STUDENT SOLUTIONS MANUAL

to accompany

CALCULUS

SINGLE AND MULTIVARIABLE   FIFTH EDITION

Deborah Hughes-Hallett
University of Arizona

Andrew M. Gleason
Harvard University

William G. McCallum
University of Arizona

et al.

Prepared by:

Rick Cangelosi
Scott Clark
Elliot J. Marks
Aaron Wootton

WILEY

John Wiley & Sons, Inc.

COVER PHOTO   © Patrick Zephyr / Patrick Zephyr Nature Photography

To order books or for customer service please, call 1-800-CALL WILEY (225-5945).

This material is based upon work supported by the National Science
Foundation under Grant No. DUE-9352905. Opinions expressed are
those of the authors and not necessarily those of the Foundation.

ISBN-13   978-0-470-41414-9

Printed in the United States of America

10 9 8 7 6 5 4 3 2 1

Printed and bound by Bind-Rite Robbinsville

Table of Contents

CHAPTER ONE

Solutions for Section 1.1

Exercises

1. Since t represents the number of years since 1970, we see that $f(35)$ represents the population of the city in 2005. In 2005, the city's population was 12 million.

5. The slope is $(3 - 2)/(2 - 0) = 1/2$. So the equation of the line is $y = (1/2)x + 2$.

9. Rewriting the equation as
$$y = -\frac{12}{7}x + \frac{2}{7}$$
shows that the line has slope $-12/7$ and vertical intercept $2/7$.

13. (a) is (V), because slope is negative, vertical intercept is 0
   (b) is (VI), because slope and vertical intercept are both positive
   (c) is (I), because slope is negative, vertical intercept is positive
   (d) is (IV), because slope is positive, vertical intercept is negative
   (e) is (III), because slope and vertical intercept are both negative
   (f) is (II), because slope is positive, vertical intercept is 0

17. $y = 5x - 3$. Since the slope of this line is 5, we want a line with slope $-\frac{1}{5}$ passing through the point $(2, 1)$. The equation is $(y - 1) = -\frac{1}{5}(x - 2)$, or $y = -\frac{1}{5}x + \frac{7}{5}$.

21. Since x goes from 1 to 5 and y goes from 1 to 6, the domain is $1 \leq x \leq 5$ and the range is $1 \leq y \leq 6$.

25. The domain is all x-values, as the denominator is never zero. The range is $0 < y \leq \frac{1}{2}$.

29. If distance is d, then $v = \dfrac{d}{t}$.

Problems

33. (a) When the car is 5 years old, it is worth $6000.
   (b) Since the value of the car decreases as the car gets older, this is a decreasing function. A possible graph is in Figure 1.1:

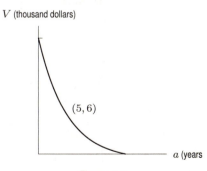

V (thousand dollars)

$(5, 6)$

a (years)

Figure 1.1

   (c) The vertical intercept is the value of V when $a = 0$, or the value of the car when it is new. The horizontal intercept is the value of a when $V = 0$, or the age of the car when it is worth nothing.

37. (a) $f(30) = 10$ means that the value of f at $t = 30$ was 10. In other words, the temperature at time $t = 30$ minutes was $10°$C. So, 30 minutes after the object was placed outside, it had cooled to 10 $°$C.
   (b) The intercept a measures the value of $f(t)$ when $t = 0$. In other words, when the object was initially put outside, it had a temperature of $a°$C. The intercept b measures the value of t when $f(t) = 0$. In other words, at time b the object's temperature is 0 $°$C.

41. (a) We have the following functions.

(i) Since a change in p of \$5 results in a decrease in q of 2, the slope of $q = D(p)$ is $-2/5$ items per dollar. So

$$q = b - \frac{2}{5}p.$$

Now we know that when $p = 550$ we have $q = 100$, so

$$100 = b - \frac{2}{5} \cdot 550$$
$$100 = b - 220$$
$$b = 320.$$

Thus a formula is

$$q = 320 - \frac{2}{5}p.$$

(ii) We can solve $q = 320 - \frac{2}{5}p$ for p in terms of q:

$$5q = 1600 - 2p$$
$$2p = 1600 - 5q$$
$$p = 800 - \frac{5}{2}q.$$

The slope of this function is $-5/2$ dollars per item, as we would expect.

(b) A graph of $p = 800 - \frac{5}{2}q$ is given in Figure 1.2.

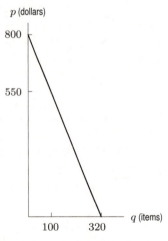

p (dollars)

800

550

100   320   *q* (items)

Figure 1.2

45. (a) The line given by $(0, 2)$ and $(1, 1)$ has slope $m = \frac{2-1}{-1} = -1$ and y-intercept 2, so its equation is

$$y = -x + 2.$$

The points of intersection of this line with the parabola $y = x^2$ are given by

$$x^2 = -x + 2$$
$$x^2 + x - 2 = 0$$
$$(x + 2)(x - 1) = 0.$$

The solution $x = 1$ corresponds to the point we are already given, so the other solution, $x = -2$, gives the x-coordinate of C. When we substitute back into either equation to get y, we get the coordinates for C, $(-2, 4)$.

(b) The line given by $(0, b)$ and $(1, 1)$ has slope $m = \frac{b-1}{-1} = 1 - b$, and y-intercept at $(0, b)$, so we can write the equation for the line as we did in part (a):

$$y = (1 - b)x + b.$$

We then solve for the points of intersection with $y = x^2$ the same way:

$$x^2 = (1 - b)x + b$$
$$x^2 - (1 - b)x - b = 0$$
$$x^2 + (b - 1)x - b = 0$$
$$(x + b)(x - 1) = 0$$

Again, we have the solution at the given point $(1, 1)$, and a new solution at $x = -b$, corresponding to the other point of intersection C. Substituting back into either equation, we can find the y-coordinate for C is b^2, and thus C is given by $(-b, b^2)$. This result agrees with the particular case of part (a) where $b = 2$.

Solutions for Section 1.2

Exercises

1. The graph shows a concave up function.

5. Initial quantity $= 5$; growth rate $= 0.07 = 7\%$.

9. Since $e^{0.25t} = \left(e^{0.25}\right)^t \approx (1.2840)^t$, we have $P = 15(1.2840)^t$. This is exponential growth since 0.25 is positive. We can also see that this is growth because $1.2840 > 1$.

13. (a) Let $Q = Q_0 a^t$. Then $Q_0 a^5 = 75.94$ and $Q_0 a^7 = 170.86$. So

$$\frac{Q_0 a^7}{Q_0 a^5} = \frac{170.86}{75.94} = 2.25 = a^2.$$

So $a = 1.5$.

(b) Since $a = 1.5$, the growth rate is $r = 0.5 = 50\%$.

Problems

17. (a) This is a linear function with slope -2 grams per day and intercept 30 grams. The function is $Q = 30 - 2t$, and the graph is shown in Figure 1.3.

Figure 1.3

Figure 1.4

(b) Since the quantity is decreasing by a constant percent change, this is an exponential function with base $1 - 0.12 = 0.88$. The function is $Q = 30(0.88)^t$, and the graph is shown in Figure 1.4.

21. (a) Advertising is generally cheaper in bulk; spending more money will give better and better marginal results initially, (Spending \$5,000 could give you a big newspaper ad reaching 200,000 people; spending \$100,000 could give you a series of TV spots reaching 50,000,000 people.) See Figure 1.5.

(b) The temperature of a hot object decreases at a rate proportional to the difference between its temperature and the temperature of the air around it. Thus, the temperature of a very hot object decreases more quickly than a cooler object. The graph is decreasing and concave up. See Figure 1.6 (We are assuming that the coffee is all at the same temperature.)

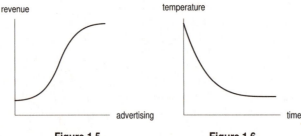

Figure 1.5   **Figure 1.6**

25. The difference, D, between the horizontal asymptote and the graph appears to decrease exponentially, so we look for an equation of the form
$$D = D_0 a^x$$
where $D_0 = 4 =$ difference when $x = 0$. Since $D = 4 - y$, we have
$$4 - y = 4a^x \quad \text{or} \quad y = 4 - 4a^x = 4(1 - a^x)$$
The point $(1, 2)$ is on the graph, so $2 = 4(1 - a^1)$, giving $a = \frac{1}{2}$.
Therefore $y = 4(1 - (\frac{1}{2})^x) = 4(1 - 2^{-x})$.

29. (a) The formula is $Q = Q_0 \left(\frac{1}{2}\right)^{(t/1620)}$.
    (b) The percentage left after 500 years is
$$\frac{Q_0(\frac{1}{2})^{(500/1620)}}{Q_0}.$$
The Q_0s cancel giving
$$\left(\frac{1}{2}\right)^{(500/1620)} \approx 0.807,$$
so 80.7% is left.

33. (a) This is the graph of a linear function, which increases at a constant rate, and thus corresponds to $k(t)$, which increases by 0.3 over each interval of 1.
    (b) This graph is concave down, so it corresponds to a function whose increases are getting smaller, as is the case with $h(t)$, whose increases are 10, 9, 8, 7, and 6.
    (c) This graph is concave up, so it corresponds to a function whose increases are getting bigger, as is the case with $g(t)$, whose increases are 1, 2, 3, 4, and 5.

37. Because the population is growing exponentially, the time it takes to double is the same, regardless of the population levels we are considering. For example, the population is 20,000 at time 3.7, and 40,000 at time 6.0. This represents a doubling of the population in a span of $6.0 - 3.7 = 2.3$ years.

How long does it take the population to double a second time, from 40,000 to 80,000? Looking at the graph once again, we see that the population reaches 80,000 at time $t = 8.3$. This second doubling has taken $8.3 - 6.0 = 2.3$ years, the same amount of time as the first doubling.

Further comparison of any two populations on this graph that differ by a factor of two will show that the time that separates them is 2.3 years. Similarly, during any 2.3 year period, the population will double. Thus, the doubling time is 2.3 years.

Suppose $P = P_0 a^t$ doubles from time t to time $t + d$. We now have $P_0 a^{t+d} = 2P_0 a^t$, so $P_0 a^t a^d = 2P_0 a^t$. Thus, canceling P_0 and a^t, d must be the number such that $a^d = 2$, no matter what t is.

Solutions for Section 1.3

Exercises

1.

(a)

(b)

(c)

(d)

(e)

(f)
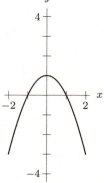

5. This graph is the graph of $m(t)$ shifted to the right by one unit. See Figure 1.7.

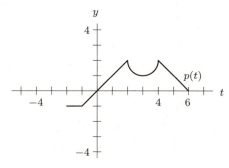

Figure 1.7

9. (a) $f(g(1)) = f(1^2) = f(1) = \sqrt{1+4} = \sqrt{5}$
   (b) $g(f(1)) = g(\sqrt{1+4}) = g(\sqrt{5}) = (\sqrt{5})^2 = 5$
   (c) $f(g(x)) = f(x^2) = \sqrt{x^2+4}$
   (d) $g(f(x)) = g(\sqrt{x+4}) = (\sqrt{x+4})^2 = x+4$
   (e) $f(t)g(t) = (\sqrt{t+4})t^2 = t^2\sqrt{t+4}$

13. (a) $f(t+1) = (t+1)^2 + 1 = t^2 + 2t + 1 + 1 = t^2 + 2t + 2.$
   (b) $f(t^2+1) = (t^2+1)^2 + 1 = t^4 + 2t^2 + 1 + 1 = t^4 + 2t^2 + 2.$
   (c) $f(2) = 2^2 + 1 = 5.$
   (d) $2f(t) = 2(t^2+1) = 2t^2 + 2.$
   (e) $[f(t)]^2 + 1 = (t^2+1)^2 + 1 = t^4 + 2t^2 + 1 + 1 = t^4 + 2t^2 + 2.$

17. $m(z) - m(z - h) = z^2 - (z - h)^2 = 2zh - h^2$.

21. $f^{-1}(75)$ is the length of the column of mercury in the thermometer when the temperature is $75°F$.

25. Since a horizontal line cuts the graph of $f(x) = x^2 + 3x + 2$ two times, f is not invertible. See Figure 1.8.

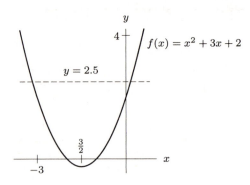

Figure 1.8

29.
$$f(-x) = (-x)^3 + (-x)^2 + (-x) = -x^3 + x^2 - x.$$

Since $f(-x) \neq f(x)$ and $f(-x) \neq -f(x)$, this function is neither even nor odd.

33. Since
$$f(-x) = e^{(-x)^2 - 1} = e^{x^2 - 1} = f(x),$$

we see f is even.

Problems

37. This looks like a shift of the graph $y = x^3$. The graph is shifted to the right 2 units and down 1 unit, so a possible formula is $y = (x - 2)^3 - 1$.

41. f is an increasing function since the amount of fuel used increases as flight time increases. Therefore f is invertible.

45. $f(g(1)) = f(2) \approx 0.4$.

49. Using the same way to compute $g(f(x))$ as in Problem 46, we get Table 1.1. Then we can plot the graph of $g(f(x))$ in Figure 1.9.

Table 1.1

x	$f(x)$	$g(f(x))$
-3	3	-2.6
-2.5	0.1	0.8
-2	-1	-1.4
-1.5	-1.3	-1.8
-1	-1.2	-1.7
-0.5	-1	-1.4
0	-0.8	-1
0.5	-0.6	-0.6
1	-0.4	-0.3
1.5	-0.1	0.3
2	0.3	1.1
2.5	0.9	2
3	1.6	2.2

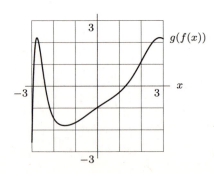

Figure 1.9

53. We have approximately $u(10) = 13$ and $v(13) = 60$ so $v(u(10)) = 60$.

57. $f(x) = \sqrt{x}, \quad g(x) = x^2 + 4$

61. (a) The function f tells us C in terms of q. To get its inverse, we want q in terms of C, which we find by solving for q:

$$C = 100 + 2q,$$
$$C - 100 = 2q,$$
$$q = \frac{C - 100}{2} = f^{-1}(C).$$

(b) The inverse function tells us the number of articles that can be produced for a given cost.

Solutions for Section 1.4

Exercises

1. Using the identity $e^{\ln x} = x$, we have $e^{\ln(1/2)} = \frac{1}{2}$.

5. Using the rules for ln, we have

$$\ln\left(\frac{1}{e}\right) + \ln AB = \ln 1 - \ln e + \ln A + \ln B$$
$$= 0 - 1 + \ln A + \ln B$$
$$= -1 + \ln A + \ln B.$$

9. Isolating the exponential term

$$20 = 50(1.04)^x$$
$$\frac{20}{50} = (1.04)^x.$$

Taking logs of both sides

$$\log \frac{2}{5} = \log(1.04)^x$$
$$\log \frac{2}{5} = x \log(1.04)$$
$$x = \frac{\log(2/5)}{\log(1.04)} = -23.4.$$

13. To solve for x, we first divide both sides by 600 and then take the natural logarithm of both sides.

$$\frac{50}{600} = e^{-0.4x}$$
$$\ln(50/600) = -0.4x$$
$$x = \frac{\ln(50/600)}{-0.4} \approx 6.212.$$

17. Using the rules for ln, we have

$$2x - 1 = x^2$$
$$x^2 - 2x + 1 = 0$$
$$(x - 1)^2 = 0$$
$$x = 1.$$

21. Taking logs of both sides yields

$$nt = \frac{\log\left(\frac{Q}{Q_0}\right)}{\log a}.$$

Hence

$$t = \frac{\log\left(\frac{Q}{Q_0}\right)}{n\log a} = \frac{\log Q - \log Q_0}{n\log a}.$$

25. Since we want $(1.5)^t = e^{kt} = (e^k)^t$, so $1.5 = e^k$, and $k = \ln 1.5 = 0.4055$. Thus, $P = 15e^{0.4055t}$. Since 0.4055 is positive, this is exponential growth.

29. If $p(t) = (1.04)^t$, then, for p^{-1} the inverse of p, we should have

$$(1.04)^{p^{-1}(t)} = t,$$
$$p^{-1}(t)\log(1.04) = \log t,$$
$$p^{-1}(t) = \frac{\log t}{\log(1.04)} \approx 58.708 \log t.$$

Problems

33. We know that the y-intercept of the line is at $(0,1)$, so we need one other point to determine the equation of the line. We observe that it intersects the graph of $f(x) = 10^x$ at the point $x = \log 2$. The y-coordinate of this point is then

$$y = 10^x = 10^{\log 2} = 2,$$

so $(\log 2, 2)$ is the point of intersection. We can now find the slope of the line:

$$m = \frac{2-1}{\log 2 - 0} = \frac{1}{\log 2}.$$

Plugging this into the point-slope formula for a line, we have

$$y - y_1 = m(x - x_1)$$
$$y - 1 = \frac{1}{\log 2}(x - 0)$$
$$y = \frac{1}{\log 2}x + 1 \approx 3.3219x + 1.$$

37. The vertical asymptote is where $x + 2 = 0$, or $x = -2$, so increasing a does not effect the vertical asymptote.

41. Since the factor by which the prices have increased after time t is given by $(1.05)^t$, the time after which the prices have doubled solves

$$2 = (1.05)^t$$
$$\log 2 = \log(1.05^t) = t\log(1.05)$$
$$t = \frac{\log 2}{\log 1.05} \approx 14.21 \text{ years.}$$

45. Let B represent the sales (in millions of dollars) at Borders bookstores t years since 2000. Since $B = 2108$ when $t = 0$ and we want the continuous growth rate, we write $B = 2108e^{kt}$. We use the information from 2005, that $B = 3880$ when $t = 5$, to find k:

$$3880 = 2108e^{k \cdot 5}$$
$$1.841 = e^{5k}$$
$$\ln(1.841) = 5k$$
$$k = 0.122.$$

We have $B = 2108e^{0.122t}$, which represents a continuous growth rate of 12.2% per year.

49. Let $t =$ number of years since 2000. Then the number of vehicles, V, in millions, at time t is given by

$$V = 213(1.04)^t$$

and the number of people, P, in millions, at time t is given by

$$P = 281(1.01)^t.$$

There is an average of one vehicle per person when $\dfrac{V}{P} = 1$, or $V = P$. Thus, we must solve for t in the equation:

$$213(1.04)^t = 281(1.01)^t,$$

which leads to

$$\left(\frac{1.04}{1.01}\right)^t = \frac{(1.04)^t}{(1.01)^t} = \frac{281}{213}$$

Taking logs on both sides, we get

$$t \log \frac{1.04}{1.01} = \log \frac{281}{213},$$

so

$$t = \frac{\log{(281/213)}}{\log{(1.04/1.01)}} = 9.5 \text{ years.}$$

This model predicts one vehicle per person in 2009.

53. We assume exponential decay and solve for k using the half-life:

$$e^{-k(5730)} = 0.5 \quad \text{so} \quad k = 1.21 \cdot 10^{-4}.$$

Now find t, the age of the painting:

$$e^{-1.21 \cdot 10^{-4} t} = 0.995, \quad \text{so} \quad t = \frac{\ln 0.995}{-1.21 \cdot 10^{-4}} = 41.43 \text{ years.}$$

Since Vermeer died in 1675, the painting is a fake.

Solutions for Section 1.5

Exercises

1. See Figure 1.10.

$$\sin\left(\frac{3\pi}{2}\right) = -1 \quad \text{is negative.}$$

$$\cos\left(\frac{3\pi}{2}\right) = 0$$

$$\tan\left(\frac{3\pi}{2}\right) \quad \text{is undefined.}$$

5. See Figure 1.11.

$$\sin\left(\frac{\pi}{6}\right) \text{ is positive.}$$

$$\cos\left(\frac{\pi}{6}\right) \text{ is positive.}$$

$$\tan\left(\frac{\pi}{6}\right) \text{ is positive.}$$

9. -1 radian $\cdot \dfrac{180°}{\pi \text{ radians}} = -\left(\dfrac{180°}{\pi}\right) \approx -60°$. See Figure 1.12.

$$\sin(-1) \quad \text{is negative}$$
$$\cos(-1) \quad \text{is positive}$$
$$\tan(-1) \quad \text{is negative.}$$

Figure 1.10   **Figure 1.11**   **Figure 1.12**

13. (a) We determine the amplitude of y by looking at the coefficient of the cosine term. Here, the coefficient is 1, so the amplitude of y is 1. Note that the constant term does not affect the amplitude.

(b) We know that the cosine function $\cos x$ repeats itself at $x = 2\pi$, so the function $\cos(3x)$ must repeat itself when $3x = 2\pi$, or at $x = 2\pi/3$. So the period of y is $2\pi/3$. Here as well the constant term has no effect.

(c) The graph of y is shown in the figure below.

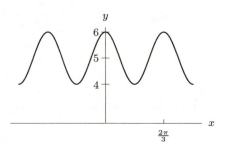

17. The period is $2\pi/\pi = 2$, since when t increases from 0 to 2, the value of πt increases from 0 to 2π. The amplitude is 0.1, since the function oscillates between 1.9 and 2.1.

21. This graph is an inverted cosine curve with amplitude 8 and period 20π, so it is given by $f(x) = -8\cos\left(\dfrac{x}{10}\right)$.

25. This can be represented by a sine function of amplitude 3 and period 18. Thus,

$$f(x) = 3\sin\left(\frac{\pi}{9}x\right).$$

29. We first isolate $\cos(2x + 1)$ and then use inverse cosine:

$$1 = 8\cos(2x + 1) - 3$$
$$4 = 8\cos(2x + 1)$$
$$0.5 = \cos(2x + 1)$$
$$\cos^{-1}(0.5) = 2x + 1$$
$$x = \frac{\cos^{-1}(0.5) - 1}{2} \approx 0.0236.$$

There are infinitely many other possible solutions since the cosine is periodic.

Problems

33. $\sin x^2$ is by convention $\sin(x^2)$, which means you square the x first and then take the sine.
$\sin^2 x = (\sin x)^2$ means find $\sin x$ and then square it.
$\sin(\sin x)$ means find $\sin x$ and then take the sine of that.
    Expressing each as a composition: If $f(x) = \sin x$ and $g(x) = x^2$, then
$\sin x^2 = f(g(x))$
$\sin^2 x = g(f(x))$
$\sin(\sin x) = f(f(x))$.

37. 200 revolutions per minute is $\frac{1}{200}$ minutes per revolution, so the period is $\frac{1}{200}$ minutes, or 0.3 seconds.

41. (a) D = the average depth of the water.
    (b) A = the amplitude = $15/2 = 7.5$.
    (c) Period = 12.4 hours. Thus $(B)(12.4) = 2\pi$ so $B = 2\pi/12.4 \approx 0.507$.
    (d) C is the time of a high tide.

45. (a) See Figure 1.13.
    (b) Average value of population = $\frac{700+900}{2} = 800$, amplitude = $\frac{900-700}{2} = 100$, and period = 12 months, so $B = 2\pi/12 = \pi/6$. Since the population is at its minimum when $t = 0$, we use a negative cosine:

$$P = 800 - 100 \cos\left(\frac{\pi t}{6}\right).$$

Figure 1.13

Figure 1.14

49. Figure 1.14 shows that the cross-sectional area is one rectangle of area hw and two triangles. Each triangle has height h and base x, where

$$\frac{h}{x} = \tan\theta \quad \text{so} \quad x = \frac{h}{\tan\theta}.$$

$$\text{Area of triangle} = \frac{1}{2}xh = \frac{h^2}{2\tan\theta}$$

$$\text{Total area} = \text{Area of rectangle} + 2(\text{Area of triangle})$$

$$= hw + 2\cdot\frac{h^2}{2\tan\theta} = hw + \frac{h^2}{\tan\theta}.$$

Solutions for Section 1.6

Exercises

1. As $x \to \infty$, $y \to \infty$.
    As $x \to -\infty$, $y \to -\infty$.

5. As $x \to \infty$, $0.25x^{1/2}$ is larger than $25{,}000x^{-3}$.

9. $f(x) = k(x+3)(x-1)(x-4) = k(x^3 - 2x^2 - 11x + 12)$, where $k < 0$. ($k \approx -\frac{1}{6}$ if the horizontal and vertical scales are equal; otherwise one can't tell how large k is.)

Problems

13. (a) II and III because in both cases, the numerator and denominator each have x^2 as the highest power, with coefficient $= 1$. Therefore,

$$y \to \frac{x^2}{x^2} = 1 \quad \text{as } x \to \pm\infty.$$

(b) I, since

$$y \to \frac{x}{x^2} = 0 \quad \text{as } x \to \pm\infty.$$

(c) II and III, since replacing x by $-x$ leaves the graph of the function unchanged.

(d) None

(e) III, since the denominator is zero and $f(x)$ tends to $\pm\infty$ when $x = \pm1$.

17. (a)   (i) The water that has flowed out of the pipe in 1 second is a cylinder of radius r and length 3 cm. Its volume is

$$V = \pi r^2 (3) = 3\pi r^2.$$

(ii) If the rate of flow is k cm/sec instead of 3 cm/sec, the volume is given by

$$V = \pi r^2 (k) = \pi r^2 k.$$

(b)   (i) The graph of V as a function of r is a quadratic. See Figure 1.15.

Figure 1.15

Figure 1.16

(ii) The graph of V as a function of k is a line. See Figure 1.16.

21. (a)   (i) If $(1, 1)$ is on the graph, we know that

$$1 = a(1)^2 + b(1) + c = a + b + c.$$

(ii) If $(1, 1)$ is the vertex, then the axis of symmetry is $x = 1$, so

$$-\frac{b}{2a} = 1,$$

and thus

$$a = -\frac{b}{2}, \text{ so } b = -2a.$$

But to be the vertex, $(1, 1)$ must also be on the graph, so we know that $a + b + c = 1$. Substituting $b = -2a$, we get $-a + c = 1$, which we can rewrite as $a = c - 1$, or $c = 1 + a$.

(iii) For $(0, 6)$ to be on the graph, we must have $f(0) = 6$. But $f(0) = a(0^2) + b(0) + c = c$, so $c = 6$.

(b) To satisfy all the conditions, we must first, from (a)(iii), have $c = 6$. From (a)(ii), $a = c - 1$ so $a = 5$. Also from (a)(ii), $b = -2a$, so $b = -10$. Thus the completed equation is

$$y = f(x) = 5x^2 - 10x + 6,$$

which satisfies all the given conditions.

25. We use the fact that at a constant speed, Time = Distance/Speed. Thus,

$$\text{Total time} = \text{Time running} + \text{Time walking}$$
$$= \frac{3}{x} + \frac{6}{x-2}.$$

Horizontal asymptote: x-axis.

Vertical asymptote: $x = 0$ and $x = 2$.

29. (a) $R(P) = kP(L - P)$, where k is a positive constant.

(b) A possible graph is in Figure 1.17.

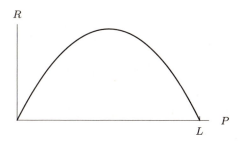

Figure 1.17

Solutions for Section 1.7

Exercises

1. Yes, because $2x + x^{2/3}$ is defined for all x.

5. Yes, because $2x - 5$ is positive for $3 \le x \le 4$.

9. No, because $e^x - 1 = 0$ at $x = 0$.

13. We have that $f(0) = -1 < 0$ and $f(1) = 1 - \cos 1 > 0$ and that f is continuous. Thus, by the Intermediate Value Theorem applied to $k = 0$, there is a number c in $[0, 1]$ such that $f(c) = k = 0$.

Problems

17. The voltage $f(t)$ is graphed in Figure 1.18.

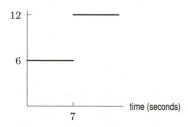

Figure 1.18: Voltage change from 6V to 12V

Using formulas, the voltage, $f(t)$, is represented by

$$f(t) = \begin{cases} 6, & 0 < t \le 7 \\ 12, & 7 < t \end{cases}$$

Although a real physical voltage is continuous, the voltage in this circuit is well-approximated by the function $f(t)$, which is not continuous on any interval around 7 seconds.

21. For $x > 0$, we have $|x| = x$, so $f(x) = 1$. For $x < 0$, we have $|x| = -x$, so $f(x) = -1$. Thus, the function is given by

$$f(x) = \begin{cases} 1 & x > 0 \\ 0 & x = 0 \\ -1 & x < 1 \end{cases},$$

so f is not continuous on any interval containing $x = 0$.

25. (a) The graphs of $y = e^x$ and $y = 4 - x^2$ cross twice in Figure 1.19. This tells us that the equation $e^x = 4 - x^2$ has two solutions.

Since $y = e^x$ increases for all x and $y = 4 - x^2$ increases for $x < 0$ and decreases for $x > 0$, these are only the two crossing points.

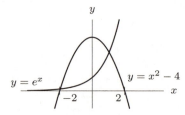

Figure 1.19

(b) Values of $f(x)$ are in Table 1.2. One solution is between $x = -2$ and $x = -1$; the second solution is between $x = 1$ and $x = 2$.

Table 1.2

x	-4	-3	-2	-1	0	1	2	3	4
$f(x)$	12.0	5.0	0.1	-2.6	-3	-0.3	7.4	25.1	66.6

29. The drug first increases linearly for half a second, at the end of which time there is 0.6 ml in the body. Thus, for $0 \le t \le 0.5$, the function is linear with slope $0.6/0.5 = 1.2$:

$$Q = 1.2t \quad \text{for} \quad 0 \le t \le 0.5.$$

At $t = 0.5$, we have $Q = 0.6$. For $t > 0.5$, the quantity decays exponentially at a continuous rate of 0.002, so Q has the form

$$Q = Ae^{-0.002t} \qquad 0.5 < t.$$

We choose A so that $Q = 0.6$ when $t = 0.5$:

$$0.6 = Ae^{-0.002(0.5)} = Ae^{-0.001}$$
$$A = 0.6e^{0.001}.$$

Thus

$$Q = \begin{cases} 1.2t & 0 \le t \le 0.5 \\ 0.6e^{0.001}e^{-.002t} & 0.5 < t. \end{cases}$$

Solutions for Section 1.8

Exercises

1. (a) As x approaches -2 from either side, the values of $f(x)$ get closer and closer to 3, so the limit appears to be about 3.

(b) As x approaches 0 from either side, the values of $f(x)$ get closer and closer to 7. (Recall that to find a limit, we are interested in what happens to the function near x but not at x.) The limit appears to be about 7.

(c) As x approaches 2 from either side, the values of $f(x)$ get closer and closer to 3 on one side of $x = 2$ and get closer and closer to 2 on the other side of $x = 2$. Thus the limit does not exist.

(d) As x approaches 4 from either side, the values of $f(x)$ get closer and closer to 8. (Again, recall that we don't care what happens right at $x = 4$.) The limit appears to be about 8.

5. For $-1 \leq x \leq 1,\ -1 \leq y \leq 1$, the graph of $y = x \ln |x|$ is in Figure 1.20. The graph suggests that

$$\lim_{x \to 0} x \ln |x| = 0.$$

9. For $-90° \leq \theta \leq 90°$, $0 \leq y \leq 0.02$, the graph of $y = \dfrac{\sin \theta}{\theta}$ is shown in Figure 1.21. Therefore, by tracing along the curve, we see that in degrees, $\lim\limits_{\theta \to 0} \dfrac{\sin \theta}{\theta} = 0.01745\ldots$.

Figure 1.20

Figure 1.21

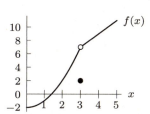

Figure 1.22

13. $f(x) = \begin{cases} x^2 - 2 & 0 < x < 3 \\ 2 & x = 3 \\ 2x + 1 & 3 < x \end{cases}$

Figure 1.22 confirms that $\lim\limits_{x \to 3^-} f(x) = \lim\limits_{x \to 3^-} (x^2 - 2) = 7$ and that $\lim\limits_{x \to 3^+} f(x) = \lim\limits_{x \to 3^+} (2x + 1) = 7$, so $\lim\limits_{x \to 3} f(x) = 7$. Note, however, that $f(x)$ is not continuous at $x = 3$ since $f(3) = 2$.

Problems

17. Since $\lim_{x \to 0^+} f(x) = 1$ and $\lim_{x \to 0^-} f(x) = -1$, we see that $\lim_{x \to 0} f(x)$ does not exist. Thus, $f(x)$ is not continuous at $x = 0$

21. When $x = 0.1$, we find $xe^{1/x} \approx 2203$. When $x = 0.01$, we find $xe^{1/x} \approx 3 \times 10^{41}$. When $x = 0.001$, the value of $xe^{1/x}$ is too big for a calculator to compute. This suggests that $\lim\limits_{x \to 0^+} xe^{1/x}$ does not exist (and in fact it does not).

25. If $x > 1$ and x approaches 1, then $p(x) = 55$. If $x < 1$ and x approaches 1, then $p(x) = 34$. There is not a single number that $p(x)$ approaches as x approaches 1, so we say that $\lim\limits_{x \to 1} p(x)$ does not exist.

29. From Table 1.3, it appears the limit is -1. This is confirmed by Figure 1.23. An appropriate window is $-0.099 < x < 0.099$, $-1.01 < y < -0.99$.

Table 1.3

x	$f(x)$	x	$f(x)$
0.1	−0.99	−0.0001	−0.99999999
0.01	−0.9999	−0.001	−0.999999
0.001	−0.999999	−0.01	−0.9999
0.0001	−0.99999999	−0.1	−0.99

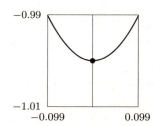

Figure 1.23

33. From Table 1.4, it appears the limit is 3. This is confirmed by Figure 1.24. An appropriate window is $-0.047 < x < 0.047$, $2.99 < y < 3.01$.

Table 1.4

x	$f(x)$
0.1	2.9552
0.01	2.9996
0.001	3.0000
0.0001	3.0000

x	$f(x)$
-0.0001	3.0000
-0.001	3.0000
-0.01	2.9996
-0.1	2.9552

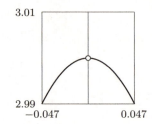

Figure 1.24

37. Divide numerator and denominator by x:

$$f(x) = \frac{\pi + 3x}{\pi x - 3} = \frac{(\pi + 3x)/x}{(\pi x - 3)/x},$$

so

$$\lim_{x \to \infty} f(x) = \lim_{x \to \infty} \frac{\pi/x + 3}{\pi - 3/x} = \frac{\lim_{x \to \infty} (\pi/x + 3)}{\lim_{x \to \infty} (\pi - 3/x)} = \frac{3}{\pi}.$$

41. Divide numerator and denominator by x^3, giving

$$f(x) = \frac{2x^3 - 16x^2}{4x^2 + 3x^3} = \frac{2 - 16/x}{4/x + 3},$$

so

$$\lim_{x \to \infty} f(x) = \lim_{x \to \infty} \frac{2 - 16/x}{4/x + 3} = \frac{\lim_{x \to \infty} (2 - 16/x)}{\lim_{x \to \infty} (4/x + 3)} = \frac{2}{3}.$$

45. $f(x) = \dfrac{2e^{-x} + 3}{3e^{-x} + 2}$, so $\lim\limits_{x \to \infty} f(x) = \dfrac{\lim_{x \to \infty} (2e^{-x} + 3)}{\lim_{x \to \infty} (3e^{-x} + 2)} = \dfrac{3}{2}$.

49. Division of numerator and denominator by x^2 yields

$$\frac{x^2 + 3x + 5}{4x + 1 + x^k} = \frac{1 + 3/x + 5/x^2}{4/x + 1/x^2 + x^{k-2}}.$$

As $x \to \infty$, the limit of the numerator is 1. The limit of the denominator depends upon k. If $k > 2$, the denominator approaches ∞ as $x \to \infty$, so the limit of the quotient is 0. If $k = 2$, the denominator approaches 1 as $x \to \infty$, so the limit of the quotient is 1. If $k < 2$ the denominator approaches 0^+ as $x \to \infty$, so the limit of the quotient is ∞. Therefore the values of k we are looking for are $k \geq 2$.

53. In the denominator, we have $\lim\limits_{x \to -\infty} 3^{2x} + 4 = 4$. In the numerator, if $k < 0$, we have $\lim\limits_{x \to -\infty} 3^{kx} + 6 = \infty$, so the quotient has a limit of ∞. If $k = 0$, we have $\lim\limits_{x \to -\infty} 3^{kx} + 6 = 7$, so the quotient has a limit of $7/4$. If $k > 0$, we have $\lim\limits_{x \to -\infty} 3^{kx} + 6 = 6$, so the quotient has a limit of $6/4$.

57. (a) Since $\sin(n\pi) = 0$ for $n = 1, 2, 3, \ldots$ the sequence of x-values

$$\frac{1}{\pi}, \frac{1}{2\pi}, \frac{1}{3\pi}, \ldots$$

works. These x-values $\to 0$ and are zeroes of $f(x)$.

(b) Since $\sin(n\pi/2) = 1$ for $n = 1, 5, 9 \ldots$ the sequence of x-values

$$\frac{2}{\pi}, \frac{2}{5\pi}, \frac{2}{9\pi}, \ldots$$

works.

(c) Since $\sin(n\pi)/2 = -1$ for $n = 3, 7, 11, \ldots$ the sequence of x-values

$$\frac{2}{3\pi}, \frac{2}{7\pi}, \frac{2}{11\pi} \ldots$$

works.

(d) Any two of these sequences of x-values show that if the limit were to exist, then it would have to have two (different) values: 0 and 1, or 0 and -1, or 1 and -1. Hence, the limit can not exist.

61. We will show $f(x) = x$ is continuous at $x = c$. Since $f(c) = c$, we need to show that

$$\lim_{x \to c} f(x) = c$$

that is, since $f(x) = x$, we need to show

$$\lim_{x \to c} x = c.$$

Pick any $\epsilon > 0$, then take $\delta = \epsilon$. Thus,

$$|f(x) - c| = |x - c| < \epsilon \quad \text{for all} \quad |x - c| < \delta = \epsilon.$$

Solutions for Chapter 1 Review

Exercises

1. The line of slope m through the point (x_0, y_0) has equation

$$y - y_0 = m(x - x_0),$$

so the line we want is

$$y - 0 = 2(x - 5)$$
$$y = 2x - 10.$$

5. A circle with center (h, k) and radius r has equation $(x - h)^2 + (y - k)^2 = r^2$. Thus $h = -1$, $k = 2$, and $r = 3$, giving

$$(x + 1)^2 + (y - 2)^2 = 9.$$

Solving for y, and taking the positive square root gives the top half, so

$$(y - 2)^2 = 9 - (x + 1)^2$$
$$y = 2 + \sqrt{9 - (x + 1)^2}.$$

See Figure 1.25.

Figure 1.25: Graph of $y = 2 + \sqrt{9 - (x + 1)^2}$

9. Since $T = f(P)$, we see that $f(200)$ is the value of T when $P = 200$; that is, the thickness of pelican eggs when the concentration of PCBs is 200 ppm.

13. (a) The equation is $y = 2x^2 + 1$. Note that its graph is narrower than the graph of $y = x^2$ which appears in gray. See Figure 1.26.

Figure 1.26

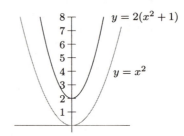

Figure 1.27

(b) $y = 2(x^2 + 1)$ moves the graph up one unit and *then* stretches it by a factor of two. See Figure 1.27.

(c) No, the graphs are not the same. Since $2(x^2 + 1) = (2x^2 + 1) + 1$, the second graph is always one unit higher than the first.

17. (a) It was decreasing from March 2 to March 5 and increasing from March 5 to March 9.

(b) From March 5 to 8, the average temperature increased, but the rate of increase went down, from $12°$ between March 5 and 6 to $4°$ between March 6 and 7 to $2°$ between March 7 and 8.

From March 7 to 9, the average temperature increased, and the rate of increase went up, from $2°$ between March 7 and 8 to $9°$ between March 8 and 9.

21. (a) A polynomial has the same end behavior as its leading term, so this polynomial behaves as $-5x^4$ globally. Thus we have:
$$f(x) \to -\infty \text{ as } x \to -\infty, \quad \text{and} \quad f(x) \to -\infty \text{ as } x \to +\infty.$$

(b) Polynomials behave globally as their leading term, so this rational function behaves globally as $(3x^2)/(2x^2)$, or $3/2$. Thus we have:
$$f(x) \to 3/2 \text{ as } x \to -\infty, \quad \text{and} \quad f(x) \to 3/2 \text{ as } x \to +\infty.$$

(c) We see from a graph of $y = e^x$ that
$$f(x) \to 0 \text{ as } x \to -\infty, \quad \text{and} \quad f(x) \to +\infty \text{ as } x \to +\infty.$$

25. Collecting similar factors yields $\left(\frac{1.04}{1.03}\right)^t = \frac{12.01}{5.02}$. Solving for t yields

$$t = \frac{\log\left(\frac{12.01}{5.02}\right)}{\log\left(\frac{1.04}{1.03}\right)} = 90.283.$$

29. The amplitude is 2. The period is $2\pi/5$. See Figure 1.28.

Figure 1.28

33. $y = -kx(x + 5) = -k(x^2 + 5x)$, where $k > 0$ is any constant.

37. $x = ky(y - 4) = k(y^2 - 4y)$, where $k > 0$ is any constant.

41. There are many solutions for a graph like this one. The simplest is $y = 1 - e^{-x}$, which gives the graph of $y = e^x$, flipped over the x-axis and moved up by 1. The resulting graph passes through the origin and approaches $y = 1$ as an upper bound, the two features of the given graph.

45. $f(x) = x^3, \quad g(x) = \ln x.$

49. $f(x) = \begin{cases} e^x & -1 < x < 0 \\ 1 & x = 0 \\ \cos x & 0 < x < 1 \end{cases}$

Figure 1.29 confirms that $\lim\limits_{x \to 0^-} f(x) = \lim\limits_{x \to 0^-} e^x = e^0 = 1$, and that $\lim\limits_{x \to 0^+} f(x) = \lim\limits_{x \to 0^+} \cos x = \cos 0 = 1$, so $\lim\limits_{x \to 0} f(x) = 1$.

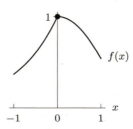

Figure 1.29

Problems

53. (a) The amplitude of the sine curve is $|A|$. Thus, increasing $|A|$ stretches the curve vertically. See Figure 1.30.
(b) The period of the wave is $2\pi/|B|$. Thus, increasing $|B|$ makes the curve oscillate more rapidly—in other words, the function executes one complete oscillation in a smaller interval. See Figure 1.31.

Figure 1.30

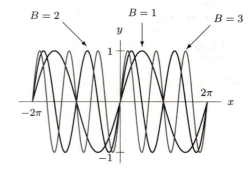

Figure 1.31

57. (a) This could be a linear function because w increases by 5 as h increases by 1.
(b) We find the slope m and the intercept b in the linear equation $w = b + mh$. We first find the slope m using the first two points in the table. Since we want w to be a function of h, we take

$$m = \frac{\Delta w}{\Delta h} = \frac{171 - 166}{69 - 68} = 5.$$

Substituting the first point and the slope $m = 5$ into the linear equation $w = b + mh$, we have $166 = b + (5)(68)$, so $b = -174$. The linear function is

$$w = 5h - 174.$$

The slope, $m = 5$, is in units of pounds per inch.
(c) We find the slope and intercept in the linear function $h = b + mw$ using $m = \Delta h/\Delta w$ to obtain the linear function

$$h = 0.2w + 34.8.$$

Alternatively, we could solve the linear equation found in part (b) for h. The slope, $m = 0.2$, has units inches per pound.

61. We can solve for the growth rate k of the bacteria using the formula $P = P_0 e^{kt}$:

$$1500 = 500 e^{k(2)}$$
$$k = \frac{\ln(1500/500)}{2}.$$

Knowing the growth rate, we can find the population P at time $t = 6$:

$$P = 500 e^{\left(\frac{\ln 3}{2}\right)6}$$
$$\approx 13{,}500 \text{ bacteria}.$$

65. We will let

$$T = \text{amount of fuel for take-off,}$$
$$L = \text{amount of fuel for landing,}$$
$$P = \text{amount of fuel per mile in the air,}$$
$$m = \text{the length of the trip in miles.}$$

Then Q, the total amount of fuel needed, is given by

$$Q(m) = T + L + Pm.$$

69. (a) Beginning at time $t = 0$, the voltage will have oscillated through a complete cycle when $\cos(120\pi t) = \cos(2\pi)$, hence when $t = \frac{1}{60}$ second. The period is $\frac{1}{60}$ second.
(b) V_0 represents the amplitude of the oscillation.
(c) See Figure 1.32.

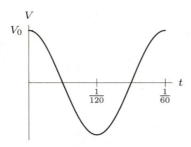

Figure 1.32

73. From Table 1.5, it appears the limit is 0. This is confirmed by Figure 1.33. An appropriate window is $-0.015 < x < 0.015$, $-0.01 < y < 0.01$.

Table 1.5

x	$f(x)$	x	$f(x)$
0.1	0.0666	−0.0001	−0.0001
0.01	0.0067	−0.001	−0.0007
0.001	0.0007	−0.01	−0.0067
0.0001	0.0001	−0.1	−0.0666

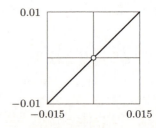

Figure 1.33

CAS Challenge Problems

77. (a) A CAS gives $f(x) = (x - a)(x + a)(x + b)(x - c)$.

(b) The graph of $f(x)$ crosses the x-axis at $x = a$, $x = -a$, $x = -b$, $x = c$; it crosses the y-axis at a^2bc. Since the coefficient of x^4 (namely 1) is positive, the graph of f looks like that shown in Figure 1.34.

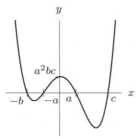

Figure 1.34: Graph of
$$f(x) =$$
$$(x-a)(x+a)(x+b)(x-c)$$

81. (a) A CAS or division gives
$$f(x) = \frac{x^3 - 30}{x - 3} = x^2 + 3x + 9 - \frac{3}{x - 3},$$
so $p(x) = x^2 + 3x + 9$, and $r(x) = -3$, and $q(x) = x - 3$.

(b) The vertical asymptote is $x = 3$. Near $x = 3$, the values of $p(x)$ are much smaller than the values of $r(x)/q(x)$. Thus
$$f(x) \approx \frac{-3}{x - 3} \qquad \text{for } x \text{ near 3.}$$

(c) For large x, the values of $p(x)$ are much larger than the value of $r(x)/q(x)$. Thus
$$f(x) \approx x^2 + 3x + 9 \qquad \text{as } x \to \infty, x \to -\infty.$$

(d) Figure 1.35 shows $f(x)$ and $y = -3/(x - 3)$ for x near 3. Figure 1.36 shows $f(x)$ and $y = x^2 + 3x + 9$ for $-20 \le x \le 20$. Note that in each case the graphs of f and the approximating function are close.

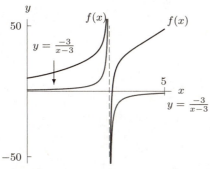

Figure 1.35: Close-up view of $f(x)$ and $y = -3/(x - 3)$

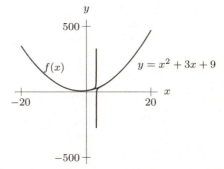

Figure 1.36: Far-away view of $f(x)$ and $y = x^2 + 3x + 9$

CHECK YOUR UNDERSTANDING

1. False. A line can be put through any two points in the plane. However, if the line is vertical, it is not the graph of a function.

5. True. The highest degree term in a polynomial determines how the polynomial behaves when x is very large in the positive or negative direction. When n is odd, x^n is positive when x is large and positive but negative when x is large and negative. Thus if a polynomial $p(x)$ has odd degree, it will be positive for some values of x and negative for other values of x. Since every polynomial is continuous, the Intermediate Value Theorem then guarantees that $p(x) = 0$ for some value of x.

9. False. Suppose $y = 5^x$. Then increasing x by 1 increases y by a factor of 5. However increasing x by 2 increases y by a factor of 25, not 10, since
$$y = 5^{x+2} = 5^x \cdot 5^2 = 25 \cdot 5^x.$$
(Other examples are possible.)

13. True. The period is $2\pi/(200\pi) = 1/100$ seconds. Thus, the function executes 100 cycles in 1 second.

17. False: When $\pi/2 < x < 3\pi/2$, we have $\cos |x| = \cos x < 0$ but $|\cos x| > 0$.

21. False. A counterexample is given by $f(x) = \sin x$, which has period 2π, and $g(x) = x^2$. The graph of $f(g(x)) = \sin(x^2)$ in Figure 1.37 is not periodic with period 2π.

Figure 1.37

25. True. If f is increasing then its reflection about the line $y = x$ is also increasing. An example is shown in Figure 1.38. The statement is true.

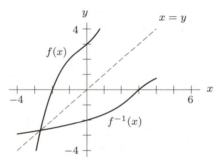

Figure 1.38

29. True. If $b > 1$, then $ab^x \to 0$ as $x \to -\infty$. If $0 < b < 1$, then $ab^x \to 0$ as $x \to \infty$. In either case, the function $y = a + ab^x$ has $y = a$ as the horizontal asymptote.

33. False. A counterexample is given by $f(x) = x^2$ and $g(x) = x + 1$. The function $f(g(x)) = (x+1)^2$ is not even because $f(g(1)) = 4$ and $f(g(-1)) = 0 \neq 4$.

37. Let $f(x) = \dfrac{1}{x + 7\pi}$. Other answers are possible.

41. Let $f(x) = x$ and $g(x) = -2x$. Then $f(x) + g(x) = -x$, which is decreasing. Note f is increasing since it has positive slope, and g is decreasing since it has negative slope.

45. False. For example, $f(x) = x/(x^2 + 1)$ has no vertical asymptote since the denominator is never 0.

49. False. For example, if $y = 4x + 1$ (so $m = 4$) and $x = 1$, then $y = 5$. Increasing x by 2 units gives 3, so $y = 4(3) + 1 = 13$. Thus, y has increased by 8 units, not $4 + 2 = 6$. (Other examples are possible.)

53. True. The constant function $f(x) = 0$ is the only function that is both even and odd. This follows, since if f is both even and odd, then, for all x, $f(-x) = f(x)$ (if f is even) and $f(-x) = -f(x)$ (if f is odd). Thus, for all x, $f(x) = -f(x)$ i.e. $f(x) = 0$, for all x. So $f(x) = 0$ is both even and odd and is the only such function.

57. True, by Property 2 of limits in Theorem 1.2.

61. True. Suppose instead that $\lim_{x \to 3} g(x)$ does not exist but $\lim_{x \to 3}(f(x)g(x))$ did exist. Since $\lim_{x \to 3} f(x)$ exists and is not zero, then $\lim_{x \to 3}((f(x)g(x))/f(x))$ exists, by Property 4 of limits in Theorem 1.2. Furthermore, $f(x) \neq 0$ for all x in some interval about 3, so $(f(x)g(x))/f(x) = g(x)$ for all x in that interval. Thus $\lim_{x \to 3} g(x)$ exists. This contradicts our assumption that $\lim_{x \to 3} g(x)$ does not exist.

65. False. Although x may be far from c, the value of $f(x)$ could be close to L. For example, suppose $f(x) = L$, the constant function.

CHAPTER TWO

Solutions for Section 2.1

Exercises

1. For t between 2 and 5, we have

$$\text{Average velocity} = \frac{\Delta s}{\Delta t} = \frac{400 - 135}{5 - 2} = \frac{265}{3} \text{ km/hr.}$$

The average velocity on this part of the trip was $265/3$ km/hr.

5. The average velocity over a time period is the change in position divided by the change in time. Since the function $s(t)$ gives the distance of the particle from a point, we read off the graph that $s(1) = 2$ and $s(3) = 6$. Thus,

$$\text{Average velocity} = \frac{\Delta s(t)}{\Delta t} = \frac{s(3) - s(1)}{3 - 1} = \frac{6 - 2}{2} = 2 \text{ meters/sec.}$$

9. (a) Let $s = f(t)$.

(i) We wish to find the average velocity between $t = 1$ and $t = 1.1$. We have

$$\text{Average velocity} = \frac{f(1.1) - f(1)}{1.1 - 1} = \frac{7.84 - 7}{0.1} = 8.4 \text{ m/sec.}$$

(ii) We have

$$\text{Average velocity} = \frac{f(1.01) - f(1)}{1.01 - 1} = \frac{7.0804 - 7}{0.01} = 8.04 \text{ m/sec.}$$

(iii) We have

$$\text{Average velocity} = \frac{f(1.001) - f(1)}{1.001 - 1} = \frac{7.008004 - 7}{0.001} = 8.004 \text{ m/sec.}$$

(b) We see in part (a) that as we choose a smaller and smaller interval around $t = 1$ the average velocity appears to be getting closer and closer to 8, so we estimate the instantaneous velocity at $t = 1$ to be 8 m/sec.

Problems

13. Using $h = 0.1, 0.01, 0.001$, we see

$$\frac{7^{0.1} - 1}{0.1} = 2.148$$

$$\frac{7^{0.01} - 1}{0.01} = 1.965$$

$$\frac{7^{0.001} - 1}{0.001} = 1.948$$

$$\frac{7^{0.0001} - 1}{0.0001} = 1.946.$$

This suggests that $\lim_{h \to 0} \dfrac{7^h - 1}{h} \approx 1.9$.

17. See Figure 2.1.

distance

time

Figure 2.1

21. Since $f(t)$ is concave down between $t = 1$ and $t = 3$, the average velocity between the two times should be less than the instantaneous velocity at $t = 1$ but greater than the instantaneous velocity at time $t = 3$, so $D < A < C$. For analogous reasons, $F < B < E$. Finally, note that f is decreasing at $t = 5$ so $E < 0$, but increasing at $t = 0$, so $D > 0$. Therefore, the ordering from smallest to greatest of the given quantities is

$$F < B < E < 0 < D < A < C.$$

25. $\displaystyle\lim_{h \to 0} \frac{(2+h)^2 - 4}{h} = \lim_{h \to 0} \frac{4 + 4h + h^2 - 4}{h} = \lim_{h \to 0}(4 + h) = 4$

Solutions for Section 2.2

Exercises

1. The derivative, $f'(2)$, is the rate of change of x^3 at $x = 2$. Notice that each time x changes by 0.001 in the table, the value of x^3 changes by 0.012. Therefore, we estimate

$$f'(2) = \begin{array}{c} \text{Rate of change} \\ \text{of } f \text{ at } x = 2 \end{array} \approx \frac{0.012}{0.001} = 12.$$

The function values in the table look exactly linear because they have been rounded. For example, the exact value of x^3 when $x = 2.001$ is 8.012006001, not 8.012. Thus, the table can tell us only that the derivative is approximately 12. Example 5 on page 89 shows how to compute the derivative of $f(x)$ exactly.

5. In Table 2.1, each x increase of 0.001 leads to an increase in $f(x)$ by about 0.031, so

$$f'(3) \approx \frac{0.031}{0.001} = 31.$$

Table 2.1

x	2.998	2.999	3.000	3.001	3.002
$x^3 + 4x$	38.938	38.969	39.000	39.031	39.062

9. Since $f'(x) = 0$ where the graph is horizontal, $f'(x) = 0$ at $x = d$. The derivative is positive at points b and c, but the graph is steeper at $x = c$. Thus $f'(x) = 0.5$ at $x = b$ and $f'(x) = 2$ at $x = c$. Finally, the derivative is negative at points a and e but the graph is steeper at $x = e$. Thus, $f'(x) = -0.5$ at $x = a$ and $f'(x) = -2$ at $x = e$. See Table 2.2.
   Thus, we have $f'(d) = 0$, $f'(b) = 0.5$, $f'(c) = 2$, $f'(a) = -0.5$, $f'(e) = -2$.

Table 2.2

x	$f'(x)$
d	0
b	0.5
c	2
a	-0.5
e	-2

Problems

13. The coordinates of A are $(4, 25)$. See Figure 2.2. The coordinates of B and C are obtained using the slope of the tangent line. Since $f'(4) = 1.5$, the slope is 1.5

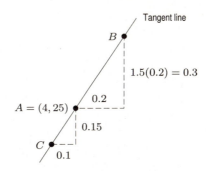

Figure 2.2

From A to B, $\Delta x = 0.2$, so $\Delta y = 1.5(0.2) = 0.3$. Thus, at C we have $y = 25 + 0.3 = 25.3$. The coordinates of B are $(4.2, 25.3)$.

From A to C, $\Delta x = -0.1$, so $\Delta y = 1.5(-0.1) = -0.15$. Thus, at C we have $y = 25 - 0.15 = 24.85$. The coordinates of C are $(3.9, 24.85)$.

17. Figure 2.3 shows the quantities in which we are interested.

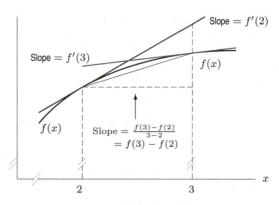

Figure 2.3

The quantities $f'(2)$, $f'(3)$ and $f(3) - f(2)$ have the following interpretations:

- $f'(2) =$ slope of the tangent line at $x = 2$
- $f'(3) =$ slope of the tangent line at $x = 3$
- $f(3) - f(2) = \frac{f(3)-f(2)}{3-2} =$ slope of the secant line from $f(2)$ to $f(3)$.

From Figure 2.3, it is clear that $0 < f(3) - f(2) < f'(2)$. By extending the secant line past the point $(3, f(3))$, we can see that it lies above the tangent line at $x = 3$.

Thus

$$0 < f'(3) < f(3) - f(2) < f'(2).$$

21. (a) Figure 2.4 shows the graph of an even function. We see that since f is symmetric about the y-axis, the tangent line at $x = -10$ is just the tangent line at $x = 10$ flipped about the y-axis, so the slope of one tangent is the negative of that of the other. Therefore, $f'(-10) = -f'(10) = -6$.

(b) From part (a) we can see that if f is even, then for any x, we have $f'(-x) = -f'(x)$. Thus $f'(-0) = -f'(0)$, so $f'(0) = 0$.

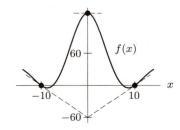

Figure 2.4

25. We want $f'(2)$. The exact answer is

$$f'(2) = \lim_{h \to 0} \frac{f(2+h) - f(2)}{h} = \lim_{h \to 0} \frac{(2+h)^{2+h} - 4}{h},$$

but we can approximate this. If $h = 0.001$, then

$$\frac{(2.001)^{2.001} - 4}{0.001} \approx 6.779$$

and if $h = 0.0001$ then

$$\frac{(2.0001)^{2.0001} - 4}{0.0001} \approx 6.773,$$

so $f'(2) \approx 6.77$.

29. The quantity $f(0)$ represents the population on October 17, 2006, so $f(0) = 300$ million.

The quantity $f'(0)$ represents the rate of change of the population (in millions per year). Since

$$\frac{1 \text{ person}}{11 \text{ seconds}} = \frac{1/10^6 \text{ million people}}{11/(60 \cdot 60 \cdot 24 \cdot 365) \text{ years}} = 2.867 \text{ million people/year},$$

so we have $f'(0) = 2.867$.

33. $\displaystyle \lim_{h \to 0} \frac{(2-h)^3 - 8}{h} = \lim_{h \to 0} \frac{8 - 12h + 6h^2 - h^3 - 8}{h} = \lim_{h \to 0} \frac{h(-12 + 6h - h^2)}{h} = \lim_{h \to 0} -12 + 6h - h^2 = -12.$

37. $\displaystyle \frac{1}{\sqrt{4+h}} - \frac{1}{2} = \frac{2 - \sqrt{4+h}}{2\sqrt{4+h}} = \frac{(2 - \sqrt{4+h})(2 + \sqrt{4+h})}{2\sqrt{4+h}(2 + \sqrt{4+h})} = \frac{4 - (4+h)}{2\sqrt{4+h}(2 + \sqrt{4+h})}.$

Therefore $\displaystyle \lim_{h \to 0} \frac{1}{h} \left(\frac{1}{\sqrt{4+h}} - \frac{1}{2} \right) = \lim_{h \to 0} \frac{-1}{2\sqrt{4+h}(2 + \sqrt{4+h})} = -\frac{1}{16}$

41.

$$f'(1) = \lim_{h \to 0} \frac{f(1+h) - f(1)}{h} = \lim_{h \to 0} \frac{((1+h)^3 + 5) - (1^3 + 5)}{h}$$

$$= \lim_{h \to 0} \frac{1 + 3h + 3h^2 + h^3 + 5 - 1 - 5}{h} = \lim_{h \to 0} \frac{3h + 3h^2 + h^3}{h}$$

$$= \lim_{h \to 0} (3 + 3h + h^2) = 3.$$

45. As we saw in the answer to Problem 39, the slope of the tangent line to $f(x) = x^3$ at $x = -2$ is 12. When $x = -2$, $f(x) = -8$ so we know the point $(-2, -8)$ is on the tangent line. Thus the equation of the tangent line is $y = 12(x + 2) - 8 = 12x + 16$.

Solutions for Section 2.3

Exercises

1. (a) We use the interval to the right of $x = 2$ to estimate the derivative. (Alternately, we could use the interval to the left of 2, or we could use both and average the results.) We have

$$f'(2) \approx \frac{f(4) - f(2)}{4 - 2} = \frac{24 - 18}{4 - 2} = \frac{6}{2} = 3.$$

We estimate $f'(2) \approx 3$.

(b) We know that $f'(x)$ is positive when $f(x)$ is increasing and negative when $f(x)$ is decreasing, so it appears that $f'(x)$ is positive for $0 < x < 4$ and is negative for $4 < x < 12$.

5. The slope of this curve is approximately -1 at $x = -4$ and at $x = 4$, approximately 0 at $x = -2.5$ and $x = 1.5$, and approximately 1 at $x = 0$. See Figure 2.5.

Figure 2.5

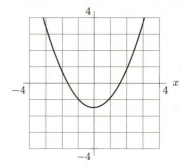

Figure 2.6

9. See Figure 2.6.

13. Since $1/x = x^{-1}$, using the power rule gives

$$k'(x) = (-1)x^{-2} = -\frac{1}{x^2}.$$

Using the definition of the derivative, we have

$$k'(x) = \lim_{h \to 0} \frac{k(x + h) - k(x)}{h} = \lim_{h \to 0} \frac{\frac{1}{x+h} - \frac{1}{x}}{h} = \lim_{h \to 0} \frac{x - (x + h)}{h(x + h)x}$$

$$= \lim_{h \to 0} \frac{-h}{h(x + h)x} = \lim_{h \to 0} \frac{-1}{(x + h)x} = -\frac{1}{x^2}.$$

17. See Figure 2.7.

Figure 2.7

21.

Problems

25. We know that $f'(x) \approx \dfrac{f(x+h) - f(x)}{h}$. For this problem, we'll take the average of the values obtained for $h = 1$

and $h = -1$; that's the average of $f(x+1) - f(x)$ and $f(x) - f(x-1)$ which equals $\dfrac{f(x+1) - f(x-1)}{2}$. Thus,

$f'(0) \approx f(1) - f(0) = 13 - 18 = -5.$
$f'(1) \approx [f(2) - f(0)]/2 = [10 - 18]/2 = -4.$
$f'(2) \approx [f(3) - f(1)]/2 = [9 - 13]/2 = -2.$
$f'(3) \approx [f(4) - f(2)]/2 = [9 - 10]/2 = -0.5.$
$f'(4) \approx [f(5) - f(3)]/2 = [11 - 9]/2 = 1.$
$f'(5) \approx [f(6) - f(4)]/2 = [15 - 9]/2 = 3.$
$f'(6) \approx [f(7) - f(5)]/2 = [21 - 11]/2 = 5.$
$f'(7) \approx [f(8) - f(6)]/2 = [30 - 15]/2 = 7.5.$
$f'(8) \approx f(8) - f(7) = 30 - 21 = 9.$

The rate of change of $f(x)$ is positive for $4 \leq x \leq 8$, negative for $0 \leq x \leq 3$. The rate of change is greatest at about $x = 8$.

29. See Figure 2.8.

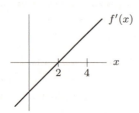

Figure 2.8

33. See Figure 2.9.

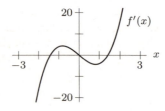

Figure 2.9

37. (a) Graph II
(b) Graph I
(c) Graph III

41. (a) The function f is increasing where f' is positive, so for $x_1 < x < x_3$.
(b) The function f is decreasing where f' is negative, so for $0 < x < x_1$ or $x_3 < x < x_5$.

45. From the given information we know that f is increasing for values of x less than -2, is decreasing between $x = -2$ and $x = 2$, and is constant for $x > 2$. Figure 2.10 shows a possible graph—yours may be different.

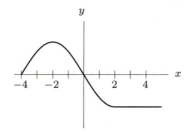

Figure 2.10

49. If $g(x)$ is odd, its graph remains the same if you rotate it $180°$ about the origin. So the tangent line to g at $x = x_0$ is the tangent line to g at $x = -x_0$, rotated $180°$.

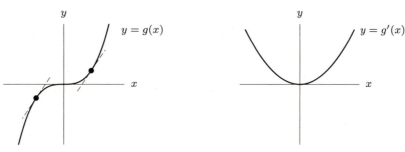

But the slope of a line stays constant if you rotate it $180°$. So $g'(x_0) = g'(-x_0)$; g' is even.

Solutions for Section 2.4

Exercises

1. (a) The function f takes quarts of ice cream to cost in dollars, so 200 is the amount of ice cream, in quarts, and \$600 is the corresponding cost, in dollars. It costs \$600 to produce 200 quarts of ice cream.
(b) Here, 200 is in quarts, but the 2 is in dollars/quart. After producing 200 quarts of ice cream, the cost to produce one additional quart is about \$2.

5. (a) The statement $f(5) = 18$ means that when 5 milliliters of catalyst are present, the reaction will take 18 minutes. Thus, the units for 5 are ml while the units for 18 are minutes.
(b) As in part (a), 5 is measured in ml. Since f' tells how fast T changes per unit a, we have f' measured in minutes/ml. If the amount of catalyst increases by 1 ml (from 5 to 6 ml), the reaction time decreases by about 3 minutes.

9. (a) This means that investing the \$1000 at 5% would yield \$1649 after 10 years.
(b) Writing $g'(r)$ as dB/dr, we see that the units of dB/dr are dollars per percent (interest). We can interpret dB as the extra money earned if interest rate is increased by dr percent. Therefore $g'(5) = \frac{dB}{dr}\big|_{r=5} \approx 165$ means that the balance, at 5% interest, would increase by about \$165 if the interest rate were increased by 1%. In other words, $g(6) \approx g(5) + 165 = 1649 + 165 = 1814$.

Problems

13. (a) Since $W = f(c)$ where W is weight in pounds and c is the number of Calories consumed per day:

$$f(1800) = 155 \quad \text{means that} \quad \text{consuming 1800 Calories per day results in a weight of 155 pounds.}$$

$$f'(2000) = 0 \quad \text{means that} \quad \text{consuming 2000 Calories per day causes neither weight gain nor loss.}$$

$$f^{-1}(162) = 2200 \quad \text{means that} \quad \text{a weight of 162 pounds is caused by a consumption of 2200 Calories per day.}$$

(b) The units of dW/dc are pounds/(Calories/day).

17. Let p be the rating points earned by the CBS Evening News, let R be the revenue earned in millions of dollars, and let $R = f(p)$. When $p = 4.3$,

$$\text{Rate of change of revenue} = \frac{\$5.5 \text{ million}}{0.1 \text{ point}} = 55 \text{ million dollars/point.}$$

Thus

$$f'(4.3) = 55.$$

21. Units of $g'(55)$ are mpg/mph. The statement $g'(55) = -0.54$ means that at 55 miles per hour the fuel efficiency (in miles per gallon, or mpg) of the car decreases at a rate of approximately one half mpg as the velocity increases by one mph.

25. (a) The company hopes that increased advertising always brings in more customers instead of turning them away. There-fore, it hopes $f'(a)$ is always positive.

(b) If $f'(100) = 2$, it means that if the advertising budget is $100,000, each extra dollar spent on advertising will bring in $2 worth of sales. If $f'(100) = 0.5$, each dollar above $100 thousand spent on advertising will bring in $0.50 worth of sales.

(c) If $f'(100) = 2$, then as we saw in part (b), spending slightly more than $100,000 will increase revenue by an amount greater than the additional expense, and thus more should be spent on advertising. If $f'(100) = 0.5$, then the increase in revenue is less than the additional expense, hence too much is being spent on advertising. The optimum amount to spend is an amount that makes $f'(a) = 1$. At this point, the increases in advertising expenditures just pay for themselves. If $f'(a) < 1$, too much is being spent; if $f'(a) > 1$, more should be spent.

29. Solving for $dp/d\delta$, we get

$$\frac{dp}{d\delta} = \left(\frac{p}{\delta + (p/c^2)} \right) \gamma.$$

(a) For $\delta \approx 10$ g/cm^3, we have $\log \delta \approx 1$, so, from Figure 2.40 in the text, we have $\gamma \approx 2.6$ and $\log p \approx 13$.
Thus $p \approx 10^{13}$, so $p/c^2 \approx 10^{13}/(9 \cdot 10^{20}) \approx 10^{-8}$, and

$$\frac{dp}{d\delta} \approx \frac{10^{13}}{10 + 10^{-8}} 2.6 \approx 2.6 \cdot 10^{12}.$$

The derivative can be interpreted as the ratio between a change in pressure and the corresponding change in density. The fact that it is so large says that a very large change in pressure brings about a very small change in density. This says that cold iron is not a very compressible material.

(b) For $\delta \approx 10^6$, we have $\log \delta \approx 6$, so, from Figure 2.40 in the text, $\gamma \approx 1.5$ and $\log p \approx 23$.
Thus $p \approx 10^{23}$, so $p/c^2 \approx 10^{23}/(9 \cdot 10^{20}) \approx 10^2$, and

$$\frac{dp}{d\delta} \approx \frac{10^{23}}{10^6 + 10^2} 1.5 \approx 1.5 \cdot 10^{17}.$$

This tells us that the matter in a white dwarf is even less compressible than cold iron.

Solutions for Section 2.5

Exercises

1. (a) Since the graph is below the x-axis at $x = 2$, the value of $f(2)$ is negative.
(b) Since $f(x)$ is decreasing at $x = 2$, the value of $f'(2)$ is negative.
(c) Since $f(x)$ is concave up at $x = 2$, the value of $f''(2)$ is positive.

5. The graph must be everywhere decreasing and concave up on some intervals and concave down on other intervals. One possibility is shown in Figure 2.11.

Figure 2.11

9. $f'(x) < 0$
$f''(x) = 0$

13. The velocity is the derivative of the distance, that is, $v(t) = s'(t)$. Therefore, we have

$$v(t) = \lim_{h \to 0} \frac{s(t+h) - s(t)}{h}$$

$$= \lim_{h \to 0} \frac{(5(t+h)^2 + 3) - (5t^2 + 3)}{h}$$

$$= \lim_{h \to 0} \frac{10th + 5h^2}{h}$$

$$= \lim_{h \to 0} \frac{h(10t + 5h)}{h} = \lim_{h \to 0} (10t + 5h) = 10t$$

The acceleration is the derivative of velocity, so $a(t) = v'(t)$:

$$a(t) = \lim_{h \to 0} \frac{10(t+h) - 10t}{h}$$

$$= \lim_{h \to 0} \frac{10h}{h} = 10.$$

Problems

17. See Figure 2.12.

Figure 2.12

21. See Figure 2.13.

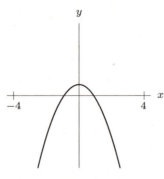

Figure 2.13

25. Suppose $p(t)$ is the average price level at time t. Then, if $t_0 = $ April 1991,

"Prices are still rising" means $p'(t_0) > 0$.

"Prices rising less fast than they were" means $p''(t_0) < 0$.

"Prices rising not as much less fast as everybody had hoped" means $H < p''(t_0)$, where H is the rate of change in rate of change of prices that people had hoped for.

29. Since f' is everywhere positive, f is everywhere increasing. Hence the greatest value of f is at x_6 and the least value of f is at x_1. Directly from the graph, we see that f' is greatest at x_3 and least at x_2. Since f'' gives the slope of the graph of f', f'' is greatest where f' is rising most rapidly, namely at x_6, and f'' is least where f' is falling most rapidly, namely at x_1.

Solutions for Section 2.6

Exercises

1. (a) Function f is not continuous at $x = 1$.
(b) Function f appears not differentiable at $x = 1, 2, 3$.

Problems

5. Yes, f is differentiable at $x = 0$, since its graph does not have a "corner" at $x = 0$. See below.

Another way to see this is by computing:

$$\lim_{h \to 0} \frac{f(h) - f(0)}{h} = \lim_{h \to 0} \frac{(h + |h|)^2}{h} = \lim_{h \to 0} \frac{h^2 + 2h|h| + |h|^2}{h}.$$

Since $|h|^2 = h^2$, we have:

$$\lim_{h \to 0} \frac{f(h) - f(0)}{h} = \lim_{h \to 0} \frac{2h^2 + 2h|h|}{h} = \lim_{h \to 0} 2(h + |h|) = 0.$$

So f is differentiable at 0 and $f'(0) = 0$.

9. We want to look at

$$\lim_{h \to 0} \frac{(h^2 + 0.0001)^{1/2} - (0.0001)^{1/2}}{h}.$$

As $h \to 0$ from positive or negative numbers, the difference quotient approaches 0. (Try evaluating it for $h = 0.001$, 0.0001, etc.) So it appears there is a derivative at $x = 0$ and that this derivative is zero. How can this be if f has a corner at $x = 0$?

The answer lies in the fact that what appears to be a corner is in fact smooth—when you zoom in, the graph of f looks like a straight line with slope 0! See Figure 2.14.

 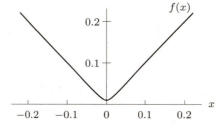

Figure 2.14: Close-ups of $f(x) = (x^2 + 0.0001)^{1/2}$ showing differentiability at $x = 0$

13. (a) Since

$$\lim_{r \to r_0^-} E = kr_0$$

and

$$\lim_{r \to r_0^+} E = \frac{kr_0^2}{r_0} = kr_0$$

and

$$E(r_0) = kr_0,$$

we see that E is continuous at r_0.

(b) The function E is not differentiable at $r = r_0$ because the graph has a corner there. The slope is positive for $r < r_0$ and the slope is negative for $r > r_0$.

(c) See Figure 2.15.

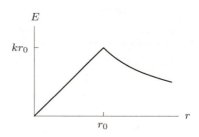

Figure 2.15

Solutions for Chapter 2 Review

Exercises

1. The average velocity over a time period is the change in position divided by the change in time. Since the function $s(t)$ gives the position of the particle, we find the values of $s(3) = 12 \cdot 3 - 3^2 = 27$ and $s(1) = 12 \cdot 1 - 1^2 = 11$. Using these values, we find

$$\text{Average velocity} = \frac{\Delta s(t)}{\Delta t} = \frac{s(3) - s(1)}{3 - 1} = \frac{27 - 11}{2} = 8 \text{ mm/sec.}$$

5. The average velocity over a time period is the change in position divided by the change in time. Since the function $s(t)$ gives the position of the particle, we find the values on the graph of $s(3) = 2$ and $s(1) = 3$. Using these values, we find

$$\text{Average velocity } = \frac{\Delta s(t)}{\Delta t} = \frac{s(3) - s(1)}{3 - 1} = \frac{2 - 3}{2} = -\frac{1}{2} \text{ mm/sec.}$$

9. See Figure 2.16.

Figure 2.16

13. See Figure 2.17.

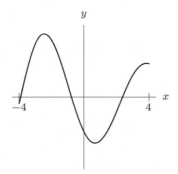

Figure 2.17

17. We need to look at the difference quotient and take the limit as h approaches zero. The difference quotient is

$$\frac{f(3 + h) - f(3)}{h} = \frac{[(3 + h)^2 + 1] - 10}{h} = \frac{9 + 6h + h^2 + 1 - 10}{h} = \frac{6h + h^2}{h} = \frac{h(6 + h)}{h}.$$

Since $h \neq 0$, we can divide by h in the last expression to get $6 + h$. Now the limit as h goes to 0 of $6 + h$ is 6, so

$$f'(3) = \lim_{h \to 0} \frac{h(6 + h)}{h} = \lim_{h \to 0} (6 + h) = 6.$$

So at $x = 3$, the slope of the tangent line is 6. Since $f(3) = 3^2 + 1 = 10$, the tangent line passes through $(3, 10)$, so its equation is

$$y - 10 = 6(x - 3), \quad \text{or} \quad y = 6x - 8.$$

21. $\lim\limits_{h \to 0} \dfrac{1}{h} \left(\dfrac{1}{(a + h)^2} - \dfrac{1}{a^2} \right) = \lim\limits_{h \to 0} \dfrac{a^2 - (a^2 + 2ah + h^2)}{(a + h)^2 a^2 h} = \lim\limits_{h \to 0} \dfrac{(-2a - h)}{(a + h)^2 a^2} = \dfrac{-2}{a^3}$

Problems

25. First note that the line $y = t$ has slope 1. From the graph, we see that

$$0 < \text{Slope at } C < \text{Slope at } B < \text{Slope between } A \text{ and } B < 1 < \text{Slope at } A.$$

Since the instantaneous velocity is represented by the slope, we have

$0 < $ Instantaneous velocity at $C <$ Instantaneous velocity at $B <$ Average velocity between A and $B < 1 <$ Instantaneous velocity at A

29. (a) A possible example is $f(x) = 1/|x - 2|$ as $\lim_{x \to 2} 1/|x - 2| = \infty$.

   (b) A possible example is $f(x) = -1/(x - 2)^2$ as $\lim_{x \to 2} -1/(x - 2)^2 = -\infty$.

33. (a)

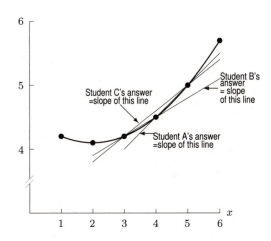

   (b) The slope of f appears to be somewhere between student A's answer and student B's, so student C's answer, halfway in between, is probably the most accurate.

   (c) Student A's estimate is $f'(x) \approx \frac{f(x+h) - f(x)}{h}$, while student B's estimate is $f'(x) \approx \frac{f(x) - f(x-h)}{h}$. Student C's estimate is the average of these two, or

$$f'(x) \approx \frac{1}{2}\left[\frac{f(x+h) - f(x)}{h} + \frac{f(x) - f(x-h)}{h}\right] = \frac{f(x+h) - f(x-h)}{2h}.$$

This estimate is the slope of the chord connecting $(x - h, f(x - h))$ to $(x + h, f(x + h))$. Thus, we estimate that the tangent to a curve is nearly parallel to a chord connecting points h units to the right and left, as shown below.

37. (a) The yam is cooling off so T is decreasing and $f'(t)$ is negative.

   (b) Since $f(t)$ is measured in degrees Fahrenheit and t is measured in minutes, df/dt must be measured in units of °F/min.

41. (a) Slope of tangent line $= \lim_{h \to 0} \frac{\sqrt{4+h} - \sqrt{4}}{h}$. Using $h = 0.001$, $\frac{\sqrt{4.001} - \sqrt{4}}{0.001} = 0.249984$. Hence the slope of the tangent line is about 0.25.

   (b)

$$y - y_1 = m(x - x_1)$$
$$y - 2 = 0.25(x - 4)$$
$$y - 2 = 0.25x - 1$$
$$y = 0.25x + 1$$

   (c) $f(x) = kx^2$

   If $(4, 2)$ is on the graph of f, then $f(4) = 2$, so $k \cdot 4^2 = 2$. Thus $k = \frac{1}{8}$, and $f(x) = \frac{1}{8}x^2$.

   (d) To find where the graph of f crosses then line $y = 0.25x + 1$, we solve:

$$\frac{1}{8}x^2 = 0.25x + 1$$

$$\begin{aligned} x^2 &= 2x + 8 \\ x^2 - 2x - 8 &= 0 \\ (x-4)(x+2) &= 0 \\ x = 4 \ \ \text{or} \ \ x &= -2 \\ f(-2) &= \frac{1}{8}(4) = 0.5 \end{aligned}$$

Therefore, $(-2, 0.5)$ is the other point of intersection. (Of course, $(4, 2)$ is a point of intersection; we know that from the start.)

45. (a) The graph looks straight because the graph shows only a small part of the curve magnified greatly.

(b) The month is March: We see that about the 21st of the month there are twelve hours of daylight and hence twelve hours of night. This phenomenon (the length of the day equaling the length of the night) occurs at the equinox, midway between winter and summer. Since the length of the days is increasing, and Madrid is in the northern hemisphere, we are looking at March, not September.

(c) The slope of the curve is found from the graph to be about 0.04 (the rise is about 0.8 hours in 20 days or 0.04 hours/day). This means that the amount of daylight is increasing by about 0.04 hours (about $2\frac{1}{2}$ minutes) per calendar day, or that each day is $2\frac{1}{2}$ minutes longer than its predecessor.

CAS Challenge Problems

49. The CAS says the derivative is zero. This can be explained by the fact that $f(x) = \sin^2 x + \cos^2 x = 1$, so $f'(x)$ is the derivative of the constant function 1. The derivative of a constant function is zero.

53. (a) The computer algebra system gives

$$\frac{d}{dx}(x^2 + 1)^2 = 4x(x^2 + 1)$$

$$\frac{d}{dx}(x^2 + 1)^3 = 6x(x^2 + 1)^2$$

$$\frac{d}{dx}(x^2 + 1)^4 = 8x(x^2 + 1)^3$$

(b) The pattern suggests that

$$\frac{d}{dx}(x^2 + 1)^n = 2nx(x^2 + 1)^{n-1}.$$

Taking the derivative of $(x^2 + 1)^n$ with a CAS confirms this.

CHECK YOUR UNDERSTANDING

1. False. For example, the car could slow down or even stop at one minute after 2 pm, and then speed back up to 60 mph at one minute before 3 pm. In this case the car would travel only a few miles during the hour, much less than 50 miles.

5. True. By definition, Average velocity = Distance traveled/Time.

9. False. If $f'(x)$ is increasing then $f(x)$ is concave up. However, $f(x)$ may be either increasing or decreasing. For example, the exponential decay function $f(x) = e^{-x}$ is decreasing but $f'(x)$ is increasing because the graph of f is concave up.

13. False. The function $f(x)$ may be discontinuous at $x = 0$, for instance $f(x) = \begin{cases} 0 \text{ if } x \le 0 \\ 1 \text{ if } x > 0 \end{cases}$. The graph of f may have a vertical tangent line at $x = 0$, for instance $f(x) = x^{1/3}$.

17. True. Instantaneous acceleration is a derivative, and all derivatives are limits of difference quotients. More precisely, instantaneous acceleration $a(t)$ is the derivative of the velocity $v(t)$, so

$$a(t) = \lim_{h \to 0} \frac{v(t+h) - v(t)}{h}.$$

21. True; $f(x) = x^3$ is increasing over any interval.

25. False. Being continuous does not imply differentiability. For example, $f(x) = |x|$ is continuous but not differentiable at $x = 0$.

CHAPTER THREE

Solutions for Section 3.1

Exercises

1. The derivative, $f'(x)$, is defined as

$$f'(x) = \lim_{h \to 0} \frac{f(x+h) - f(x)}{h}.$$

If $f(x) = 7$, then

$$f'(x) = \lim_{h \to 0} \frac{7 - 7}{h} = \lim_{h \to 0} \frac{0}{h} = 0.$$

5. $y' = \pi x^{\pi - 1}$.   (power rule)

9. $y' = 11x^{-12}$.

13. $y' = -\frac{3}{4} x^{-7/4}$.

17. Since $y = \dfrac{1}{r^{7/2}} = r^{-7/2}$, we have $\dfrac{dy}{dx} = -\dfrac{7}{2} r^{-9/2}$.

21. Since $f(x) = \sqrt{\dfrac{1}{x^3}} = \dfrac{1}{x^{3/2}} = x^{-3/2}$, we have $f'(x) = -\dfrac{3}{2} x^{-5/2}$.

25. $y' = 17 + 12x^{-1/2}$.

29. $y' = 18x^2 + 8x - 2$.

33. Since $y = t^{3/2}(2 + \sqrt{t}) = 2t^{3/2} + t^{3/2}t^{1/2} = 2t^{3/2} + t^2$, we have $\frac{dy}{dx} = 3t^{1/2} + 2t$.

37. $f(z) = \dfrac{z}{3} + \dfrac{1}{3}z^{-1} = \dfrac{1}{3}\left(z + z^{-1}\right)$, so $f'(z) = \dfrac{1}{3}\left(1 - z^{-2}\right) = \dfrac{1}{3}\left(\dfrac{z^2 - 1}{z^2}\right)$.

41. Since $f(x) = \dfrac{ax + b}{x} = \dfrac{ax}{x} + \dfrac{b}{x} = a + bx^{-1}$, we have $f'(x) = -bx^{-2}$.

45. Since w is a constant times q, we have $dw/dq = 3ab^2$.

Problems

49. The x is in the exponent and we have not learned how to handle that yet.

53. $y' = -2/3z^3$.   (power rule and sum rule)

57. The slopes of the tangent lines to $y = x^2 - 2x + 4$ are given by $y' = 2x - 2$. A line through the origin has equation $y = mx$. So, at the tangent point, $x^2 - 2x + 4 = mx$ where $m = y' = 2x - 2$.

$$x^2 - 2x + 4 = (2x - 2)x$$
$$x^2 - 2x + 4 = 2x^2 - 2x$$
$$-x^2 + 4 = 0$$
$$-(x + 2)(x - 2) = 0$$
$$x = 2, -2.$$

Thus, the points of tangency are $(2, 4)$ and $(-2, 12)$. The lines through these points and the origin are $y = 2x$ and $y = -6x$, respectively. Graphically, this can be seen in Figure 3.1.

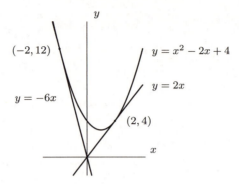

Figure 3.1

61. (a) Since the power of x will go down by one every time you take a derivative (until the exponent is zero after which the derivative will be zero), we can see immediately that $f^{(8)}(x) = 0$.

(b) $f^{(7)}(x) = 7 \cdot 6 \cdot 5 \cdot 4 \cdot 3 \cdot 2 \cdot 1 \cdot x^0 = 5040$.

65. (a) The average velocity between $t = 0$ and $t = 2$ is given by

$$\text{Average velocity} = \frac{f(2) - f(0)}{2 - 0} = \frac{-4.9(2^2) + 25(2) + 3 - 3}{2 - 0} = \frac{33.4 - 3}{2} = 15.2 \text{ m/sec.}$$

(b) Since $f'(t) = -9.8t + 25$, we have

$$\text{Instantaneous velocity} = f'(2) = -9.8(2) + 25 = 5.4 \text{ m/sec.}$$

(c) Acceleration is given $f''(t) = -9.8$. The acceleration at $t = 2$ (and all other times) is the acceleration due to gravity, which is -9.8 m/sec^2.

(d) We can use a graph of height against time to estimate the maximum height of the tomato. See Figure 3.2. Alternately, we can find the answer analytically. The maximum height occurs when the velocity is zero and $v(t) = -9.8t + 25 = 0$ when $t = 2.6$ sec. At this time the tomato is at a height of $f(2.6) = 34.9$. The maximum height is 34.9 meters.

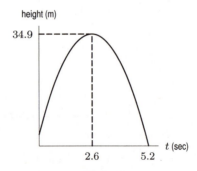

Figure 3.2

(e) We see in Figure 3.2 that the tomato hits ground at about $t = 5.2$ seconds. Alternately, we can find the answer analytically. The tomato hits the ground when

$$f(t) = -4.9t^2 + 25t + 3 = 0.$$

We solve for t using the quadratic formula:

$$t = \frac{-25 \pm \sqrt{(25)^2 - 4(-4.9)(3)}}{2(-4.9)}$$

$$t = \frac{-25 \pm \sqrt{683.8}}{-9.8}$$

$$t = -0.12 \quad \text{and} \quad t = 5.2.$$

We use the positive values, so the tomato hits the ground at $t = 5.2$ seconds.

69. $V = \frac{4}{3}\pi r^3$. Differentiating gives $\frac{dV}{dr} = 4\pi r^2$ = surface area of a sphere.

The difference quotient $\frac{V(r+h)-V(r)}{h}$ is the volume between two spheres divided by the change in radius. Furthermore, when h is very small, the difference between volumes, $V(r+h) - V(r)$, is like a coating of paint of depth h applied to the surface of the sphere. The volume of the paint is about $h \cdot$ (Surface Area) for small h: dividing by h gives back the surface area.

Thinking about the derivative as the rate of change of the function for a small change in the variable gives another way of seeing the result. If you increase the radius of a sphere a small amount, the volume increases by a very thin layer whose volume is the surface area at that radius multiplied by that small amount.

73. (a)

$$\frac{d(x^{-1})}{dx} = \lim_{h \to 0} \frac{(x+h)^{-1} - x^{-1}}{h} = \lim_{h \to 0} \frac{1}{h}\left[\frac{1}{x+h} - \frac{1}{x}\right]$$

$$= \lim_{h \to 0} \frac{1}{h}\left[\frac{x - (x+h)}{x(x+h)}\right] = \lim_{h \to 0} \frac{1}{h}\left[\frac{-h}{x(x+h)}\right]$$

$$= \lim_{h \to 0} \frac{-1}{x(x+h)} = \frac{-1}{x^2} = -1x^{-2}.$$

$$\frac{d(x^{-3})}{dx} = \lim_{h \to 0} \frac{(x+h)^{-3} - x^{-3}}{h}$$

$$= \lim_{h \to 0} \frac{1}{h}\left[\frac{1}{(x+h)^3} - \frac{1}{x^3}\right]$$

$$= \lim_{h \to 0} \frac{1}{h}\left[\frac{x^3 - (x+h)^3}{x^3(x+h)^3}\right]$$

$$= \lim_{h \to 0} \frac{1}{h}\left[\frac{x^3 - (x^3 + 3hx^2 + 3h^2x + h^3)}{x^3(x+h)^3}\right]$$

$$= \lim_{h \to 0} \frac{1}{h}\left[\frac{-3hx^2 - 3xh^2 - h^3}{x^3(x+h)^3}\right]$$

$$= \lim_{h \to 0} \frac{-3x^2 - 3xh - h^2}{x^3(x+h)^3}$$

$$= \frac{-3x^2}{x^6} = -3x^{-4}.$$

(b) For clarity, let $n = -k$, where k is a positive integer. So $x^n = x^{-k}$.

$$\frac{d(x^{-k})}{dx} = \lim_{h \to 0} \frac{(x+h)^{-k} - x^{-k}}{h}$$

$$= \lim_{h \to 0} \frac{1}{h}\left[\frac{1}{(x+h)^k} - \frac{1}{x^k}\right]$$

$$= \lim_{h \to 0} \frac{1}{h}\left[\frac{x^k - (x+h)^k}{x^k(x+h)^k}\right]$$

terms involving h^2 and higher powers of h

$$= \lim_{h \to 0} \frac{1}{h}\left[\frac{x^k - x^k - khx^{k-1} - \overbrace{\cdots - h^k}}{x^k(x+h)^k}\right]$$

$$= \frac{-kx^{k-1}}{x^k(x)^k} = \frac{-k}{x^{k+1}} = -kx^{-(k+1)} = -kx^{-k-1}.$$

Solutions for Section 3.2

Exercises

1. $f'(x) = 2e^x + 2x$.

5. $y' = 10x + (\ln 2)2^x$.

9. Since $y = 2^x + \dfrac{2}{x^3} = 2^x + 2x^{-3}$, we have $\dfrac{dy}{dx} = (\ln 2)2^x - 6x^{-4}$.

13. $f'(t) = (\ln(\ln 3))(\ln 3)^t$.

17. $f'(x) = 3x^2 + 3^x \ln 3$

21. Since e and k are constants, e^k is constant, so we have $f'(x) = (\ln k)k^x$.

25. $y'(x) = a^x \ln a + ax^{a-1}$.

29. We can take the derivative of the sum $x^2 + 2^x$, but not the product.

33. The exponent is x^2, and we haven't learned what to do about that yet.

Problems

37. $f(s) = 5^s e^s = (5e)^s$, so $f'(s) = \ln(5e) \cdot (5e)^s = (1 + \ln 5)5^s e^s$.

41. (a) $P = 10.186(0.997)^{20} \approx 9.592$ million.

(b) Differentiating, we have

$$\frac{dP}{dt} = 10.186(\ln 0.997)(0.997)^t$$

so $\left.\dfrac{dP}{dt}\right|_{t=20} = 10.186(\ln 0.997)(0.997)^{20} \approx -0.0325$ million/year.

Thus in 2020, Hungary's population will be decreasing by 32,500 people per year.

45. Since $y = 2^x$, $y' = (\ln 2)2^x$. At $(0, 1)$, the tangent line has slope $\ln 2$ so its equation is $y = (\ln 2)x + 1$. At c, $y = 0$, so $0 = (\ln 2)c + 1$, thus $c = -\dfrac{1}{\ln 2}$.

49. We are interested in when the derivative $\dfrac{d(a^x)}{dx}$ is positive and when it is negative. The quantity a^x is always positive. However $\ln a > 0$ for $a > 1$ and $\ln a < 0$ for $0 < a < 1$. Thus the function a^x is increasing for $a > 1$ and decreasing for $a < 1$.

Solutions for Section 3.3

Exercises

1. By the product rule, $f'(x) = 2x(x^3 + 5) + x^2(3x^2) = 2x^4 + 3x^4 + 10x = 5x^4 + 10x$. Alternatively, $f'(x) = (x^5 + 5x^2)' = 5x^4 + 10x$. The two answers should, and do, match.

5. $y' = \frac{1}{2\sqrt{x}}2^x + \sqrt{x}(\ln 2)2^x$.

9. $f'(y) = (\ln 4)4^y(2 - y^2) + 4^y(-2y) = 4^y((\ln 4)(2 - y^2) - 2y)$.

13. $\dfrac{dy}{dx} = \dfrac{1 \cdot 2^t - (t + 1)(\ln 2)2^t}{(2^t)^2} = \dfrac{2^t(1 - (t + 1)\ln 2)}{(2^t)^2} = \dfrac{1 - (t + 1)\ln 2}{2^t}$

17. $\dfrac{dz}{dt} = \dfrac{3(5t + 2) - (3t + 1)5}{(5t + 2)^2} = \dfrac{15t + 6 - 15t - 5}{(5t + 2)^2} = \dfrac{1}{(5t + 2)^2}$.

21. $w = y^2 - 6y + 7$. $w' = 2y - 6, y \neq 0$.

25. $h'(r) = \dfrac{d}{dr}\left(\dfrac{r^2}{2r + 1}\right) = \dfrac{(2r)(2r + 1) - 2r^2}{(2r + 1)^2} = \dfrac{2r(r + 1)}{(2r + 1)^2}$.

29.

$$f'(x) = \frac{(2 + 3x + 4x^2)(1) - (1 + x)(3 + 8x)}{(2 + 3x + 4x^2)^2}$$

$$= \frac{2 + 3x + 4x^2 - 3 - 11x - 8x^2}{(2 + 3x + 4x^2)^2}$$

$$= \frac{-4x^2 - 8x - 1}{(2 + 3x + 4x^2)^2}.$$

Problems

33. Using the quotient rule, we know that $j'(x) = (g'(x) \cdot f(x) - g(x) \cdot f'(x))/(f(x))^2$. We use slope to compute the derivatives. Since $f(x)$ is linear on the interval $0 < x < 2$, we compute the slope of the line to see that $f'(x) = 2$ on this interval. Similarly, we compute the slope on the interval $2 < x < 4$ to see that $f'(x) = -2$ on the interval $2 < x < 4$. Since $f(x)$ has a corner at $x = 2$, we know that $f'(2)$ does not exist.

Similarly, $g(x)$ is linear on the interval shown, and we see that the slope of $g(x)$ on this interval is -1 so we have $g'(x) = -1$ on this interval.

(a) We have
$$j'(1) = \frac{g'(1) \cdot f(1) - g(1) \cdot f'(1)}{(f(1))^2} = \frac{(-1)2 - 3 \cdot 2}{2^2} = \frac{-2 - 6}{4} = \frac{-8}{4} = -2.$$

(b) We have $j'(2) = (g'(2) \cdot f(2) - g(2) \cdot f'(2))/(f(2)^2)$. Since $f(x)$ has a corner at $x = 2$, we know that $f'(2)$ does not exist. Therefore, $j'(2)$ does not exist.

(c) We have
$$j'(3) = \frac{g'(3) \cdot f(3) - g(3) \cdot f'(3)}{(f(3))^2} = \frac{(-1)2 - 1(-2)}{2^2} = \frac{-2 + 2}{4} = 0.$$

37. Estimates may vary. From the graphs, we estimate $f(2) \approx 0.3$, $f'(2) \approx 1.1$, $g(2) \approx 1.6$, and $g'(2) \approx -0.5$. By the quotient rule, to one decimal place
$$k'(2) = \frac{f'(2) \cdot g(2) - f(2) \cdot g'(2)}{(g(2))^2} \approx \frac{1.1(1.6) - 0.3(-0.5)}{(1.6)^2} = 0.7.$$

41. $f(x) = e^x \cdot e^x$
$f'(x) = e^x \cdot e^x + e^x \cdot e^x = 2e^{2x}.$

45. Since $f(0) = -5/1 = -5$, the tangent line passes through the point $(0, -5)$, so its vertical intercept is -5. To find the slope of the tangent line, we find the derivative of $f(x)$ using the quotient rule:
$$f'(x) = \frac{(x + 1) \cdot 2 - (2x - 5) \cdot 1}{(x + 1)^2} = \frac{7}{(x + 1)^2}.$$

At $x = 0$, the slope of the tangent line is $m = f'(0) = 7$. The equation of the tangent line is $y = 7x - 5$.

49. (a) $G'(z) = F'(z)H(z) + H'(z)F(z)$, so
$G'(3) = F'(3)H(3) + H'(3)F(3) = 4 \cdot 1 + 3 \cdot 5 = 19.$

(b) $G'(w) = \dfrac{F'(w)H(w) - H'(w)F(w)}{[H(w)]^2}$, so $G'(3) = \dfrac{4(1) - 3(5)}{1^2} = -11.$

53. (a) If the museum sells the painting and invests the proceeds $P(t)$ at time t, then t years have elapsed since 2000, and the time span up to 2020 is $20 - t$. This is how long the proceeds $P(t)$ are earning interest in the bank. Each year the money is in the bank it earns 5% interest, which means the amount in the bank is multiplied by a factor of 1.05. So, at the end of $(20 - t)$ years, the balance is given by
$$B(t) = P(t)(1 + 0.05)^{20-t} = P(t)(1.05)^{20-t}.$$

(b)
$$B(t) = P(t)(1.05)^{20}(1.05)^{-t} = (1.05)^{20}\frac{P(t)}{(1.05)^t}.$$

(c) By the quotient rule,
$$B'(t) = (1.05)^{20}\left[\frac{P'(t)(1.05)^t - P(t)(1.05)^t \ln 1.05}{(1.05)^{2t}}\right].$$

So,

$$B'(10) = (1.05)^{20} \left[\frac{5000(1.05)^{10} - 150,000(1.05)^{10} \ln 1.05}{(1.05)^{20}} \right]$$

$$= (1.05)^{10}(5000 - 150,000 \ln 1.05)$$

$$\approx -3776.63.$$

57. From the answer to Problem 56, we find that

$$f'(x) = (x - r_1)(x - r_2) \cdots (x - r_{n-1}) \cdot 1$$
$$+ (x - r_1)(x - r_2) \cdots (x - r_{n-2}) \cdot 1 \cdot (x - r_n)$$
$$+ (x - r_1)(x - r_2) \cdots (x - r_{n-3}) \cdot 1 \cdot (x - r_{n-1})(x - r_n)$$
$$+ \cdots + 1 \cdot (x - r_2)(x - r_3) \cdots (x - r_n)$$
$$= f(x) \left(\frac{1}{x - r_1} + \frac{1}{x - r_2} + \cdots + \frac{1}{x - r_n} \right).$$

Solutions for Section 3.4

Exercises

1. $f'(x) = 99(x + 1)^{98} \cdot 1 = 99(x + 1)^{98}$.

5. $\dfrac{d}{dx}(\sqrt{e^x + 1}) = \dfrac{d}{dx}(e^x + 1)^{1/2} = \dfrac{1}{2}(e^x + 1)^{-1/2} \dfrac{d}{dx}(e^x + 1) = \dfrac{e^x}{2\sqrt{e^x + 1}}$.

9. $k'(x) = 4(x^3 + e^x)^3(3x^2 + e^x)$.

13. $f(\theta) = (2^{-1})^\theta = (\frac{1}{2})^\theta$ so $f'(\theta) = (\ln \frac{1}{2})2^{-\theta}$.

17. $p'(t) = 4e^{4t+2}$.

21. $y' = -4e^{-4t}$.

25. $z'(x) = \dfrac{(\ln 2)2^x}{3\sqrt[3]{(2^x + 5)^2}}$.

29. We can write this as $f(z) = \sqrt{z}e^{-z}$, in which case it is the same as problem 24. So $f'(z) = \dfrac{1}{2\sqrt{z}}e^{-z} - \sqrt{z}e^{-z}$.

33. $\dfrac{dy}{dx} = \dfrac{2e^{2x}(x^2 + 1) - e^{2x}(2x)}{(x^2 + 1)^2} = \dfrac{2e^{2x}(x^2 + 1 - x)}{(x^2 + 1)^2}$

37. $w' = (2t + 3)(1 - e^{-2t}) + (t^2 + 3t)(2e^{-2t})$.

41.
$$f'(w) = (e^{w^2})(10w) + (5w^2 + 3)(e^{w^2})(2w)$$
$$= 2we^{w^2}(5 + 5w^2 + 3)$$
$$= 2we^{w^2}(5w^2 + 8).$$

45. $f'(y) = e^{e^{(y^2)}} \left[(e^{y^2})(2y) \right] = 2ye^{[e^{(y^2)} + y^2]}$.

49. We use the product rule. We have

$$f'(x) = (ax)(e^{-bx}(-b)) + (a)(e^{-bx}) = -abxe^{-bx} + ae^{-bx}.$$

Problems

53. Using the chain rule, we know that $v'(x) = f'(f(x)) \cdot f'(x)$. We use slope to compute the derivatives. Since $f(x)$ is linear on the interval $0 < x < 2$, we compute the slope of the line to see that $f'(x) = 2$ on this interval. Similarly, we compute the slope on the interval $2 < x < 4$ to see that $f'(x) = -2$ on the interval $2 < x < 4$. Since $f(x)$ has a corner at $x = 2$, we know that $f'(2)$ does not exist.

    (a) We have $v'(1) = f'(f(1)) \cdot f'(1) = f'(2) \cdot 2$. Since $f(x)$ has a corner at $x = 2$, we know that $f'(2)$ does not exist. Therefore, $v'(1)$ does not exist.

    (b) We have $v'(2) = f'(f(2)) \cdot f'(2)$. Since $f(x)$ has a corner at $x = 2$, we know that $f'(2)$ does not exist. Therefore, $v'(2)$ does not exist.

    (c) We have $v'(3) = f'(f(3)) \cdot f'(3) = (f'(2))(-2)$. Since $f(x)$ has a corner at $x = 2$, we know that $f'(2)$ does not exist. Therefore, $v'(3)$ does not exist.

57. The chain rule gives

$$\frac{d}{dx} g(f(x)) \bigg|_{x=30} = g'(f(30))f'(30) = g'(20)f'(30) = (1/2)(-2) = -1.$$

61. The graph is concave down when $f''(x) < 0$.

$$f'(x) = e^{-x^2}(-2x)$$
$$f''(x) = \left[e^{-x^2}(-2x) \right](-2x) + e^{-x^2}(-2)$$
$$= \frac{4x^2}{e^{x^2}} - \frac{2}{e^{x^2}}$$
$$= \frac{4x^2 - 2}{e^{x^2}} < 0$$

The graph is concave down when $4x^2 < 2$. This occurs when $x^2 < \frac{1}{2}$, or $-\frac{1}{\sqrt{2}} < x < \frac{1}{\sqrt{2}}$.

65. (a) The rate of change of the population is $P'(t)$. If $P'(t)$ is proportional to $P(t)$, we have

$$P'(t) = kP(t).$$

    (b) If $P(t) = Ae^{kt}$, then $P'(t) = kAe^{kt} = kP(t)$.

69. We see that $m'(x)$ is nearly of the form $f'(g(x)) \cdot g'(x)$ where

$$f(g) = e^g \quad \text{and} \quad g(x) = x^6,$$

but $g'(x)$ is off by a multiple of 6. Therefore, using the chain rule, let

$$m(x) = \frac{f(g(x))}{6} = \frac{e^{(x^6)}}{6}.$$

73. We have $h(-c) = f(g(-c)) = f(-b) = 0$. From the chain rule,

$$h'(-c) = f'(g(-c))g'(-c).$$

Since g is increasing at $x = -c$, we know that $g'(-c) > 0$. We have

$$f'(g(-c)) = f'(-b),$$

and since f is decreasing at $x = -b$, we have $f'(g(-c)) < 0$. Thus,

$$h'(-c) = \underbrace{f'(g(-c))}_{-} \cdot \underbrace{g'(-c)}_{+} < 0,$$

so h is decreasing at $x = -c$.

77. We have $f(0) = 6$ and $f(10) = 6e^{0.013(10)} = 6.833$. The derivative of $f(t)$ is

$$f'(t) = 6e^{0.013t} \cdot 0.013 = 0.078e^{0.013t},$$

and so $f'(0) = 0.078$ and $f'(10) = 0.089$.

These values tell us that in 1999 (at $t = 0$), the population of the world was 6 billion people and the population was growing at a rate of 0.078 billion people per year. In the year 2009 (at $t = 10$), this model predicts that the population of the world will be 6.833 billion people and growing at a rate of 0.089 billion people per year.

81. (a) For $t < 0$, $I = \dfrac{dQ}{dt} = 0$.

For $t > 0$, $I = \dfrac{dQ}{dt} = -\dfrac{Q_0}{RC}e^{-t/RC}$.

(b) For $t > 0$, $t \to 0$ (that is, as $t \to 0^+$),

$$I = -\frac{Q_0}{RC}e^{-t/RC} \to -\frac{Q_0}{RC}.$$

Since $I = 0$ just to the left of $t = 0$ and $I = -Q_0/RC$ just to the right of $t = 0$, it is not possible to define I at $t = 0$.

(c) Q is not differentiable at $t = 0$ because there is no tangent line at $t = 0$.

85. Using the chain and product rule:

$$\frac{d^2}{dx^2}\left(f(g(x))\right) = \frac{d}{dx}\left(\frac{d}{dx}\left(f(g(x))\right)\right) = \frac{d}{dx}\left(f'(g(x)) \cdot g'(x)\right)$$

$$= f''(g(x)) \cdot g'(x) \cdot g'(x) + f'(g(x)) \cdot g''(x)$$

$$= f''(g(x)) \cdot \left(g'(x)\right)^2 + f'(g(x)) \cdot g''(x).$$

Solutions for Section 3.5

Exercises

1.

Table 3.1

x	$\cos x$	Difference Quotient	$-\sin x$
0	1.0	−0.0005	0.0
0.1	0.995	−0.10033	−0.099833
0.2	0.98007	−0.19916	−0.19867
0.3	0.95534	−0.296	−0.29552
0.4	0.92106	−0.38988	−0.38942
0.5	0.87758	−0.47986	−0.47943
0.6	0.82534	−0.56506	−0.56464

5. $f'(x) = \cos(3x) \cdot 3 = 3\cos(3x)$.

9. $f'(x) = (2x)(\cos x) + x^2(-\sin x) = 2x\cos x - x^2\sin x$.

13. $z' = e^{\cos\theta} - \theta(\sin\theta)e^{\cos\theta}$.

17.

$$f(x) = (1 - \cos x)^{\frac{1}{2}}$$

$$f'(x) = \frac{1}{2}(1 - \cos x)^{-\frac{1}{2}}(-(-\sin x))$$

$$= \frac{\sin x}{2\sqrt{1 - \cos x}}.$$

21. $f'(x) = 2 \cdot [\sin(3x)] + 2x[\cos(3x)] \cdot 3 = 2\sin(3x) + 6x\cos(3x)$

25. $y' = 5\sin^4\theta\cos\theta$.

29. $h'(t) = 1 \cdot (\cos t) + t(-\sin t) + \frac{1}{\cos^2 t} = \cos t - t\sin t + \frac{1}{\cos^2 t}$.

33. $y' = -2\cos w\sin w - \sin(w^2)(2w) = -2(\cos w\sin w + w\sin(w^2))$

37. Using the power and quotient rules gives

$$f'(x) = \frac{1}{2}\left(\frac{1-\sin x}{1-\cos x}\right)^{-1/2}\left[\frac{-\cos x(1-\cos x) - (1-\sin x)\sin x}{(1-\cos x)^2}\right]$$

$$= \frac{1}{2}\sqrt{\frac{1-\cos x}{1-\sin x}}\left[\frac{-\cos x(1-\cos x) - (1-\sin x)\sin x}{(1-\cos x)^2}\right]$$

$$= \frac{1}{2}\sqrt{\frac{1-\cos x}{1-\sin x}}\left[\frac{1-\cos x - \sin x}{(1-\cos x)^2}\right].$$

Problems

41. The pattern in the table below allows us to generalize and say that the $(4n)^{\text{th}}$ derivative of $\cos x$ is $\cos x$, i.e.,

$$\frac{d^4y}{dx^4} = \frac{d^8y}{dx^8} = \cdots = \frac{d^{4n}y}{dx^{4n}} = \cos x.$$

Thus we can say that $d^{48}y/dx^{48} = \cos x$. From there we differentiate twice more to obtain $d^{50}y/dx^{50} = -\cos x$.

n	1	2	3	4	$\cdots$	48	49	50
n^{th} derivative	$-\sin x$	$-\cos x$	$\sin x$	$\cos x$		$\cos x$	$-\sin x$	$-\cos x$

45. (a) $v(t) = \dfrac{dy}{dt} = \dfrac{d}{dt}(15 + \sin(2\pi t)) = 2\pi\cos(2\pi t)$.

(b)

49. (a) Using triangle OPD in Figure 3.3, we see

$$\frac{OD}{a} = \cos\theta \quad \text{so} \quad OD = a\cos\theta$$

$$\frac{PD}{a} = \sin\theta \quad \text{so} \quad PD = a\sin\theta.$$

Figure 3.3

Using triangle PQD, we have

$$(PD)^2 + d^2 = l^2$$

so

$$a^2 \sin^2 \theta + d^2 = l^2, \quad d = \sqrt{l^2 - a^2 \sin^2 \theta}.$$

Thus,

$$x = OD + DQ$$
$$= a \cos \theta + \sqrt{l^2 - a^2 \sin^2 \theta}.$$

(b) Differentiating, regarding a and l as constants,

$$\frac{dx}{d\theta} = -a \sin \theta + \frac{1}{2} \frac{(-2a^2 \sin \theta \cos \theta)}{\sqrt{l^2 - a^2 \sin^2 \theta}}$$

$$= -a \sin \theta - \frac{a^2 \sin \theta \cos \theta}{\sqrt{l^2 - a^2 \sin^2 \theta}}.$$

We want to find dx/dt. Using the chain rule and the fact that $d\theta/dt = 2$, we have

$$\frac{dx}{dt} = \frac{dx}{d\theta} \cdot \frac{d\theta}{dt} = 2 \frac{dx}{d\theta}.$$

(i) Substituting $\theta = \pi/2$, we have

$$\frac{dx}{dt} = 2 \frac{dx}{d\theta}\bigg|_{\theta=\pi/2} = -2a \sin \left(\frac{\pi}{2} \right) - 2 \frac{a^2 \sin(\frac{\pi}{2}) \cos(\frac{\pi}{2})}{\sqrt{l^2 - a^2 \sin^2(\frac{\pi}{2})}}$$

$$= -2a \text{ cm/sec.}$$

(ii) Substituting $\theta = \pi/4$, we have

$$\frac{dx}{dt} = 2 \frac{dx}{d\theta}\bigg|_{\theta=\pi/4} = -2a \sin \left(\frac{\pi}{4} \right) - 2 \frac{a^2 \sin(\frac{\pi}{4}) \cos(\frac{\pi}{4})}{\sqrt{l^2 - a^2 \sin^2(\frac{\pi}{4})}}$$

$$= -a\sqrt{2} - \frac{a^2}{\sqrt{l^2 - a^2/2}} \text{ cm/sec.}$$

53. (a) If $f(x) = \sin x$, then

$$f'(x) = \lim_{h \to 0} \frac{\sin(x + h) - \sin x}{h}$$

$$= \lim_{h \to 0} \frac{(\sin x \cos h + \sin h \cos x) - \sin x}{h}$$

$$= \lim_{h \to 0} \frac{\sin x (\cos h - 1) + \sin h \cos x}{h}$$

$$= \sin x \lim_{h \to 0} \frac{\cos h - 1}{h} + \cos x \lim_{h \to 0} \frac{\sin h}{h}.$$

(b) $\frac{\cos h - 1}{h} \to 0$ and $\frac{\sin h}{h} \to 1$, as $h \to 0$. Thus, $f'(x) = \sin x \cdot 0 + \cos x \cdot 1 = \cos x$.

(c) Similarly,

$$g'(x) = \lim_{h \to 0} \frac{\cos(x + h) - \cos x}{h}$$

$$= \lim_{h \to 0} \frac{(\cos x \cos h - \sin x \sin h) - \cos x}{h}$$

$$= \lim_{h \to 0} \frac{\cos x (\cos h - 1) - \sin x \sin h}{h}$$

$$= \cos x \lim_{h \to 0} \frac{\cos h - 1}{h} - \sin x \lim_{h \to 0} \frac{\sin h}{h}$$

$$= -\sin x.$$

Solutions for Section 3.6

Exercises

1. $f'(t) = \dfrac{2t}{t^2+1}$.

5. $f'(x) = \dfrac{1}{1-e^{-x}} \cdot (-e^{-x})(-1) = \dfrac{e^{-x}}{1-e^{-x}}$.

9. $j'(x) = \dfrac{ae^{ax}}{(e^{ax}+b)}$

13. $f'(w) = \dfrac{1}{\cos(w-1)}[-\sin(w-1)] = -\tan(w-1)$.

    [This could be done easily using the answer from Problem 10 and the chain rule.]

17. $g(\alpha) = \alpha$, so $g'(\alpha) = 1$.

21. $h'(w) = \arcsin w + \dfrac{w}{\sqrt{1-w^2}}$.

25. $f'(x) = -\sin(\arctan 3x)\left(\dfrac{1}{1+(3x)^2}\right)(3) = \dfrac{-3\sin(\arctan 3x)}{1+9x^2}$.

29. Using the chain rule gives $f'(x) = \dfrac{\cos x - \sin x}{\sin x + \cos x}$.

33. $f'(x) = -\sin(\arcsin(x+1))(\dfrac{1}{\sqrt{1-(x+1)^2}}) = \dfrac{-(x+1)}{\sqrt{1-(x+1)^2}}$.

Problems

37. Estimates may vary. From the graphs, we estimate $f(2) \approx 0.3$, $f'(2) \approx 1.1$, and $g'(0.3) \approx 1.7$. Thus, by the chain rule,

$$k'(2) = g'(f(2)) \cdot f'(2) \approx g'(0.3) \cdot f'(2) \approx 1.7 \cdot 1.1 \approx 1.9.$$

41. $pH = 2 = -\log x$ means $\log x = -2$ so $x = 10^{-2}$. Rate of change of pH with hydrogen ion concentration is

$$\frac{d}{dx}pH = -\frac{d}{dx}(\log x) = \frac{-1}{x(\ln 10)} = -\frac{1}{(10^{-2})\ln 10} = -43.4$$

45. (a)

$$f'(x) = \frac{1}{1+x^2} + \frac{1}{1+\frac{1}{x^2}} \cdot (-\frac{1}{x^2})$$
$$= \frac{1}{1+x^2} + \left(-\frac{1}{x^2+1}\right)$$
$$= \frac{1}{1+x^2} - \frac{1}{1+x^2}$$
$$= 0$$

    (b) f is a constant function. Checking at a few values of x,

Table 3.2

x	$\arctan x$	$\arctan x^{-1}$	$f(x) = \arctan x + \arctan x^{-1}$
1	0.785392	0.7853982	1.5707963
2	1.1071487	0.4636476	1.5707963
3	1.2490458	0.3217506	1.5707963

49. Since the chain rule gives $h'(x) = n'(m(x))m'(x) = 1$ we must find values a and x such that $a = m(x)$ and $n'(a)m'(x) = 1$.

Calculating slopes from the graph of n gives

$$n'(a) = \begin{cases} 1 & \text{if} \quad 0 < a < 50 \\ 1/2 & \text{if} \quad 50 < a < 100. \end{cases}$$

Calculating slopes from the graph of m gives

$$m'(x) = \begin{cases} -2 & \text{if} \quad 0 < x < 50 \\ 2 & \text{if} \quad 50 < x < 100. \end{cases}$$

The only values of the derivative n' are 1 and $1/2$ and the only values of the derivative m' are 2 and -2. In order to have $n'(a)m'(x) = 1$ we must therefore have $n'(a) = 1/2$ and $m'(x) = 2$. Thus $50 < a < 100$ and $50 < x < 100$.

Now $a = m(x)$ and from the graph of m we see that $50 < m(x) < 100$ for $0 < x < 25$ or $75 < x < 100$.

The two conditions on x we have found are both satisfied when $75 < x < 100$. Thus $h'(x) = 1$ for all x in the interval $75 < x < 100$. The question asks for just one of these x values, for example $x = 80$.

53. Since the point $(2, 5)$ is on the curve, we know $f(2) = 5$. The point $(2.1, 5.3)$ is on the tangent line, so

$$\text{Slope tangent} = \frac{5.3 - 5}{2.1 - 2} = \frac{0.3}{0.1} = 3.$$

Thus, $f'(2) = 3$. Since g is the inverse function of f and $f(2) = 5$, we know $f^{-1}(5) = 2$, so $g(5) = 2$.

Differentiating, we have

$$g'(2) = \frac{1}{f'(g(5))} = \frac{1}{f'(2)} = \frac{1}{3}.$$

57. To find $(f^{-1})'(3)$, we first look in the table to find that $3 = f(9)$, so $f^{-1}(3) = 9$. Thus,

$$(f^{-1})'(3) = \frac{1}{f'(f^{-1}(3))} = \frac{1}{f'(9)} = \frac{1}{5}.$$

61. We must have

$$(f^{-1})'(5) = \frac{1}{f'(f^{-1}(5))} = \frac{1}{f'(10)} = \frac{1}{8}.$$

Solutions for Section 3.7

Exercises

1. We differentiate implicitly both sides of the equation with respect to x.

$$2x + 2y\frac{dy}{dx} = 0,$$

$$\frac{dy}{dx} = -\frac{2x}{2y} = -\frac{x}{y}.$$

5. We differentiate implicitly both sides of the equation with respect to x.

$$x^{1/2} = 5y^{1/2}$$

$$\frac{1}{2}x^{-1/2} = \frac{5}{2}y^{-1/2}\frac{dy}{dx}$$

$$\frac{dy}{dx} = \frac{\frac{1}{2}x^{-1/2}}{\frac{5}{2}y^{-1/2}} = \frac{1}{5}\sqrt{\frac{y}{x}} = \frac{1}{25}.$$

We can also obtain this answer by realizing that the original equation represents part of the line $x = 25y$ which has slope $1/25$.

9.

$$2ax - 2by\frac{dy}{dx} = 0$$

$$\frac{dy}{dx} = \frac{-2ax}{-2by} = \frac{ax}{by}$$

13. Using the relation $\cos^2 y + \sin^2 y = 1$, the equation becomes:
$1 = y + 2$ or $y = -1$. Hence, $\frac{dy}{dx} = 0$.

17. We differentiate implicitly both sides of the equation with respect to x.

$$(x - a)^2 + y^2 = a^2$$

$$2(x - a) + 2y\frac{dy}{dx} = 0$$

$$2y\frac{dy}{dx} = 2a - 2x$$

$$\frac{dy}{dx} = \frac{2a - 2x}{2y} = \frac{a - x}{y}.$$

21. Differentiating with respect to x gives

$$3x^2 + 2xy' + 2y + 2yy' = 0$$

so that

$$y' = -\frac{3x^2 + 2y}{2x + 2y}$$

At the point $(1, 1)$ the slope is $-\frac{5}{4}$.

25. First, we must find the slope of the tangent, $\left.\frac{dy}{dx}\right|_{(4,2)}$. Implicit differentiation yields:

$$2y\frac{dy}{dx} = \frac{2x(xy - 4) - x^2\left(x\frac{dy}{dx} + y\right)}{(xy - 4)^2}.$$

Given the complexity of the above equation, we first want to substitute 4 for x and 2 for y (the coordinates of the point where we are constructing our tangent line), then solve for $\frac{dy}{dx}$. Substitution yields:

$$2 \cdot 2\frac{dy}{dx} = \frac{(2 \cdot 4)(4 \cdot 2 - 4) - 4^2\left(4\frac{dy}{dx} + 2\right)}{(4 \cdot 2 - 4)^2} = \frac{8(4) - 16(4\frac{dy}{dx} + 2)}{16} = -4\frac{dy}{dx}.$$

$$4\frac{dy}{dx} = -4\frac{dy}{dx},$$

Solving for $\frac{dy}{dx}$, we have:

$$\frac{dy}{dx} = 0.$$

The tangent is a horizontal line through $(4, 2)$, hence its equation is $y = 2$.

Problems

29. (a) Taking derivatives implicitly, we get

$$\frac{2}{25}x + \frac{2}{9}y\frac{dy}{dx} = 0$$

$$\frac{dy}{dx} = \frac{-9x}{25y}.$$

(b) The slope is not defined anywhere along the line $y = 0$. This ellipse intersects that line in two places, $(-5, 0)$ and $(5, 0)$. (These are the "ends" of the ellipse where the tangent is vertical.)

33. (a) Differentiating both sides of the equation with respect to P gives

$$\frac{d}{dP}\left(\frac{4f^2P}{1-f^2}\right) = \frac{dK}{dP} = 0.$$

By the product rule

$$\frac{d}{dP}\left(\frac{4f^2P}{1-f^2}\right) = \frac{d}{dP}\left(\frac{4f^2}{1-f^2}\right)P + \left(\frac{4f^2}{1-f^2}\right)\cdot 1$$

$$= \left(\frac{(1-f^2)(8f) - 4f^2(-2f)}{(1-f^2)^2}\right)\frac{df}{dP}P + \left(\frac{4f^2}{1-f^2}\right)$$

$$= \left(\frac{8f}{(1-f^2)^2}\right)\frac{df}{dP}P + \left(\frac{4f^2}{1-f^2}\right) = 0.$$

So

$$\frac{df}{dP} = \frac{-4f^2/(1-f^2)}{8fP/(1-f^2)^2} = \frac{-1}{2P}f(1-f^2).$$

(b) Since f is a fraction of a gas, $0 \le f \le 1$. Also, in the equation relating f and P we can't have $f = 0$, since that would imply $K = 0$, and we can't have $f = 1$, since the left side is undefined there. So $0 < f < 1$. Thus $1 - f^2 > 0$. Also, pressure can't be negative, and from the equation relating f and P, we see that P can't be zero either, so $P > 0$. Therefore $df/dP = -(1/2P)f(1-f^2) < 0$ always. This means that at larger pressures less of the gas decomposes.

Solutions for Section 3.8

Exercises

1. Using the chain rule, $\dfrac{d}{dx}\left(\cosh(2x)\right) = \left(\sinh(2x)\right)\cdot 2 = 2\sinh(2x)$.

5. Using the chain rule,

$$\frac{d}{dt}\left(\cosh^2 t\right) = 2\cosh t \cdot \sinh t.$$

9. Using the chain rule twice,

$$\frac{d}{dy}\left(\sinh\left(\sinh(3y)\right)\right) = \cosh\left(\sinh(3y)\right)\cdot\cosh(3y)\cdot 3$$

$$= 3\cosh(3y)\cdot\cosh\left(\sinh(3y)\right).$$

13. Substituting $-x$ for x in the formula for $\sinh x$ gives

$$\sinh(-x) = \frac{e^{-x} - e^{-(-x)}}{2} = \frac{e^{-x} - e^{x}}{2} = -\frac{e^{x} - e^{-x}}{2} = -\sinh x.$$

Problems

17. The graph of $\sinh x$ in the text suggests that

$$\text{As } x \to \infty, \quad \sinh x \to \frac{1}{2}e^{x}.$$

$$\text{As } x \to -\infty, \quad \sinh x \to -\frac{1}{2}e^{-x}.$$

Using the facts that

$$\text{As } x \to \infty, \quad e^{-x} \to 0,$$
$$\text{As } x \to -\infty, \quad e^{x} \to 0,$$

we can obtain the same results analytically:

$$\text{As } x \to \infty, \quad \sinh x = \frac{e^x - e^{-x}}{2} \to \frac{1}{2}e^x.$$
$$\text{As } x \to -\infty, \quad \sinh x = \frac{e^x - e^{-x}}{2} \to -\frac{1}{2}e^{-x}.$$

21. Recall that

$$\sinh A = \frac{1}{2}(e^A - e^{-A}) \quad \text{and} \quad \cosh A = \frac{1}{2}(e^A + e^{-A}).$$

Now substitute, expand and collect terms:

$$\cosh A \cosh B + \sinh B \sinh A = \frac{1}{2}(e^A + e^{-A}) \cdot \frac{1}{2}(e^B + e^{-B}) + \frac{1}{2}(e^B - e^{-B}) \cdot \frac{1}{2}(e^A - e^{-A})$$

$$= \frac{1}{4}\left(e^{A+B} + e^{A-B} + e^{-A+B} + e^{-(A+B)}\right.$$

$$\left. +e^{B+A} - e^{B-A} - e^{-B+A} + e^{-A-B}\right)$$

$$= \frac{1}{2}\left(e^{A+B} + e^{-(A+B)}\right)$$

$$= \cosh(A + B).$$

25. Using the definition of $\cosh x$ and $\sinh x$, we have $\cosh x^2 = \dfrac{e^{x^2} + e^{-x^2}}{2}$ and $\sinh x^2 = \dfrac{e^{x^2} - e^{-x^2}}{2}$. Therefore

$$\lim_{x \to \infty} \frac{\sinh(x^2)}{\cosh(x^2)} = \lim_{x \to \infty} \frac{e^{x^2} - e^{-x^2}}{e^{x^2} + e^{-x^2}}$$

$$= \lim_{x \to \infty} \frac{e^{x^2}\left(1 - e^{-2x^2}\right)}{e^{x^2}\left(1 + e^{-2x^2}\right)}$$

$$= \lim_{x \to \infty} \frac{1 - e^{-2x^2}}{1 + e^{-2x^2}}$$

$$= 1.$$

29. We want to show that for any A, B with $A > 0$, $B > 0$, we can find K and c such that

$$y = Ae^x + Be^{-x} = \frac{Ke^{(x-c)} + Ke^{-(x-c)}}{2}$$

$$= \frac{K}{2}e^x e^{-c} + \frac{K}{2}e^{-x}e^c$$

$$= \left(\frac{Ke^{-c}}{2}\right)e^x + \left(\frac{Ke^c}{2}\right)e^{-x}.$$

Thus, we want to find K and c such that

$$\frac{Ke^{-c}}{2} = A \quad \text{and} \quad \frac{Ke^c}{2} = B.$$

Dividing, we have

$$\frac{Ke^c}{Ke^{-c}} = \frac{B}{A}$$

$$e^{2c} = \frac{B}{A}$$

$$c = \frac{1}{2}\ln\left(\frac{B}{A}\right).$$

If $A > 0,\ B > 0$, then there is a solution for c. Substituting to find K, we have

$$\frac{Ke^{-c}}{2} = A$$

$$K = 2Ae^c = 2Ae^{(\ln(B/A))/2}$$

$$= 2Ae^{\ln\sqrt{B/A}} = 2A\sqrt{\frac{B}{A}} = 2\sqrt{AB}.$$

Thus, if $A > 0,\ B > 0$, there is a solution for K also.

The fact that $y = Ae^x + Be^{-x}$ can be rewritten in this way shows that the graph of $y = Ae^x + Be^{-x}$ is the graph of $\cosh x$, shifted over by c and stretched (or shrunk) vertically by a factor of K.

Solutions for Section 3.9

Exercises

1. Since $f(1) = 1$ and we showed that $f'(1) = 2$, the local linearization is

$$f(x) \approx 1 + 2(x - 1) = 2x - 1.$$

5. With $f(x) = 1/(\sqrt{1+x})$, we see that the tangent line approximation to f near $x = 0$ is

$$f(x) \approx f(0) + f'(0)(x - 0),$$

which becomes

$$\frac{1}{\sqrt{1+x}} \approx 1 + f'(0)x.$$

Since $f'(x) = (-1/2)(1+x)^{-3/2}$, $f'(0) = -1/2$. Thus our formula reduces to

$$\frac{1}{\sqrt{1+x}} \approx 1 - x/2.$$

This is the local linearization of $\dfrac{1}{\sqrt{1+x}}$ near $x = 0$.

9. From Figure 3.4, we see that the error has its maximum magnitude at the end points of the interval, $x = \pm 1$. The magnitude of the error can be read off the graph as less than 0.2 or estimated as

$$|\text{Error}| \leq |1 - \sin 1| = 0.159 < 0.2.$$

The approximation is an overestimate for $x > 0$ and an underestimate for $x < 0$.

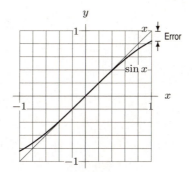

Figure 3.4

Problems

13. (a) Since

$$\frac{d}{dx}(\cos x) = -\sin x,$$

the slope of the tangent line is $-\sin(\pi/4) = -1/\sqrt{2}$. Since the tangent line passes through the point $(\pi/4, \cos(\pi/4)) = (\pi/4, 1/\sqrt{2})$, its equation is

$$y - \frac{1}{\sqrt{2}} = -\frac{1}{\sqrt{2}}\left(x - \frac{\pi}{4}\right)$$

$$y = -\frac{1}{\sqrt{2}}x + \frac{1}{\sqrt{2}}\left(\frac{\pi}{4} + 1\right).$$

Thus, the tangent line approximation to $\cos x$ is

$$\cos x \approx -\frac{1}{\sqrt{2}}x + \frac{1}{\sqrt{2}}\left(\frac{\pi}{4} + 1\right).$$

(b) From Figure 3.5, we see that the tangent line approximation is an overestimate.

(c) From Figure 3.5, we see that the maximum error for $0 \le x \le \pi/2$ is either at $x = 0$ or at $x = \pi/2$. The error can either be estimated from the graph, or as follows. At $x = 0$,

$$|\text{Error}| = \left|\cos 0 - \frac{1}{\sqrt{2}}\left(\frac{\pi}{4} + 1\right)\right| = 0.262 < 0.3.$$

At $x = \pi/2$,

$$|\text{Error}| = \left|\cos\frac{\pi}{2} + \frac{1}{\sqrt{2}}\frac{\pi}{2} - \frac{1}{\sqrt{2}}\left(\frac{\pi}{4} + 1\right)\right| = 0.152 < 0.2.$$

Thus, for $0 \le x \le \pi/2$, we have

$$|\text{Error}| < 0.3.$$

Figure 3.5

17. We have $f(x) = x + \ln(1 + x)$, so $f'(x) = 1 + 1/(1 + x)$. Thus $f'(0) = 2$ so

Local linearization near $x = 0$ is $f(x) \approx f(0) + f'(0)x = 2x$.

We get an approximate solution using the local linearization instead of $f(x)$, so the equation becomes

$2x = 0.2$,   with solution $x = 0.1$.

A computer or calculator gives the actual value as $x = 0.102$.

21. (a) Suppose g is a constant and

$$T = f(l) = 2\pi\sqrt{\frac{l}{g}}.$$

Then

$$f'(l) = \frac{2\pi}{\sqrt{g}}\frac{1}{2}l^{-1/2} = \frac{\pi}{\sqrt{gl}}.$$

Thus, local linearity tells us that

$$f(l + \Delta l) \approx f(l) + \frac{\pi}{\sqrt{gl}}\Delta l.$$

Now $T = f(l)$ and $\Delta T = f(l + \Delta l) - f(l)$, so

$$\Delta T \approx \frac{\pi}{\sqrt{gl}}\Delta l = 2\pi\sqrt{\frac{l}{g}}\cdot\frac{1}{2}\frac{\Delta l}{l} = \frac{T}{2}\frac{\Delta l}{l}.$$

(b) Knowing that the length of the pendulum increases by 2% tells us that

$$\frac{\Delta l}{l} = 0.02.$$

Thus,

$$\Delta T \approx \frac{T}{2}(0.02) = 0.01T.$$

So

$$\frac{\Delta T}{T} \approx 0.01.$$

Thus, T increases by 1%.

25. (a) The range is $f(20) = 16{,}398$ meters.

(b) We have

$$f'(\theta) = \frac{\pi}{90}25510\cos\frac{\pi\theta}{90}$$
$$f'(20) = 682.$$

Thus, for angles, θ, near $20°$, we have

Range $= f(\theta) \approx f(20) + f'(20)(\theta - 20) = 16398 + 682(\theta - 20)$ meters.

(c) The true range for $21°$ is $f(21) = 17{,}070$ meters. The linear approximation gives

Approximate range $= 16398 + 682(21 - 20) = 17080$ meters

which is a little too high.

29. We have $f(0) = 1$ and $f'(0) = 0$. Thus

$$E(x) = \cos x - 1.$$

Values for $E(x)/(x - 0)$ near $x = 0$ are in Table 3.3.

Table 3.3

x	0.1	0.01	0.001
$E(x)/(x - 0)$	−0.050	−0.0050	−0.00050

From the table, we can see that

$$\frac{E(x)}{(x-0)} \approx -0.5(x-0),$$

so $k = -1/2$ and

$$E(x) \approx -\frac{1}{2}(x-0)^2 = -\frac{1}{2}x^2.$$

In addition, $f''(0) = -1$, so

$$E(x) \approx -\frac{1}{2}x^2 = \frac{f''(0)}{2}x^2.$$

33. The local linearization of e^x near $x = 0$ is $1 + 1x$ so

$$e^x \approx 1 + x.$$

Squaring this yields, for small x,

$$e^{2x} = (e^x)^2 \approx (1+x)^2 = 1 + 2x + x^2.$$

Local linearization of e^{2x} directly yields

$$e^{2x} \approx 1 + 2x$$

for small x. The two approximations are consistent because they agree: the tangent line approximation to $1 + 2x + x^2$ is just $1 + 2x$.

The first approximation is more accurate. One can see this numerically or by noting that the approximation for e^{2x} given by $1 + 2x$ is really the same as approximating e^y at $y = 2x$. Since the other approximation approximates e^y at $y = x$, which is twice as close to 0 and therefore a better general estimate, it's more likely to be correct.

37. Note that

$$[f(g(x))]' = \lim_{h \to 0} \frac{f(g(x+h)) - f(g(x))}{h}.$$

Using the local linearizations of f and g, we get that

$$f(g(x+h)) - f(g(x)) \approx f\left(g(x) + g'(x)h\right) - f(g(x))$$
$$\approx f(g(x)) + f'(g(x))g'(x)h - f(g(x))$$
$$= f'(g(x))g'(x)h.$$

Therefore,

$$[f(g(x))]' = \lim_{h \to 0} \frac{f(g(x+h)) - f(g(x))}{h}$$
$$= \lim_{h \to 0} \frac{f'(g(x))g'(x)h}{h}$$
$$= \lim_{h \to 0} f'(g(x))g'(x) = f'(g(x))g'(x).$$

A more complete derivation can be given using the error term discussed in the section on Differentiability and Linear Approximation in Chapter 2. Adapting the notation of that section to this problem, we write

$$f(z+k) = f(z) + f'(z)k + E_f(k) \quad \text{and} \quad g(x+h) = g(x) + g'(x)h + E_g(h),$$

where $\lim_{h \to 0} \frac{E_g(h)}{h} = \lim_{k \to 0} \frac{E_f(k)}{k} = 0$.

Now we let $z = g(x)$ and $k = g(x+h) - g(x)$. Then we have $k = g'(x)h + E_g(h)$. Thus,

$$\frac{f(g(x+h)) - f(g(x))}{h} = \frac{f(z+k) - f(z)}{h}$$
$$= \frac{f(z) + f'(z)k + E_f(k) - f(z)}{h} = \frac{f'(z)k + E_f(k)}{h}$$
$$= \frac{f'(z)g'(x)h + f'(z)E_g(h)}{h} + \frac{E_f(k)}{k} \cdot \left(\frac{k}{h}\right)$$
$$= f'(z)g'(x) + \frac{f'(z)E_g(h)}{h} + \frac{E_f(k)}{k}\left[\frac{g'(x)h + E_g(h)}{h}\right]$$
$$= f'(z)g'(x) + \frac{f'(z)E_g(h)}{h} + \frac{g'(x)E_f(k)}{k} + \frac{E_g(h) \cdot E_f(k)}{h \cdot k}$$

Now, if $h \to 0$ then $k \to 0$ as well, and all the terms on the right except the first go to zero, leaving us with the term $f'(z)g'(x)$. Substituting $g(x)$ for z, we obtain

$$[f(g(x))]' = \lim_{h \to 0} \frac{f(g(x+h)) - f(g(x))}{h} = f'(g(x))g'(x).$$

Solutions for Section 3.10

Exercises

1. False. The derivative, $f'(x)$, is not equal to zero everywhere, because the function is not continuous at integral values of x, so $f'(x)$ does not exist there. Thus, the Constant Function Theorem does not apply.

5. False. Let $f(x) = x^3$ on $[-1, 1]$. Then $f(x)$ is increasing but $f'(x) = 0$ for $x = 0$.

9. No. This function does not satisfy the hypotheses of the Mean Value Theorem, as it is not continuous.

    However, the function has a point c such that

$$f'(c) = \frac{f(b) - f(a)}{b - a}.$$

Thus, this satisfies the conclusion of the theorem.

Problems

13. A polynomial $p(x)$ satisfies the conditions of Rolle's Theorem for all intervals $a \le x \le b$.

    Suppose $a_1, a_2, a_3, a_4, a_5, a_6, a_7$ are the seven distinct zeros of $p(x)$ in increasing order. Thus $p(a_1) = p(a_2) = 0$, so by Rolle's Theorem, $p'(x)$ has a zero, c_1, between a_1 and a_2.

    Similarly, $p'(x)$ has 6 distinct zeros, $c_1, c_2, c_3, c_4, c_5, c_6$, where

$$a_1 < c_1 < a_2$$
$$a_2 < c_2 < a_3$$
$$a_3 < c_3 < a_4$$
$$a_4 < c_4 < a_5$$
$$a_5 < c_5 < a_6$$
$$a_6 < c_6 < a_7.$$

    The polynomial $p'(x)$ is of degree 6, so $p'(x)$ cannot have more than 6 zeros.

17. The Decreasing Function Theorem is: Suppose that f is continuous on $[a, b]$ and differentiable on (a, b). If $f'(x) < 0$ on (a, b), then f is decreasing on $[a, b]$. If $f'(x) \le 0$ on (a, b), then f is nonincreasing on $[a, b]$.

    To prove the theorem, we note that if f is decreasing then $-f$ is increasing and vice-versa. Similarly, if f is non-increasing, then $-f$ is nondecreasing. Thus if $f'(x) < 0$, then $-f'(x) > 0$, so $-f$ is increasing, which means f is decreasing. And if $f'(x) \le 0$, then $-f'(x) \ge 0$, so $-f$ is nondecreasing, which means f is nonincreasing.

21. By the Mean Value Theorem, Theorem 3.7, there is a number c, with $0 < c < 1$, such that

$$f'(c) = \frac{f(1) - f(0)}{1 - 0}.$$

Since $f(1) - f(0) > 0$, we have $f'(c) > 0$.

    Alternatively if $f'(c) \le 0$ for all c in $(0, 1)$, then by the Increasing Function Theorem, $f(0) \ge f(1)$.

25. Let $h(x) = f(x) - g(x)$. Then $h'(x) = f'(x) - g'(x) = 0$ for all x in (a, b). Hence, by the Constant Function Theorem, there is a constant C such that $h(x) = C$ on (a, b). Thus $f(x) = g(x) + C$.

Solutions for Chapter 3 Review

Exercises

1. $w' = 100(t^2 + 1)^{99}(2t) = 200t(t^2 + 1)^{99}$.

5. Using the quotient rule,
$$h'(t) = \frac{(-1)(4+t) - (4-t)}{(4+t)^2} = -\frac{8}{(4+t)^2}.$$

9. Since $h(\theta) = \theta(\theta^{-1/2} - \theta^{-2}) = \theta\theta^{-1/2} - \theta\theta^{-2} = \theta^{1/2} - \theta^{-1}$, we have $h'(\theta) = \frac{1}{2}\theta^{-1/2} + \theta^{-2}$.

13. $y' = 0$

17. $s'(\theta) = \frac{d}{d\theta}\sin^2(3\theta - \pi) = 6\cos(3\theta - \pi)\sin(3\theta - \pi)$.

21. $f'(\theta) = -1(1 + e^{-\theta})^{-2}(e^{-\theta})(-1) = \frac{e^{-\theta}}{(1 + e^{-\theta})^2}$.

25. Using the chain rule and simplifying,
$$q'(\theta) = \frac{1}{2}(4\theta^2 - \sin^2(2\theta))^{-1/2}(8\theta - 2\sin(2\theta)(2\cos(2\theta))) = \frac{4\theta - 2\sin(2\theta)\cos(2\theta)}{\sqrt{4\theta^2 - \sin^2(2\theta)}}.$$

29. Using the chain rule, we get:
$$m'(n) = \cos(e^n) \cdot (e^n)$$

33. $\dfrac{d}{dx}xe^{\tan x} = e^{\tan x} + xe^{\tan x}\dfrac{1}{\cos^2 x}$.

37. $h(x) = ax \cdot \ln e = ax$, so $h'(x) = a$.

41. Using the product rule gives
$$H'(t) = 2ate^{-ct} - c(at^2 + b)e^{-ct}$$
$$= (-cat^2 + 2at - bc)e^{-ct}.$$

45. Using the quotient rule gives
$$w'(r) = \frac{2ar(b + r^3) - 3r^2(ar^2)}{(b + r^3)^2}$$
$$= \frac{2abr - ar^4}{(b + r^3)^2}.$$

49. Since $g(w) = 5(a^2 - w^2)^{-2}$, $g'(w) = -10(a^2 - w^2)^{-3}(-2w) = \dfrac{20w}{(a^2 - w^2)^3}$

53. Using the quotient rule gives
$$g'(t) = \frac{\left(\frac{k}{kt} + 1\right)(\ln(kt) - t) - (\ln(kt) + t)\left(\frac{k}{kt} - 1\right)}{(\ln(kt) - t)^2}$$
$$g'(t) = \frac{\left(\frac{1}{t} + 1\right)(\ln(kt) - t) - (\ln(kt) + t)\left(\frac{1}{t} - 1\right)}{(\ln(kt) - t)^2}$$
$$g'(t) = \frac{\ln(kt)/t - 1 + \ln(kt) - t - \ln(kt)/t - 1 + \ln(kt) + t}{(\ln(kt) - t)^2}$$
$$g'(t) = \frac{2\ln(kt) - 2}{(\ln(kt) - t)^2}.$$

57. $g'(x) = -\dfrac{1}{2}(5x^4 + 2)$.

61. $g'(x) = \dfrac{d}{dx}(2x - x^{-1/3} + 3^x - e) = 2 + \dfrac{1}{3x^{\frac{4}{3}}} + 3^x \ln 3$.

65. $r'(\theta) = \dfrac{d}{d\theta}\sin[(3\theta - \pi)^2] = \cos[(3\theta - \pi)^2] \cdot 2(3\theta - \pi) \cdot 3 = 6(3\theta - \pi)\cos[(3\theta - \pi)^2]$.

69. $f'(x) = \dfrac{d}{dx}(2 - 4x - 3x^2)(6x^e - 3\pi) = (-4 - 6x)(6x^e - 3\pi) + (2 - 4x - 3x^2)(6ex^{e-1})$.

73.

$$
\begin{aligned}
h'(x) &= \left(-\frac{1}{x^2} + \frac{2}{x^3}\right)\left(2x^3 + 4\right) + \left(\frac{1}{x} - \frac{1}{x^2}\right)\left(6x^2\right) \\
&= -2x + 4 - \frac{4}{x^2} + \frac{8}{x^3} + 6x - 6 \\
&= 4x - 2 - 4x^{-2} + 8x^{-3}
\end{aligned}
$$

77. We wish to find the slope $m = dy/dx$. To do this, we can implicitly differentiate the given formula in terms of x:

$$
\begin{aligned}
x^2 + 3y^2 &= 7 \\
2x + 6y\frac{dy}{dx} &= \frac{d}{dx}(7) = 0 \\
\frac{dy}{dx} &= \frac{-2x}{6y} = \frac{-x}{3y}.
\end{aligned}
$$

Thus, at $(2, -1)$, $m = -(2)/3(-1) = 2/3$.

Problems

81.

$$
\begin{aligned}
f'(x) &= -8 + 2\sqrt{2}x \\
f'(r) &= -8 + 2\sqrt{2}r = 4 \\
r &= \frac{12}{2\sqrt{2}} = 3\sqrt{2}.
\end{aligned}
$$

85. Since $r(x) = s(t(x))$, the chain rule gives $r'(x) = s'(t(x)) \cdot t'(x)$. Thus,

$$
r'(0) = s'(t(0)) \cdot t'(0) \approx s'(2) \cdot (-2) \approx (-2)(-2) = 4.
$$

Note that since $t(x)$ is a linear function whose slope looks like -2 from the graph, $t'(x) \approx -2$ everywhere. To find $s'(2)$, draw a line tangent to the curve at the point $(2, s(2))$, and estimate the slope.

89. We have $r(1) = s(t(1)) \approx s(0) \approx 2$. By the chain rule, $r'(x) = s'(t(x)) \cdot t'(x)$, so

$$
r'(1) = s'(t(1)) \cdot t'(1) \approx s'(0) \cdot (-2) \approx 2(-2) = -4.
$$

Thus the equation of the tangent line is

$$
\begin{aligned}
y - 2 &= -4(x - 1) \\
y &= -4x + 6.
\end{aligned}
$$

Note that since $t(x)$ is a linear function whose slope looks like -2 from the graph, $t'(x) \approx -2$ everywhere. To find $s'(0)$, draw a line tangent to the curve at the point $(0, s(0))$, and estimate the slope.

93. Since W is proportional to r^3, we have $W = kr^3$ for some constant k. Thus, $dW/dr = k(3r^2) = 3kr^2$. Thus, dW/dr is proportional to r^2.

97. (a) $f(x) = x^2 - 4g(x)$
   $f(2) = 4 - 4(3) = -8$
   $f'(2) = 20$
   Thus, we have a point $(2, -8)$ and slope $m = 20$. This gives

$$
\begin{aligned}
-8 &= 2(20) + b \\
b &= -48, \quad \text{so} \\
y &= 20x - 48.
\end{aligned}
$$

(b) $f(x) = \dfrac{x}{g(x)}$

$f(2) = \dfrac{2}{3}$

$f'(2) = \dfrac{11}{9}$

Thus, we have point $(2, \frac{2}{3})$ and slope $m = \frac{11}{9}$. This gives

$$\frac{2}{3} = (\frac{11}{9})(2) + b$$

$$b = \frac{2}{3} - \frac{22}{9} = \frac{-16}{9}, \quad \text{so}$$

$$y = \frac{11}{9}x - \frac{16}{9}.$$

(c) $f(x) = x^2 g(x)$

$f(2) = 4 \cdot g(2) = 4(3) = 12$

$f'(2) = -4$

Thus, we have point $(2, 12)$ and slope $m = -4$. This gives

$$12 = 2(-4) + b$$

$$b = 20, \quad \text{so}$$

$$y = -4x + 20.$$

(d) $f(x) = (g(x))^2$

$f(2) = (g(2))^2 = (3)^2 = 9$

$f'(2) = -24$

Thus, we have point $(2, 9)$ and slope $m = -24$. This gives

$$9 = 2(-24) + b$$

$$b = 57, \quad \text{so}$$

$$y = -24x + 57.$$

(e) $f(x) = x \sin(g(x))$

$f(2) = 2 \sin(g(2)) = 2 \sin 3$

$f'(2) = \sin 3 - 8 \cos 3$

We will use a decimal approximation for $f(2)$ and $f'(2)$, so the point $(2, 2 \sin 3) \approx (2, 0.28)$ and $m \approx 8.06$. Thus,

$$0.28 = 2(8.06) + b$$

$$b = -15.84, \quad \text{so}$$

$$y = 8.06x - 15.84.$$

(f) $f(x) = x^2 \ln g(x)$

$f(2) = 4 \ln g(2) = 4 \ln 3 \approx 4.39$

$f'(2) = 4 \ln 3 - \dfrac{16}{3} \approx -0.94.$

Thus, we have point $(2, 4.39)$ and slope $m = -0.94$. This gives

$$4.39 = 2(-0.94) + b$$

$$b = 6.27, \quad \text{so}$$

$$y = -0.94x + 6.27.$$

101. The curves meet when $1 - x^3/3 = x - 1$, that is when $x^3 + 3x - 6 = 0$. So the roots of this equation give us the x-coordinates of the intersection point. By numerical methods, we see there is one solution near $x = 1.3$. See Figure 3.6. Let

$$y_1(x) = 1 - \frac{x^3}{3} \quad \text{and} \quad y_2(x) = x - 1.$$

So we have

$$y_1' = -x^2 \quad \text{and} \quad y_2' = 1.$$

However, $y_2'(x) = +1$, so if the curves are to be perpendicular when they cross, then y_1' must be -1. Since $y_1' = -x^2$, $y_1' = -1$ only at $x = \pm 1$ which is not the point of intersection. The curves are therefore not perpendicular when they cross.

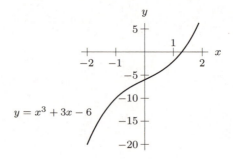

$y = x^3 + 3x - 6$

Figure 3.6

105. Using the definition of $\cosh x$ and $\sinh x$, we have $\cosh x^2 = \dfrac{e^{x^2} + e^{-x^2}}{2}$ and $\sinh x^2 = \dfrac{e^{x^2} - e^{-x^2}}{2}$. Therefore

$$\lim_{x \to -\infty} \frac{\sinh(x^2)}{\cosh(x^2)} = \lim_{x \to -\infty} \frac{e^{x^2} - e^{-x^2}}{e^{x^2} + e^{-x^2}}$$

$$= \lim_{x \to -\infty} \frac{e^{x^2}(1 - e^{-2x^2})}{e^{x^2}(1 + e^{-2x^2})}$$

$$= \lim_{x \to -\infty} \frac{1 - e^{-2x^2}}{1 + e^{-2x^2}}$$

$$= 1.$$

109. (a) $\dfrac{dg}{dr} = GM \dfrac{d}{dr}\left(\dfrac{1}{r^2}\right) = GM \dfrac{d}{dr}\left(r^{-2}\right) = GM(-2)r^{-3} = -\dfrac{2GM}{r^3}.$

(b) $\dfrac{dg}{dr}$ is the rate of change of acceleration due to the pull of gravity. The further away from the center of the earth, the weaker the pull of gravity is. So g is decreasing and therefore its derivative, $\dfrac{dg}{dr}$, is negative.

(c) By part (a),

$$\left.\frac{dg}{dr}\right|_{r=6400} = \left.-\frac{2GM}{r^3}\right|_{r=6400} = -\frac{2(6.67 \times 10^{-20})(6 \times 10^{24})}{(6400)^3} \approx -3.05 \times 10^{-6}.$$

(d) It is reasonable to assume that g is a constant near the surface of the earth.

113. (a) Differentiating, we see

$$v = \frac{dy}{dt} = -2\pi \omega y_0 \sin(2\pi \omega t)$$

$$a = \frac{dv}{dt} = -4\pi^2 \omega^2 y_0 \cos(2\pi \omega t).$$

(b) We have

$$y = y_0 \cos(2\pi \omega t)$$
$$v = -2\pi \omega y_0 \sin(2\pi \omega t)$$
$$a = -4\pi^2 \omega^2 y_0 \cos(2\pi \omega t).$$

So

Amplitude of y is $|y_0|$,

Amplitude of v is $|2\pi \omega y_0| = 2\pi \omega |y_0|$,

Amplitude of a is $|4\pi^2 \omega^2 y_0| = 4\pi^2 \omega^2 |y_0|$.

The amplitudes are different (provided $2\pi \omega \neq 1$). The periods of the three functions are all the same, namely $1/\omega$.

(c) Looking at the answer to part (a), we see

$$\frac{d^2y}{dt^2} = a = -4\pi^2\omega^2\left(y_0\cos(2\pi\omega t)\right)$$
$$= -4\pi^2\omega^2 y.$$

So we see that

$$\frac{d^2y}{dt^2} + 4\pi^2\omega^2 y = 0.$$

117. (a) The statement $f(2) = 4023$ tells us that when the price is \$2 per gallon, 4023 gallons of gas are sold.
   (b) Since $f(2) = 4023$, we have $f^{-1}(4023) = 2$. Thus, 4023 gallons are sold when the price is \$2 per gallon.
   (c) The statement $f'(2) = -1250$ tells us that if the price increases from \$2 per gallon, the sales decrease at a rate of 1250 gallons per \$1 increase in price.
   (d) The units of $(f^{-1})'(4023)$ are dollars per gallon. We have

$$(f^{-1})'(4023) = \frac{1}{f'(f^{-1}(4023))} = \frac{1}{f'(2)} = -\frac{1}{1250} = -0.0008.$$

Thus, when 4023 gallons are already sold, sales decrease at the rate of one gallon per price increase of 0.0008 dollars. In others words, an additional gallon is sold if the price drops by 0.0008 dollars.

121. (a) We multiply through by $h = f \cdot g$ and cancel as follows:

$$\frac{f'}{f} + \frac{g'}{g} = \frac{h'}{h}$$
$$\left(\frac{f'}{f} + \frac{g'}{g}\right) \cdot fg = \frac{h'}{h} \cdot fg$$
$$\frac{f'}{f} \cdot fg + \frac{g'}{g} \cdot fg = \frac{h'}{h} \cdot h$$
$$f' \cdot g + g' \cdot f = h',$$

which is the product rule.
   (b) We start with the product rule, multiply through by $1/(fg)$ and cancel as follows:

$$f' \cdot g + g' \cdot f = h'$$
$$(f' \cdot g + g' \cdot f) \cdot \frac{1}{fg} = h' \cdot \frac{1}{fg}$$
$$(f' \cdot g) \cdot \frac{1}{fg} + (g' \cdot f) \cdot \frac{1}{fg} = h' \cdot \frac{1}{fg}$$
$$\frac{f'}{f} + \frac{g'}{g} = \frac{h'}{h},$$

which is the additive rule shown in part (a).

CAS Challenge Problems

125. (a) A CAS gives $g'(r) = 0$.
   (b) Using the product rule,

$$g'(r) = \frac{d}{dr}(2^{-2r}) \cdot 4^r + 2^{-2r}\frac{d}{dr}(4^r) = -2\ln 2 \cdot 2^{-2r}4^r + 2^{-2r}\ln 4 \cdot 4^r$$
$$= -\ln 4 \cdot 2^{-2r}4^r + \ln 4 \cdot 2^{-2r}4^r = (-\ln 4 + \ln 4)2^{-2r}4^r = 0 \cdot 2^{-2r}4^r = 0.$$

   (c) By the laws of exponents, $4^r = (2^2)^r = 2^{2r}$, so $2^{-2r}4^r = 2^{-2r}2^{2r} = 2^0 = 1$. Therefore, its derivative is zero.

CHECK YOUR UNDERSTANDING

1. True. Since $d(x^n)/dx = nx^{n-1}$, the derivative of a power function is a power function, so the derivative of a polynomial is a polynomial.

5. True. Since $f'(x)$ is the limit

$$f'(x) = \lim_{h \to 0} \frac{f(x+h) - f(x)}{h},$$

the function f must be defined for all x.

9. True; differentiating the equation with respect to x, we get

$$2y\frac{dy}{dx} + y + x\frac{dy}{dx} = 0.$$

Solving for dy/dx, we get that

$$\frac{dy}{dx} = \frac{-y}{2y + x}.$$

Thus dy/dx exists where $2y + x \neq 0$. Now if $2y + x = 0$, then $x = -2y$. Substituting for x in the original equation, $y^2 + xy - 1 = 0$, we get

$$y^2 - 2y^2 - 1 = 0.$$

This simplifies to $y^2 + 1 = 0$, which has no solutions. Thus dy/dx exists everywhere.

13. False. Since $(\sinh x)' = \cosh x > 0$, the function $\sinh x$ is increasing everywhere so can never repeat any of its values.

17. False; the fourth derivative of $\cos t + C$, where C is any constant, is indeed $\cos t$. But any function of the form $\cos t + p(t)$, where $p(t)$ is a polynomial of degree less than or equal to 3, also has its fourth derivative equal to $\cos t$. So $\cos t + t^2$ will work.

21. False; for example, if both f and g are constant functions, then the derivative of $f(g(x))$ is zero, as is the derivative of $f(x)$. Another example is $f(x) = 5x + 7$ and $g(x) = x + 2$.

25. False. Let $f(x) = e^{-x}$ and $g(x) = x^2$. Let $h(x) = f(g(x)) = e^{-x^2}$. Then $h'(x) = -2xe^{-x^2}$ and $h''(x) = (-2 + 4x^2)e^{-x^2}$. Since $h''(0) < 0$, clearly h is not concave up for all x.

29. False. For example, let $f(x) = x + 5$, and $g(x) = 2x - 3$. Then $f'(x) \leq g'(x)$ for all x, but $f(0) > g(0)$.

33. Let f be defined by

$$f(x) = \begin{cases} x & \text{if } 0 \leq x < 2 \\ 19 & \text{if } x = 2 \end{cases}$$

Then f is differentiable on $(0, 2)$ and $f'(x) = 1$ for all x in $(0, 2)$. Thus there is no c in $(0, 2)$ such that

$$f'(c) = \frac{f(2) - f(0)}{2 - 0} = \frac{19}{2}.$$

The reason that this function does not satisfy the conclusion of the Mean Value Theorem is that it is not continuous at $x = 2$.

CHAPTER FOUR

Solutions for Section 4.1

Exercises

1. $f'(x) = 6x^2 + 6x - 36$. To find critical points, we set $f'(x) = 0$. Then

$$6(x^2 + x - 6) = 6(x + 3)(x - 2) = 0.$$

Therefore, the critical points of f are $x = -3$ and $x = 2$. To the left of $x = -3$, $f'(x) > 0$. Between $x = -3$ and $x = 2$, $f'(x) < 0$. To the right of $x = 2$, $f'(x) > 0$. Thus $f(-3)$ is a local maximum, $f(2)$ a local minimum. See Figure 4.1.

Figure 4.1

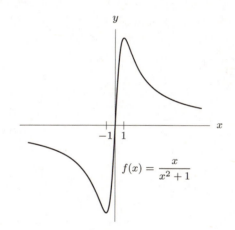

Figure 4.2

5.

$$f'(x) = \frac{x^2 + 1 - x \cdot 2x}{(x^2 + 1)^2} = \frac{1 - x^2}{(x^2 + 1)^2} = \frac{(1 - x)(1 + x)}{(x^2 + 1)^2}.$$

Critical points are $x = \pm 1$. To the left of $x = -1$, $f'(x) < 0$.
Between $x = -1$ and $x = 1$, $f'(x) > 0$.
To the right of $x = 1$, $f'(x) < 0$.
So, $f(-1)$ is a local minimum, $f(1)$ a local maximum. See Figure 4.2.

9. There are many possible answers. One possible graph is shown in Figure 4.3.

Figure 4.3

13. From the graph of $f(x)$ in the figure below, we see that the function must have two inflection points. We calculate $f'(x) = 4x^3 + 3x^2 - 6x$, and $f''(x) = 12x^2 + 6x - 6$. Solving $f''(x) = 0$ we find that:

$$x_1 = -1 \quad \text{and} \quad x_2 = \frac{1}{2}.$$

Since $f''(x) > 0$ for $x < x_1$, $f''(x) < 0$ for $x_1 < x < x_2$, and $f''(x) > 0$ for $x_2 < x$, it follows that both points are inflection points.

17. (a) Decreasing for $x < -1$, increasing for $-1 < x < 0$, decreasing for $0 < x < 1$, and increasing for $x > 1$.

(b) $f(-1)$ and $f(1)$ are local minima, $f(0)$ is a local maximum.

21. The inflection points of f are the points where f'' changes sign. See Figure 4.4.

Figure 4.4

Problems

25. To find the critical points, we set the derivative equal to zero and solve for t.

$$F'(t) = Ue^t + Ve^{-t}(-1) = 0$$
$$Ue^t - \frac{V}{e^t} = 0$$
$$Ue^t = \frac{V}{e^t}$$
$$Ue^{2t} = V$$
$$e^{2t} = \frac{V}{U}$$
$$2t = \ln(V/U)$$
$$t = \frac{\ln(V/U)}{2}.$$

The derivative $F'(t)$ is never undefined, so the only critical point is $t = 0.5 \ln(V/U)$.

29. See Figure 4.5.

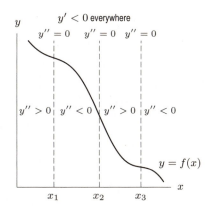

Figure 4.5

33. (a) It appears that this function has a local maximum at about $x = 1$, a local minimum at about $x = 4$, and a local maximum at about $x = 8$.

(b) The table now gives values of the derivative, so critical points occur where $f'(x) = 0$. Since f' is continuous, this occurs between 2 and 3, so there is a critical point somewhere around 2.5. Since f' is positive for values less than 2.5 and negative for values greater than 2.5, it appears that f has a local maximum at about $x = 2.5$. Similarly, it appears that f has a local minimum at about $x = 6.5$ and another local maximum at about $x = 9.5$.

37. Differentiating using the product rule gives

$$f'(x) = 3x^2(1 - x)^4 - 4x^3(1 - x)^3 = x^2(1 - x)^3(3(1 - x) - 4x) = x^2(1 - x)^3(3 - 7x).$$

The critical points are the solutions to

$$f'(x) = x^2(1 - x)^3(3 - 7x) = 0$$
$$x = 0, 1, \frac{3}{7}.$$

For $x < 0$, since $1 - x > 0$ and $3 - 7x > 0$, we have $f'(x) > 0$.
For $0 < x < \frac{3}{7}$, since $1 - x > 0$ and $3 - 7x > 0$, we have $f'(x) > 0$.
For $\frac{3}{7} < x < 1$, since $1 - x > 0$ and $3 - 7x < 0$, we have $f'(x) < 0$.
For $1 < x$, since $1 - x < 0$ and $3 - 7x < 0$, we have $f'(x) > 0$.
Thus, $x = 0$ is neither a local maximum nor a local minimum; $x = 3/7$ is a local maximum; $x = 1$ is a local minimum.

41. First, we wish to have $f'(6) = 0$, since $f(6)$ should be a local minimum:

$$f'(x) = 2x + a = 0$$
$$x = -\frac{a}{2} = 6$$
$$a = -12.$$

Next, we need to have $f(6) = -5$, since the point $(6, -5)$ is on the graph of $f(x)$. We can substitute $a = -12$ into our equation for $f(x)$ and solve for b:

$$f(x) = x^2 - 12x + b$$
$$f(6) = 36 - 72 + b = -5$$
$$b = 31.$$

Thus, $f(x) = x^2 - 12x + 31$.

45. From the first condition, we get that $x = 2$ is a local minimum for f. From the second condition, it follows that $x = 4$ is an inflection point. A possible graph is shown in Figure 4.6.

Figure 4.6

49. Since the derivative of an even function is odd and the derivative of an odd function is even, f and f'' are either both odd or both even, and f' is the opposite. Graphs I and II represent odd functions; III represents an even function, so III is f'. Since the maxima and minima of III occur where I crosses the x-axis, I must be the derivative of f', that is, f''. In addition, the maxima and minima of II occur where III crosses the x-axis, so II is f.

53. (a) Since $f''(x) > 0$ and $g''(x) > 0$ for all x, then $f''(x) + g''(x) > 0$ for all x, so $f(x) + g(x)$ is concave up for all x.

(b) Nothing can be concluded about the concavity of $(f + g)(x)$. For example, if $f(x) = ax^2$ and $g(x) = bx^2$ with $a > 0$ and $b < 0$, then $(f + g)''(x) = a + b$. So $f + g$ is either always concave up, always concave down, or a straight line, depending on whether $a > |b|$, $a < |b|$, or $a = |b|$. More generally, it is even possible that $(f + g)(x)$ may have one or more changes in concavity.

(c) It is possible to have infinitely many changes in concavity. Consider $f(x) = x^2 + \cos x$ and $g(x) = -x^2$. Since $f''(x) = 2 - \cos x$, we see that $f(x)$ is concave up for all x. Clearly $g(x)$ is concave down for all x. However, $f(x) + g(x) = \cos x$, which changes concavity an infinite number of times.

Solutions for Section 4.2

Exercises

1. See Figure 4.7.

Figure 4.7

5. Since $f(x) = x^3 - 3x^2 + 20$ is continuous and the interval $-1 \leq x \leq 3$ is closed, there must be a global maximum and minimum. The candidates are critical points in the interval and endpoints. Since there are no points where $f'(x)$ is undefined, we solve $f'(x) = 0$ to find all the critical points:

$$f'(x) = 3x^2 - 6x = 3x(x - 2) = 0,$$

so $x = 0$ and $x = 2$ are the critical points; both are in the interval. We then compare the values of f at the critical points and the endpoints:

$$f(-1) = 16, \quad f(0) = 20, \quad f(2) = 16, \quad f(3) = 20.$$

Thus the global maximum is 20 at $x = 0$ and $x = 3$, and the global minimum is 16 at $x = -1$ and $x = 2$.

9. Since $f(x) = e^{-x} \sin x$ is continuous and the interval $0 \leq x \leq 2\pi$ is closed, there must be a global maximum and minimum. The possible candidates are critical points in the interval and endpoints. Since there are no points where f' is undefined, we solve $f'(x) = 0$ to find all critical points:

$$f'(x) = -e^{-x} \sin x + e^{-x} \cos x = e^{-x}(-\sin x + \cos x) = 0.$$

Since $e^{-x} \neq 0$, the critical points are when $\sin x = \cos x$; the only solutions in the given interval are $x = \pi/4$ and $x = 5\pi/4$. We then compare the values of f at the critical points and the endpoints:

$$f(0) = 0, \quad f(\pi/4) = e^{-\pi/4}\left(\frac{\sqrt{2}}{2}\right) = 0.322, \quad f(5\pi/4) = e^{-5\pi/4}\left(\frac{-\sqrt{2}}{2}\right) = -0.0139, \quad f(2\pi) = 0.$$

Thus the global maximum is 0.322 at $x = \pi/4$, and the global minimum is -0.0139 at $x = 5\pi/4$.

13. (a) We have $f'(x) = 10x^9 - 10 = 10(x^9 - 1)$. This is zero when $x = 1$, so $x = 1$ is a critical point of f. For values of x less than 1, x^9 is less than 1, and thus $f'(x)$ is negative when $x < 1$. Similarly, $f'(x)$ is positive for $x > 1$. Thus $f(1) = -9$ is a local minimum.
   We also consider the endpoints $f(0) = 0$ and $f(2) = 1004$. Since $f'(0) < 0$ and $f'(2) > 0$, we see $x = 0$ and $x = 2$ are local maxima.
   (b) Comparing values of f shows that the global minimum is at $x = 1$, and the global maximum is at $x = 2$.

Problems

17. Differentiating gives

$$f'(x) = 1 - \frac{1}{x^2},$$

so the critical points satisfy

$$1 - \frac{1}{x^2} = 0$$
$$x^2 = 1$$
$$x = 1 \quad \text{(We want } x > 0\text{)}.$$

Since f' is negative for $0 < x < 1$ and f' is positive for $x > 1$, there is a local minimum at $x = 1$.
   Since $f(x) \to \infty$ as $x \to 0^+$ and as $x \to \infty$, the local minimum at $x = 1$ is a global minimum; there is no global maximum. See Figure 4.8. The the global minimum is $f(1) = 2$.

$f(x) = x + 1/x$

Figure 4.8

21. Differentiating using the product rule gives

$$f'(t) = 2 \sin t \cos t \cdot \cos t - (\sin^2 t + 2) \sin t = 0$$
$$\sin t (2 \cos^2 t - \sin^2 t - 2) = 0$$
$$\sin t (2(1 - \sin^2 t) - \sin^2 t - 2) = 0$$
$$\sin t (-3 \sin^2 t) = -3 \sin^3 t = 0.$$

Thus, the critical points are where $\sin t = 0$, so

$$t = 0, \pm\pi, \pm 2\pi, \pm 3\pi, \ldots.$$

Since $f'(t) = -3\sin^3 t$ is negative for $-\pi < t < 0$, positive for $0 < t < \pi$, negative for $\pi < t < 2\pi$, and so on, we find that $t = 0, \pm 2\pi, \ldots$ give local minima, while $t = \pm\pi, \pm 3\pi, \ldots$ give local maxima. Evaluating gives

$$f(0) = f(\pm 2\pi) = (0 + 2)1 = 2$$
$$f(\pm\pi) = f(\pm 3\pi) = (0 + 2)(-1) = -2.$$

Thus, the global maximum of $f(t)$ is 2, occurring at $t = 0, \pm 2\pi, \ldots$, and the global minimum of $f(t)$ is -2, occurring at $t = \pm\pi, \pm 3\pi, \ldots$. See Figure 4.9.

Figure 4.9

25. We want to maximize the height, y, of the grapefruit above the ground, as shown in the figure below. Using the derivative we can find exactly when the grapefruit is at the highest point. We can think of this in two ways. By common sense, at the peak of the grapefruit's flight, the velocity, dy/dt, must be zero. Alternately, we are looking for a global maximum of y, so we look for critical points where $dy/dt = 0$. We have

$$\frac{dy}{dt} = -32t + 50 = 0 \qquad \text{and so} \qquad t = \frac{-50}{-32} \approx 1.56 \text{ sec.}$$

Thus, we have the time at which the height is a maximum; the maximum value of y is then

$$y \approx -16(1.56)^2 + 50(1.56) + 5 = 44.1 \text{ feet.}$$

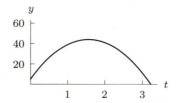

29. We set $f'(r) = 0$ to find the critical points:

$$\frac{2A}{r^3} - \frac{3B}{r^4} = 0$$
$$\frac{2Ar - 3B}{r^4} = 0$$
$$2Ar - 3B = 0$$
$$r = \frac{3B}{2A}.$$

The only critical point is at $r = 3B/(2A)$. If $r > 3B/(2A)$, we have $f' > 0$ and if $r < 3B/(2A)$, we have $f' < 0$. Thus, the force between the atoms is minimized at $r = 3B/(2A)$.

33. For $x > 0$, the line in Figure 4.10 has

$$\text{Slope} = \frac{y}{x} = \frac{x^2 e^{-3x}}{x} = xe^{-3x}.$$

If the slope has a maximum, it occurs where

$$\frac{d}{dx} (\text{Slope}) = 1 \cdot e^{-3x} - 3xe^{-3x} = 0$$
$$e^{-3x} (1 - 3x) = 0$$
$$x = \frac{1}{3}.$$

For this x-value,

$$\text{Slope} = \frac{1}{3}e^{-3(1/3)} = \frac{1}{3}e^{-1} = \frac{1}{3e}.$$

Figure 4.10 shows that the slope tends toward 0 as $x \to \infty$; the formula for the slope shows that the slope tends toward 0 as $x \to 0$. Thus the only critical point, $x = 1/3$, must give a local and global maximum.

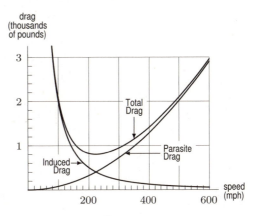

Figure 4.10   Figure 4.11

37. (a) Figure 4.11 contains the graph of total drag, plotted on the same coordinate system with induced and parasite drag. It was drawn by adding the vertical coordinates of Induced and Parasite drag.
(b) Airspeeds of approximately 160 mph and 320 mph each result in a total drag of 1000 pounds. Since two distinct airspeeds are associated with a single total drag value, the total drag function does not have an inverse. The parasite and induced drag functions do have inverses, because they are strictly increasing and strictly decreasing functions, respectively.
(c) To conserve fuel, fly the at the airspeed which minimizes total drag. This is the airspeed corresponding to the lowest point on the total drag curve in part (a): that is, approximately 220 mph.

41. (a) We know that $h''(x) < 0$ for $-2 \le x < -1$, $h''(-1) = 0$, and $h''(x) > 0$ for $x > -1$. Thus, $h'(x)$ decreases to its minimum value at $x = -1$, which we know to be zero, and then increases; it is never negative.
(b) Since $h'(x)$ is non-negative for $-2 \le x \le 1$, we know that $h(x)$ is never decreasing on $[-2, 1]$. So a global maximum must occur at the right hand endpoint of the interval.
(c) The graph below shows a function that is increasing on the interval $-2 \le x \le 1$ with a horizontal tangent and an inflection point at $(-1, 2)$.

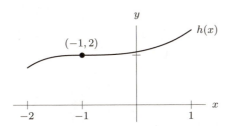

Solutions for Section 4.3

Exercises

1. (a) See Figure 4.12.

(b) We see in Figure 4.12 that in each case the graph of f is a parabola with one critical point, its vertex, on the positive x-axis. The critical point moves to the right along the x-axis as a increases.

(c) To find the critical points, we set the derivative equal to zero and solve for x.

$$f'(x) = 2(x - a) = 0$$
$$x = a.$$

The only critical point is at $x = a$. As we saw in the graph, and as a increases, the critical point moves to the right.

Figure 4.12

Figure 4.13

5. (a) See Figure 4.13.

(b) We see in Figure 4.13 that in each case f appears to have two critical points. One critical point is a local minimum at the origin and the other is a local maximum in quadrant I. As the parameter a increases, the critical point in quadrant I appears to move down and to the left, closer to the origin.

(c) To find the critical points, we set the derivative equal to zero and solve for x. Using the product rule, we have:

$$f'(x) = x^2 \cdot e^{-ax}(-a) + 2x \cdot e^{-ax} = 0$$
$$xe^{-ax}(-ax + 2) = 0$$
$$x = 0 \quad \text{and} \quad x = \frac{2}{a}.$$

There are two critical points, at $x = 0$ and $x = 2/a$. As we saw in the graph, as a increases the nonzero critical point moves to the left.

9. (a) Figure 4.14 shows the effect of varying a with $b = 1$.

(b) Figure 4.15 shows the effect of varying b with $a = 1$.

Figure 4.14

Figure 4.15

(c) In each case f appears to have two critical points, a local maximum in quadrant I and a local minimum on the positive x-axis. From Figure 4.14 it appears that increasing a moves the local maximum up and does not move the local minimum. From Figure 4.15 it appears that increasing b moves the local maximum up and to the right and moves the local minimum to the right along the x-axis.

(d) To find the critical points, we set the derivative equal to zero and solve for x. Using the product rule, we have:

$$f'(x) = ax \cdot 2(x - b) + a \cdot (x - b)^2 = 0$$
$$a(x - b)(2x + (x - b)) = 0$$
$$a(x - b)(3x - b) = 0$$
$$x = b \quad \text{or} \quad x = \frac{b}{3}.$$

There are two critical points, at $x = b$ and at $x = b/3$. Increasing a does not move either critical point horizontally. Increasing b moves both critical points to the right. This confirms what we saw in the graphs.

Problems

13. (a) Writing $y = L(1 + Ae^{-kt})^{-1}$, we find the first derivative by the chain rule

$$\frac{dy}{dt} = -L(1 + Ae^{-kt})^{-2}(-Ake^{-kt}) = \frac{LAke^{-kt}}{(1 + Ae^{-kt})^2}.$$

Using the quotient rule to calculate the second derivative gives

$$\frac{d^2y}{dt^2} = \frac{\left(-LAk^2e^{-kt}(1 + Ae^{-kt})^2 - 2LAke^{-kt}(1 + Ae^{-kt})(-Ake^{-kt})\right)}{(1 + Ae^{-kt})^4}$$

$$= \frac{LAk^2e^{-kt}(-1 + Ae^{-kt})}{(1 + Ae^{-kt})^3}.$$

(b) Since $L, A > 0$ and $e^{-kt} > 0$ for all t, the factor LAk^2e^{-kt} and the denominator are never zero. Thus, possible inflection points occur where

$$-1 + Ae^{-kt} = 0.$$

Solving for t gives

$$t = \frac{\ln A}{k}.$$

(c) The second derivative is positive to the left of $t = \ln(A)/k$ and negative to the right, so the function changes from concave up to concave down at $t = \ln(A)/k$.

17. Cubic polynomials are all of the form $f(x) = Ax^3 + Bx^2 + Cx + D$. There is an inflection point at the origin $(0,0)$ if $f''(0) = 0$ and $f(0) = 0$. Since $f(0) = D$, we must have $D = 0$. Since $f''(x) = 6Ax + 2B$, giving $f''(0) = 2B$, we must have $B = 0$. The family of cubic polynomials with inflection point at the origin is the two parameter family $f(x) = Ax^3 + Cx$.

21. (a) See Figure 4.16.

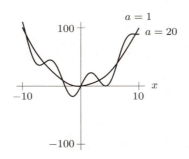

Figure 4.16

(b) The function $f(x) = x^2 + a\sin x$ is concave up for all x if $f''(x) > 0$ for all x. We have $f''(x) = 2 - a\sin x$. Because $\sin x$ varies between -1 and 1, we have $2 - a\sin x > 0$ for all x if $-2 < a < 2$ but not otherwise. Thus $f(x)$ is concave up for all x if $-2 < a < 2$.

25. Since the horizontal asymptote is $y = 5$, we know $a = 5$. The value of b can be any number. Thus $y = 5(1 - e^{-bx})$ for any $b > 0$.

29. Since $y(0) = a/(1 + b) = 2$, we have $a = 2 + 2b$. To find a point of inflection, we calculate

$$\frac{dy}{dt} = \frac{abe^{-t}}{(1 + be^{-t})^2},$$

and using the quotient rule,

$$\frac{d^2y}{dx^2} = \frac{-abe^{-t}(1 + be^{-t})^2 - abe^{-t}2(1 + be^{-t})(-be^{-t})}{(1 + be^{-t})^4}$$

$$= \frac{abe^{-t}(-1 + be^{-t})}{(1 + be^{-t})^3}.$$

The second derivative is equal to 0 when $be^{-t} = 1$, or $b = e^t$. When $t = 1$, we have $b = e$. The second derivative changes sign at this point, so we have an inflection point. Thus

$$y = \frac{2 + 2e}{1 + e^{1-t}}.$$

33. Differentiating $y = ae^{-x} + bx$ gives

$$\frac{dy}{dx} = -ae^{-x} + b.$$

Since the global minimum occurs for $x = 1$, we have $-a/e + b = 0$, so $b = a/e$.

The value of the function at $x = 1$ is 2, so we have $2 = a/e + b$, which gives

$$2 = \frac{a}{e} + \frac{a}{e} = \frac{2a}{e},$$

so $a = e$ and $b = 1$. Thus

$$y = e^{1-x} + x.$$

We compute $d^2y/dx^2 = e^{1-x}$, which is always positive, so this confirms that $x = 1$ is a local minimum. Because the value of $e^{1-x} + x$ as $x \to \pm\infty$ grows without bound, this local minimum is a global minimum.

37. Since $\lim_{t\to\infty} N = a$, we have $a = 200{,}000$. Note that while $N(t)$ will never actually reach 200,000, it will become arbitrarily close to 200,000. Since N represents the number of people, it makes sense to round up long before $t \to \infty$. When $t = 1$, we have $N = 0.1(200{,}000) = 20{,}000$ people, so plugging into our formula gives

$$N(1) = 20{,}000 = 200{,}000 \left(1 - e^{-k(1)}\right).$$

Solving for k gives

$$0.1 = 1 - e^{-k}$$
$$e^{-k} = 0.9$$
$$k = -\ln 0.9 \approx 0.105.$$

41. (a) The graph of r has a vertical asymptote if the denominator is zero. Since $(x - b)^2$ is nonnegative, the denominator can only be zero if $a \le 0$. Then

$$a + (x - b)^2 = 0$$
$$(x - b)^2 = -a$$
$$x - b = \pm\sqrt{-a}$$
$$x = b \pm \sqrt{-a}.$$

In order for there to be a vertical asymptote, a must be less than or equal to zero. There are no restrictions on b.

(b) Differentiating gives

$$r'(x) = \frac{-1}{(a + (x - b)^2)^2} \cdot 2(x - b),$$

so $r' = 0$ when $x = b$. If $a \le 0$, then r' is undefined at the same points at which r is undefined. Thus the only critical point is $x = b$. Since we want $r(x)$ to have a maximum at $x = 3$, we choose $b = 3$. Also, since $r(3) = 5$, we have

$$r(3) = \frac{1}{a + (3 - 3)^2} = \frac{1}{a} = 5 \qquad \text{so} \qquad a = \frac{1}{5}.$$

45. Let $f(x) = Ae^{-Bx^2}$. Since

$$f(x) = Ae^{-Bx^2} = Ae^{-\frac{(x-0)^2}{(1/B)}},$$

this is just the family of curves $y = e^{-\frac{(x-a)^2}{b}}$ multiplied by a constant A. This family of curves is discussed in the text; here, $a = 0$, $b = \frac{1}{B}$. When $x = 0$, $y = Ae^0 = A$, so A determines the y-intercept. A also serves to flatten or stretch the graph of e^{-Bx^2} vertically. Since $f'(x) = -2ABxe^{-Bx^2}$, $f(x)$ has a critical point at $x = 0$. For $B > 0$, the graphs are bell-shaped curves centered at $x = 0$, and $f(0) = A$ is a global maximum.

To find the inflection points of f, we solve $f''(x) = 0$. Since $f'(x) = -2ABxe^{-Bx^2}$,

$$f''(x) = -2ABe^{-Bx^2} + 4AB^2x^2e^{-Bx^2}.$$

Since e^{-Bx^2} is always positive, $f''(x) = 0$ when

$$-2AB + 4AB^2x^2 = 0$$
$$x^2 = \frac{2AB}{4AB^2}$$
$$x = \pm\sqrt{\frac{1}{2B}}.$$

These are points of inflection, since the second derivative changes sign here. Thus for large values of B, the inflection points are close to $x = 0$, and for smaller values of B the inflection points are further from $x = 0$. Therefore B affects the width of the graph.

In the graphs in Figure 4.17, A is held constant, and variations in B are shown.

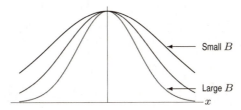

Small B

Large B

Figure 4.17: $f(x) = Ae^{-Bx^2}$ for varying B

49.

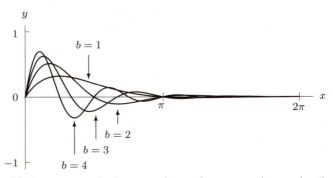

The larger the value of b, the narrower the humps and more humps per given region there are in the graph.

53. (a) To find $\lim_{r \to 0^+} V(r)$, first rewrite $V(r)$ with a common denominator:

$$\lim_{r \to 0^+} V(r) = \lim_{r \to 0^+} \frac{A}{r^{12}} - \frac{B}{r^6}$$
$$= \lim_{r \to 0^+} \frac{A - Br^6}{r^{12}}$$
$$\to \frac{A}{0^+} \to +\infty.$$

As the distance between the two atoms becomes small, the potential energy diverges to $+\infty$.

(b) The critical point of $V(r)$ will occur where $V'(r) = 0$:

$$V'(r) = -\frac{12A}{r^{13}} + \frac{6B}{r^7} = 0$$
$$\frac{-12A + 6Br^6}{r^{13}} = 0$$
$$-12A + 6Br^6 = 0$$
$$r^6 = \frac{2A}{B}$$
$$r = \left(\frac{2A}{B}\right)^{1/6}$$

To determine whether this is a local maximum or minimum, i we can use the first derivative test. Since r is positive, the sign of $V'(r)$ is determined by the sign of $-12A + 6Br^6$. Notice that this is an increasing function of r for $r > 0$, so $V'(r)$ changes sign from $-$ to $+$ at $r = (2A/B)^{1/6}$. The first derivative test yields

r	$\leftarrow$	$\left(\frac{2A}{B}\right)^{1/6}$	$\rightarrow$
$V'(r)$	neg.	zero	pos.

Thus $V(r)$ goes from decreasing to increasing at the critical point $r = (2A/B)^{1/6}$, so this is a local minimum.

(c) Since $F(r) = -V'(r)$, the force is zero exactly where $V'(r) = 0$, i.e. at the critical points of V. The only critical point was the one found in part (b), so the only such point is $r = (2A/B)^{1/6}$.

(d) Since the numerator in $r = (2A/B)^{1/6}$ is proportional to $A^{1/6}$, the equilibrium size of the molecule increases when the parameter A is increased. Conversely, since B is in the denominator, when B is increased the equilibrium size of the molecules decrease.

57. (a) The vertical intercept is $W = Ae^{-e^{b-c\cdot 0}} = Ae^{-e^b}$. There is no horizontal intercept since the exponential function is always positive. There is a horizontal asymptote. As $t \to \infty$, we see that $e^{b-ct} = e^b/e^{ct} \to 0$, since t is positive. Therefore $W \to Ae^0 = A$, so there is a horizontal asymptote at $W = A$.

(b) The derivative is
$$\frac{dW}{dt} = Ae^{-e^{b-ct}}(-e^{b-ct})(-c) = Ace^{-e^{b-ct}}e^{b-ct}.$$

Thus, dW/dt is always positive, so W is always increasing and has no critical points. The second derivative is
$$\frac{d^2W}{dt^2} = \frac{d}{dt}(Ace^{-e^{b-ct}})e^{b-ct} + Ace^{-e^{b-ct}}\frac{d}{dt}(e^{b-ct})$$
$$= Ac^2e^{-e^{b-ct}}e^{b-ct}e^{b-ct} + Ace^{-e^{b-ct}}(-c)e^{b-ct}$$
$$= Ac^2e^{-e^{b-ct}}e^{b-ct}(e^{b-ct} - 1).$$

Now e^{b-ct} decreases from $e^b > 1$ when $t = 0$ toward 0 as $t \to \infty$. The second derivative changes sign from positive to negative when $e^{b-ct} = 1$, i.e., when $b - ct = 0$, or $t = b/c$. Thus the curve has an inflection point at $t = b/c$, where $W = Ae^{-e^{b-(b/c)c}} = Ae^{-1}$.

(c) See Figure 4.18.

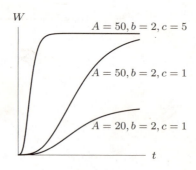

W

$A = 50, b = 2, c = 5$

$A = 50, b = 2, c = 1$

$A = 20, b = 2, c = 1$

t

Figure 4.18

(d) The final size of the organism is given by the horizontal asymptote $W = A$. The curve is steepest at its inflection point, which occurs at $t = b/c$, $W = Ae^{-1}$. Since $e = 2.71828\ldots \approx 3$, the size the organism when it is growing fastest is about $A/3$, one third its final size. So yes, the Gompertz growth function is useful in modeling such growth.

Solutions for Section 4.4

Exercises

1. We look for critical points of M:
$$\frac{dM}{dx} = \frac{1}{2}wL - wx.$$

Now $dM/dx = 0$ when $x = L/2$. At this point $d^2M/dx^2 = -w$ so this point is a local maximum. The graph of $M(x)$ is a parabola opening downward, so the local maximum is also the global maximum.

5. We only consider $\lambda > 0$. For such λ, the value of $v \to \infty$ as $\lambda \to \infty$ and as $\lambda \to 0^+$. Thus, v does not have a maximum velocity. It will have a minimum velocity. To find it, we set $dv/d\lambda = 0$:

$$\frac{dv}{d\lambda} = k\frac{1}{2}\left(\frac{\lambda}{c} + \frac{c}{\lambda}\right)^{-1/2}\left(\frac{1}{c} - \frac{c}{\lambda^2}\right) = 0.$$

Solving, and remembering that $\lambda > 0$, we obtain

$$\frac{1}{c} - \frac{c}{\lambda^2} = 0$$
$$\frac{1}{c} = \frac{c}{\lambda^2}$$
$$\lambda^2 = c^2,$$

so

$$\lambda = c.$$

Thus, we have one critical point. Since

$$\frac{dv}{d\lambda} < 0 \quad \text{for } \lambda < c$$

and

$$\frac{dv}{d\lambda} > 0 \quad \text{for } \lambda > c,$$

the first derivative test tells us that we have a local minimum of v at $x = c$. Since $\lambda = c$ is the only critical point, it gives the global minimum. Thus the minimum value of v is

$$v = k\sqrt{\frac{c}{c} + \frac{c}{c}} = \sqrt{2}k.$$

9. A graph of F against θ is shown in Figure 4.19.

Taking the derivative:

$$\frac{dF}{d\theta} = -\frac{mg\mu(\cos\theta - \mu\sin\theta)}{(\sin\theta + \mu\cos\theta)^2}.$$

At a critical point, $dF/d\theta = 0$, so

$$\cos\theta - \mu\sin\theta = 0$$
$$\tan\theta = \frac{1}{\mu}$$
$$\theta = \arctan\left(\frac{1}{\mu}\right).$$

If $\mu = 0.15$, then $\theta = \arctan(1/0.15) = 1.422 \approx 81.5°$. To calculate the maximum and minimum values of F, we evaluate at this critical point and the endpoints:

$$\text{At } \theta = 0, \quad F = \frac{0.15mg}{\sin 0 + 0.15\cos 0} = 1.0mg \text{ newtons.}$$

$$\text{At } \theta = 1.422, \ F = \frac{0.15mg}{\sin(1.422) + 0.15\cos(1.422)} = 0.148mg \text{ newtons.}$$

$$\text{At } \theta = \pi/2, \quad F = \frac{0.15mg}{\sin(\frac{\pi}{2}) + 0.15\cos(\frac{\pi}{2})} = 0.15mg \text{ newtons.}$$

Thus, the maximum value of F is $1.0mg$ newtons when $\theta = 0$ (her arm is vertical) and the minimum value of F is $0.148mg$ newtons is when $\theta = 1.422$ (her arm is close to horizontal). See Figure 4.20.

Figure 4.19

Figure 4.20

13. Examination of the graph suggests that $0 \leq x^3 e^{-x} \leq 2$. The lower bound of 0 is the best possible lower bound since

$$f(0) = (0)^3 e^{-0} = 0.$$

To find the best possible upper bound, we find the critical points. Differentiating, using the product rule, yields

$$f'(x) = 3x^2 e^{-x} - x^3 e^{-x}$$

Setting $f'(x) = 0$ and factoring gives

$$3x^2 e^{-x} - x^3 e^{-x} = 0$$
$$x^2 e^{-x} (3 - x) = 0$$

So the critical points are $x = 0$ and $x = 3$. Note that $f'(x) < 0$ for $x > 3$ and $f'(x) > 0$ for $x < 3$, so $f(x)$ has a local maximum at $x = 3$. Examination of the graph tells us that this is the global maximum. So $0 \leq x^3 e^{-x} \leq f(3)$.

$$f(3) = 3^3 e^{-3} \approx 1.34425$$

So $0 \leq x^3 e^{-x} \leq 3^3 e^{-3} \approx 1.34425$ are the best possible bounds for the function.

Figure 4.21

Problems

17. (a) The rectangle on the left has area xy, and the semicircle on the right with radius $y/2$ has area $(1/2)\pi(y/2)^2 = \pi y^2/8$. We have

$$\text{Area of entire region} = xy + \frac{\pi}{8}y^2.$$

(b) The perimeter of the figure is made of two horizontal line segments of length x each, one vertical segment of length y, and a semicircle of radius $y/2$ of length $\pi y/2$. We have

$$\text{Perimeter of entire region} = 2x + y + \frac{\pi}{2}y = 2x + (1 + \frac{\pi}{2})y.$$

(c) We want to maximize the area $xy + \pi y^2/8$, with the perimeter condition $2x + (1 + \pi/2)y = 100$. Substituting

$$x = 50 - \left(\frac{1}{2} + \frac{\pi}{4}\right) y = 50 - \frac{2 + \pi}{4} y$$

into the area formula, we must maximize

$$A(y) = \left(50 - \left(\frac{1}{2} + \frac{\pi}{4}\right)y\right)y + \frac{\pi}{8}y^2 = 50y - \left(\frac{1}{2} + \frac{\pi}{8}\right)y^2$$

on the interval

$$0 \le y \le 200/(2 + \pi) = 38.8985$$

where $y \ge 0$ because y is a length and $y \le 200/(2 + \pi)$ because $x \ge 0$ is a length.

The critical point of A occurs where

$$A'(y) = 50 - \left(1 + \frac{\pi}{4}\right)y = 0$$

at

$$y = \frac{200}{4 + \pi} = 28.005.$$

A maximum for A must occur at the critical point $y = 200/(4 + \pi)$ or at one of the endpoints $y = 0$ or $y = 200/(2 + \pi)$. Since

$$A(0) = 0 \qquad A\left(\frac{200}{4 + \pi}\right) = 700.124 \qquad A\left(\frac{200}{2 + \pi}\right) = 594.2$$

the maximum is at

$$y = \frac{200}{4 + \pi}.$$

Hence

$$x = 50 - \frac{2 + \pi}{4}y = 50 - \frac{2 + \pi}{4}\frac{200}{4 + \pi} = \frac{100}{4 + \pi}.$$

The dimensions giving maximum area with perimeter 100 are

$$x = \frac{100}{4 + \pi} = 14.0 \qquad y = \frac{200}{4 + \pi} = 28.0.$$

The length y is twice as great as x.

21. We want to minimize the surface area S of the box, shown in Figure 4.22. The box has 5 faces: the bottom which has area x^2 and the four sides, each of which has area xh. Thus we want to minimize

$$S = x^2 + 4xh.$$

The volume of the box is $8 = x^2h$, so $h = 8/x^2$. Substituting this expression in for h in the formula for S gives

$$S = x^2 + 4x \cdot \frac{8}{x^2} = x^2 + \frac{32}{x}.$$

Differentiating gives

$$\frac{dS}{dx} = 2x - \frac{32}{x^2}.$$

To minimize S we look for critical points, so we solve $0 = 2x - 32/x^2$. Multiplying by x^2 gives

$$0 = 2x^3 - 32,$$

so $x = 16^{1/3}$ cm. Then we can find

$$h = \frac{8}{x^2} = \frac{8}{16^{2/3}} = \frac{16}{2 \cdot 16^{2/3}} = \frac{16^{1/3}}{2}$$

cm.

We can check that this critical point is a minimum of S by checking the sign of

$$\frac{d^2S}{dx^2} = 2 + \frac{64}{x^3}$$

which is positive when $x > 0$. So S is concave up at the critical point and therefore $x = 16^{1/3}$ gives a minimum value of S.

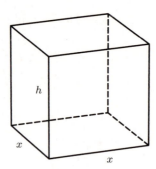

Figure 4.22

25. The geometry shows that as x increases from 0 to 10 the length y decreases. The minimum value of y is 5 when $x = 10$, and the maximum value of y is $\sqrt{10^2 + 5^2} = \sqrt{125}$ when $x = 0$.

We can also use calculus. By the Pythagorean theorem we have

$$y = \sqrt{5^2 + (10 - x)^2}.$$

The extreme values of

$$y = f(x) = \sqrt{5^2 + (10 - x)^2}$$

for $0 \le x \le 10$ occur at the endpoints $x = 0$ or $x = 10$ or at a critical point of f. The only solution of the equation

$$f'(x) = \frac{x - 10}{\sqrt{5^2 + (10 - x)^2}} = 0$$

is $x = 10$, which is the only critical point of f. We have

$$f(0) = \sqrt{125} = 11.18 \qquad f(10) = 5.$$

Therefore the minimum value of y is 5 and the maximum value is $\sqrt{125} = 5\sqrt{5}$.

29. The rectangle in Figure 4.23 has area, A, given by

$$A = 2xy = \frac{2x}{1 + x^2} \qquad \text{for } x \ge 0.$$

At a critical point,

$$\frac{dA}{dx} = \frac{2}{1 + x^2} + 2x \left(\frac{-2x}{(1 + x^2)^2} \right) = 0$$

$$\frac{2(1 + x^2 - 2x^2)}{(1 + x^2)^2} = 0$$

$$1 - x^2 = 0$$

$$x = \pm 1.$$

Since $A = 0$ for $x = 0$ and $A \to 0$ as $x \to \infty$, the critical point $x = 1$ is a local and global maximum for the area. Then $y = 1/2$, so the vertices are

$$(-1, 0), \ (1, 0), \ \left(1, \frac{1}{2}\right), \ \left(-1, \frac{1}{2}\right).$$

Figure 4.23

33. From the triangle shown in Figure 4.24, we see that

$$\left(\frac{w}{2}\right)^2 + \left(\frac{h}{2}\right)^2 = 30^2$$
$$w^2 + h^2 = 4(30)^2 = 3600.$$

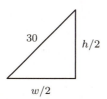

Figure 4.24

The strength, S, of the beam is given by

$$S = kwh^2,$$

for some constant k. To make S a function of only one variable, substitute for h^2, giving

$$S = kw(3600 - w^2) = k(3600w - w^3).$$

Differentiating and setting $dS/dw = 0$,

$$\frac{dS}{dw} = k(3600 - 3w^2) = 0.$$

Solving for w gives

$$w = \sqrt{1200} = 34.64 \text{ cm},$$

so

$$h^2 = 3600 - w^2 = 3600 - 1200 = 2400$$
$$h = \sqrt{2400} = 48.99 \text{ cm}.$$

Thus, $w = 34.64$ cm and $h = 48.99$ cm give a critical point. To check that this is a local maximum, we compute

$$\frac{d^2S}{dw^2} = -6w < 0 \quad \text{for} \quad w > 0.$$

Since $d^2S/dw^2 < 0$, we see that $w = 34.64$ cm is a local maximum. It is the only critical point, so it is a global maximum.

37. Any point on the curve can be written (x, x^2). The distance between such a point and $(3, 0)$ is given by

$$s(x) = \sqrt{(3 - x)^2 + (0 - x^2)^2} = \sqrt{(3 - x)^2 + x^4}.$$

Plotting this function in Figure 4.25, we see that there is a minimum near $x = 1$.

To find the value of x that minimizes the distance we can instead minimize the function $Q = s^2$ (the derivative is simpler). Then we have

$$Q(x) = (3 - x)^2 + x^4.$$

Differentiating $Q(x)$ gives

$$\frac{dQ}{dx} = -6 + 2x + 4x^3.$$

Plotting the function $4x^3 + 2x - 6$ shows that there is one real solution at $x = 1$, which can be verified by substitution; the required coordinates are therefore $(1, 1)$. Because $Q''(x) = 2 + 12x^2$ is always positive, $x = 1$ is indeed the minimum. See Figure 4.26.

Figure 4.25

Figure 4.26

41. (a) Suppose n passengers sign up for the cruise. If $n \leq 100$, then the cruise's revenue is $R = 2000n$, so the maximum revenue is
$$R = 2000 \cdot 100 = 200,000.$$
If $n > 100$, then the price is
$$p = 2000 - 10(n - 100)$$
and hence the revenue is
$$R = n(2000 - 10(n - 100)) = 3000n - 10n^2.$$
To find the maximum revenue, we set $dR/dn = 0$, giving $20n = 3000$ or $n = 150$. Then the revenue is
$$R = (2000 - 10 \cdot 50) \cdot 150 = 225,000.$$
Since this is more than the maximum revenue when $n \leq 100$, the boat maximizes its revenue with 150 passengers, each paying $1500.

(b) We approach this problem in a similar way to part (a), except now we are dealing with the profit function π. If $n \leq 100$, we have
$$\pi = 2000n - 80,000 - 400n,$$
so π is maximized with 100 passengers yielding a profit of
$$\pi = 1600 \cdot 100 - 80,000 = \$80,000.$$
If $n > 100$, we have
$$\pi = n(2000 - 10(n - 100)) - (80,000 + 400n).$$
We again set $d\pi/dn = 0$, giving $2600 = 20n$, so $n = 130$. The profit is then $89,000. So the boat maximizes profit by boarding 130 passengers, each paying $1700. This gives the boat $89,000 in profit.

45. (a) The line in the left-hand figure has slope equal to the rate worms arrive. To understand why, see line (1) in the right-hand figure. (This is the same line.) For any point Q on the loading curve, the line PQ has slope
$$\frac{QT}{PT} = \frac{QT}{PO + OT} = \frac{\text{load}}{\text{traveling time} + \text{searching time}}.$$

(b) The slope of the line PQ is maximized when the line is tangent to the loading curve, which happens with line (2). The load is then approximately 7 worms.

(c) If the traveling time is increased, the point P moves to the left, to point P', say. If line (3) is tangent to the curve, it will be tangent to the curve further to the right than line (2), so the optimal load is larger. This makes sense: if the bird has to fly further, you'd expect it to bring back more worms each time.

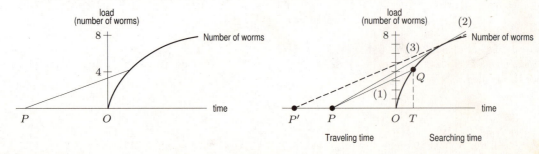

49. (a) Since the speed of light is a constant, the time of travel is minimized when the distance of travel is minimized. From Figure 4.27,

$$\text{Distance } \overrightarrow{OP} = \sqrt{x^2 + 1^2} = \sqrt{x^2 + 1}$$
$$\text{Distance } \overrightarrow{PQ} = \sqrt{(2 - x)^2 + 1^2} = \sqrt{(2 - x)^2 + 1}$$

Thus,

$$\text{Total distance traveled} = s = \sqrt{x^2 + 1} + \sqrt{(2 - x)^2 + 1}.$$

The total distance is a minimum if

$$\frac{ds}{dx} = \frac{1}{2}(x^2 + 1)^{-1/2} \cdot 2x + \frac{1}{2}((2 - x)^2 + 1)^{-1/2} \cdot 2(2 - x)(-1) = 0,$$

giving

$$\frac{x}{\sqrt{x^2 + 1}} - \frac{2 - x}{\sqrt{(2 - x)^2 + 1}} = 0$$
$$\frac{x}{\sqrt{x^2 + 1}} = \frac{2 - x}{\sqrt{(2 - x)^2 + 1}}$$

Squaring both sides gives

$$\frac{x^2}{x^2 + 1} = \frac{(2 - x)^2}{(2 - x)^2 + 1}.$$

Cross multiplying gives

$$x^2((2 - x)^2 + 1) = (2 - x)^2(x^2 + 1).$$

Multiplying out

$$x^2(4 - 4x + x^2 + 1) = (4 - 4x + x^2)(x^2 + 1)$$
$$4x^2 - 4x^3 + x^4 + x^2 = 4x^2 - 4x^3 + x^4 + 4 - 4x + x^2.$$

Collecting terms and canceling gives

$$0 = 4 - 4x$$
$$x = 1.$$

We can see that this value of x gives a minimum by comparing the value of s at this point and at the endpoints, $x = 0, x = 2$.

At $x = 1$,

$$s = \sqrt{1^2 + 1} + \sqrt{(2 - 1)^2 + 1} = 2.83.$$

At $x = 0$,

$$s = \sqrt{0^2 + 1} + \sqrt{(2 - 0)^2 + 1} = 3.24.$$

At $x = 2$,

$$s = \sqrt{2^2 + 1} + \sqrt{(2 - 2)^2 + 1} = 3.24.$$

Thus the shortest travel time occurs when $x = 1$; that is, when P is at the point $(1, 1)$.

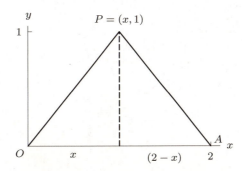

Figure 4.27

(b) Since $x = 1$ is halfway between $x = 0$ and $x = 2$, the angles θ_1 and θ_2 are equal.

Solutions for Section 4.5

Exercises

1. The fixed costs are $5000, the marginal cost per item is $2.40, and the price per item is $4.

5. (a) Total cost, in millions of dollars, $C(q) = 3 + 0.4q$.
   (b) Revenue, in millions of dollars, $R(q) = 0.5q$.
   (c) Profit, in millions of dollars, $\pi(q) = R(q) - C(q) = 0.5q - (3 + 0.4q) = 0.1q - 3$.

9. (a) At $q = 5000$, $MR > MC$, so the marginal revenue to produce the next item is greater than the marginal cost. This means that the company will make money by producing additional units, and production should be increased.
   (b) Profit is maximized where $MR = MC$, and where the profit function is going from increasing ($MR > MC$) to decreasing ($MR < MC$). This occurs at $q = 8000$.

Problems

13. (a) $\pi(q)$ is maximized when $R(q) > C(q)$ and they are as far apart as possible. See Figure 4.28.
    (b) $\pi'(q_0) = R'(q_0) - C'(q_0) = 0$ implies that $C'(q_0) = R'(q_0) = p$.
       Graphically, the slopes of the two curves at q_0 are equal. This is plausible because if $C'(q_0)$ were greater than p or less than p, the maximum of $\pi(q)$ would be to the left or right of q_0, respectively. In economic terms, if the cost were rising more quickly than revenues, the profit would be maximized at a lower quantity (and if the cost were rising more slowly, at a higher quantity).
    (c) See Figure 4.29.

Figure 4.28

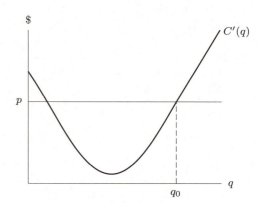

Figure 4.29

17. For each month,

$$\text{Profit} = \text{Revenue} - \text{Cost}$$
$$\pi = pq - wL = pcK^{\alpha}L^{\beta} - wL$$

The variable on the right is L, so at the maximum

$$\frac{d\pi}{dL} = \beta pcK^{\alpha}L^{\beta-1} - w = 0$$

Now $\beta - 1$ is negative, since $0 < \beta < 1$, so $1 - \beta$ is positive and we can write

$$\frac{\beta pcK^{\alpha}}{L^{1-\beta}} = w$$

giving

$$L = \left(\frac{\beta pcK^{\alpha}}{w}\right)^{\frac{1}{1-\beta}}$$

Since $\beta - 1$ is negative, when L is just above 0, the quantity $L^{\beta-1}$ is huge and positive, so $d\pi/dL > 0$. When L is large, $L^{\beta-1}$ is small, so $d\pi/dL < 0$. Thus the value of L we have found gives a global maximum, since it is the only critical point.

21. (a) $a(q) = C(q)/q$, so $C(q) = 0.01q^3 - 0.6q^2 + 13q$.

(b) Taking the derivative of $C(q)$ gives an expression for the marginal cost:

$$C'(q) = MC(q) = 0.03q^2 - 1.2q + 13.$$

To find the smallest MC we take its derivative and find the value of q that makes it zero. So: $MC'(q) = 0.06q - 1.2 = 0$ when $q = 1.2/0.06 = 20$. This value of q must give a minimum because the graph of $MC(q)$ is a parabola opening upward. Therefore the minimum marginal cost is $MC(20) = 1$. So the marginal cost is at a minimum when the additional cost per item is $1.

(c) $a'(q) = 0.02q - 0.6$

Setting $a'(q) = 0$ and solving for q gives $q = 30$ as the quantity at which the average is minimized, since the graph of a is a parabola which opens upward. The minimum average cost is $a(30) = 4$ dollars per item.

(d) The marginal cost at $q = 30$ is $MC(30) = 0.03(30)^2 - 1.2(30) + 13 = 4$. This is the same as the average cost at this quantity. Note that since $a(q) = C(q)/q$, we have $a'(q) = (qC'(q) - C(q))/q^2$. At a critical point, q_0, of $a(q)$, we have

$$0 = a'(q_0) = \frac{q_0 C'(q_0) - C(q_0)}{q_0^2},$$

so $C'(q_0) = C(q_0)/q_0 = a(q_0)$. Therefore $C'(30) = a(30) = 4$ dollars per item.

Another way to see why the marginal cost at $q = 30$ must equal the minimum average cost $a(30) = 4$ is to view $C'(30)$ as the approximate cost of producing the 30th or 31st good. If $C'(30) < a(30)$, then producing the 31st good would lower the average cost, i.e. $a(31) < a(30)$. If $C'(30) > a(30)$, then producing the 30th good would raise the average cost, i.e. $a(30) > a(29)$. Since $a(30)$ is the global minimum, we must have $C'(30) = a(30)$.

25. (a) See Figure 4.30.

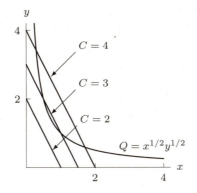

Figure 4.30

(b) Comparing the lines $C = 2$, $C = 3$, $C = 4$, we see that the cost increases as we move away from the origin. The line $C = 2$ does not cut the curve $Q = 1$; the lines $C = 3$ and $C = 4$ cut twice.

The minimum cost occurs where a cost line is tangent to the production curve.

(c) Using implicit differentiation, the slope of $x^{1/2}y^{1/2} = 1$ is given by

$$\frac{1}{2}x^{-1/2}y^{1/2} + \frac{1}{2}x^{1/2}y^{-1/2}y' = 0$$

$$y' = \frac{-x^{-1/2}y^{1/2}}{x^{1/2}y^{-1/2}} = -\frac{y}{x}.$$

The cost lines all have slope -2. Thus, if the curve is tangent to a line, we have

$$-\frac{y}{x} = -2$$
$$y = 2x.$$

Substituting into $Q = x^{1/2}y^{1/2} = 1$ gives

$$x^{1/2}(2x)^{1/2} = 1$$
$$\sqrt{2}x = 1$$
$$x = \frac{1}{\sqrt{2}}$$
$$y = 2 \cdot \frac{1}{\sqrt{2}} = \sqrt{2}.$$

Thus the minimum cost is

$$C = 2\frac{1}{\sqrt{2}} + \sqrt{2} = 2\sqrt{2}.$$

Solutions for Section 4.6

Exercises

1. The rate of growth, in billions of people per year, was

$$\frac{dP}{dt} = 6.7(0.011)e^{0.011t}.$$

On January 1, 2007, we have $t = 0$, so

$$\frac{dP}{dt} = 6.7(0.011)e^0 = 0.0737 \text{ billion/year} = 73.7 \text{ million people/year}.$$

5. The rate of change of the power dissipated is given by

$$\frac{dP}{dR} = -\frac{81}{R^2}.$$

9. The rate of change of velocity is given by

$$\frac{dv}{dt} = -\frac{mg}{k}\left(-\frac{k}{m}e^{-kt/m}\right) = ge^{-kt/m}.$$

When $t = 0$,

$$\left.\frac{dv}{dt}\right|_{t=0} = g.$$

When $t = 1$,

$$\left.\frac{dv}{dt}\right|_{t=1} = ge^{-k/m}.$$

These answers give the acceleration at $t = 0$ and $t = 1$. The acceleration at $t = 0$ is g, the acceleration due to gravity, and at $t = 1$, the acceleration is $ge^{-k/m}$, a smaller value.

13. We know $dR/dt = 0.2$ when $R = 5$ and $V = 9$ and we want to know dI/dt. Differentiating $I = V/R$ with V constant gives

$$\frac{dI}{dt} = V\left(-\frac{1}{R^2}\frac{dR}{dt}\right),$$

so substituting gives

$$\frac{dI}{dt} = 9\left(-\frac{1}{5^2}\cdot 0.2\right) = -0.072 \text{ amps per second}.$$

17. We have

$$\frac{dA}{dt} = \frac{9}{16}[4 - 4\cos(4\theta)]\frac{d\theta}{dt}.$$

So

$$\left.\frac{dA}{dt}\right|_{\theta=\pi/4} = \frac{9}{16}\left(4 - 4\cos\left(4\cdot\frac{\pi}{4}\right)\right)0.2 = \frac{9}{16}(4+4)0.2 = 0.9 \text{ cm}^2/\text{min}.$$

Problems

21. Since the radius is 3 feet, the volume of the gas when the depth is h is given by

$$V = \pi 3^2 h = 9\pi h.$$

We want to find dV/dt when $h = 4$ and $dh/dt = 0.2$. Differentiating gives

$$\frac{dV}{dt} = 9\pi\frac{dh}{dt} = 9\pi(0.2) = 5.655 \text{ ft}^3/\text{sec}.$$

Notice that the value $H = 4$ is not used. This is because V is proportional to h so dV/dt does not depend on h.

25. The rate of change of temperature with distance, dH/dy, at altitude 4000 ft approximated by

$$\frac{dH}{dy} \approx \frac{\Delta H}{\Delta y} = \frac{38 - 52}{6 - 4} = -7°\text{F/thousand ft.}$$

A speed of 3000 ft/min tells us $dy/dt = 3000$, so

$$\text{Rate of change of temperature with time} = \frac{dH}{dy} \cdot \frac{dy}{dt} \approx -7\frac{°\text{F}}{\text{thousand ft}} \cdot \frac{3 \text{ thousand ft}}{\text{min}} = -21°\text{F/min.}$$

Other estimates can be obtained by estimating the derivative as

$$\frac{dH}{dy} \approx \frac{\Delta H}{\Delta y} = \frac{52 - 60}{4 - 2} = -4°\text{F/thousand ft}$$

or by averaging the two estimates

$$\frac{dH}{dy} \approx \frac{-7 - 4}{2} = -5.5°\text{F/thousand ft.}$$

If the rate of change of temperature with distance is $-4°$/thousand ft, then

$$\text{Rate of change of temperature with time} = \frac{dH}{dy} \cdot \frac{dy}{dt} \approx -4\frac{°\text{F}}{\text{thousand ft}} \cdot \frac{3 \text{ thousand ft}}{\text{min}} = -12°\text{F/min.}$$

Thus, estimates for the rate at which temperature was decreasing range from $12°$F/min to $21°$F/min.

29. (a) The surface of the water is circular with radius r cm. Applying Pythagoras' Theorem to the triangle in Figure 4.31 shows that

$$(10 - h)^2 + r^2 = 10^2$$

so

$$r = \sqrt{10^2 - (10 - h)^2} = \sqrt{20h - h^2} \text{ cm.}$$

(b) We know $dh/dt = -0.1$ cm/hr and we want to know dr/dt when $h = 5$ cm. Differentiating

$$r = \sqrt{20h - h^2}$$

gives

$$\frac{dr}{dt} = \frac{1}{2}(20h - h^2)^{-1/2}\left(20\frac{dh}{dt} - 2h\frac{dh}{dt}\right) = \frac{10 - h}{\sqrt{20h - h^2}} \cdot \frac{dh}{dt}.$$

Substituting $dh/dt = -0.1$ and $h = 5$ gives

$$\left.\frac{dr}{dt}\right|_{h=5} = \frac{5}{\sqrt{20 \cdot 5 - 5^2}} \cdot (-0.1) = -\frac{1}{2\sqrt{75}} = -0.0577 \text{ cm/hr.}$$

Thus, the radius is decreasing at 0.0577 cm per hour.

Figure 4.31

33. Let the volume of clay be V. The clay is in the shape of a cylinder, so $V = \pi r^2 L$. We know $dL/dt = 0.1$ cm/sec and we want to know dr/dt when $r = 1$ cm and $L = 5$ cm. Differentiating with respect to time t gives

$$\frac{dV}{dt} = \pi 2 r L \frac{dr}{dt} + \pi r^2 \frac{dL}{dt}.$$

However, the amount of clay is unchanged, so $dV/dt = 0$ and

$$2rL\frac{dr}{dt} = -r^2 \frac{dL}{dt},$$

therefore

$$\frac{dr}{dt} = -\frac{r}{2L}\frac{dL}{dt}.$$

When the radius is 1 cm and the length is 5 cm, and the length is increasing at 0.1 cm per second, the rate at which the radius is changing is

$$\frac{dr}{dt} = -\frac{1}{2 \cdot 5} \cdot 0.1 = -0.01 \text{ cm/sec.}$$

Thus, the radius is decreasing at 0.01 cm/sec.

37. From Figure 4.32, Pythagoras' Theorem shows that the ground distance, d, between the train and the point, B, vertically below the plane is given by

$$d^2 = x^2 + y^2.$$

Figure 4.33 shows that

$$z^2 = d^2 + 4^2$$

so

$$z^2 = x^2 + y^2 + 4^2.$$

We know that when $x = 1$, $dx/dt = 80$, $y = 5$, $dy/dt = 500$, and we want to know dz/dt. First, we find z:

$$z^2 = 1^2 + 5^2 + 4^2 = 42, \text{ so } z = \sqrt{42}.$$

Differentiating $z^2 = x^2 + y^2 + 4^2$ gives

$$2z\frac{dz}{dt} = 2x\frac{dx}{dt} + 2y\frac{dy}{dt}.$$

Canceling 2s and substituting gives

$$\sqrt{42}\frac{dz}{dt} = 1(80) + 5(500)$$

$$\frac{dz}{dt} = \frac{2580}{\sqrt{42}} = 398.103 \text{ mph.}$$

Figure 4.32: View from air

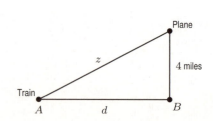

Figure 4.33: Vertical view

41. The volume of a cube is $V = x^3$. So

$$\frac{dV}{dt} = 3x^2\frac{dx}{dt},$$

and

$$\frac{1}{V}\frac{dV}{dt} = \frac{3}{x}\frac{dx}{dt}.$$

The surface area of a cube is $A = 6x^2$. So

$$\frac{dA}{dt} = 12x\frac{dx}{dt},$$

and

$$\frac{1}{A}\frac{dA}{dt} = \frac{2}{x}\frac{dx}{dt}.$$

Thus the percentage rate of change of the volume of the cube, $\frac{1}{V}\frac{dV}{dt}$, is larger.

45. (a) Since the elevator is descending at 30 ft/sec, its height from the ground is given by $h(t) = 300 - 30t$,   for $0 \leq t \leq 10$.

(b) From the triangle in the figure,

$$\tan\theta = \frac{h(t) - 100}{150} = \frac{300 - 30t - 100}{150} = \frac{200 - 30t}{150}.$$

Therefore

$$\theta = \arctan\left(\frac{200 - 30t}{150}\right)$$

and

$$\frac{d\theta}{dt} = \frac{1}{1 + \left(\frac{200-30t}{150}\right)^2} \cdot \left(\frac{-30}{150}\right) = -\frac{1}{5}\left(\frac{150^2}{150^2 + (200 - 30t)^2}\right).$$

Notice that $\frac{d\theta}{dt}$ is always negative, which is reasonable since θ decreases as the elevator descends.

(c) If we want to know when θ changes (decreases) the fastest, we want to find out when $d\theta/dt$ has the largest magnitude. This will occur when the denominator, $150^2 + (200 - 30t)^2$, in the expression for $d\theta/dt$ is the smallest, or when $200 - 30t = 0$. This occurs when $t = \frac{200}{30}$ seconds, and so $h(\frac{200}{30}) = 100$ feet, i.e., when the elevator is at the level of the observer.

Solutions for Section 4.7

Exercises

1. Since $f'(a) > 0$ and $g'(a) < 0$, l'Hopital's rule tells us that

$$\lim_{x \to a}\frac{f(x)}{g(x)} = \frac{f'(a)}{g'(a)} < 0.$$

5. The denominator approaches zero as x goes to zero and the numerator goes to zero even faster, so you should expect that the limit to be 0. You can check this by substituting several values of x close to zero. Alternatively, using l'Hopital's rule, we have

$$\lim_{x \to 0}\frac{x^2}{\sin x} = \lim_{x \to 0}\frac{2x}{\cos x} = 0.$$

9. The larger power dominates. Using l'Hopital's rule

$$\lim_{x \to \infty}\frac{x^5}{0.1x^7} = \lim_{x \to \infty}\frac{5x^4}{0.7x^6} = \lim_{x \to \infty}\frac{20x^3}{4.2x^5}$$

$$= \lim_{x \to \infty}\frac{60x^2}{21x^4} = \lim_{x \to \infty}\frac{120x}{84x^3} = \lim_{x \to \infty}\frac{120}{252x^2} = 0$$

so $0.1x^7$ dominates.

Problems

13. We want to find $\lim\limits_{x\to\infty} f(x)$, which we do by three applications of l'Hopital's rule:

$$\lim_{x\to\infty} \frac{2x^3 + 5x^2}{3x^3 - 1} = \lim_{x\to\infty} \frac{6x^2 + 10x}{9x^2} = \lim_{x\to\infty} \frac{12x + 10}{18x} = \lim_{x\to\infty} \frac{12}{18} = \frac{2}{3}.$$

So the line $y = 2/3$ is the horizontal asymptote.

17. We have $\lim_{x\to 1} x = 1$ and $\lim_{x\to 1}(x - 1) = 0$, so l'Hopital's rule does not apply.

21. This is an ∞^0 form. With $y = \lim_{x\to\infty}(1 + x)^{1/x}$, we take logarithms to get

$$\ln y = \lim_{x\to\infty} \frac{1}{x} \ln(1 + x).$$

This limit is a $0 \cdot \infty$ form,

$$\lim_{x\to\infty} \frac{1}{x} \ln(1 + x),$$

which can be rewritten as the ∞/∞ form

$$\lim_{x\to\infty} \frac{\ln(1 + x)}{x},$$

to which l'Hopital's rule applies.

25. Let $f(x) = \ln x$ and $g(x) = 1/x$ so $f'(x) = 1/x$ and $g'(x) = -1/x^2$ and

$$\lim_{x\to 0^+} \frac{\ln x}{1/x} = \lim_{x\to 0^+} \frac{1/x}{-1/x^2} = \lim_{x\to 0^+} \frac{x}{-1} = 0.$$

29. To get this expression in a form in which l'Hopital's rule applies, we combine the fractions:

$$\frac{1}{x} - \frac{1}{\sin x} = \frac{\sin x - x}{x \sin x}.$$

Letting $f(x) = \sin x - x$ and $g(x) = x \sin x$, we have $f(0) = 0$ and $g(0) = 0$ so l'Hopital's rule can be used. Differentiating gives $f'(x) = \cos x - 1$ and $g'(x) = x \cos x + \sin x$, so $f'(0) = 0$ and $g'(0) = 0$, so $f'(0)/g'(0)$ is undefined. Therefore, to apply l'Hopital's rule we differentiate again to obtain $f''(x) = -\sin x$ and $g''(x) = 2\cos x - x \sin x$, for which $f''(0) = 0$ and $g''(0) = 2 \neq 0$. Then

$$\lim_{x\to 0} \left(\frac{1}{x} - \frac{1}{\sin x}\right) = \lim_{x\to 0} \left(\frac{\sin x - x}{x \sin x}\right)$$
$$= \lim_{x\to 0} \left(\frac{\cos x - 1}{x \cos x + \sin x}\right)$$
$$= \lim_{x\to 0} \left(\frac{-\sin x}{2\cos x - x \sin x}\right)$$
$$= \frac{0}{2} = 0.$$

33. Since $\lim_{t\to 0} \sin^2 At = 0$ and $\lim_{t\to 0} \cos At - 1 = 1 - 1 = 0$, this is a $0/0$ form. Applying l'Hopital's rule we get

$$\lim_{t\to 0} \frac{\sin^2 At}{\cos At - 1} = \lim_{t\to 0} \frac{2A \sin At \cos At}{-A \sin At} = \lim_{t\to 0} -2\cos At = -2.$$

37. Let $f(x) = \cos x$ and $g(x) = x$. Observe that since $f(0) = 1$, l'Hopital's rule does not apply. But since $g(0) = 0$,

$$\lim_{x\to 0} \frac{\cos x}{x} \quad \text{does not exist.}$$

41. Let $k = n/2$, so $k \to \infty$ as $n \to \infty$. Thus,

$$\lim_{n\to\infty} \left(1 + \frac{2}{n}\right)^n = \lim_{k\to\infty} \left(1 + \frac{1}{k}\right)^{2k} = \lim_{k\to\infty} \left(\left(1 + \frac{1}{k}\right)^k\right)^2 = e^2.$$

45. This limit is of the form 0^0 so we apply l'Hopital's rule to

$$\ln f(t) = \frac{\ln\left((3^t + 5^t)/2\right)}{t}.$$

We have

$$\lim_{t\to-\infty} \ln f(t) = \lim_{t\to-\infty} \frac{\left((\ln 3)3^t + (\ln 5)5^t\right) / \left(3^t + 5^t\right)}{1}$$

$$= \lim_{t\to-\infty} \frac{(\ln 3)3^t + (\ln 5)5^t}{3^t + 5^t}$$

$$= \lim_{t\to-\infty} \frac{\ln 3 + (\ln 5)(5/3)^t}{1 + (5/3)^t}$$

$$= \frac{\ln 3 + 0}{1 + 0} = \ln 3.$$

Thus

$$\lim_{t\to-\infty} f(t) = \lim_{t\to-\infty} e^{\ln f(t)} = e^{\lim_{t\to-\infty} \ln f(t)} = e^{\ln 3} = 3.$$

49. To evaluate, we use l'Hopital's Rule:

$$\lim_{x\to 0} \frac{1 - \cosh 3x}{x} = \lim_{x\to 0} \frac{-3\sinh 3x}{1} = 0.$$

53. Since the limit is of the form $0/0$, we can apply l'Hopital's rule. We have

$$\lim_{x\to\pi/2} \frac{1 - \sin x + \cos x}{\sin x + \cos x - 1} = \lim_{x\to\pi/2} \frac{-\cos x - \sin x}{\cos x - \sin x} = \frac{-1}{-1} = 1.$$

Solutions for Section 4.8

Exercises

1. Between times $t = 0$ and $t = 1$, x goes at a constant rate from 0 to 1 and y goes at a constant rate from 1 to 0. So the particle moves in a straight line from $(0, 1)$ to $(1, 0)$. Similarly, between times $t = 1$ and $t = 2$, it goes in a straight line to $(0, -1)$, then to $(-1, 0)$, then back to $(0, 1)$. So it traces out the diamond shown in Figure 4.34.

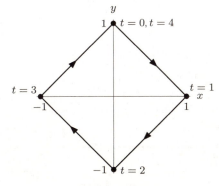

Figure 4.34

5. For $0 \leq t \leq \frac{\pi}{2}$, we have $x = \sin t$ increasing and $y = \cos t$ decreasing, so the motion is clockwise for $0 \leq t \leq \frac{\pi}{2}$. Similarly, we see that the motion is clockwise for the time intervals $\frac{\pi}{2} \leq t \leq \pi, \pi \leq t \leq \frac{3\pi}{2}$, and $\frac{3\pi}{2} \leq t \leq 2\pi$.

9. Let $f(t) = \ln t$. Then $f'(t) = \frac{1}{t}$. The particle is moving counterclockwise when $f'(t) > 0$, that is, when $t > 0$. Any other time, when $t \leq 0$, the position is not defined.

13. We have
$$\frac{dx}{dt} = -2\sin 2t, \frac{dy}{dt} = \cos t.$$
The speed is
$$v = \sqrt{4\sin^2(2t) + \cos^2 t}.$$
Thus, $v = 0$ when $\sin(2t) = \cos t = 0$, and so the particle stops when $t = \pm\pi/2, \pm3\pi/2, \ldots$ or $t = (2n+1)\frac{\pi}{2}$, for any integer n.

17. One possible answer is $x = -2, y = t$.

21. The ellipse $x^2/25 + y^2/49 = 1$ can be parameterized by $x = 5\cos t, y = 7\sin t, 0 \leq t \leq 2\pi$.

25. We have
$$\frac{dy}{dx} = \frac{dy/dt}{dx/dt} = \frac{4\cos(4t)}{3\cos(3t)}.$$
Thus when $t = \pi$, the slope of the tangent line is $-4/3$. Since $x = 0$ and $y = 0$ when $t = \pi$, the equation of the tangent line is $y = -(4/3)x$.

29. We see from the parametric equations that the particle moves along a line. It suffices to plot two points: at $t = 0$, the particle is at point $(1, -4)$, and at $t = 1$, the particle is at point $(4, -3)$. Since x increases as t increases, the motion is left to right on the line as shown in Figure 4.35.

Alternately, we can solve the first equation for t, giving $t = (x - 1)/3$, and substitute this into the second equation to get
$$y = \frac{x-1}{3} - 4 = \frac{1}{3}x - \frac{13}{3}.$$
The line is $y = \frac{1}{3}x - \frac{13}{3}$.

Figure 4.35

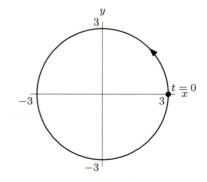

Figure 4.36

33. The graph is a circle centered at the origin with radius 3. The equation is
$$x^2 + y^2 = (3\cos t)^2 + (3\sin t)^2 = 9.$$
The particle is at the point $(3, 0)$ when $t = 0$, and motion is counterclockwise. See Figure 4.36.

Problems

37. (a) Eliminating t between
$$x = 2 + t, \quad y = 4 + 3t$$
gives
$$y - 4 = 3(x - 2),$$
$$y = 3x - 2.$$

Eliminating t between

$$x = 1 - 2t, \quad y = 1 - 6t$$

gives

$$y - 1 = 3(x - 1),$$
$$y = 3x - 2.$$

Since both parametric equations give rise to the same equation in x and y, they both parameterize the same line.

(b) Slope $= 3$, y-intercept $= -2$.

41. It is a straight line through the point $(3, 5)$ with slope -1. A linear parameterization of the same line is $x = 3 + t$, $y = 5 - t$.

45. (a) (i) A horizontal tangent occurs when $dy/dt = 0$ and $dx/dt \neq 0$. Thus,

$$\frac{dy}{dt} = 6e^{2t} - 2e^{-2t} = 0$$
$$6e^{2t} = 2e^{-2t}$$
$$e^{4t} = \frac{1}{3}$$
$$4t = \ln \frac{1}{3}$$
$$t = \frac{1}{4} \ln \frac{1}{3} = -0.25 \ln 3 = -0.275.$$

We need to check that $dx/dt \neq 0$ when $t = -0.25 \ln 3$. Since $dx/dt = 2e^{2t} + 2e^{-2t}$ is always positive, dx/dt is never zero.

(ii) A vertical tangent occurs when $dx/dt = 0$ and $dy/dt \neq 0$. Since $dx/dt = 2e^{2t} + 2e^{-2t}$ is always positive, there is no vertical tangent.

(b) The chain rule gives

$$\frac{dy}{dx} = \frac{dy/dt}{dx/dt} = \frac{6e^{2t} - 2e^{-2t}}{2e^{2t} + 2e^{-2t}} = \frac{3e^{2t} - e^{-2t}}{e^{2t} + e^{-2t}}.$$

(c) As $t \to \infty$, we have $e^{-2t} \to 0$. Thus,

$$\lim_{t \to \infty} \frac{dy}{dx} = \lim_{t \to \infty} \frac{3e^{2t} - e^{-2t}}{e^{2t} + e^{-2t}} = \lim_{t \to \infty} \frac{3e^{2t}}{e^{2t}} = 3.$$

As $t \to \infty$, the fraction gets closer and closer to 3.

49. (a) To determine if the particles collide, we check whether they are ever at the same point at the same time. We first set the two x-coordinates equal to each other:

$$4t - 4 = 3t$$
$$t = 4.$$

When $t = 4$, both x-coordinates are 12. Now we check whether the y-coordinates are also equal at $t = 4$:

$$y_A(4) = 2 \cdot 4 - 5 = 3$$
$$y_B(4) = 4^2 - 2 \cdot 4 - 1 = 7.$$

Thus, the particles do not collide since they are not at the same point at the same time.

(b) For the particles to collide, we need both x- and y-coordinates to be equal. Since the x-coordinates are equal at $t = 4$, we find the k value making $y_A(4) = y_B(4)$.

Substituting $t = 4$ into $y_A(t) = 2t - k$ and $y_B(t) = t^2 - 2t - 1$, we have

$$8 - k = 16 - 8 - 1$$
$$k = 1.$$

(c) To find the speed of the particles, we differentiate.

For particle A,
$x(t) = 4t - 4$, so $x'(t) = 4$, and $x'(4) = 4$
$y(t) = 2t - 1$, so $y'(t) = 2$, and $y'(4) = 2$

$$\text{Speed}_A = \sqrt{(x'(t))^2 + (y'(t))^2} = \sqrt{4^2 + 2^2} = \sqrt{20}.$$

For particle B,
$x(t) = 3t$, so $x'(t) = 3$, and $x'(4) = 3$
$y(t) = t^2 - 2t - 1$, so $y'(t) = 2t - 2$, and $y'(4) = 6$

$$\text{Speed}_B = \sqrt{(x'(t))^2 + (y'(t))^2} = \sqrt{3^2 + 6^2} = \sqrt{45}.$$

Thus, when $t = 4$, particle B is moving faster.

53. (a) The particle touches the x-axis when $y = 0$. Since $y = \cos(2t) = 0$ for the first time when $2t = \pi/2$, we have $t = \pi/4$. To find the speed of the particle at that time, we use the formula

$$\text{Speed} = \sqrt{\left(\frac{dx}{dt}\right)^2 + \left(\frac{dy}{dt}\right)^2} = \sqrt{(\cos t)^2 + (-2\sin(2t))^2}.$$

When $t = \pi/4$,

$$\text{Speed} = \sqrt{(\cos(\pi/4))^2 + (-2\sin(\pi/2))^2} = \sqrt{(\sqrt{2}/2)^2 + (-2 \cdot 1)^2} = \sqrt{9/2}.$$

(b) The particle is at rest when its speed is zero. Since $\sqrt{(\cos t)^2 + (-2\sin(2t))^2} \geq 0$, the speed is zero when

$$\cos t = 0 \quad \text{and} \quad -2\sin(2t) = 0.$$

Now $\cos t = 0$ when $t = \pi/2$ or $t = 3\pi/2$. Since $-2\sin(2t) = -4\sin t \cos t$, we see that this expression also equals zero when $t = \pi/2$ or $t = 3\pi/2$.

(c) We need to find d^2y/dx^2. First, we must determine dy/dx. We know

$$\frac{dy}{dx} = \frac{dy/dt}{dx/dt} = \frac{-2\sin 2t}{\cos t} = \frac{-4\sin t \cos t}{\cos t} = -4\sin t.$$

Since $dy/dx = -4\sin t$, we can now use the formula:

$$\frac{d^2y}{dx^2} = \frac{dw/dt}{dx/dt} \quad \text{where} \quad w = \frac{dy}{dx}$$

$$\frac{d^2y}{dx^2} = \frac{-4\cos t}{\cos t} = -4.$$

Since d^2y/dx^2 is always negative, our graph is concave down everywhere.

Using the identity $y = \cos(2t) = 1 - 2\sin^2 t$, we can eliminate the parameter and write the original equation as $y = 1 - 2x^2$, which is a parabola that is concave down everywhere.

57. For $0 \leq t \leq 2\pi$, we get Figure 4.37.

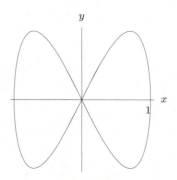

Figure 4.37

Solutions for Chapter 4 Review

Exercises

1. See Figure 4.38.

Figure 4.38

5. (a) First we find f' and f'':

$$f'(x) = -e^{-x}\sin x + e^{-x}\cos x$$
$$f''(x) = e^{-x}\sin x - e^{-x}\cos x$$
$$-e^{-x}\cos x - e^{-x}\sin x$$
$$= -2e^{-x}\cos x$$

  (b) The critical points are $x = \pi/4, 5\pi/4$, since $f'(x) = 0$ here.

  (c) The inflection points are $x = \pi/2, 3\pi/2$, since f'' changes sign at these points.

  (d) At the endpoints, $f(0) = 0$, $f(2\pi) = 0$. So we have $f(\pi/4) = (e^{-\pi/4})(\sqrt{2}/2)$ as the global maximum; $f(5\pi/4) = -e^{-5\pi/4}(\sqrt{2}/2)$ as the global minimum.

  (e) See Figure 4.39.

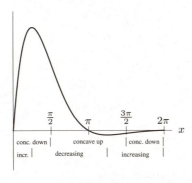

Figure 4.39

9. As $x \to -\infty$, $e^{-x} \to \infty$, so $xe^{-x} \to -\infty$. Thus $\lim_{x\to-\infty} xe^{-x} = -\infty$.

As $x \to \infty$, $\frac{x}{e^x} \to 0$, since e^x grows much more quickly than x. Thus $\lim_{x\to\infty} xe^{-x} = 0$.

Using the product rule,

$$f'(x) = e^{-x} - xe^{-x} = (1-x)e^{-x},$$

which is zero when $x = 1$, negative when $x > 1$, and positive when $x < 1$. Thus $f(1) = 1/e^1 = 1/e$ is a local maximum.

Again, using the product rule,

$$f''(x) = -e^{-x} - e^{-x} + xe^{-x}$$
$$= xe^{-x} - 2e^{-x}$$
$$= (x-2)e^{-x},$$

which is zero when $x = 2$, positive when $x > 2$, and negative when $x < 2$, giving an inflection point at $(2, \frac{2}{e^2})$. With the above, we have the following diagram:

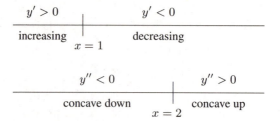

The graph of f is shown in Figure 4.40.

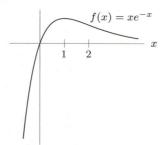

Figure 4.40

and $f(x)$ has one global maximum at $1/e$ and no local or global minima.

13. Since $g(t)$ is always decreasing for $t \geq 0$, we expect it to a global maximum at $t = 0$ but no global minimum. At $t = 0$, we have $g(0) = 1$, and as $t \to \infty$, we have $g(t) \to 0$.

Alternatively, rewriting as $g(t) = (t^3 + 1)^{-1}$ and differentiating using the chain rule gives

$$g'(t) = -(t^3 + 1)^{-2} \cdot 3t^2.$$

Since $3t^2 = 0$ when $t = 0$, there is a critical point at $t = 0$, and g decreases for all $t > 0$. See Figure 4.41.

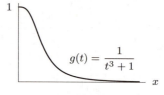

Figure 4.41

17. $\lim\limits_{x \to +\infty} f(x) = +\infty$, and $\lim\limits_{x \to 0^+} f(x) = +\infty$.

Hence, $x = 0$ is a vertical asymptote.

$f'(x) = 1 - \dfrac{2}{x} = \dfrac{x - 2}{x}$, so $x = 2$ is the only critical point.

$f''(x) = \dfrac{2}{x^2}$, which can never be zero. So there are no inflection points.

x		2	
f'	$-$	0	$+$
f''	$+$	$+$	$+$
f	$\searsmile$		$\nearsmile$

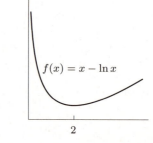

Thus, $f(2)$ is a local and global minimum.

21. We see from the parametric equations that the particle moves along a line. It suffices to plot two points: at $t = 0$, the particle is at point $(4, 1)$, and at $t = 1$, the particle is at point $(2, 5)$. Since x decreases as t increases, the motion is right to left and the curve is shown in Figure 4.42.

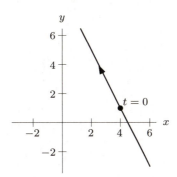

Figure 4.42

Alternately, we can solve the first equation for t, giving $t = -(x - 4)/2$, and substitute this into the second equation to get

$$y = 4 \left(\frac{-(x - 4)}{2} \right) + 1 = -2x + 9.$$

The line is $y = -2x + 9$.

25. We have

$$\frac{dM}{dt} = (3x^2 + 0.4x^3)\frac{dx}{dt}.$$

If $x = 5$, then

$$\frac{dM}{dt} = [3(5^2) + 0.4(5^3)](0.02) = 2.5 \text{ gm/hr}.$$

Problems

29. The local maxima and minima of f correspond to places where f' is zero and changes sign or, possibly, to the endpoints of intervals in the domain of f. The points at which f changes concavity correspond to local maxima and minima of f'. The change of sign of f', from positive to negative corresponds to a maximum of f and change of sign of f' from negative to positive corresponds to a minimum of f.

33. Since the x^3 term has coefficient of 1, the cubic polynomial is of the form $y = x^3 + ax^2 + bx + c$. We now find a, b, and c. Differentiating gives

$$\frac{dy}{dx} = 3x^2 + 2ax + b.$$

The derivative is 0 at local maxima and minima, so

$$\left. \frac{dy}{dx} \right|_{x=1} = 3(1)^2 + 2a(1) + b = 3 + 2a + b = 0$$

$$\left. \frac{dy}{dx} \right|_{x=3} = 3(3)^2 + 2a(3) + b = 27 + 6a + b = 0$$

Subtracting the first equation from the second and solving for a and b gives

$$24 + 4a = 0 \qquad \text{so} \qquad a = -6$$
$$b = -3 - 2(-6) = 9.$$

Since the y-intercept is 5, the cubic is

$$y = x^3 - 6x^2 + 9x + 5.$$

Since the coefficient of x^3 is positive, $x = 1$ is the maximum and $x = 3$ is the minimum. See Figure 4.43. To confirm that $x = 1$ gives a maximum and $x = 3$ gives a minimum, we calculate

$$\frac{d^2y}{dx^2} = 6x + 2a = 6x - 12.$$

At $x = 1$, $\frac{d^2y}{dx^2} = -6 < 0$, so we have a maximum.

At $x = 3$, $\frac{d^2y}{dx^2} = 6 > 0$, so we have a minimum.

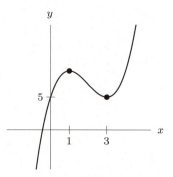

Figure 4.43: Graph of $y = x^3 - 6x^2 + 9x + 5$

37. First notice that since this function approaches 0 as x approaches either plus or minus infinity, any local extrema that we find are also global extrema.

Differentiating $y = axe^{-bx^2}$ gives

$$\frac{dy}{dx} = ae^{-bx^2} - 2abx^2e^{-bx^2} = ae^{-bx^2}(1 - 2bx^2).$$

Since we have a critical points at $x = 1$ and $x = -1$, we know $1 - 2b = 0$, so $b = 1/2$.

The global maximum is 2 at $x = 1$, so we have $2 = ae^{-1/2}$ which gives $a = 2e^{1/2}$. Notice that this value of a also gives the global minimum at $x = -1$.

Thus,

$$y = 2xe^{\left(\frac{1-x^2}{2}\right)}.$$

41. We want to maximize the volume $V = x^2h$ of the box, shown in Figure 4.44. The box has 5 faces: the bottom, which has area x^2 and the four sides, each of which has area xh. Thus $8 = x^2 + 4xh$, so

$$h = \frac{8 - x^2}{4x}.$$

Substituting this expression in for h in the formula for V gives

$$V = x^2 \cdot \frac{8 - x^2}{4x} = \frac{1}{4}(8x - x^3).$$

Differentiating gives

$$\frac{dV}{dx} = \frac{1}{4}(8 - 3x^2).$$

To maximize V we look for critical points, so we solve $0 = (8 - 3x^2)/4$, getting $x = \pm\sqrt{8/3}$. We discard the negative solution, since x is a positive length. Then we can find

$$h = \frac{8 - x^2}{4x} = \frac{8 - \frac{8}{3}}{4\sqrt{\frac{8}{3}}} = \frac{\frac{16}{3}}{4\sqrt{\frac{8}{3}}} = \frac{\frac{4}{3}}{2\sqrt{\frac{2}{3}}} = \sqrt{\frac{2}{3}}.$$

Thus $x = \sqrt{8/3}$ cm and $h = \sqrt{2/3}$ cm.

We can check that this critical point is a maximum of V by checking the sign of

$$\frac{d^2V}{dx^2} = -\frac{3}{2}x,$$

which is negative when $x > 0$. So V is concave down at the critical point and therefore $x = \sqrt{8/3}$ gives a maximum value of V.

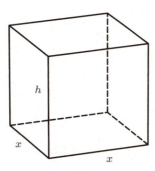

Figure 4.44

45. Let $f(x) = x \sin x$. Then $f'(x) = x \cos x + \sin x$.

$f'(x) = 0$ when $x = 0$, $x \approx 2$, and $x \approx 5$. The latter two estimates we can get from the graph of $f'(x)$.

Zooming in (or using some other approximation method), we can find the zeros of $f'(x)$ with more precision. They are (approximately) 0, 2.029, and 4.913. We check the endpoints and critical points for the global maximum and minimum.

$$f(0) = 0, \qquad\qquad f(2\pi) = 0,$$
$$f(2.029) \approx 1.8197, \quad f(4.914) \approx -4.814.$$

Thus for $0 \le x \le 2\pi$, $-4.81 \le f(x) \le 1.82$.

49. (a) We set the derivative equal to zero and solve for t to find critical points. Using the product rule, we have:

$$f'(t) = (at^2)(e^{-bt}(-b)) + (2at)e^{-bt} = 0$$
$$ate^{-bt}(-bt + 2) = 0$$
$$t = 0 \quad \text{or} \quad t = \frac{2}{b}.$$

There are two critical points: $t = 0$ and $t = 2/b$.

(b) Since we want a critical point at $t = 5$, we substitute and solve for b:

$$5 = 2/b$$
$$b = \frac{2}{5} = 0.4.$$

To find the value of a, we use the fact that $f(5) = 12$, so we have:

$$a(5^2)e^{-0.4(5)} = 12$$
$$a \cdot 25e^{-2} = 12$$
$$a = \frac{12e^2}{25} = 3.547.$$

(c) To show that $f(t)$ has a local minimum at $t = 0$ and a local maximum at $t = 5$, we can use the first derivative test or the second derivative test. Using the first derivative test, we evaluate f' at values on either side of $t = 0$ and $t = 5$. Since $f'(t) = 3.547te^{-0.4t}(-0.4t + 2)$, we have

$$f'(-1) = -3.547e^{0.4}(2.4) = -12.700 < 0$$

and

$$f'(1) = 3.547e^{-0.4}(1.6) = 3.804 > 0,$$

and
$$f'(6) = 3.547(6)e^{-2.4}(-0.4) = -0.772 < 0.$$

The function f is decreasing to the left of $t = 0$, increasing between $t = 0$ and $t = 5$, and decreasing to the right of $t = 5$. Therefore, $f(t)$ has a local minimum at $t = 0$ and a local maximum at $t = 5$. See Figure 4.45.

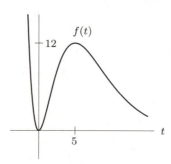

Figure 4.45

53. Let V be the volume of the ice, so that $V = 3\pi(r^2 - 1^2)$. Now,

$$\frac{dV}{dt} = 6\pi r \frac{dr}{dt}.$$

Thus for $r = 1.5$, we have
$$\frac{dV}{dt} = 6\pi(1.5)(0.03) = 0.848 \, \text{cm}^3/\text{hr}.$$

57. (a) The business must reorder often enough to keep pace with sales. If reordering is done every t months, then,

$$\text{Quantity sold in } t \text{ months} = \text{Quantity reordered in each batch}$$
$$rt = q$$
$$t = \frac{q}{r} \text{ months.}$$

(b) The amount spent on each order is $a + bq$, which is spent every q/r months. To find the monthly expenditures, divide by q/r. Thus, on average,

$$\text{Amount spent on ordering per month} = \frac{a + bq}{q/r} = \frac{ra}{q} + rb \text{ dollars.}$$

(c) The monthly cost of storage is $kq/2$ dollars, so

$$C = \text{Ordering costs} + \text{Storage costs}$$
$$C = \frac{ra}{q} + rb + \frac{kq}{2} \text{ dollars.}$$

(d) The optimal batch size minimizes C, so

$$\frac{dC}{dq} = \frac{-ra}{q^2} + \frac{k}{2} = 0$$
$$\frac{ra}{q^2} = \frac{k}{2}$$
$$q^2 = \frac{2ra}{k}$$

so

$$q = \sqrt{\frac{2ra}{k}} \text{ items per order.}$$

61. Since the volume is fixed at 200 ml (i.e. 200 cm^3), we can solve the volume expression for h in terms of r to get (with h and r in centimeters)

$$h = \frac{200 \cdot 3}{7\pi r^2}.$$

Using this expression in the surface area formula we arrive at

$$S = 3\pi r \sqrt{r^2 + \left(\frac{600}{7\pi r^2}\right)^2}$$

By plotting $S(r)$ we see that there is a minimum value near $r = 2.7$ cm.

65. (a) The length of the piece of wire made into a circle is x cm, so the length of the piece made into a square is $(L - x)$ cm. See Figure 4.46.

Figure 4.46

The circumference of the circle is x, so its radius, r cm, is given by

$$r = \frac{x}{2\pi} \text{ cm.}$$

The perimeter of the square is $(L - x)$, so the side length, s cm, is given by

$$s = \frac{L - x}{4} \text{ cm.}$$

Thus, the sum of areas is given by

$$A = \pi r^2 + s^2 = \pi \left(\frac{x}{2\pi}\right)^2 + \left(\frac{L - x}{4}\right)^2 = \frac{x^2}{4\pi} + \frac{(L - x)^2}{16}, \quad \text{for } 0 \le x \le L.$$

Setting $dA/dx = 0$ to find the critical points gives

$$\frac{dA}{dx} = \frac{x}{2\pi} - \frac{(L - x)}{8} = 0$$

$$8x = 2\pi L - 2\pi x$$

$$(8 + 2\pi)x = 2\pi L$$

$$x = \frac{2\pi L}{8 + 2\pi} = \frac{\pi L}{4 + \pi} \approx 0.44L.$$

To find the maxima and minima, we substitute the critical point and the endpoints, $x = 0$ and $x = L$, into the area function.

For $x = 0$, we have $A = \dfrac{L^2}{16}$.

For $x = \dfrac{\pi L}{4 + \pi}$, we have $L - x = L - \dfrac{\pi L}{4 + \pi} = \dfrac{4L}{4 + \pi}$. Then

$$A = \frac{\pi^2 L^2}{4\pi(4 + \pi)^2} + \frac{1}{16}\left(\frac{4L}{4 + \pi}\right)^2 = \frac{\pi L^2}{4(4 + \pi)^2} + \frac{L^2}{(4 + \pi)^2}$$

$$= \frac{\pi L^2 + 4L^2}{4(4 + \pi)^2} = \frac{L^2}{4(4 + \pi)} = \frac{L^2}{16 + 4\pi}.$$

For $x = L$, we have $A = \dfrac{L^2}{4\pi}$.

Thus, $x = \dfrac{\pi L}{4 + \pi}$ gives the minimum value of $A = \dfrac{L^2}{16 + 4\pi}$.

Since $4\pi < 16$, we see that $x = L$ gives the maximum value of $A = \dfrac{L^2}{4\pi}$.

This corresponds to the situation in which we do not cut the wire at all and use the single piece to make a circle.

(b) At the maximum, $x = L$, so

$$\frac{\text{Length of wire in square}}{\text{Length of wire in circle}} = \frac{0}{L} = 0.$$

$$\frac{\text{Area of square}}{\text{Area of circle}} = \frac{0}{L^2/4\pi} = 0.$$

At the minimum, $x = \dfrac{\pi L}{4 + \pi}$, so $L - x = L - \dfrac{\pi L}{4 + \pi} = \dfrac{4L}{4 + \pi}$.

$$\frac{\text{Length of wire in square}}{\text{Length of wire in circle}} = \frac{4L/(4 + \pi)}{\pi L/(4 + \pi)} = \frac{4}{\pi}.$$

$$\frac{\text{Area of square}}{\text{Area of circle}} = \frac{L^2/(4 + \pi)^2}{\pi L^2/(4(4 + \pi)^2)} = \frac{4}{\pi}.$$

(c) For a general value of x,

$$\frac{\text{Length of wire in square}}{\text{Length of wire in circle}} = \frac{L - x}{x}.$$

$$\frac{\text{Area of square}}{\text{Area of circle}} = \frac{(L - x)^2/16}{x^2/(4\pi)} = \frac{\pi}{4} \cdot \frac{(L - x)^2}{x^2}.$$

If the ratios are equal, we have

$$\frac{L - x}{x} = \frac{\pi}{4} \cdot \frac{(L - x)^2}{x^2}.$$

So either $L - x = 0$, giving $x = L$, or we can cancel $(L - x)$ and multiply through by $4x^2$, giving

$$4x = \pi(L - x)$$

$$x = \frac{\pi L}{4 + \pi}.$$

Thus, the two values of x found in part (a) are the only values of x in $0 \leq x \leq L$ making the ratios in part (b) equal. (The ratios are not defined if $x = 0$.)

69. Evaluating the limits in the numerator and the denominator we get $0/e^0 = 0/1 = 0$, so this is not an indeterminate form. l'Hopital's rule does not apply.

73. If $f(x) = 1 - \cosh(5x)$ and $g(x) = x^2$, then $f(0) = g(0) = 0$, so we use l'Hopital's Rule:

$$\lim_{x \to 0} \frac{1 - \cosh 5x}{x^2} = \lim_{x \to 0} \frac{-5 \sinh 5x}{2x} = \lim_{x \to 0} \frac{-25 \cosh 5x}{2} = -\frac{25}{2}.$$

77. The radius r is related to the volume by the formula $V = \frac{4}{3}\pi r^3$. By implicit differentiation, we have

$$\frac{dV}{dt} = \frac{4}{3}\pi 3r^2 \frac{dr}{dt} = 4\pi r^2 \frac{dr}{dt}.$$

The surface area of a sphere is $4\pi r^2$, so we have

$$\frac{dV}{dt} = s \cdot \frac{dr}{dt},$$

but since $\dfrac{dV}{dt} = \dfrac{1}{3}s$ was given, we have

$$\frac{dr}{dt} = \frac{1}{3}.$$

81. We want to find dP/dV. Solving $PV = k$ for P gives

$$P = k/V$$

so,

$$\frac{dP}{dV} = -\frac{k}{V^2}.$$

CAS Challenge Problems

85. (a) A CAS gives

$$\frac{d}{dx}\operatorname{arcsinh} x = \frac{1}{\sqrt{1+x^2}}$$

(b) Differentiating both sides of $\sinh(\operatorname{arcsinh} x) = x$, we get

$$\cosh(\operatorname{arcsinh} x)\frac{d}{dx}(\operatorname{arcsinh} x) = 1$$

$$\frac{d}{dx}(\operatorname{arcsinh} x) = \frac{1}{\cosh(\operatorname{arcsinh} x)}.$$

Since $\cosh^2 x - \sinh^2 x = 1$, $\cosh x = \pm\sqrt{1+\sinh^2 x}$. Furthermore, since $\cosh x > 0$ for all x, we take the positive square root, so $\cosh x = \sqrt{1+\sinh^2 x}$. Therefore, $\cosh(\operatorname{arcsinh} x) = \sqrt{1+(\sinh(\operatorname{arcsinh} x))^2} = \sqrt{1+x^2}$. Thus

$$\frac{d}{dx}\operatorname{arcsinh} x = \frac{1}{\sqrt{1+x^2}}.$$

89. (a)

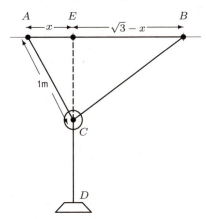

Figure 4.47

We want to maximize the sum of the lengths EC and CD in Figure 4.47. Let x be the distance AE. Then x can be between 0 and 1, the length of the left rope. By the Pythagorean theorem,

$$EC = \sqrt{1-x^2}.$$

The length of the rope from B to C can also be found by the Pythagorean theorem:

$$BC = \sqrt{EC^2 + EB^2} = \sqrt{1-x^2+(\sqrt{3}-x)^2} = \sqrt{4-2\sqrt{3}x}.$$

Since the entire rope from B to D has length 3 m, the length from C to D is

$$CD = 3 - \sqrt{4-2\sqrt{3}x}.$$

The distance we want to maximize is

$$f(x) = EC + CD = \sqrt{1-x^2} + 3 - \sqrt{4-2\sqrt{3}x}, \quad \text{for} \quad 0 \le x \le 1.$$

Differentiating gives

$$f'(x) = \frac{-2x}{2\sqrt{1-x^2}} - \frac{-2\sqrt{3}}{2\sqrt{4-2\sqrt{3}x}}.$$

Setting $f'(x) = 0$ gives the cubic equation

$$2\sqrt{3}x^3 - 7x^2 + 3 = 0.$$

Using a computer algebra system to solve the equation gives three roots: $x = -1/\sqrt{3}, x = \sqrt{3}/2, x = \sqrt{3}$. We discard the negative root. Since x cannot be larger than 1 meter (the length of the left rope), the only critical point of interest is $x = \sqrt{3}/2$, that is, halfway between A and B.

To find the global maximum, we calculate the distance of the weight from the ceiling at the critical point and at the endpoints:

$$f(0) = \sqrt{1} + 3 - \sqrt{4} = 2$$

$$f\left(\frac{\sqrt{3}}{2}\right) = \sqrt{1 - \frac{3}{4}} + 3 - \sqrt{4 - 2\sqrt{3} \cdot \frac{\sqrt{3}}{2}} = 2.5$$

$$f(1) = \sqrt{0} + 3 - \sqrt{4 - 2\sqrt{3}} = 4 - \sqrt{3} = 2.27.$$

Thus, the weight is at the maximum distance from the ceiling when $x = \sqrt{3}/2$; that is, the weight comes to rest at a point halfway between points A and B.

(b) No, the equilibrium position depends on the length of the rope. For example, suppose that the left-hand rope was 1 cm long. Then there is no way for the pulley at its end to move to a point halfway between the anchor points.

CHECK YOUR UNDERSTANDING

1. True. Since the domain of f is all real numbers, all local minima occur at critical points.

5. False. For example, if $f(x) = x^3$, then $f'(0) = 0$, but $f(x)$ does not have either a local maximum or a local minimum at $x = 0$.

9. Let $f(x) = ax^2$, with $a \neq 0$. Then $f'(x) = 2ax$, so f has a critical point only at $x = 0$.

13. False. For example, if $f(x) = x^3$, then $f'(0) = 0$, so $x = 0$ is a critical point, but $x = 0$ is neither a local maximum nor a local minimum.

17. True. If the maximum is not at an endpoint, then it must be at critical point of f. But $x = 0$ is the only critical point of $f(x) = x^2$ and it gives a minimum, not a maximum.

21. False. The circumference A and radius r are related by $A = \pi r^2$, so $dA/dt = 2\pi r\, dr/dt$. Thus dA/dt depends on r and since r is not constant, neither is dA/dt.

25. $f(x) = x^2 + 1$ is positive for all x and concave up.

29. This is impossible. Since f'' exists, so must f', which means that f is differentiable and hence continuous. If $f(x)$ were positive for some values of x and negative for other values, then by the Intermediate Value Theorem, $f(x)$ would have to be zero somewhere, but this is impossible since $f(x)f''(x) < 0$ for all x. Thus either $f(x) > 0$ for all values of x, in which case $f''(x) < 0$ for all values of x, that is f is concave down. But this is impossible by Problem 26. Or else $f(x) < 0$ for all x, in which case $f''(x) > 0$ for all x, that is f is concave up. But this is impossible by Problem 28.

CHAPTER FIVE

Solutions for Section 5.1

Exercises

1. (a) (i) Since the velocity is increasing, for an upper estimate we use a right sum. Using $n = 4$, we have $\Delta t = 3$, so

$$\text{Upper estimate} = (37)(3) + (38)(3) + (40)(3) + (45)(3) = 480.$$

  (ii) Using $n = 2$, we have $\Delta t = 6$, so

$$\text{Upper estimate} = (38)(6) + (45)(6) = 498.$$

(b) The answer using $n = 4$ is more accurate as it uses the values of $v(t)$ when $t = 3$ and $t = 9$.

(c) Since the velocity is increasing, for a lower estimate we use a left sum. Using $n = 4$, we have $\Delta t = 3$, so

$$\text{Lower estimate} = (34)(3) + (37)(3) + (38)(3) + (40)(3) = 447.$$

5. (a) The velocity is always positive, so the particle is moving in the same direction throughout. However, the particle is speeding up until shortly before $t = 0$, and slowing down thereafter.

(b) The distance traveled is represented by the area under the curve. Using whole grid squares, we can overestimate the area as $3 + 3 + 3 + 3 + 2 + 1 = 15$, and we can underestimate the area as $1 + 2 + 2 + 1 + 0 + 0 = 6$.

9. Using $\Delta t = 0.2$, our upper estimate is

$$\frac{1}{1+0}(0.2) + \frac{1}{1+0.2}(0.2) + \frac{1}{1+0.4}(0.2) + \frac{1}{1+0.6}(0.2) + \frac{1}{1+0.8}(0.2) \approx 0.75.$$

The lower estimate is

$$\frac{1}{1+0.2}(0.2) + \frac{1}{1+0.4}(0.2) + \frac{1}{1+0.6}(0.2) + \frac{1}{1+0.8}(0.2)\frac{1}{1+1}(0.2) \approx 0.65.$$

Since v is a decreasing function, the bug has crawled more than 0.65 meters, but less than 0.75 meters. We average the two to get a better estimate:

$$\frac{0.65 + 0.75}{2} = 0.70 \text{ meters.}$$

13. From $t = 0$ to $t = 5$ the velocity is positive so the change in position is to the right. The area under the velocity graph gives the distance traveled. The region is a triangle, and so has area $(1/2)bh = (1/2)5 \cdot 10 = 25$. Thus the change in position is 25 cm to the right.

Problems

17. Since f is increasing, the right-hand sum is the upper estimate and the left-hand sum is the lower estimate. We have $f(a) = 13$, $f(b) = 23$ and $\Delta t = (b - a)/n = 2/100$. Thus,

$$|\text{Difference in estimates}| = |f(b) - f(a)|\Delta t$$
$$= |23 - 13|\frac{1}{50} = \frac{1}{5}.$$

21. The change in position is calculated from the area between the velocity graph and the t-axis, with the region below the axis corresponding to negatives velocities and counting negatively.

Figure 5.1 shows the graph of $f(t)$. From $t = 0$ to $t = 3$ the velocity is positive. The region under the graph of $f(t)$ is a triangle with height 6 cm/sec and base 3 seconds. Thus, from $t = 0$ to $t = 3$, the particle moves

$$\text{Distance moved to right} = \frac{1}{2} \cdot 3 \cdot 6 = 9 \text{ centimeters.}$$

From $t = 3$ to $t = 4$, the velocity is negative. The region between the graph of $f(t)$ and the t-axis is a triangle with height 2 cm/sec and base 1 second, so in this interval the particle moves

$$\text{Distance moved to left} = \frac{1}{2} \cdot 1 \cdot 2 = 1 \text{ centimeter.}$$

Thus, the total change in position is $9 - 1 = 8$ centimeters to the right.

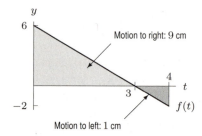

Figure 5.1

25. The graph of her velocity against time is a straight line from 0 mph to 60 mph; see Figure 5.2. Since the distance traveled is the area under the curve, we have

$$\text{Shaded area} = \frac{1}{2} \cdot t \cdot 60 = 10 \text{ miles}$$

Solving for t gives

$$t = \frac{1}{3}\text{hr} = 20 \text{ minutes .}$$

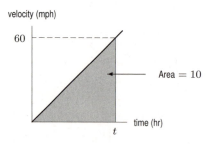

Figure 5.2

Solutions for Section 5.2

Exercises

1. (a) Left-hand sum. Right-hand sum would be smaller.
  (b) We have $a = 0$, $b = 2$, $n = 6$, $\Delta x = \frac{2}{6} = \frac{1}{3}$.

5. With $\Delta x = 5$, we have

$$\text{Left-hand sum} = 5(0 + 100 + 200 + 100 + 200 + 250 + 275) = 5625,$$

$$\text{Right-hand sum} = 5(100 + 200 + 100 + 200 + 250 + 275 + 300) = 7125.$$

The average of these two sums is our best guess for the value of the integral;

$$\int_{-15}^{20} f(x)\,dx \approx \frac{5625 + 7125}{2} = 6375.$$

9. With $\Delta x = 3$, we have

$$\text{Left-hand sum} = 3(32 + 22 + 15 + 11) = 240,$$

$$\text{Right-hand sum} = 3(22 + 15 + 11 + 9) = 171.$$

The average of these two sums is our best guess for the value of the integral;

$$\int_{0}^{12} f(x)\,dx \approx \frac{240 + 171}{2} = 205.5.$$

13. We use a calculator or computer to see that $\displaystyle\int_{-1}^{1} e^{-x^2}\,dx = 1.4936.$

17. A graph of $y = \ln x$ shows that this function is non-negative on the interval $x = 1$ to $x = 4$. Thus,

$$\text{Area} = \int_{1}^{4} \ln x\,dx = 2.545.$$

The integral was evaluated on a calculator.

21. The graph of $y = x^4 - 8$ has intercepts $x = \pm\sqrt[4]{8}$. See Figure 5.3. Since the region is below the x-axis, the integral is negative, so

$$\text{Area} = -\int_{-\sqrt[4]{8}}^{\sqrt[4]{8}} (x^4 - 8)\,dx = 21.527.$$

The integral was evaluated on a calculator.

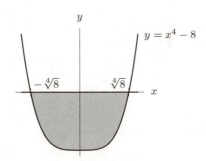

Figure 5.3

Problems

25. Left-hand sum gives: $1^2(1/4) + (1.25)^2(1/4) + (1.5)^2(1/4) + (1.75)^2(1/4) = 1.96875.$
Right-hand sum gives: $(1.25)^2(1/4) + (1.5)^2(1/4) + (1.75)^2(1/4) + (2)^2(1/4) = 2.71875.$

We estimate the value of the integral by taking the average of these two sums, which is 2.34375. Since x^2 is monotonic on $1 \le x \le 2$, the true value of the integral lies between 1.96875 and 2.71875. Thus the most our estimate could be off is 0.375. We expect it to be much closer. (And it is—the true value of the integral is $7/3 \approx 2.333$.)

29. Looking at the graph of $e^{-x}\sin x$ for $0 \le x \le 2\pi$ in Figure 5.4, we see that the area, A_1, below the curve for $0 \le x \le \pi$ is much greater than the area, A_2, above the curve for $\pi \le x \le 2\pi$. Thus, the integral is

$$\int_0^{2\pi} e^{-x}\sin x\,dx = A_1 - A_2 > 0.$$

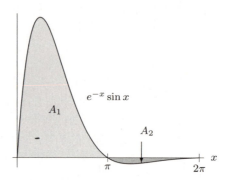

Figure 5.4

33. (a) $\displaystyle\int_{-3}^0 f(x)\,dx = -2.$

(b) $\displaystyle\int_{-3}^4 f(x)\,dx = \int_{-3}^0 f(x)\,dx + \int_0^3 f(x)\,dx + \int_3^4 f(x)\,dx = -2 + 2 - \frac{A}{2} = -\frac{A}{2}.$

37. (a) If the interval $1 \le t \le 2$ is divided into n equal subintervals of length $\Delta t = 1/n$, the subintervals are given by

$$1 \le t \le 1+\frac{1}{n},\ 1+\frac{1}{n} \le t \le 1+\frac{2}{n},\ \dots,\ 1+\frac{n-1}{n} \le t \le 2.$$

The left-hand sum is given by

$$\text{Left sum} = \sum_{r=0}^{n-1} f\left(1+\frac{r}{n}\right)\frac{1}{n} = \sum_{r=0}^{n-1}\frac{1}{1+r/n}\cdot\frac{1}{n} = \sum_{r=0}^{n-1}\frac{1}{n+r}$$

and the right-hand sum is given by

$$\text{Right sum} = \sum_{r=1}^{n} f\left(1+\frac{r}{n}\right)\frac{1}{n} = \sum_{r=1}^{n}\frac{1}{n+r}.$$

Since $f(t) = 1/t$ is decreasing in the interval $1 \le t \le 2$, we know that the right-hand sum is less than $\int_1^2 1/t\,dt$ and the left-hand sum is larger than this integral. Thus we have

$$\sum_{r=1}^{n}\frac{1}{n+r} \quad < \quad \int_1^t \frac{1}{t}\,dt \quad < \quad \sum_{r=0}^{n-1}\frac{1}{n+r}.$$

(b) Subtracting the sums gives

$$\sum_{r=0}^{n-1}\frac{1}{n+r} - \sum_{r=1}^{n}\frac{1}{n+r} = \frac{1}{n} - \frac{1}{2n} = \frac{1}{2n}.$$

(c) Here we need to find n such that

$$\frac{1}{2n} \le 5 \times 10^{-6},\quad \text{so}\quad n \ge \frac{1}{10}\times 10^6 = 10^5.$$

Solutions for Section 5.3

Exercises

1. The units of measurement are dollars.

5. The integral $\int_0^6 a(t)\, dt$ represents the change in velocity between times $t = 0$ and $t = 6$ seconds; it is measured in km/hr.

9. Average value $= \dfrac{1}{2-0} \displaystyle\int_0^2 (1+t)\, dt = \dfrac{1}{2}(4) = 2.$

13. Since the average value is given by

$$\text{Average value} = \frac{1}{b-a} \int_a^b f(x)\, dx,$$

the units for dx inside the integral are canceled by the units for $1/(b-a)$ outside the integral, leaving only the units for $f(x)$. This is as it should be, since the average value of f should be measured in the same units as $f(x)$.

Problems

17. (a) Using rectangles under the curve, we get

$$\text{Acres defaced} \approx (1)(0.2 + 0.4 + 1 + 2) = 3.6 \text{ acres.}$$

(b) Using rectangles above the curve, we get

$$\text{Acres defaced} \approx (1)(0.4 + 1 + 2 + 3.5) = 6.9 \text{ acres.}$$

(c) The number of acres defaced is between 3.6 and 6.9, so we estimate the average, 5.25 acres.

21. (a) An overestimate is 7 tons. An underestimate is 5 tons.
(b) An overestimate is $7 + 8 + 10 + 13 + 16 + 20 = 74$ tons. An underestimate is $5 + 7 + 8 + 10 + 13 + 16 = 59$ tons.
(c) If measurements are made every Δt months, then the error is $|f(6) - f(0)| \cdot \Delta t$. So for this to be less than 1 ton, we need $(20 - 5) \cdot \Delta t < 1$, or $\Delta t < 1/15$. So measurements every 2 days or so will guarantee an error in over- and underestimates of less than 1 ton.

25. The area under the curve represents the number of cubic feet of storage times the number of days the storage was used. This area is given by

$$\text{Area under graph} = \text{Area of rectangle} + \text{Area of triangle}$$
$$= 30 \cdot 10{,}000 + \frac{1}{2} \cdot 30(30{,}000 - 10{,}000)$$
$$= 600{,}000.$$

Since the warehouse charges \$5 for every 10 cubic feet of storage used for a day, the company will have to pay $(5)(60{,}000) = \$300{,}000$.

29. We know that the the integral of F, and therefore the work, can be obtained by computing the areas in Figure 5.5.

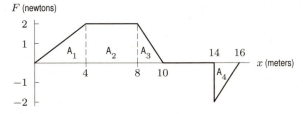

Figure 5.5

$$W = \int_0^{16} F(x)\, dx = \text{Area above } x\text{-axis} - \text{Area below } x\text{-axis}$$

$$= A_1 + A_2 + A_3 - A_4$$
$$= \frac{1}{2} \cdot 4 \cdot 2 + 4 \cdot 2 + \frac{1}{2} \cdot 2 \cdot 2 - \frac{1}{2} \cdot 2 \cdot 2$$
$$= 12 \text{ newton} \cdot \text{meters}.$$

33. (a) Average value of $f = \frac{1}{5} \int_0^5 f(x)\,dx$.

(b) Average value of $|f| = \frac{1}{5} \int_0^5 |f(x)|\,dx = \frac{1}{5}(\int_0^2 f(x)\,dx - \int_2^5 f(x)\,dx)$.

37. Suppose $F(t)$ represents the total quantity of water in the water tower at time t, where t is in days since April 1. Then the graph shown in the problem is a graph of $F'(t)$. By the Fundamental Theorem,

$$F(30) - F(0) = \int_0^{30} F'(t)dt.$$

We can calculate the change in the quantity of water by calculating the area under the curve. If each box represents about 300 liters, there is about one box, or -300 liters, from $t = 0$ to $t = 12$, and 6 boxes, or about $+1800$ liters, from $t = 12$ to $t = 30$. Thus

$$\int_0^{30} F'(t)dt = 1800 - 300 = 1500,$$

so the final amount of water is given by

$$F(30) = F(0) + \int_0^{30} F'(t)dt = 12{,}000 + 1500 = 13{,}500 \text{ liters}.$$

41. (a) Since $t = 0$ to $t = 31$ covers January:

$$\begin{array}{l}\text{Average number of} \\ \text{daylight hours in January}\end{array} = \frac{1}{31} \int_0^{31} [12 + 2.4\sin(0.0172(t - 80))]\,dt.$$

Using left and right sums with $n = 100$ gives

$$\text{Average} \approx \frac{306}{31} \approx 9.9 \text{ hours}.$$

(b) Assuming it is not a leap year, the last day of May is $t = 151 (= 31 + 28 + 31 + 30 + 31)$ and the last day of June is $t = 181 (= 151 + 30)$. Again finding the integral numerically:

$$\begin{array}{l}\text{Average number of} \\ \text{daylight hours in June}\end{array} = \frac{1}{30} \int_{151}^{181} [12 + 2.4\sin(0.0172(t - 80))]\,dt$$
$$\approx \frac{431}{30} \approx 14.4 \text{ hours}.$$

(c)

$$\text{Average for whole year} = \frac{1}{365} \int_0^{365} [12 + 2.4\sin(0.0172(t - 80))]\,dt$$
$$\approx \frac{4381}{365} \approx 12.0 \text{ hours}.$$

(d) The average over the whole year should be 12 hours, as computed in (c). Since Madrid is in the northern hemisphere, the average for a winter month, such as January, should be less than 12 hours (it is 9.9 hours) and the average for a summer month, such as June, should be more than 12 hours (it is 14.4 hours).

Solutions for Section 5.4

Exercises

1. (a) A graph of $f'(x) = \sin(x^2)$ is shown in Figure 5.6. Since the derivative $f'(x)$ is positive between $x = 0$ and $x = 1$, the change in $f(x)$ is positive, so $f(1)$ is larger than $f(0)$. Between $x = 2$ and $x = 2.5$, we see that $f'(x)$ is negative, so the change in $f(x)$ is negative; thus, $f(2)$ is greater than $f(2.5)$.

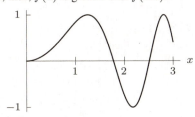

Figure 5.6: Graph of $f'(x) = \sin(x^2)$

(b) The change in $f(x)$ between $x = 0$ and $x = 1$ is given by the Fundamental Theorem of Calculus:

$$f(1) - f(0) = \int_0^1 \sin(x^2)dx = 0.310.$$

Since $f(0) = 2$, we have

$$f(1) = 2 + 0.310 = 2.310.$$

Similarly, since

$$f(2) - f(0) = \int_0^2 \sin(x^2)dx = 0.805,$$

we have

$$f(2) = 2 + 0.805 = 2.805.$$

Since

$$f(3) - f(0) = \int_0^3 \sin(x^2)dx = 0.774,$$

we have

$$f(3) = 2 + 0.774 = 2.774.$$

The results are shown in the table.

x	0	1	2	3
$f(x)$	2	2.310	2.805	2.774

5. The graph of $y = 5\ln(2x)$ is above the line $y = 3$ for $3 \le x \le 5$. See Figure 5.7. Therefore

$$\text{Area} = \int_3^5 (5\ln(2x) - 3)\, dx = 14.688.$$

The integral was evaluated on a calculator.

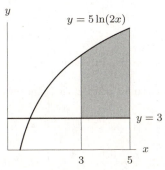

Figure 5.7

9. The graph of $y = \cos t$ is above the graph of $y = \sin t$ for $0 \leq t \leq \pi/4$ and $y = \cos t$ is below $y = \sin t$ for $\pi/4 < t < \pi$. See Figure 5.8. Therefore, we find the area in two pieces:

$$\text{Area} = \int_0^{\pi/4} (\cos t - \sin t)\, dt + \int_{\pi/4}^{\pi} (\sin t - \cos t)\, dt = 2.828.$$

The integral was evaluated on a calculator.

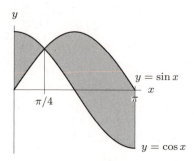

Figure 5.8

13. We have $f(t) = F'(t) = 6t + 4$, so by the Fundamental Theorem of Calculus,

$$\int_2^5 (6t + 4)\, dt = F(5) - F(2) = 95 - 20 = 75.$$

17. We have $f(t) = F'(t) = 1/(\cos^2 t)$, so by the Fundamental Theorem of Calculus,

$$\int_0^{\pi} 1/(\cos^2 t)\, dt = F(\pi) - F(0) = 0 - 0 = 0.$$

Problems

21. Note that $\int_a^b g(x)\, dx = \int_a^b g(t)\, dt$. Thus, we have

$$\int_a^b (f(x) + g(x))\, dx = \int_a^b f(x)\, dx + \int_a^b g(x)\, dx = 8 + 2 = 10.$$

25. We write

$$\int_a^b \left(c_1 g(x) + (c_2 f(x))^2 \right)\, dx = \int_a^b \left(c_1 g(x) + c_2^2 (f(x))^2 \right)\, dx$$

$$= \int_a^b c_1 g(x)\, dx + \int_a^b c_2^2 (f(x))^2\, dx$$

$$= c_1 \int_a^b g(x)\, dx + c_2^2 \int_a^b (f(x))^2\, dx$$

$$= c_1(2) + c_2^2(12) = 2c_1 + 12c_2^2.$$

29. Since f is even, $\int_0^2 f(x)\, dx = (1/2)6 = 3$ and $\int_0^5 f(x)\, dx = (1/2)14 = 7$. Therefore

$$\int_2^5 f(x)\, dx = \int_0^5 f(x)\, dx - \int_0^2 f(x)\, dx = 7 - 3 = 4.$$

33. (a) 0, since the integrand is an odd function and the limits are symmetric around 0.

(b) 0, since the integrand is an odd function and the limits are symmetric around 0.

37. See Figure 5.9. Since $\int_0^1 f(x)\,dx = A_1$ and $\int_1^2 f(x)\,dx = -A_2$ and $\int_2^3 f(x)\,dx = A_3$, we know that

$$0 < \int_0^1 f(x)\,dx < -\int_1^2 f(x)\,dx < \int_2^3 f(x)\,dx.$$

In addition, $\int_0^2 f(x)\,dx = A_1 - A_2$, which is negative, but smaller in magnitude than $\int_1^2 f(x)\,dx$. Thus

$$\int_1^2 f(x)\,dx < \int_0^2 f(x)\,dx < 0.$$

The area A_3 lies inside a rectangle of height 20 and base 1, so $A_3 < 20$. The area A_2 lies inside a rectangle below the x-axis of height 10 and width 1, so $-10 < A_2$. Thus:

$$(\text{viii}) < (\text{ii}) < (\text{iii}) < (\text{vi}) < (\text{i}) < (\text{v}) < (\text{iv}) < (\text{vii}).$$

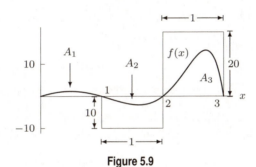

Figure 5.9

41. (a) Yes.

(b) No, because the sum of the left sums has 20 subdivisions. The result is the left sum approximation with 20 subdivisions to $\int_1^3 f(x)\,dx$.

45. See Figure 5.10.

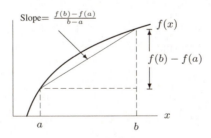

Figure 5.10

49. By the given property, $\int_a^a f(x)\,dx = -\int_a^a f(x)\,dx$, so $2\int_a^a f(x)\,dx = 0$. Thus $\int_a^a f(x)\,dx = 0$.

Solutions for Chapter 5 Review

Exercises

1. (a) Suppose $f(t)$ is the flow rate in m³/hr at time t. We are only given two values of the flow rate, so in making our estimates of the flow, we use one subinterval, with $\Delta t = 3/1 = 3$:

$$\text{Left estimate} = 3[f(6 \text{ am})] = 3 \cdot 100 = 300 \text{ m}^3 \quad \text{(an underestimate)}$$
$$\text{Right estimate} = 3[f(9 \text{ am})] = 3 \cdot 280 = 840 \text{ m}^3 \quad \text{(an overestimate)}.$$

The best estimate is the average of these two estimates,

$$\text{Best estimate} = \frac{\text{Left} + \text{Right}}{2} = \frac{300 + 840}{2} = 570 \text{ m}^3.$$

(b) Since the flow rate is increasing throughout, the error, i.e., the difference between over- and under-estimates, is given by

$$\text{Error} \le \Delta t \left[f(9 \text{ am}) - f(6 \text{ am}) \right] = \Delta t[280 - 100] = 180\Delta t.$$

We wish to choose Δt so that the the error $180\Delta t \le 6$, or $\Delta t \le 6/180 = 1/30$. So the flow rate gauge should be read every $1/30$ of an hour, or every 2 minutes.

5. We know that

$$\int_{-3}^{5} f(x)\,dx = \text{Area above the axis} - \text{Area below the axis}.$$

The area above the axis is about 3 boxes. Since each box has area $(1)(5) = 5$, the area above the axis is about $(3)(5) = 15$. The area below the axis is about 11 boxes, giving an area of about $(11)(5) = 55$. We have

$$\int_{-3}^{5} f(x)\,dx \approx 15 - 55 = -40.$$

9. Since x intercepts are $x = 0, \pi, 2\pi, \ldots$,

$$\text{Area} = \int_{0}^{\pi} \sin x \, dx = 2.$$

The integral was evaluated on a calculator.

13. The graph of $y = -e^x + e^{2(x-1)}$ has intercepts where $e^x = e^{2(x-1)}$, or where $x = 2(x-1)$, so $x = 2$. See Figure 5.11. Since the region is below the x-axis, the integral is negative, so

$$\text{Area} = -\int_{0}^{2} -e^x + e^{2(x-1)} \, dx = 2.762.$$

The integral was evaluated on a calculator.

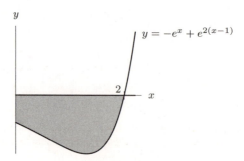

Figure 5.11

17. Distance traveled $= \displaystyle\int_0^{1.1} \sin(t^2)\, dt \approx 0.40$ miles.

Problems

21. (a) By the chain rule,
$$\frac{d}{dx}\left(\frac{1}{2}\sin^2 t\right) = \frac{1}{2}\cdot 2\sin t\cos t = \sin t\cos t.$$

  (i) Using a calculator, $\int_{0.2}^{0.4}\sin t\cos t\, dt = 0.056$

  (ii) The Fundamental Theorem of Calculus tells us that the integral is
$$\int_{0.2}^{0.4}\sin t\cos t\, dt = F(0.4) - F(0.2) = \frac{1}{2}\left(\sin^2(0.4) - \sin^2(0.2)\right) = 0.05609.$$

25. On the interval $2 \le x \le 5$,
$$\begin{array}{c}\text{Average value}\\ \text{of } f\end{array} = \frac{1}{5-2}\int_2^5 f(x)\, dx = 4,$$

so
$$\int_2^5 f(x)\, dx = 12.$$

Thus
$$\int_2^5 (3f(x)+2)\, dx = 3\int_2^5 f(x)\, dx + 2\int_2^5 1\, dx = 3(12) + 2(5-2) = 42.$$

29. (a) Train A starts earlier than Train B, and stops later. At every moment Train A is going faster than Train B. Both trains increase their speed at a constant rate through the first half of their trip and slow down during the second half. Both trains reach their maximum speed at the same time. The area under the velocity graph for Train A is larger than the area under the velocity graph for Train B, meaning that Train A travels farther—as would be expected, given that its speed is always higher than B's.

  (b) (i) The maximum velocity is read off the vertical axis. The graph for Train A appears to go about twice as high as the graph for Train B; see Figure 5.12. So
$$\frac{\text{Maximum velocity of Train } A}{\text{Maximum velocity of Train } B} = \frac{v_A}{v_B} \approx 2.$$

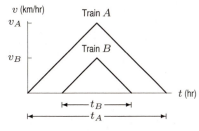

Figure 5.12

    (ii) The time of travel is the horizontal distance between the start and stop times (the two t-intercepts). The horizontal distance for Train A appears to be about twice the corresponding distance for Train B; see Figure 5.12. So
$$\frac{\text{Time traveled by Train } A}{\text{Time traveled by Train } B} = \frac{t_A}{t_B} \approx 2.$$

    (iii) The distance traveled by each train is given by the area under its graph. Since the area of triangle is $\frac{1}{2}\cdot$ Base $\cdot$ Height, and since the base and height for Train A is approximately twice that for Train B, we have
$$\frac{\text{Distance traveled by Train } A}{\text{Distance traveled by Train } B} = \frac{\frac{1}{2}\cdot v_A \cdot t_A}{\frac{1}{2}\cdot v_B \cdot t_B} \approx 2\cdot 2 = 4.$$

33. (a) At $t = 20$ minutes, she stops moving toward the lake (with $v > 0$) and starts to move away from the lake (with $v < 0$). So at $t = 20$ minutes the cyclist turns around.

(b) The cyclist is going the fastest when v has the greatest magnitude, either positive or negative. Looking at the graph, we can see that this occurs at $t = 40$ minutes, when $v = -25$ and the cyclist is pedaling at 25 km/hr away from the lake.

(c) From $t = 0$ to $t = 20$ minutes, the cyclist comes closer to the lake, since $v > 0$; thereafter, $v < 0$ so the cyclist moves away from the lake. So at $t = 20$ minutes, the cyclist comes the closest to the lake. To find out how close she is, note that between $t = 0$ and $t = 20$ minutes the distance she has come closer is equal to the area under the graph of v. Each box represents 5/6 of a kilometer, and there are about 2.5 boxes under the graph, giving a distance of about 2 km. Since she was originally 5 km away, she then is about $5 - 2 = 3$ km from the lake.

(d) At $t = 20$ minutes she turns around , since v changes sign then. Since the area below the t-axis is greater than the area above, the farthest she is from the lake is at $t = 60$ minutes. Between $t = 20$ and $t = 60$ minutes, the area under the graph is about 10.8 km. (Since 13 boxes $\cdot 5/6 = 10.8$.) So at $t = 60$ she will be about $3 + 10.8 = 13.8$ km from the lake.

37. All the integrals have positive values, since $f \geq 0$. The integral in (ii) is about one-half the integral in (i), due to the apparent symmetry of f. The integral in (iv) will be much larger than the integral in (i), since the two peaks of f^2 rise to 10,000. The integral in (iii) will be smaller than half of the integral in (i), since the peaks in $f^{1/2}$ will only rise to 10. So

$$\int_0^2 (f(x))^{1/2}\, dx < \int_0^1 f(x)\, dx < \int_0^2 f(x)\, dx < \int_0^2 (f(x))^2\, dx.$$

41. (a) V, since the slope is constant.

(b) IV, since the net area under this curve is the most negative.

(c) III, since the area under the curve is largest.

(d) II, since the steepest ascent at $t = 0$ occurs on this curve.

(e) III, since average velocity is (total distance)/5, and III moves the largest total distance.

(f) I, since average acceleration is $\dfrac{1}{5}\displaystyle\int_0^5 v'(t)\, dt = \dfrac{1}{5}(v(5) - v(0))$, and in I, the velocity increases the most from start $(t = 0)$ to finish $(t = 5)$.

45. (a) About 300 meter^3/sec.

(b) About 250 meter^3/sec.

(c) Looking at the graph, we can see that the 1996 flood reached its maximum just between March and April, for a high of about 1250 meter^3/sec. Similarly, the 1957 flood reached its maximum in mid-June, for a maximum flow rate of 3500 meter^3/sec.

(d) The 1996 flood lasted about 1/3 of a month, or about 10 days. The 1957 flood lasted about 4 months.

(e) The area under the controlled flood graph is about 2/3 box. Each box represents 500 meter^3/sec for one month. Since

$$1 \text{ month} = 30\frac{\text{days}}{\text{month}} \cdot 24\frac{\text{hours}}{\text{day}} \cdot 60\frac{\text{minutes}}{\text{hour}} \cdot 60\frac{\text{seconds}}{\text{minute}}$$
$$= 2.592 \cdot 10^6 \approx 2.6 \cdot 10^6 \text{ seconds,}$$

each box represents

$$\text{Flow} \approx (500 \text{ meter}^3/\text{sec}) \cdot (2.6 \cdot 10^6 \text{ sec}) = 13 \cdot 10^8 \text{ meter}^3 \text{of water.}$$

So, for the artificial flood,

$$\text{Additional flow} \approx \frac{2}{3} \cdot 13 \cdot 10^8 = 8.7 \cdot 10^8 \text{ meter}^3 \approx 10^9 \text{ meter}^3.$$

(f) The 1957 flood released a volume of water represented by about 12 boxes above the 250 meter/sec baseline. Thus, for the natural flood,

$$\text{Additional flow} \approx 12 \cdot 13 \cdot 10^8 = 1.95 \cdot 10^{10} \approx 2 \cdot 10^{10} \text{ meter}^3.$$

So, the natural flood was nearly 20 times larger than the controlled flood and lasted much longer.

49. In (a), $f'(1)$ is the slope of a tangent line at $x = 1$, which is negative. As for (c), the rate of change in $f(x)$ is given by $f'(x)$, and the average value of this over $0 \le x \le a$ is

$$\frac{1}{a - 0} \int_0^a f'(x)\,dx = \frac{f(a) - f(0)}{a - 0}.$$

This is the slope of the line through the points $(0, 1)$ and $(a, 0)$, which is less negative that the tangent line at $x = 1$. Therefore, $(a) < (c) < 0$. The quantity (b) is $\left(\int_0^a f(x)\,dx\right)/a$ and (d) is $\int_0^a f(x)\,dx$, which is the net area under the graph of f (counting the area as negative for f below the x-axis). Since $a > 1$ and $\int_0^a f(x)\,dx > 0$, we have $0 < \text{(b)} < \text{(d)}$. Therefore

$$\text{(a)} < \text{(c)} < \text{(b)} < \text{(d)}.$$

CAS Challenge Problems

53. (a) Since the length of the interval of integration is $2 - 1 = 1$, the width of each subdivision is $\Delta t = 1/n$. Thus the endpoints of the subdivision are

$$t_0 = 1, \quad t_1 = 1 + \Delta t = 1 + \frac{1}{n}, \quad t_2 = 1 + 2\Delta t = 1 + \frac{2}{n}, \ldots,$$

$$t_i = 1 + i\Delta t = 1 + \frac{i}{n}, \ldots, \quad t_{n-1} = 1 + (n-1)\Delta t = 1 + \frac{n-1}{n}.$$

Thus, since the integrand is $f(t) = t$,

$$\text{Left-hand sum} = \sum_{i=0}^{n-1} f(t_i)\Delta t = \sum_{i=0}^{n-1} t_i \Delta t = \sum_{i=0}^{n-1} \left(1 + \frac{i}{n}\right)\frac{1}{n} = \sum_{i=0}^{n-1} \frac{n+i}{n^2}.$$

(b) The CAS finds the formula for the Riemann sum

$$\sum_{i=0}^{n-1} \frac{n+i}{n^2} = \frac{\frac{(-1+n)\,n}{2} + n^2}{n^2} = \frac{3}{2} - \frac{1}{2n}.$$

(c) Taking the limit as $n \to \infty$

$$\lim_{n\to\infty}\left(\frac{3}{2} - \frac{1}{2n}\right) = \lim_{n\to\infty}\frac{3}{2} - \lim_{n\to\infty}\frac{1}{2n} = \frac{3}{2} + 0 = \frac{3}{2}.$$

(d) The shape under the graph of $y = t$ between $t = 1$ and $t = 2$ is a trapezoid of width 1, height 1 on the left and 2 on the right. So its area is $1 \cdot (1 + 2)/2 = 3/2$. This is the same answer we got by computing the definite integral.

57. (a) Different systems may give different answers. A typical answer is

$$\int_a^c \frac{x}{1 + bx^2}\,dx = \frac{\ln\left(\frac{|c^2 b + 1|}{|a^2 b + 1|}\right)}{2b}.$$

Some CASs may not have the absolute values in the answer; since $b > 0$, the answer is correct without the absolute values.

(b) Using the properties of logarithms, we can rewrite the answer to part (a) as

$$\int_a^c \frac{x}{1 + bx^2}\,dx = \frac{\ln\left|c^2 b + 1\right| - \ln\left|a^2 b + 1\right|}{2b} = \frac{\ln\left|c^2 b + 1\right|}{2b} - \frac{\ln\left|a^2 b + 1\right|}{2b}.$$

If $F(x)$ is an antiderivative of $x/(1 + bx^2)$, then the Fundamental Theorem of Calculus says that

$$\int_a^c \frac{x}{1 + bx^2}\,dx = F(c) - F(a).$$

Thus

$$F(c) - F(a) = \frac{\ln\left|c^2 b + 1\right|}{2b} - \frac{\ln\left|a^2 b + 1\right|}{2b}.$$

This suggests that

$$F(x) = \frac{\ln\left|1 + bx^2\right|}{2b}.$$

(Since $b > 0$, we know $\left|1 + bx^2\right| = 1 + bx^2$.) Taking the derivative confirms this:

$$\frac{d}{dx}\left(\frac{\ln(1 + bx^2)}{2b}\right) = \frac{x}{1 + bx^2}.$$

CHECK YOUR UNDERSTANDING

1. False. The units of the integral are the product of the units for $f(x)$ times the units for x.

5. False. The integral is the change in position from $t = a$ to $t = b$. If the velocity changes sign in the interval, the total distance traveled and the change in position will not be the same.

9. True, since $\int_0^2 2f(x)dx = 2\int_0^2 f(x)dx$.

13. False. Let $f(x) = 7$ and $g(x) = 9$ for all x.
Then $\int_1^2 f(x)\,dx + \int_2^3 g(x)\,dx = 7 + 9 = 16$, but $\int_1^3 (f(x) + g(x))\,dx = \int_1^3 16\,dx = 32$.

17. False. Let $f(x) = x$ and $g(x) = 5$. Then $\int_2^6 f(x)\,dx = 16$ and $\int_2^6 g(x)\,dx = 20$, so $\int_2^6 f(x)\,dx \leq \int_2^6 g(x)\,dx$, but $f(x) > g(x)$ for $5 < x < 6$.

21. True. We have by the properties of integrals in Theorem 5.3,

$$\int_1^9 f(x)dx = \int_1^4 f(x)dx + \int_4^9 f(x)dx.$$

Since $(1/(4 - 1)) \int_1^4 f(x)dx = A$ and $(1/(9 - 4)) \int_4^9 f(x)dx = B$, we have

$$\int_1^9 f(x)dx = 3A + 5B.$$

Dividing this equation through by 8, we get that the average value of f on the interval $[1, 9]$ is $(3/8)A + (5/8)B$.

25. False. A counterexample is given by the functions f and g in Figure 5.13. The function f is decreasing, g is increasing, and we have

$$\int_1^2 f(x)\,dx = \int_1^2 g(x)\,dx,$$

because both integrals equal $1/2$, the area of of the same sized triangle.

Figure 5.13

CHAPTER SIX

Solutions for Section 6.1

Exercises

1. See Figure 6.1.

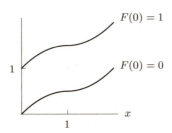

Figure 6.1

5. Since dP/dt is negative for $t < 3$ and positive for $t > 3$, we know that P is decreasing for $t < 3$ and increasing for $t > 3$. Between each two integer values, the magnitude of the change is equal to the area between the graph dP/dt and the t-axis. For example, between $t = 0$ and $t = 1$, we see that the change in P is -1. Since $P = 2$ at $t = 0$, we must have $P = 1$ at $t = 1$. The other values are found similarly, and are shown in Table 6.1.

Table 6.1

t	1	2	3	4	5
P	1	0	$-1/2$	0	1

Problems

9. See Figure 6.2. Note that since $f(x_1) = 0$ and $f'(x_1) < 0$, $F(x_1)$ is a local maximum; since $f(x_3) = 0$ and $f'(x_3) > 0$, $F(x_3)$ is a local minimum. Also, since $f'(x_2) = 0$ and f changes from decreasing to increasing about $x = x_2$, F has an inflection point at $x = x_2$.

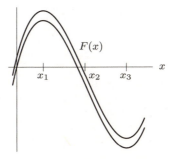

Figure 6.2

13. Between $t = 0$ and $t = 1$, the particle moves at 10 km/hr for 1 hour. Since it starts at $x = 5$, the particle is at $x = 15$ when $t = 1$. See Figure 6.3. The graph of distance is a straight line between $t = 0$ and $t = 1$ because the velocity is constant then.

Between $t = 1$ and $t = 2$, the particle moves 10 km to the left, ending at $x = 5$. Between $t = 2$ and $t = 3$, it moves 10 km to the right again. See Figure 6.3.

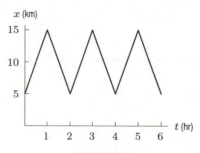

Figure 6.3

As an aside, note that the original velocity graph is not entirely realistic as it suggests the particle reverses direction instantaneously at the end of each hour. In practice this means the reversal of direction occurs over a time interval that is short in comparison to an hour.

17. Looking at the graph of g' in Figure 6.4, we see that the critical points of g occur when $x = 15$ and $x = 40$, since $g'(x) = 0$ at these values. Inflection points of g occur when $x = 10$ and $x = 20$, because $g'(x)$ has a local maximum or minimum at these values. Knowing these four key points, we sketch the graph of $g(x)$ in Figure 6.5.

We start at $x = 0$, where $g(0) = 50$. Since g' is negative on the interval $[0, 10]$, the value of $g(x)$ is decreasing there. At $x = 10$ we have

$$g(10) = g(0) + \int_0^{10} g'(x)\,dx$$
$$= 50 - (\text{area of shaded trapezoid } T_1)$$
$$= 50 - \left(\frac{10 + 20}{2} \cdot 10\right) = -100.$$

Similarly,

$$g(15) = g(10) + \int_{10}^{15} g'(x)\,dx$$
$$= -100 - (\text{area of triangle } T_2)$$
$$= -100 - \frac{1}{2}(5)(20) = -150.$$

Continuing,

$$g(20) = g(15) + \int_{15}^{20} g'(x)\,dx = -150 + \frac{1}{2}(5)(10) = -125,$$

and

$$g(40) = g(20) + \int_{20}^{40} g'(x)\,dx = -125 + \frac{1}{2}(20)(10) = -25.$$

We now find concavity of $g(x)$ in the intervals $[0, 10]$, $[10, 15]$, $[15, 20]$, $[20, 40]$ by checking whether $g'(x)$ increases or decreases in these same intervals. If $g'(x)$ increases, then $g(x)$ is concave up; if $g'(x)$ decreases, then $g(x)$ is concave down. Thus we finally have the graph of $g(x)$ in Figure 6.5.

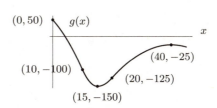

Figure 6.4

Figure 6.5

21. (a) Critical points of $F(x)$ are the zeros of f: $x = 1$ and $x = 3$.
   (b) $F(x)$ has a local minimum at $x = 1$ and a local maximum at $x = 3$.
   (c) See Figure 6.6.

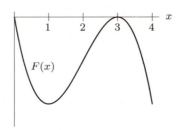

Figure 6.6

Notice that the graph could also be above or below the x-axis at $x = 3$.

25. Both $F(x)$ and $G(x)$ have roots at $x = 0$ and $x = 4$. Both have a critical point (which is a local maximum) at $x = 2$. However, since the area under $g(x)$ between $x = 0$ and $x = 2$ is larger than the area under $f(x)$ between $x = 0$ and $x = 2$, the y-coordinate of $G(x)$ at 2 will be larger than the y-coordinate of $F(x)$ at 2. See below.

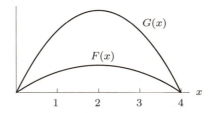

Solutions for Section 6.2

Exercises

1. $5x$

5. $\sin t$

9. $-\dfrac{1}{2z^2}$

13. $\dfrac{t^4}{4} - \dfrac{t^3}{6} - \dfrac{t^2}{2}$

17. $F(t) = \displaystyle\int 6t\, dt = 3t^2 + C$

21. $F(z) = \displaystyle\int (z + e^z)\, dz = \dfrac{z^2}{2} + e^z + C$

25. $P(t) = \displaystyle\int (2 + \sin t)\, dt = 2t - \cos t + C$

29. $f(x) = 3$, so $F(x) = 3x + C$. $F(0) = 0$ implies that $3 \cdot 0 + C = 0$, so $C = 0$. Thus $F(x) = 3x$ is the only possibility.

33. $f(x) = x^2$, so $F(x) = \dfrac{x^3}{3} + C$. $F(0) = 0$ implies that $\dfrac{0^3}{3} + C = 0$, so $C = 0$. Thus $F(x) = \dfrac{x^3}{3}$ is the only possibility.

37. $\displaystyle\int 5x\, dx = \dfrac{5}{2}x^2 + C.$

41. $\displaystyle\int \left(t^2 + \dfrac{1}{t^2}\right) dt = \dfrac{t^3}{3} - \dfrac{1}{t} + C$

45. $2t^2 + 7t + C$

49. $-\cos t + C$

53. $\displaystyle\int_0^3 (x^2 + 4x + 3)\, dx = \left(\dfrac{x^3}{3} + 2x^2 + 3x\right)\Big|_0^3 = (9 + 18 + 9) - 0 = 36$

57. $\displaystyle\int_2^5 (x^3 - \pi x^2)\, dx = \left(\dfrac{x^4}{4} - \dfrac{\pi x^3}{3}\right)\Big|_2^5 = \dfrac{609}{4} - 39\pi \approx 29.728.$

61. $\displaystyle\int_0^{\pi/4} (\sin t + \cos t)\, dt = (-\cos t + \sin t)\Big|_0^{\pi/4} = \left(-\dfrac{\sqrt{2}}{2} + \dfrac{\sqrt{2}}{2}\right) - (-1 + 0) = 1.$

Problems

65. The graph crosses the x-axis where

$$7 - 8x + x^2 = 0$$
$$(x - 7)(x - 1) = 0;$$

so $x = 1$ and $x = 7$. See Figure 6.7. The parabola opens upward and the region is below the x-axis, so

$$\text{Area} = -\int_1^7 (7 - 8x + x^2)\, dx$$
$$= -\left(7x - 4x^2 + \dfrac{x^3}{3}\right)\Big|_1^7 = 36.$$

$$y = 7 - 8x + x^2$$

Figure 6.7

69. The graph is shown in Figure 6.8. Since $\cos\theta \geq \sin\theta$ for $0 \leq \theta \leq \pi/4$, we have

$$\text{Area} = \int_0^{\pi/4} (\cos\theta - \sin\theta)\, d\theta$$
$$= (\sin\theta + \cos\theta)\Big|_0^{\pi/4}$$
$$= \dfrac{1}{\sqrt{2}} + \dfrac{1}{\sqrt{2}} - 1 = \sqrt{2} - 1.$$

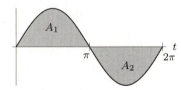

Figure 6.8

73. The area under $f(x) = 8x$ between $x = 1$ and $x = b$ is given by $\int_1^b (8x)dx$. Using the Fundamental Theorem to evaluate the integral:

$$\text{Area } = 4x^2 \Big|_1^b = 4b^2 - 4.$$

Since the area is 192, we have

$$4b^2 - 4 = 192$$
$$4b^2 = 196$$
$$b^2 = 49$$
$$b = \pm 7.$$

Since b is larger than 1, we have $b = 7$.

77. (a) The average value of $f(t) = \sin t$ over $0 \leq t \leq 2\pi$ is given by the formula

$$\text{Average } = \frac{1}{2\pi - 0} \int_0^{2\pi} \sin t \, dt$$
$$= \frac{1}{2\pi}(-\cos t)\Big|_0^{2\pi}$$
$$= \frac{1}{2\pi}(-\cos 2\pi - (-\cos 0)) = 0.$$

We can check this answer by looking at the graph of $\sin t$ in Figure 6.9. The area below the curve and above the t-axis over the interval $0 \leq t \leq \pi$, A_1, is the same as the area above the curve but below the t-axis over the interval $\pi \leq t \leq 2\pi$, A_2. When we take the integral of $\sin t$ over the entire interval $0 \leq t \leq 2\pi$, we get $A_1 - A_2 = 0$.

Figure 6.9

(b) Since

$$\int_0^\pi \sin t \, dt = -\cos t \Big|_0^\pi = -\cos \pi - (-\cos 0) = -(-1) - (-1) = 2,$$

the average value of $\sin t$ on $0 \leq t \leq \pi$ is given by

$$\text{Average value } = \frac{1}{\pi} \int_0^\pi \sin t \, dt = \frac{2}{\pi}.$$

81. The rate at which water is entering the tank is

$$\frac{dV}{dt} = 120 - 6t.$$

Thus, the total quantity of water in the tank at time $t = 4$ is

$$V = \int_0^4 (120 - 6t)\,dt.$$

Since an antiderivative to $120 - 6t$ is

$$120t - 3t^2,$$

we have

$$
\begin{aligned}
V = \int_0^4 (120 - 6t)\,dt &= \left.(120t - 3t^2)\right|_0^4 \\
&= (120 \cdot 4 - 6 \cdot 4^2) - (120 \cdot 0 - 3 \cdot 0^2) \\
&= 384 \text{ ft}^3.
\end{aligned}
$$

The radius is 5 feet, so if the height is h ft, the volume is $V = \pi 5^2 h = 25\pi h$. Thus, at time $t = 4$, we have $V = 384$, so

$$
\begin{aligned}
384 &= 25\pi h \\
h &= \frac{384}{25\pi} = 4.889 \text{ ft.}
\end{aligned}
$$

85. (a) See Figure 6.10.

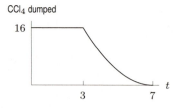

CCl$_4$ dumped

Figure 6.10

(b) 7 years, because $t^2 - 14t + 49 = (t-7)^2$ indicates that the rate of flow was zero after 7 years.

(c)

$$
\begin{aligned}
\text{Area under the curve} &= 3(16) + \int_3^7 (t^2 - 14t + 49)\,dt \\
&= 48 + \left.\left(\frac{1}{3}t^3 - 7t^2 + 49t\right)\right|_3^7 \\
&= 48 + \frac{343}{3} - 343 + 343 - 9 + 63 - 147 \\
&= \frac{208}{3} = 69.333 \text{ cubic yards.}
\end{aligned}
$$

Solutions for Section 6.3

Exercises

1. $y = \int (x^3 + 5)\,dx = \dfrac{x^4}{4} + 5x + C$

5. Since $y = x + \sin x - \pi$, we differentiate to see that $dy/dx = 1 + \cos x$, so y satisfies the differential equation. To show that it also satisfies the initial condition, we check that $y(\pi) = 0$:

$$
\begin{aligned}
y &= x + \sin x - \pi \\
y(\pi) &= \pi + \sin \pi - \pi = 0.
\end{aligned}
$$

9. Integrating gives

$$\int \frac{dq}{dz}\, dz = \int (2 + \sin z)\, dz = 2z - \cos z + C.$$

If $q = 5$ when $z = 0$, then $2(0) - \cos(0) + C = 5$ so $C = 6$. Thus $q = 2z - \cos z + 6$.

Problems

13. (a) We find F for each piece, $0 \le x \le 1$ and $1 \le x \le 2$.

For $0 \le x \le 1$, we have $f(x) = -x + 1$, so F is of the form

$$\int (-x + 1)\, dx = -\frac{x^2}{2} + x + C.$$

Since we want $F(1) = 1$, we need $C = 1/2$. See Figure 6.11.

For $1 \le x \le 2$, we have $f(x) = x - 1$, so F is of the form

$$\int (x - 1)\, dx = \frac{x^2}{2} - x + C.$$

Again, since we want $F(1) = 1$, we have $C = 3/2$. See Figure 6.11.

(b) Evaluating

$$F(2) - F(0) = \left(\frac{2^2}{2} - 2 + \frac{3}{2}\right) - \left(-\frac{0^2}{2} + 0 + \frac{1}{2}\right) = 1.$$

The region under the graph of f consists of two triangles, whose area is

$$\text{Area} = \frac{1}{2} + \frac{1}{2} = 1.$$

(c) The Fundamental Theorem of Calculus says

$$\int_0^2 f(x)\, dx = F(2) - F(0).$$

Since the value of the integral is just the area under the curve, we have shown this in part (b).

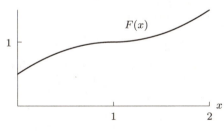

Figure 6.11

17. (a) To find the height of the balloon, we integrate its velocity with respect to time:

$$h(t) = \int v(t)\, dt$$

$$= \int (-32t + 40)\, dt$$

$$= -32\frac{t^2}{2} + 40t + C.$$

Since at $t = 0$, we have $h = 30$, we can solve for C to get $C = 30$, giving us a height of

$$h(t) = -16t^2 + 40t + 30.$$

(b) To find the average velocity between $t = 1.5$ and $t = 3$, we find the total displacement and divide by time.

$$\text{Average velocity} = \frac{h(3) - h(1.5)}{3 - 1.5} = \frac{6 - 54}{1.5} = -32 \text{ ft/sec.}$$

The balloon's average velocity is 32 ft/sec downward.

(c) First, we must find the time when $h(t) = 6$. Solving the equation $-16t^2 + 40t + 30 = 6$, we get

$$6 = -16t^2 + 40t + 30$$
$$0 = -16t^2 + 40t + 24$$
$$0 = 2t^2 - 5t - 3$$
$$0 = (2t + 1)(t - 3).$$

Thus, $t = -1/2$ or $t = 3$. Since $t = -1/2$ makes no physical sense, we use $t = 3$ to calculate the balloon's velocity. At $t = 3$, we have a velocity of $v(3) = -32(3) + 40 = -56$ ft/sec. So the balloon's velocity is 56 ft/sec downward at the time of impact.

21. Since the acceleration is constant, a graph of the velocity versus time looks like this:

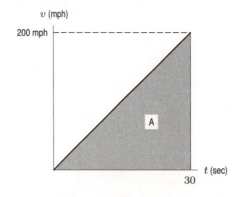

The distance traveled in 30 seconds, which is how long the runway must be, is equal to the area represented by A. We have $A = \frac{1}{2}(\text{base})(\text{height})$. First we convert the required velocity into miles per second.

$$200 \text{ mph} = \frac{200 \text{ miles}}{\text{hour}} \left(\frac{1 \text{ hour}}{60 \text{ minutes}} \right) \left(\frac{1 \text{ minute}}{60 \text{ seconds}} \right)$$
$$= \frac{200}{3600} \frac{\text{miles}}{\text{second}}$$
$$= \frac{1}{18} \text{ miles/second.}$$

Therefore $A = \frac{1}{2}(30 \text{ sec})(200 \text{ mph}) = \frac{1}{2}(30 \text{ sec}) \left(\frac{1}{18} \text{ miles/sec} \right) = \frac{5}{6} \text{ miles.}$

25. (a) $a(t) = 1.6$, so $v(t) = 1.6t + v_0 = 1.6t$, since the initial velocity is 0.

(b) $s(t) = 0.8t^2 + s_0$, where s_0 is the rock's initial height.

Solutions for Section 6.4

Exercises

1. By the Fundamental Theorem, $f(x) = F'(x)$. Since f is positive and increasing, F is increasing and concave up. Since $F(0) = \int_0^0 f(t)dt = 0$, the graph of F must start from the origin. See Figure 6.12.

Figure 6.12

5. $(1+x)^{200}$.

9. $\arctan(x^2)$.

13.

Table 6.2

x	0	0.5	1	1.5	2
$I(x)$	0	0.50	1.09	2.03	3.65

17. If $f'(x) = \dfrac{\sin x}{x}$, then $f(x)$ is of the form

$$f(x) = C + \int_a^x \frac{\sin t}{t}\, dt.$$

Since $f(1) = 5$, we take $a = 1$ and $C = 5$, giving

$$f(x) = 5 + \int_1^x \frac{\sin t}{t}\, dt.$$

Problems

21. Using the solution to Problem 20, we see that $F(x)$ is increasing for $x < \sqrt{\pi}$ and $F(x)$ is decreasing for $x > \sqrt{\pi}$. Thus $F(x)$ has its maximum value when $x = \sqrt{\pi}$.

By the Fundamental Theorem of Calculus,

$$F(\sqrt{\pi}) - F(0) = \int_0^{\sqrt{\pi}} f(t)\, dt = \int_0^{\sqrt{\pi}} \sin(t^2)\, dt.$$

Since $F(0) = 0$, calculating Riemann sums gives

$$F(\sqrt{\pi}) = \int_0^{\sqrt{\pi}} \sin(t^2)\, dt = 0.895.$$

25. Since $G'(x) = \cos(x^2)$ and $G(0) = -3$, we have

$$G(x) = G(0) + \int_0^x \cos(t^2)\, dt = -3 + \int_0^x \cos(t^2)\, dt.$$

Substituting $x = -1$ and evaluating the integral numerically gives

$$G(-1) = -3 + \int_0^{-1} \cos(t^2)\, dt = -3.905.$$

29. If we let $f(x) = \int_2^x \sin(t^2)\, dt$ and $g(x) = x^3$, using the chain rule gives

$$\frac{d}{dx} \int_2^{x^3} \sin(t^2)\, dt = f'(g(x)) \cdot g'(x) = \sin((x^3)^2) \cdot 3x^2 = 3x^2 \sin(x^6).$$

33. Since $\int_{\cos x}^3 e^{t^2}\, dt = -\int_3^{\cos x} e^{t^2}\, dt$, if we let $f(x) = \int_3^x e^{t^2}\, dt$ and $g(x) = \cos x$, using the chain rule gives

$$\frac{d}{dx} \int_{\cos x}^3 e^{t^2}\, dt = -\frac{d}{dx} \int_3^{\cos x} e^{t^2}\, dt = -f'(g(x)) \cdot g'(x) = -e^{(\cos x)^2}(-\sin x) = \sin x\, e^{\cos^2 x}.$$

37. (a) The definition of P gives

$$P(0) = \int_0^0 \arctan(t^2)\, dt = 0$$

and

$$P(-x) = \int_0^{-x} \arctan(t^2)\, dt.$$

Changing the variable of integration by letting $t = -z$ gives

$$\int_0^{-x} \arctan(t^2)\, dt = \int_0^x \arctan((-z)^2)(-dz) = -\int_0^x \arctan(z^2)\, dz.$$

Thus P is an odd function.

(b) Using the Second Fundamental Theorem gives $P'(x) = \arctan(x^2)$, which is greater than 0 for $x \neq 0$. Thus P is increasing everywhere.

(c) Since

$$P''(x) = \frac{2x}{1 + x^4},$$

we have P concave up if $x > 0$ and concave down if $x < 0$.

(d) See Figure 6.13.

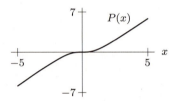

Figure 6.13

41. The derivative of $s(t - t_0)$ is $s'(t - t_0)$, so $s'(t - t_0) = v(t)$. Substituting $w = t - t_0$, so that $t = w + t_0$, we get $s'(w) = v(w + t_0)$. Renaming w to t, we get $s'(t) = v(t + t_0)$.

45. If we let $f(x) = \operatorname{erf}(x)$ and $g(x) = \sqrt{x}$, then we are looking for $\frac{d}{dx}[f(g(x))]$. By the chain rule, this is the same as $g'(x)f'(g(x))$. Since

$$f'(x) = \frac{d}{dx}\left(\frac{2}{\sqrt{\pi}} \int_0^x e^{-t^2}\, dt\right)$$
$$= \frac{2}{\sqrt{\pi}} e^{-x^2}$$

and $g'(x) = \dfrac{1}{2\sqrt{x}}$, we have

$$f'(g(x)) = \frac{2}{\sqrt{\pi}} e^{-x},$$

and so

$$\frac{d}{dx}[\operatorname{erf}(\sqrt{x})] = \frac{1}{2\sqrt{x}} \frac{2}{\sqrt{\pi}} e^{-x} = \frac{1}{\sqrt{\pi x}} e^{-x}.$$

Solutions for Section 6.5

Exercises

1. (a) The object is thrown from an initial height of $y = 1.5$ meters.

(b) The velocity is obtained by differentiating, which gives $v = -9.8t + 7$ m/sec. The initial velocity is $v = 7$ m/sec upward.

(c) The acceleration due to gravity is obtained by differentiating again, giving $g = -9.8$ m/sec^2, or 9.8 m/sec^2 downward.

Problems

5. In Problem 4 we used the equation $0 = -16t^2 + 400$ to learn that the object hits the ground after 5 seconds. In a more general form this is the equation $y = -\frac{g}{2}t^2 + v_0 t + y_0$, and we know that $v_0 = 0$, $y_0 = 400$ ft. So the moment the object hits the ground is given by $0 = -\frac{g}{2}t^2 + 400$. In Problem 4 we used $g = 32$ ft/sec^2, but in this case we want to find a g that results in the object hitting the ground after only 5/2 seconds. We put in 5/2 for t and solve for g:

$$0 = -\frac{g}{2}\left(\frac{5}{2}\right)^2 + 400, \text{ so } g = \frac{2(400)}{(5/2)^2} = 128 \text{ ft/sec}^2.$$

9. (a) $t = \dfrac{s}{\frac{1}{2}v_{\max}}$, where t is the time it takes for an object to travel the distance s, starting from rest with uniform acceleration a. $v_{\max}$ is the highest velocity the object reaches. Since its initial velocity is 0, the mean of its highest velocity and initial velocity is $\frac{1}{2}v_{\max}$.

(b) By Problem 8, $s = \frac{1}{2}gt^2$, where g is the acceleration due to gravity, so it takes $\sqrt{200/32} = 5/2$ seconds for the body to hit the ground. Since $v = gt$, $v_{\max} = 32(\frac{5}{2}) = 80$ ft/sec. Galileo's statement predicts $(100 \text{ ft})/(40 \text{ ft/sec}) = 5/2$ seconds, and so Galileo's result is verified.

(c) If the acceleration is a constant a, then $s = \frac{1}{2}at^2$, and $v_{\max} = at$. Thus

$$\frac{s}{\frac{1}{2}v_{\max}} = \frac{\frac{1}{2}at^2}{\frac{1}{2}at} = t.$$

Solutions for Chapter 6 Review

Exercises

1. See Figure 6.14

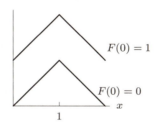

$F(0) = 1$

$F(0) = 0$

x

1

Figure 6.14

5. $\displaystyle\int (2 + \cos t)\, dt = 2t + \sin t + C$

9. $\displaystyle\int \frac{8}{\sqrt{x}}\, dx = 16x^{1/2} + C$

13. $\tan x + C$

17. $\frac{1}{10}(x+1)^{10} + C$

21. $3\sin x + 7\cos x + C$

25. $F(x) = \displaystyle\int \frac{1}{x^2}\, dx = -\frac{1}{x} + C$

29. $F(x) = \displaystyle\int 5e^x\, dx = 5e^x + C$

33. We have $F(x) = \dfrac{x^4}{4} + 2x^3 - 4x + C$. Since $F(0) = 4$, we have $4 = 0 + C$, so $C = 4$. So $F(x) = \dfrac{x^4}{4} + 2x^3 - 4x + 4$.

37. $F(x) = \displaystyle\int \cos x\, dx = \sin x + C$. If $F(0) = 4$, then $F(0) = 0 + C = 4$ and thus $C = 4$. So $F(x) = \sin x + 4$.

Problems

41. $\displaystyle\int_0^3 x^2\, dx = \dfrac{x^3}{3}\bigg|_0^3 = 9 - 0 = 9$.

45. (a) See Figure 6.15. Since $f(x) > 0$ for $0 < x < 2$ and $f(x) < 0$ for $2 < x < 5$, we have

$$
\begin{aligned}
\text{Area} &= \int_0^2 f(x)\, dx - \int_2^5 f(x)\, dx \\
&= \int_0^2 (x^3 - 7x^2 + 10x)\, dx - \int_2^5 (x^3 - 7x^2 + 10x)\, dx \\
&= \left(\frac{x^4}{4} - \frac{7x^3}{3} + 5x^2 \right)\bigg|_0^2 - \left(\frac{x^4}{4} - \frac{7x^3}{3} + 5x^2 \right)\bigg|_2^5 \\
&= \left[\left(4 - \frac{56}{3} + 20 \right) - (0 - 0 + 0) \right] - \left[\left(\frac{625}{4} - \frac{875}{3} + 125 \right) - \left(4 - \frac{56}{3} + 20 \right) \right] \\
&= \frac{253}{12}.
\end{aligned}
$$

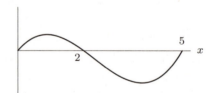

Figure 6.15: Graph of $f(x) = x^3 - 7x^2 + 10x$

(b) Calculating $\int_0^5 f(x)\, dx$ gives

$$
\begin{aligned}
\int_0^5 f(x)\, dx &= \int_0^5 (x^3 - 7x^2 + 10x)\, dx \\
&= \left(\frac{x^4}{4} - \frac{7x^3}{3} + 5x^2 \right)\bigg|_0^5 \\
&= \left(\frac{625}{4} - \frac{875}{3} + 125 \right) - (0 - 0 + 0) \\
&= -\frac{125}{12}.
\end{aligned}
$$

This integral measures the difference between the area above the x-axis and the area below the x-axis. Since the definite integral is negative, the graph of $f(x)$ lies more below the x-axis than above it. Since the function crosses the axis at $x = 2$,

$$
\int_0^5 f(x)\, dx = \int_0^2 f(x)\, dx + \int_2^5 f(x)\, dx = \frac{16}{3} - \frac{63}{4} = \frac{-125}{12},
$$

whereas

$$
\text{Area} = \int_0^2 f(x)\, dx - \int_2^5 f(x)\, dx = \frac{16}{3} + \frac{64}{4} = \frac{253}{12}.
$$

49. The graph of $y = c(1 - x^2)$ has x-intercepts of $x = \pm 1$. See Figure 6.16. Since it is symmetric about the y-axis, we have

$$\text{Area} = \int_{-1}^{1} c(1 - x^2)\,dx = 2c \int_{0}^{1} (1 - x^2)\,dx$$

$$= 2c \left(x - \frac{x^3}{3} \right) \Big|_{0}^{1} = \frac{4c}{3}.$$

We want the area to be 1, so

$$\frac{4c}{3} = 1, \quad \text{giving} \quad c = \frac{3}{4}.$$

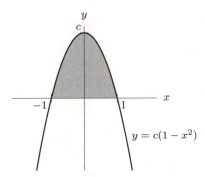

Figure 6.16

53. Since $A'(r) = C(r)$ and $C(r) = 2\pi r$, we have

$$A'(r) = 2\pi r.$$

Thus, we have, for some arbitrary constant K:

$$A(r) = \int 2\pi r\,dr = 2\pi \int r\,dr = 2\pi \frac{r^2}{2} + K = \pi r^2 + K.$$

Since a circle of radius $r = 0$ has area $= 0$, we substitute to find K:

$$0 = \pi 0^2 + K$$
$$K = 0.$$

Thus

$$A(r) = \pi r^2.$$

57. See Figure 6.17.

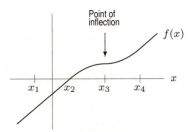

Figure 6.17

61. We have

$$\frac{d}{dx} \int_{2}^{x} \arccos(t^7)\,dt = \arccos x^7.$$

65. If we let $f(x) = \int_5^x \cos(t^3)\,dt$ and $g(x) = e^x$, using the chain rule gives

$$\frac{d}{dx}\int_5^{e^x} \cos(t^3)\,dt = f'(g(x))\cdot g'(x) = \cos((e^x)^3)\cdot e^x = e^x\cos(e^{3x}).$$

69. $F(x)$ represents the net area between $(\sin t)/t$ and the t-axis from $t = \frac{\pi}{2}$ to $t = x$, with area counted as negative for $(\sin t)/t$ below the t-axis. As long as the integrand is positive $F(x)$ is increasing. Therefore, the global maximum of $F(x)$ occurs at $x = \pi$ and is given by the area

$$A_1 = \int_{\pi/2}^{\pi} \frac{\sin t}{t}\,dt.$$

At $x = \pi/2$, $F(x) = 0$. Figure 6.18 shows that the area A_1 is larger than the area A_2. Thus $F(x) > 0$ for $\frac{\pi}{2} < x \leq \frac{3\pi}{2}$. Therefore the global minimum is $F(\frac{\pi}{2}) = 0$.

Figure 6.18

73. Let v be the velocity and s be the position of the particle at time t. We know that $a = dv/dt$, so acceleration is the slope of the velocity graph. Similarly, velocity is the slope of the position graph. Graphs of v and s are shown in Figures 6.19 and 6.20, respectively.

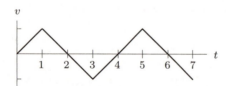

Figure 6.19: Velocity against time

Figure 6.20: Position against time

77. (a) Using $g = -32$ ft/sec^2, we have

t (sec)	0	1	2	3	4	5
$v(t)$ (ft/sec)	80	48	16	-16	-48	-80

(b) The object reaches its highest point when $v = 0$, which appears to be at $t = 2.5$ seconds. By symmetry, the object should hit the ground again at $t = 5$ seconds.

(c) Left sum $= 80(1) + 48(1) + 16(\frac{1}{2}) = 136$ ft , which is an overestimate.
Right sum $= 48(1) + 16(1) + (-16)(\frac{1}{2}) = 56$ ft , which is an underestimate.
Note that we used a smaller third rectangle of width $1/2$ to end our sum at $t = 2.5$.

(d) We have $v(t) = 80 - 32t$, so antidifferentiation yields $s(t) = 80t - 16t^2 + s_0$.
But $s_0 = 0$, so $s(t) = 80t - 16t^2$.
At $t = 2.5$, $s(t) = 100$ ft., so 100 ft. is the highest point.

CAS Challenge Problems

81. (a) We have $\Delta x = \dfrac{(b-a)}{n}$ and $x_i = a + i(\Delta x) = a + i\left(\dfrac{b-a}{n}\right)$, so, since $f(x_i) = {x_i}^3$,

$$\text{Riemann sum } = \sum_{i=1}^{n} f(x_i)\Delta x = \sum_{i=1}^{n} \left[a + i\left(\frac{b-a}{n}\right)\right]^3 \left(\frac{b-a}{n}\right).$$

(b) A CAS gives

$$\sum_{i=1}^{n} \left[a + \frac{i(b-a)}{n}\right]^3 \frac{(b-a)}{n} = -\frac{(a-b)(a^3(n-1)^2 + (a^2b + ab^2)(n^2-1) + b^3(n+1)^3)}{4n^2}.$$

Taking the limit as $n \to \infty$ gives

$$\lim_{n \to \infty} \sum_{i=1}^{n} \left[a + i\left(\frac{b-a}{n}\right)\right]^3 \left(\frac{b-a}{n}\right) = -\frac{(a+b)(a-b)(a^2+b^2)}{4}.$$

(c) The answer to part (b) simplifies to $\dfrac{b^4}{4} - \dfrac{a^4}{4}$. Since $\dfrac{d}{dx}\left(\dfrac{x^4}{4}\right) = x^3$, the Fundamental Theorem of Calculus says that

$$\int_a^b x^3\,dx = \left.\frac{x^4}{4}\right|_a^b = \frac{b^4}{4} - \frac{a^4}{4}.$$

85. (a) A CAS gives

$$\int \frac{1}{(x-1)(x-3)}\,dx = \frac{1}{2}(\ln|x-3| - \ln|x-1|)$$

$$\int \frac{1}{(x-1)(x-4)}\,dx = \frac{1}{3}(\ln|x-4| - \ln|x-1|)$$

$$\int \frac{1}{(x-1)(x+3)}\,dx = \frac{1}{4}(\ln|x+3| - \ln|x-1|).$$

Although the absolute values are needed in the answer, some CASs may not include them.

(b) The three integrals in part (a) obey the rule

$$\int \frac{1}{(x-a)(x-b)}\,dx = \frac{1}{b-a}(\ln|x-b| - \ln|x-a|).$$

(c) Checking the formula by calculating the derivative

$$\frac{d}{dx}\left(\frac{1}{b-a}(\ln|x-b| - \ln|x-a|)\right) = \frac{1}{b-a}\left(\frac{1}{x-b} - \frac{1}{x-a}\right)$$

$$= \frac{1}{b-a}\left(\frac{(x-a) - (x-b)}{(x-a)(x-b)}\right)$$

$$= \frac{1}{b-a}\left(\frac{b-a}{(x-a)(x-b)}\right) = \frac{1}{(x-a)(x-b)}.$$

CHECK YOUR UNDERSTANDING

1. True. A function can have only one derivative.

5. False. Differentiating using the product and chain rules gives

$$\frac{d}{dx}\left(\frac{-1}{2x}e^{-x^2}\right) = \frac{1}{2x^2}e^{-x^2} + e^{-x^2}.$$

9. True. If $y = F(x)$ is a solution to the differential equation $dy/dx = f(x)$, then $F'(x) = f(x)$, so $F(x)$ is an antiderivative of $f(x)$.

13. False. The solution of the initial value problem $dy/dx = 1$ with $y(0) = -5$ is a solution of the differential equation that is not positive at $x = 0$.

17. True. All solutions of the differential equation $dy/dt = 3t^2$ are in the family $y(t) = t^3 + C$ of antiderivatives of $3t^2$. The initial condition $y(1) = \pi$ tells us that $y(1) = \pi = 1^3 + C$, so $C = \pi - 1$. Thus $y(t) = t^3 + \pi - 1$ is the only solution of the initial value problem.

21. True. Suppose t is measured in seconds from when the ball was thrown. The acceleration $a = dv/dt$ is -32 ft/sec^2, so the velocity of the ball is $v = -32t + C$ feet/second at time t. At $t = 0$ the velocity is -10, so $v = -32t - 10$. Since $v = ds/dt$, an antiderivative gives the height $s = -16t^2 - 10t + K$ feet of the ball at time t. Since the ball starts at the top of the building, $s = 100$ when $t = 0$. Substituting gives $s = -16t^2 - 10t + 100$. The ball hits the ground when $s = 0$, so we solve $0 = -16t^2 - 10t + 100$. The positive solution $t = 2.2$ tells us that the ball hits the ground after 2.2 seconds.

25. True. Since F and G are both antiderivatives of f, they must differ by a constant. In fact, we can see that the constant C is equal to $\int_0^2 f(t)dt$ since

$$F(x) = \int_0^x f(t)dt = \int_2^x f(t)dt + \int_0^2 f(t)dt = G(x) + C.$$

CHAPTER SEVEN

Solutions for Section 7.1

Exercises

1. (a) We substitute $w = 1 + x^2$, $dw = 2x \, dx$.

$$\int_{x=0}^{x=1} \frac{x}{1+x^2} \, dx = \frac{1}{2} \int_{w=1}^{w=2} \frac{1}{w} \, dw = \frac{1}{2} \ln|w| \Big|_1^2 = \frac{1}{2} \ln 2.$$

(b) We substitute $w = \cos x$, $dw = -\sin x \, dx$.

$$\int_{x=0}^{x=\frac{\pi}{4}} \frac{\sin x}{\cos x} \, dx = -\int_{w=1}^{w=\sqrt{2}/2} \frac{1}{w} \, dw$$

$$= -\ln|w| \Big|_1^{\sqrt{2}/2} = -\ln \frac{\sqrt{2}}{2} = \frac{1}{2} \ln 2.$$

5. We use the substitution $w = -x$, $dw = -dx$.

$$\int e^{-x} dx = -\int e^w dw = -e^w + C = -e^{-x} + C.$$

Check: $\frac{d}{dx}(-e^{-x} + C) = -(-e^{-x}) = e^{-x}$.

9. We use the substitution $w = 3 - t$, $dw = -dt$.

$$\int \sin(3-t) dt = -\int \sin(w) dw = -(-\cos(w)) + C = \cos(3-t) + C.$$

Check: $\frac{d}{dt}(\cos(3-t) + C) = -\sin(3-t)(-1) = \sin(3-t)$.

13. We use the substitution $w = t^3 - 3$, $dw = 3t^2 \, dt$.

$$\int t^2 (t^3 - 3)^{10} \, dt = \frac{1}{3} \int (t^3 - 3)^{10} (3t^2 dt) = \int w^{10} \left(\frac{1}{3} dw\right)$$

$$= \frac{1}{3} \frac{w^{11}}{11} + C = \frac{1}{33}(t^3 - 3)^{11} + C.$$

Check: $\frac{d}{dt}[\frac{1}{33}(t^3 - 3)^{11} + C] = \frac{1}{3}(t^3 - 3)^{10}(3t^2) = t^2(t^3 - 3)^{10}$.

17. In this case, it seems easier not to substitute.

$$\int y^2 (1+y)^2 \, dy = \int y^2 (y^2 + 2y + 1) \, dy = \int (y^4 + 2y^3 + y^2) \, dy$$

$$= \frac{y^5}{5} + \frac{y^4}{2} + \frac{y^3}{3} + C.$$

Check: $\frac{d}{dy}\left(\frac{y^5}{5} + \frac{y^4}{2} + \frac{y^3}{3} + C\right) = y^4 + 2y^3 + y^2 = y^2(y+1)^2$.

21. In this case, it seems easier not to substitute.

$$\int (x^2 + 3)^2 \, dx = \int (x^4 + 6x^2 + 9) \, dx = \frac{x^5}{5} + 2x^3 + 9x + C.$$

Check: $\dfrac{d}{dx}\left[\dfrac{x^5}{5} + 2x^3 + 9x + C\right] = x^4 + 6x^2 + 9 = (x^2 + 3)^2.$

25. We use the substitution $w = \sin\theta$, $dw = \cos\theta\,d\theta$.

$$\int \sin^6\theta \cos\theta \, d\theta = \int w^6 \, dw = \frac{w^7}{7} + C = \frac{\sin^7\theta}{7} + C.$$

Check: $\dfrac{d}{d\theta}\left[\dfrac{\sin^7\theta}{7} + C\right] = \sin^6\theta \cos\theta.$

29. We use the substitution $w = \ln z$, $dw = \frac{1}{z}\,dz$.

$$\int \frac{(\ln z)^2}{z} \, dz = \int w^2 \, dw = \frac{w^3}{3} + C = \frac{(\ln z)^3}{3} + C.$$

Check: $\dfrac{d}{dz}\left[\dfrac{(\ln z)^3}{3} + C\right] = 3 \cdot \dfrac{1}{3}(\ln z)^2 \cdot \dfrac{1}{z} = \dfrac{(\ln z)^2}{z}.$

33. We use the substitution $w = \sqrt{y}$, $dw = \dfrac{1}{2\sqrt{y}}\,dy$.

$$\int \frac{e^{\sqrt{y}}}{\sqrt{y}} \, dy = 2 \int e^w \, dw = 2e^w + C = 2e^{\sqrt{y}} + C.$$

Check: $\dfrac{d}{dy}(2e^{\sqrt{y}} + C) = 2e^{\sqrt{y}} \cdot \dfrac{1}{2\sqrt{y}} = \dfrac{e^{\sqrt{y}}}{\sqrt{y}}.$

37. We use the substitution $w = 1 + 3t^2$, $dw = 6t\,dt$.

$$\int \frac{t}{1 + 3t^2} \, dt = \int \frac{1}{w}\left(\frac{1}{6}\,dw\right) = \frac{1}{6}\ln|w| + C = \frac{1}{6}\ln(1 + 3t^2) + C.$$

(We can drop the absolute value signs since $1 + 3t^2 > 0$ for all t).

Check: $\dfrac{d}{dt}\left[\dfrac{1}{6}\ln(1 + 3t^2) + C\right] = \dfrac{1}{6}\dfrac{1}{1 + 3t^2}(6t) = \dfrac{t}{1 + 3t^2}.$

41. Since $d(\sinh x)/dx = \cosh x$, we have

$$\int \cosh x \, dx = \sinh x + C.$$

45. We use the substitution $w = x^2$ and $dw = 2x\,dx$ so

$$\int x\cosh x^2 \, dx = \frac{1}{2}\int \cosh w \, dw = \frac{1}{2}\sinh w + C = \frac{1}{2}\sinh x^2 + C.$$

Check this answer by taking the derivative: $\dfrac{d}{dx}\left[\dfrac{1}{2}\sinh x^2 + C\right] = x\cosh x^2.$

49. Make the substitution $w = x^2$, $dw = 2x\,dx$. We have

$$\int 2x\cos(x^2) \, dx = \int \cos w \, dw = \sin w + C = \sin x^2 + C.$$

53. Make the substitution $w = x^2 + 1$, $dw = 2x\,dx$. We have

$$\int \frac{x}{x^2 + 1} \, dx = \frac{1}{2}\int \frac{dw}{w} = \frac{1}{2}\ln|w| + C = \frac{1}{2}\ln(x^2 + 1) + C.$$

(Notice that since $x^2 + 1 \geq 0$, $|x^2 + 1| = x^2 + 1$.)

57. $\displaystyle\int_0^{\pi/2} e^{-\cos\theta}\sin\theta \, d\theta = e^{-\cos\theta}\bigg|_0^{\pi/2} = e^{-\cos(\pi/2)} - e^{-\cos(0)} = 1 - \frac{1}{e}$

61. We substitute $w = \sqrt{x}$. Then $dw = \dfrac{1}{2}x^{-1/2}dx$.

$$\int_{x=1}^{x=4} \frac{\cos\sqrt{x}}{\sqrt{x}}\,dx = \int_{w=1}^{w=2} \cos w\,(2\,dw)$$

$$= 2(\sin w)\Big|_1^2 = 2(\sin 2 - \sin 1).$$

65. $\displaystyle\int_1^3 \frac{1}{x}\,dx = \ln x\,\Big|_1^3 = \ln 3.$

69. Let $w = \sqrt{y+1}$, so $y = w^2 - 1$ and $dy = 2w\,dw$. Thus

$$\int y\sqrt{y+1}\,dy = \int (w^2 - 1)w2w\,dw = 2\int w^4 - w^2\,dw$$

$$= \frac{2}{5}w^5 - \frac{2}{3}w^3 + C = \frac{2}{5}(y+1)^{5/2} - \frac{2}{3}(y+1)^{3/2} + C.$$

73. Let $w = \sqrt{x-2}$, so $x = w^2 + 2$ and $dx = 2w\,dw$. Thus

$$\int x^2\sqrt{x-2}\,dx = \int (w^2+2)^2 w2w\,dw = 2\int w^6 + 4w^4 + 4w^2\,dw$$

$$= \frac{2}{7}w^7 + \frac{8}{5}w^5 + \frac{8}{3}w^3 + C$$

$$= \frac{2}{7}(x-2)^{7/2} + \frac{8}{5}(x-2)^{5/2} + \frac{8}{3}(x-2)^{3/2} + C.$$

Problems

77. If we let $y = 3x$ in the first integral, we get $dy = 3dx$. Also, the limits $x = 0$ and $x = \pi/3$ become $y = 0$ and $y = 3(\pi/3) = \pi$. Thus

$$\int_0^{\pi/3} 3\sin^2(3x)\,dx = \int_0^{\pi/3} \sin^2(3x)\,3dx = \int_0^\pi \sin^2(y)\,dy.$$

81. As x goes from $\sqrt{a}$ to $\sqrt{b}$ the values of $w = x^2$ increase from a to b. Since $x = \sqrt{w}$ we have $dx = dw/(2\sqrt{w})$. Hence

$$\int_{\sqrt{a}}^{\sqrt{b}} dx = \int_a^b \frac{1}{2\sqrt{w}}dw = \int_a^b g(w)dw$$

with

$$g(w) = \frac{1}{2\sqrt{w}}.$$

85. For the first integral, let $w = \sin x$, $dw = \cos x\,dx$. Then

$$\int e^{\sin x}\cos x\,dx = \int e^w\,dw.$$

For the second integral, let $w = \arcsin x$, $dw = \dfrac{1}{\sqrt{1-x^2}}dx$. Then

$$\int \frac{e^{\arcsin x}}{\sqrt{1-x^2}}\,dx = \int e^w\,dw.$$

89. (a) This integral can be evaluated using integration by substitution. We use $w = x^2$, $dw = 2xdx$.

$$\int x\sin x^2\,dx = \frac{1}{2}\int \sin(w)dw = -\frac{1}{2}\cos(w) + C = -\frac{1}{2}\cos(x^2) + C.$$

(b) This integral cannot be evaluated using a simple integration by substitution.

(c) This integral cannot be evaluated using a simple integration by substitution.

(d) This integral can be evaluated using integration by substitution. We use $w = 1 + x^2$, $dw = 2x\,dx$.

$$\int \frac{x}{(1+x^2)^2}\,dx = \frac{1}{2}\int \frac{1}{w^2}\,dw = \frac{1}{2}\left(\frac{-1}{w}\right) + C = \frac{-1}{2(1+x^2)} + C.$$

(e) This integral cannot be evaluated using a simple integration by substitution.

(f) This integral can be evaluated using integration by substitution. We use $w = 2 + \cos x$, $dw = -\sin x\,dx$.

$$\int \frac{\sin x}{2 + \cos x}\,dx = -\int \frac{1}{w}\,dw = -\ln|w| + C = -\ln|2 + \cos x| + C.$$

93. Since $(e^{\theta+1})^3 = e^{3\theta+3} = e^{3\theta} \cdot e^3$, we have

$$\text{Area} = \int_0^2 (e^{\theta+1})^3\,d\theta = \int_0^2 e^{3\theta} \cdot e^3\,d\theta$$

$$= e^3 \int_0^2 e^{3\theta}\,d\theta = e^3 \frac{1}{3} e^{3\theta} \bigg|_0^2 = \frac{e^3}{3}(e^6 - 1).$$

97. If $f(x) = \dfrac{1}{x+1}$, the average value of f on the interval $0 \le x \le 2$ is defined to be

$$\frac{1}{2-0}\int_0^2 f(x)\,dx = \frac{1}{2}\int_0^2 \frac{dx}{x+1}.$$

We'll integrate by substitution. We let $w = x + 1$ and $dw = dx$, and we have

$$\int_{x=0}^{x=2} \frac{dx}{x+1} = \int_{w=1}^{w=3} \frac{dw}{w} = \ln w \bigg|_1^3 = \ln 3 - \ln 1 = \ln 3.$$

Thus, the average value of $f(x)$ on $0 \le x \le 2$ is $\frac{1}{2}\ln 3 \approx 0.5493$. See Figure 7.1.

$$f(x) = \tfrac{1}{1+x}$$

0.54931

2

x

Figure 7.1

101. (a) The Fundamental Theorem gives

$$\int_{-\pi}^{\pi} \cos^2\theta \sin\theta\,d\theta = -\frac{\cos^3\theta}{3}\bigg|_{-\pi}^{\pi} = \frac{-(-1)^3}{3} - \frac{-(-1)^3}{3} = 0.$$

This agrees with the fact that the function $f(\theta) = \cos^2\theta \sin\theta$ is odd and the interval of integration is centered at $x = 0$, thus we must get 0 for the definite integral.

(b) The area is given by

$$\text{Area} = \int_0^{\pi} \cos^2\theta \sin\theta\,d\theta = -\frac{\cos^3\theta}{3}\bigg|_0^{\pi} = \frac{-(-1)^3}{3} - \frac{-(1)^3}{3} = \frac{2}{3}.$$

105. We substitute $w = 1 - x$ into $I_{m,n}$. Then $dw = -dx$, and $x = 1 - w$.

When $x = 0$, $w = 1$, and when $x = 1$, $w = 0$, so

$$I_{m,n} = \int_0^1 x^m (1-x)^n dx = \int_1^0 (1-w)^m w^n (-dw)$$

$$= -\int_1^0 w^n (1-w)^m dw = \int_0^1 w^n (1-w)^m dw = I_{n,m}.$$

109. (a) At time $t = 0$, the rate of oil leakage $= r(0) = 50$ thousand liters/minute.

At $t = 60$, rate $= r(60) = 15.06$ thousand liters/minute.

(b) To find the amount of oil leaked during the first hour, we integrate the rate from $t = 0$ to $t = 60$:

$$\text{Oil leaked} = \int_0^{60} 50 e^{-0.02t}\, dt = \left(-\frac{50}{0.02} e^{-0.02t} \right) \Big|_0^{60}$$

$$= -2500 e^{-1.2} + 2500 e^0 = 1747 \text{ thousand liters.}$$

113. Since v is given as the velocity of a falling body, the height h is decreasing, so $v = -\frac{dh}{dt}$, and it follows that $h(t) = -\int v(t)\, dt$ and $h(0) = h_0$. Let $w = e^{t\sqrt{gk}} + e^{-t\sqrt{gk}}$. Then

$$dw = \sqrt{gk} \left(e^{t\sqrt{gk}} - e^{-t\sqrt{gk}} \right) dt,$$

so $\dfrac{dw}{\sqrt{gk}} = (e^{t\sqrt{gk}} - e^{-t\sqrt{gk}})\, dt$. Therefore,

$$-\int v(t)dt = -\int \sqrt{\frac{g}{k}} \left(\frac{e^{t\sqrt{gk}} - e^{-t\sqrt{gk}}}{e^{t\sqrt{gk}} + e^{-t\sqrt{gk}}} \right) dt$$

$$= -\sqrt{\frac{g}{k}} \int \frac{1}{e^{t\sqrt{gk}} + e^{-t\sqrt{gk}}} \left(e^{t\sqrt{gk}} - e^{-t\sqrt{gk}} \right) dt$$

$$= -\sqrt{\frac{g}{k}} \int \left(\frac{1}{w} \right) \frac{dw}{\sqrt{gk}}$$

$$= -\sqrt{\frac{g}{gk^2}} \ln|w| + C$$

$$= -\frac{1}{k} \ln \left(e^{t\sqrt{gk}} + e^{-t\sqrt{gk}} \right) + C.$$

Since

$$h(0) = -\frac{1}{k} \ln(e^0 + e^0) + C = -\frac{\ln 2}{k} + C = h_0,$$

we have $C = h_0 + \dfrac{\ln 2}{k}$. Thus,

$$h(t) = -\frac{1}{k} \ln \left(e^{t\sqrt{gk}} + e^{-t\sqrt{gk}} \right) + \frac{\ln 2}{k} + h_0 = -\frac{1}{k} \ln \left(\frac{e^{t\sqrt{gk}} + e^{-t\sqrt{gk}}}{2} \right) + h_0.$$

Solutions for Section 7.2

Exercises

1. (a) Since we can change x^2 into a multiple of x^3 by integrating, let $v' = x^2$ and $u = e^x$. Using $v = x^3/3$ and $u' = e^x$ we get

$$\int x^2 e^x dx = \int uv' dx = uv - \int u'v dx$$

$$= \frac{x^3 e^x}{3} - \frac{1}{3} \int x^3 e^x dx.$$

(b) Since we can change x^2 into a multiple of x by differentiating, let $u = x^2$ and $v' = e^x$. Using $u' = 2x$ and $v = e^x$ we have

$$\int x^2 e^x dx = \int uv' dx = uv - \int u'v dx$$

$$= x^2 e^x - 2 \int x e^x dx.$$

5. Let $u = t$ and $v' = e^{5t}$, so $u' = 1$ and $v = \frac{1}{5} e^{5t}$.
Then $\int t e^{5t} dt = \frac{1}{5} t e^{5t} - \int \frac{1}{5} e^{5t} dt = \frac{1}{5} t e^{5t} - \frac{1}{25} e^{5t} + C.$

9. Let $u = \ln y$, $v' = y$. Then, $v = \frac{1}{2} y^2$ and $u' = \dfrac{1}{y}$. Integrating by parts, we get:

$$\int y \ln y \, dy = \frac{1}{2} y^2 \ln y - \int \frac{1}{2} y^2 \cdot \frac{1}{y} \, dy$$

$$= \frac{1}{2} y^2 \ln y - \frac{1}{2} \int y \, dy$$

$$= \frac{1}{2} y^2 \ln y - \frac{1}{4} y^2 + C.$$

13. Let $u = \sin \theta$ and $v' = \sin \theta$, so $u' = \cos \theta$ and $v = -\cos \theta$. Then

$$\int \sin^2 \theta \, d\theta = -\sin \theta \cos \theta + \int \cos^2 \theta \, d\theta$$

$$= -\sin \theta \cos \theta + \int (1 - \sin^2 \theta) \, d\theta$$

$$= -\sin \theta \cos \theta + \int 1 \, d\theta - \int \sin^2 \theta \, d\theta.$$

By adding $\int \sin^2 \theta \, d\theta$ to both sides of the above equation, we find that $2 \int \sin^2 \theta \, d\theta = -\sin \theta \cos \theta + \theta + C$, so $\int \sin^2 \theta \, d\theta = -\frac{1}{2} \sin \theta \cos \theta + \frac{\theta}{2} + C'.$

17. Let $u = t + 2$ and $v' = \sqrt{2 + 3t}$, so $u' = 1$ and $v = \frac{2}{9}(2 + 3t)^{3/2}$. Then

$$\int (t + 2)\sqrt{2 + 3t} \, dt = \frac{2}{9}(t + 2)(2 + 3t)^{3/2} - \frac{2}{9} \int (2 + 3t)^{3/2} \, dt$$

$$= \frac{2}{9}(t + 2)(2 + 3t)^{3/2} - \frac{4}{135}(2 + 3t)^{5/2} + C.$$

21. Let $u = y$ and $v' = \frac{1}{\sqrt{5-y}}$, so $u' = 1$ and $v = -2(5 - y)^{1/2}$.

$$\int \frac{y}{\sqrt{5 - y}} \, dy = -2y(5 - y)^{1/2} + 2 \int (5 - y)^{1/2} \, dy = -2y(5 - y)^{1/2} - \frac{4}{3}(5 - y)^{3/2} + C.$$

25. Let $u = \arctan 7z$ and $v' = 1$, so $u' = \frac{7}{1+49z^2}$ and $v = z$. Now $\int \frac{7z\,dz}{1+49z^2}$ can be evaluated by the substitution $w = 1 + 49z^2$, $dw = 98z\,dz$, so

$$\int \frac{7z\,dz}{1+49z^2} = 7 \int \frac{\frac{1}{98}\,dw}{w} = \frac{1}{14} \int \frac{dw}{w} = \frac{1}{14} \ln|w| + C = \frac{1}{14} \ln(1 + 49z^2) + C$$

So

$$\int \arctan 7z\,dz = z\arctan 7z - \frac{1}{14} \ln(1 + 49z^2) + C.$$

29. Let $u = x, u' = 1$ and $v' = \sinh x, v = \cosh x$. Integrating by parts, we get

$$\int x\sinh x\,dx = x\cosh x - \int \cosh x\,dx$$
$$= x\cosh x - \sinh x + C.$$

33. We use integration by parts. Let $u = z$ and $v' = e^{-z}$, so $u' = 1$ and $v = -e^{-z}$. Then

$$\int_0^{10} ze^{-z}\,dz = -ze^{-z}\Big|_0^{10} + \int_0^{10} e^{-z}\,dz$$
$$= -10e^{-10} + (-e^{-z})\Big|_0^{10}$$
$$= -11e^{-10} + 1$$
$$\approx 0.9995.$$

37. We use integration by parts. Let $u = \arcsin z$ and $v' = 1$, so $u' = \dfrac{1}{\sqrt{1-z^2}}$ and $v = z$. Then

$$\int_0^1 \arcsin z\,dz = z\arcsin z\Big|_0^1 - \int_0^1 \frac{z}{\sqrt{1-z^2}}\,dz = \frac{\pi}{2} - \int_0^1 \frac{z}{\sqrt{1-z^2}}\,dz.$$

To find $\displaystyle\int_0^1 \frac{z}{\sqrt{1-z^2}}\,dz$, we substitute $w = 1 - z^2$, so $dw = -2z\,dz$.
Then

$$\int_{z=0}^{z=1} \frac{z}{\sqrt{1-z^2}}\,dz = -\frac{1}{2}\int_{w=1}^{w=0} w^{-\frac{1}{2}}\,dw = \frac{1}{2}\int_{w=0}^{w=1} w^{-\frac{1}{2}}\,dw = w^{\frac{1}{2}}\Big|_0^1 = 1.$$

Thus our final answer is $\frac{\pi}{2} - 1 \approx 0.571$.

Problems

41. Using integration by parts with $u' = e^{-t}$, $v = t$, so $u = -e^{-t}$ and $v' = 1$, we have

$$\text{Area} = \int_0^2 te^{-t}\,dt = -te^{-t}\Big|_0^2 - \int_0^2 -1\cdot e^{-t}\,dt$$
$$= (-te^{-t} - e^{-t})\Big|_0^2 = -2e^{-2} - e^{-2} + 1 = 1 - 3e^{-2}.$$

45. Since the graph of $f(t) = \ln(t^2 - 1)$ is above the graph of $g(t) = \ln(t - 1)$ for $t > 1$, we have

$$\text{Area} = \int_2^3 (\ln(t^2 - 1) - \ln(t - 1)) \, dt = \int_2^3 \ln\left(\frac{t^2 - 1}{t - 1}\right) dt = \int_2^3 \ln(t + 1) \, dt.$$

We can cancel the factor of $(t - 1)$ in the last step above because the integral is over $2 \le t \le 3$, where $(t - 1)$ is not zero.

We use $\int \ln x \, dx = x \ln x - 1$ with the substitution $x = t + 1$. The limits $t = 2$, $t = 3$ become $x = 3$, $x = 4$, respectively. Thus

$$\text{Area} = \int_2^3 \ln(t + 1) \, dt = \int_3^4 \ln x \, dx = (x \ln x - x) \Big|_3^4$$
$$= 4 \ln 4 - 4 - (3 \ln 3 - 3) = 4 \ln 4 - 3 \ln 3 - 1.$$

49. First, let $u = e^x$ and $v' = \sin x$, so $u' = e^x$ and $v = -\cos x$.

Thus $\int e^x \sin x \, dx = -e^x \cos x + \int e^x \cos x \, dx$. To calculate $\int e^x \cos x \, dx$, we again need to use integration by parts. Let $u = e^x$ and $v' = \cos x$, so $u' = e^x$ and $v = \sin x$.

Thus

$$\int e^x \cos x \, dx = e^x \sin x - \int e^x \sin x \, dx.$$

This gives

$$\int e^x \sin x \, dx = e^x \sin x - e^x \cos x - \int e^x \sin x \, dx.$$

By adding $\int e^x \sin x \, dx$ to both sides, we obtain

$$2 \int e^x \sin x \, dx = e^x (\sin x - \cos x) + C.$$

$$\text{Thus} \quad \int e^x \sin x \, dx = \frac{1}{2} e^x (\sin x - \cos x) + C.$$

This problem could also be done in other ways; for example, we could have started with $u = \sin x$ and $v' = e^x$ as well.

53. We integrate by parts. Since we know what the answer is supposed to be, it's easier to choose u and v'. Let $u = x^n$ and $v' = e^x$, so $u' = nx^{n-1}$ and $v = e^x$. Then

$$\int x^n e^x \, dx = x^n e^x - n \int x^{n-1} e^x \, dx.$$

57. (a) One way to avoid integrating by parts is to take the derivative of the right hand side instead. Since $\int e^{ax} \sin bx \, dx$ is the antiderivative of $e^{ax} \sin bx$,

$$e^{ax} \sin bx = \frac{d}{dx} [e^{ax} (A \sin bx + B \cos bx) + C]$$
$$= a e^{ax} (A \sin bx + B \cos bx) + e^{ax} (Ab \cos bx - Bb \sin bx)$$
$$= e^{ax} [(aA - bB) \sin bx + (aB + bA) \cos bx].$$

Thus $aA - bB = 1$ and $aB + bA = 0$. Solving for A and B in terms of a and b, we get

$$A = \frac{a}{a^2 + b^2}, \quad B = -\frac{b}{a^2 + b^2}.$$

Thus

$$\int e^{ax} \sin bx = e^{ax} \left(\frac{a}{a^2 + b^2} \sin bx - \frac{b}{a^2 + b^2} \cos bx\right) + C.$$

(b) If we go through the same process, we find

$$ae^{ax}[(aA - bB)\sin bx + (aB + bA)\cos bx] = e^{ax}\cos bx.$$

Thus $aA - bB = 0$, and $aB + bA = 1$. In this case, solving for A and B yields

$$A = \frac{b}{a^2 + b^2}, \ B = \frac{a}{a^2 + b^2}.$$

Thus $\int e^{ax}\cos bx = e^{ax}(\frac{b}{a^2+b^2}\sin bx + \frac{a}{a^2+b^2}\cos bx) + C.$

61. We have

$$\text{Bioavailability} = \int_0^3 15te^{-0.2t}\,dt.$$

We first use integration by parts to evaluate the indefinite integral of this function. Let $u = 15t$ and $v' = e^{-0.2t}\,dt$, so $u' = 15dt$ and $v = -5e^{-0.2t}$. Then,

$$\int 15te^{-0.2t}\,dt = (15t)(-5e^{-0.2t}) - \int (-5e^{-0.2t})(15dt)$$

$$= -75te^{-0.2t} + 75\int e^{-0.2t}\,dt = -75te^{-0.2} - 375e^{-0.2t} + C.$$

Thus,

$$\int_0^3 15te^{-0.2t}\,dt = \left(-75te^{-0.2t} - 375e^{-0.2t}\right)\Big|_0^3 = -329.29 + 375 = 45.71.$$

The bioavailability of the drug over this time interval is 45.71 (ng/ml)-hours.

65. (a) We want to compute C_1, with $C_1 > 0$, such that

$$\int_0^1 (\Psi_1(x))^2 \, dx = \int_0^1 (C_1\sin(\pi x))^2 \, dx = C_1^2 \int_0^1 \sin^2(\pi x)\,dx = 1.$$

We use integration by parts with $u = v' = \sin(\pi x)$.
So $u' = \pi\cos(\pi x)$ and $v = -\frac{1}{\pi}\cos(\pi x)$. Thus

$$\int_0^1 \sin^2(\pi x)\,dx = -\frac{1}{\pi}\sin(\pi x)\cos(\pi x)\Big|_0^1 + \int_0^1 \cos^2(\pi x)\,dx$$

$$= -\frac{1}{\pi}\sin(\pi x)\cos(\pi x)\Big|_0^1 + \int_0^1 (1 - \sin^2(\pi x))\,dx.$$

Moving $\int_0^1 \sin^2(\pi x)\,dx$ from the right side to the left side of the equation and solving, we get

$$2\int_0^1 \sin^2(\pi x)\,dx = -\frac{1}{\pi}\sin(\pi x)\cos(\pi x)\Big|_0^1 + \int_0^1 1\,dx = 0 + 1 = 1,$$

so

$$\int_0^1 \sin^2(\pi x)\,dx = \frac{1}{2}.$$

Thus, we have

$$\int_0^1 (\Psi_1(x))^2 \, dx = C_1^2 \int_0^1 \sin^2(\pi x)\,dx = \frac{C_1^2}{2}.$$

So, to normalize Ψ_1, we take $C_1 > 0$ such that

$$\frac{C_1^2}{2} = 1 \quad \text{so} \quad C_1 = \sqrt{2}.$$

(b) To normalize Ψ_n, we want to compute C_n, with $C_n > 0$, such that

$$\int_0^1 (\Psi_n(x))^2 \, dx = C_n^2 \int_0^1 \sin^2(n\pi x)\,dx = 1.$$

The solution to part (a) shows us that

$$\int \sin^2(\pi t)\, dt = -\frac{1}{2\pi} \sin(\pi t) \cos(\pi t) + \frac{1}{2} \int 1\, dt.$$

In the integral for Ψ_n, we make the substitution $t = nx$, so $dx = \frac{1}{n} dt$. Since $t = 0$ when $x = 0$ and $t = n$ when $x = 1$, we have

$$\int_0^1 \sin^2(n\pi x)\, dx = \frac{1}{n} \int_0^n \sin^2(\pi t)\, dt$$

$$= \frac{1}{n} \left(-\frac{1}{2\pi} \sin(\pi t)\cos(\pi t) \Big|_0^n + \frac{1}{2} \int_0^n 1\, dt \right)$$

$$= \frac{1}{n} \left(0 + \frac{n}{2} \right) = \frac{1}{2}.$$

Thus, we have

$$\int_0^1 (\Psi_n(x))^2\, dx = C_n^2 \int_0^1 \sin^2(n\pi x)\, dx = \frac{C_n^2}{2}.$$

So to normalize Ψ_n, we take C_n such that

$$\frac{C_n^2}{2} = 1 \quad \text{so} \quad C_n = \sqrt{2}.$$

Solutions for Section 7.3

Exercises

1. $\dfrac{1}{10} e^{(-3\theta)} (-3\cos\theta + \sin\theta) + C.$
(Let $a = -3, b = 1$ in II-9.)

5. Note that you can't use substitution here: letting $w = x^3 + 5$ does not work, since there is no $dw = 3x^2\, dx$ in the integrand. What will work is simply multiplying out the square: $(x^3 + 5)^2 = x^6 + 10x^3 + 25$. Then use I-1:

$$\int (x^3 + 5)^2\, dx = \int x^6\, dx + 10 \int x^3\, dx + 25 \int 1\, dx = \frac{1}{7}x^7 + 10 \cdot \frac{1}{4}x^4 + 25x + C.$$

9. $\dfrac{1}{\sqrt{3}} \arctan \dfrac{y}{\sqrt{3}} + C.$
(Let $a = \sqrt{3}$ in V-24).

13. $\dfrac{5}{16} \sin 3\theta \sin 5\theta + \dfrac{3}{16} \cos 3\theta \cos 5\theta + C.$
(Let $a = 3, b = 5$ in II-12.)

17. $\left(\dfrac{1}{3}x^4 - \dfrac{4}{9}x^3 + \dfrac{4}{9}x^2 - \dfrac{8}{27}x + \dfrac{8}{81} \right) e^{3x} + C.$
(Let $a = 3, p(x) = x^4$ in III-14.)

21. Substitute $w = x^2, dw = 2x\, dx$. Then $\int x^3 \sin x^2\, dx = \dfrac{1}{2} \int w \sin w\, dw$. By III-15, we have

$$\int w \sin w\, dw = -\frac{1}{2} w \cos w + \frac{1}{2} \sin w + C = -\frac{1}{2}x^2 \cos x^2 + \frac{1}{2} \sin x^2 + C.$$

25. Use IV-21 twice to get the exponent down to 1:

$$\int \frac{1}{\cos^5 x}\,dx = \frac{1}{4}\frac{\sin x}{\cos^4 x} + \frac{3}{4}\int \frac{1}{\cos^3 x}\,dx$$

$$\int \frac{1}{\cos^3 x}\,dx = \frac{1}{2}\frac{\sin x}{\cos^2 x} + \frac{1}{2}\int \frac{1}{\cos x}\,dx.$$

Now use IV-22 to get

$$\int \frac{1}{\cos x}\,dx = \frac{1}{2}\ln\left|\frac{(\sin x)+1}{(\sin x)-1}\right| + C.$$

Putting this all together gives

$$\int \frac{1}{\cos^5 x}\,dx = \frac{1}{4}\frac{\sin x}{\cos^4 x} + \frac{3}{8}\frac{\sin x}{\cos^2 x} + \frac{3}{16}\ln\left|\frac{(\sin x)+1}{(\sin x)-1}\right| + C.$$

29.

$$\int \frac{1}{x^2+4x+3}\,dx = \int \frac{1}{(x+1)(x+3)}\,dx = \frac{1}{2}(\ln|x+1| - \ln|x+3|) + C.$$

(Let $a = -1$ and $b = -3$ in V-26).

33. $\arctan(z+2) + C.$
(Substitute $w = z + 2$ and use V-24, letting $a = 1$.)

37.

$$\int \sin^3 3\theta \cos^2 3\theta\,d\theta = \int (\sin 3\theta)(\cos^2 3\theta)(1 - \cos^2 3\theta)\,d\theta$$

$$= \int \sin 3\theta(\cos^2 3\theta - \cos^4 3\theta)\,d\theta.$$

Using an extension of the tip given in rule IV-23, we let $w = \cos 3\theta$, $dw = -3\sin 3\theta\,d\theta$.

$$\int \sin 3\theta(\cos^2 3\theta - \cos^4 3\theta)\,d\theta = -\frac{1}{3}\int (w^2 - w^4)\,dw$$

$$= -\frac{1}{3}\left(\frac{w^3}{3} - \frac{w^5}{5}\right) + C$$

$$= -\frac{1}{9}(\cos^3 3\theta) + \frac{1}{15}(\cos^5 3\theta) + C.$$

41. Let $a = \sqrt{3}$ in VI-30 and VI-28:

$$\int_0^1 \sqrt{3-x^2}\,dx = \left(\frac{1}{2}x\sqrt{3-x^2} + \frac{3}{2}\arcsin\frac{x}{\sqrt{3}}\right)\Big|_0^1 \approx 1.630.$$

45. $\displaystyle\int_0^1 \frac{1}{x^2+2x+1}\,dx = \int_0^1 \frac{1}{(x+1)^2}\,dx.$
We substitute $w = x + 1$, so $dw = dx$. Note that when $x = 1$, we have $w = 2$, and when $x = 0$, we have $w = 1$.

$$\int_{x=0}^{x=1} \frac{1}{(x+1)^2}\,dx = \int_{w=1}^{w=2} \frac{1}{w^2}\,dw = -\frac{1}{w}\Big|_{w=1}^{w=2} = -\frac{1}{2} + 1 = \frac{1}{2}.$$

49. Let $w = x^2$, $dw = 2x\,dx$. When $x = 0$, $w = 0$, and when $x = \frac{1}{\sqrt{2}}$, $w = \frac{1}{2}$. Then

$$\int_0^{\frac{1}{\sqrt{2}}} \frac{x\,dx}{\sqrt{1-x^4}} = \int_0^{\frac{1}{2}} \frac{\frac{1}{2}\,dw}{\sqrt{1-w^2}} = \frac{1}{2}\arcsin w\Big|_0^{\frac{1}{2}} = \frac{1}{2}\left(\arcsin\frac{1}{2} - \arcsin 0\right) = \frac{\pi}{12}.$$

Problems

53. (a)

$$\frac{1}{1-0} \int_0^1 V_0 \cos(120\pi t)dt = \frac{V_0}{120\pi} \sin(120\pi t)\bigg|_0^1$$

$$= \frac{V_0}{120\pi} [\sin(120\pi) - \sin(0)]$$

$$= \frac{V_0}{120\pi} [0 - 0] = 0.$$

(b) Let's find the average of V^2 first.

$$\overline{V^2} = \text{Average of } V^2 = \frac{1}{1-0} \int_0^1 V^2 dt$$

$$= \frac{1}{1-0} \int_0^1 (V_0 \cos(120\pi t))^2 dt$$

$$= V_0^2 \int_0^1 \cos^2(120\pi t)dt$$

Now, let $120\pi t = x$, and $dt = \dfrac{dx}{120\pi}$. So

$$\overline{V^2} = \frac{V_0^2}{120\pi} \int_0^{120\pi} \cos^2 x\, dx.$$

$$= \frac{V_0^2}{120\pi} \left(\frac{1}{2} \cos x \sin x + \frac{1}{2}x\right)\bigg|_0^{120\pi} \quad \text{II-18}$$

$$= \frac{V_0^2}{120\pi} 60\pi = \frac{V_0^2}{2}.$$

So, the average of V^2 is $\dfrac{V_0^2}{2}$ and $\overline{V} = \sqrt{\text{average of } V^2} = \dfrac{V_0}{\sqrt{2}}$.

(c) $V_0 = \sqrt{2} \cdot \overline{V} = 110\sqrt{2} \approx 156$ volts.

Solutions for Section 7.4

Exercises

1. Since $25 - x^2 = (5 - x)(5 + x)$, we take

$$\frac{20}{25 - x^2} = \frac{A}{5 - x} + \frac{B}{5 + x}.$$

So,

$$20 = A(5 + x) + B(5 - x)$$
$$20 = (A - B)x + 5A + 5B,$$

giving

$$A - B = 0$$
$$5A + 5B = 20.$$

Thus $A = B = 2$ and

$$\frac{20}{25 - x^2} = \frac{2}{5 - x} + \frac{2}{5 + x}.$$

5. Since $s^4 - 1 = (s^2 - 1)(s^2 + 1) = (s - 1)(s + 1)(s^2 + 1)$, we have

$$\frac{2}{s^4 - 1} = \frac{A}{s - 1} + \frac{B}{s + 1} + \frac{Cs + D}{s^2 + 1}.$$

Thus,

$$2 = A(s + 1)(s^2 + 1) + B(s - 1)(s^2 + 1) + (Cs + D)(s - 1)(s + 1)$$
$$2 = (A + B + C)s^3 + (A - B + D)s^2 + (A + B - C)s + (A - B - D),$$

giving

$$A + B + C = 0$$
$$A - B + D = 0$$
$$A + B - C = 0$$
$$A - B - D = 2.$$

From the first and third equations we find $A + B = 0$ and $C = 0$. From the second and fourth we find $A - B = 1$ and $D = -1$. Thus $A = 1/2$ and $B = -1/2$ and

$$\frac{2}{s^4 - 1} = \frac{1}{2(s - 1)} - \frac{1}{2(s + 1)} - \frac{1}{s^2 + 1}.$$

9. Using the result of Problem 2, we have

$$\int \frac{x + 1}{6x + x^2} \, dx = \int \frac{1/6}{x} \, dx + \int \frac{5/6}{6 + x} \, dx = \frac{1}{6} \left(\ln |x| + 5 \ln |6 + x| \right) + C.$$

13. Using the result of Problem 6, we have

$$\int \frac{2y}{y^3 - y^2 + y - 1} \, dy = \int \frac{1}{y - 1} \, dy + \int \frac{1 - y}{y^2 + 1} \, dy = \ln |y - 1| + \arctan y - \frac{1}{2} \ln \left| y^2 + 1 \right| + C.$$

17. We let

$$\frac{10x + 2}{x^3 - 5x^2 + x - 5} = \frac{10x + 2}{(x - 5)(x^2 + 1)} = \frac{A}{x - 5} + \frac{Bx + C}{x^2 + 1}$$

giving

$$10x + 2 = A(x^2 + 1) + (Bx + C)(x - 5)$$
$$10x + 2 = (A + B)x^2 + (C - 5B)x + A - 5C$$

so

$$A + B = 0$$
$$C - 5B = 10$$
$$A - 5C = 2.$$

Thus, $A = 2$, $B = -2$, $C = 0$, so

$$\int \frac{10x + 2}{x^3 - 5x^2 + x - 5} \, dx = \int \frac{2}{x - 5} \, dx - \int \frac{2x}{x^2 + 1} \, dx = 2 \ln |x - 5| - \ln \left| x^2 + 1 \right| + K.$$

21. Completing the square gives $x^2 + 4x + 5 = 1 + (x + 2)^2$. Since $x + 2 = \tan t$ and $dx = (1/\cos^2 t)dt$, we have

$$\int \frac{1}{x^2 + 4x + 5} \, dx = \int \frac{1}{1 + \tan^2 t} \cdot \frac{1}{\cos^2 t} dt = \int dt = t + C = \arctan(x + 2) + C.$$

Problems

25. (a) We have

$$\frac{3x+6}{x^2+3x} = \frac{2(x+3)}{x(x+3)} + \frac{x}{x(x+3)} = \frac{2}{x} + \frac{1}{x+3}.$$

Thus

$$\int \frac{3x+6}{x^2+3x}\,dx = \int \left(\frac{2}{x} + \frac{1}{x+3}\right)\,dx = 2\ln|x| + \ln|x+3| + C.$$

(b) Let $a = 0, b = -3, c = 3$, and $d = 6$ in V-27.

$$\int \frac{3x+6}{x^2+3x}\,dx = \int \frac{3x+6}{x(x+3)}\,dx$$

$$= \frac{1}{3}(6\ln|x| + 3\ln|x+3|) + C = 2\ln|x| + \ln|x+3| + C.$$

29. Since $y^2 + 3y + 3 = (y+3/2)^2 + (3 - 9/4) = (y+3/2)^2 + 3/4$, we have

$$\int \frac{dy}{y^2 + 3y + 3} = \int \frac{dy}{(y+3/2)^2 + 3/4}.$$

Substitute $y + 3/2 = \tan\theta$, so $y = (\tan\theta) - 3/2$.

33. Since $t^2 + 4t + 7 = (t+2)^2 + 3$, we have

$$\int (t+2)\sin(t^2 + 4t + 7)\,dt = \int (t+2)\sin((t+2)^2 + 3)\,dt.$$

Substitute $w = (t+2)^2 + 3$, so $dw = 2(t+2)\,dt$.

This integral can also be computed without completing the square, by substituting $w = t^2 + 4t + 7$, so $dw = (2t+4)\,dt$.

37. We write

$$\frac{1}{(x+7)(x-2)} = \frac{A}{x+7} + \frac{B}{x-2},$$

giving

$$1 = A(x-2) + B(x+7)$$
$$1 = (A+B)x + (-2A + 7B)$$

so

$$A + B = 0$$
$$-2A + 7B = 1.$$

Thus, $A = -1/9$, $B = 1/9$, so

$$\int \frac{1}{(x+7)(x-2)}\,dx = -\int \frac{1/9}{x+7}\,dx + \int \frac{1/9}{x-2}\,dx = -\frac{1}{9}\ln|x+7| + \frac{1}{9}\ln|x-2| + C.$$

41. We use partial fractions and write

$$\frac{1}{3P - 3P^2} = \frac{A}{3P} + \frac{B}{1-P},$$

multiply through by $3P(1-P)$, and then solve for A and B, getting $A = 1$ and $B = 1/3$. So

$$\int \frac{dP}{3P - 3P^2} = \int \left(\frac{1}{3P} + \frac{1}{3(1-P)}\right)dP = \frac{1}{3}\int \frac{dP}{P} + \frac{1}{3}\int \frac{dP}{1-P}$$

$$= \frac{1}{3}\ln|P| - \frac{1}{3}\ln|1 - P| + C = \frac{1}{3}\ln\left|\frac{P}{1-P}\right| + C.$$

45. Since $x^2 + x^4 = x^2(1+x^2)$ cannot be factored further, we write

$$\frac{x-2}{x^2+x^4} = \frac{A}{x} + \frac{B}{x^2} + \frac{Cx+D}{1+x^2}.$$

Multiplying by $x^2(1+x^2)$ gives

$$x - 2 = Ax(1+x^2) + B(1+x^2) + (Cx+D)x^2$$
$$x - 2 = (A+C)x^3 + (B+D)x^2 + Ax + B,$$

so

$$A + C = 0$$
$$B + D = 0$$
$$A = 1$$
$$B = -2.$$

Thus, $A = 1$, $B = -2$, $C = -1$, $D = 2$, and we have

$$\int \frac{x-2}{x^2+x^4}\,dx = \int \left(\frac{1}{x} - \frac{2}{x^2} + \frac{-x+2}{1+x^2}\right)dx = \int \frac{dx}{x} - 2\int \frac{dx}{x^2} - \int \frac{x\,dx}{1+x^2} + 2\int \frac{dx}{1+x^2}$$
$$= \ln|x| + \frac{2}{x} - \frac{1}{2}\ln\left|1+x^2\right| + 2\arctan x + K.$$

We use K as the constant of integration, since we already used C in the problem.

49. Since $(4-z^2)^{3/2} = (\sqrt{4-z^2})^3$, we substitute $z = 2\sin\theta$, so $dz = 2\cos\theta\,d\theta$. We get

$$\int \frac{dz}{(4-z^2)^{3/2}} = \int \frac{2\cos\theta\,d\theta}{(4-4\sin^2\theta)^{3/2}} = \int \frac{2\cos\theta\,d\theta}{8\cos^3\theta} = \frac{1}{4}\int \frac{d\theta}{\cos^2\theta} = \frac{1}{4}\tan\theta + C$$

Since $\sin\theta = z/2$, we have $\cos\theta = \sqrt{1-(z/2)^2} = (\sqrt{4-z^2})/2$, so

$$\int \frac{dz}{(4-z^2)^{3/2}} = \frac{1}{4}\tan\theta + C = \frac{1}{4}\frac{\sin\theta}{\cos\theta} + C = \frac{1}{4}\frac{z/2}{(\sqrt{4-z^2})/2} + C = \frac{z}{4\sqrt{4-z^2}} + C$$

53. Notice that because $\frac{3x}{(x-1)(x-4)}$ is negative for $2 \le x \le 3$,

$$\text{Area} = -\int_2^3 \frac{3x}{(x-1)(x-4)}\,dx.$$

Using partial fractions gives

$$\frac{3x}{(x-1)(x-4)} = \frac{A}{x-1} + \frac{B}{x-4} = \frac{(A+B)x - B - 4A}{(x-1)(x-4)}.$$

Multiplying through by $(x-1)(x-4)$ gives

$$3x = (A+B)x - B - 4A$$

so $A = -1$ and $B = 4$. Thus

$$-\int_2^3 \frac{3x}{(x-1)(x-4)}\,dx = -\int_2^3 \left(\frac{-1}{x-1} + \frac{4}{x-4}\right)dx = (\ln|x-1| - 4\ln|x-4|)\Big|_2^3 = 5\ln 2.$$

57. We have

$$\text{Area} = \int_0^3 \frac{1}{\sqrt{x^2 + 9}}\, dx.$$

Let $x = 3\tan\theta$ so $dx = (3/\cos^2\theta)d\theta$ and

$$\sqrt{x^2 + 9} = \sqrt{\frac{9\sin^2\theta}{\cos^2\theta} + 9} = \frac{3}{\cos\theta}.$$

When $x = 0, \theta = 0$ and when $x = 3, \theta = \pi/4$. Thus

$$\int_0^3 \frac{1}{\sqrt{x^2+9}}\,dx = \int_0^{\pi/4} \frac{1}{\sqrt{9\tan^2\theta+9}}\,\frac{3}{\cos^2\theta}\,d\theta = \int_0^{\pi/4} \frac{1}{3/\cos\theta}\cdot\frac{3}{\cos^2\theta}\,d\theta = \int_0^{\pi/4}\frac{1}{\cos\theta}\,d\theta$$

$$= \frac{1}{2}\ln\left|\frac{\sin\theta+1}{\sin\theta-1}\right|\Bigg|_0^{\pi/4} = \frac{1}{2}\ln\left|\frac{1/\sqrt{2}+1}{1/\sqrt{2}-1}\right| = \frac{1}{2}\ln\left(\frac{1+\sqrt{2}}{\sqrt{2}-1}\right).$$

This answer can be simplified to $\ln(1+\sqrt{2})$ by multiplying the numerator and denominator of the fraction by $(\sqrt{2}+1)$ and using the properties of logarithms. The integral $\int(1/\cos\theta)d\theta$ is done using the Table of Integrals.

61. Using partial fractions, we write

$$\frac{3x^2+1}{x^3+x} = \frac{3x^2+1}{x(x^2+1)} = \frac{A}{x} + \frac{Bx+C}{x^2+1}$$

$$3x^2 + 1 = A(x^2+1) + (Bx+C)x = (A+B)x^2 + Cx + A.$$

So, $A + B = 3, C = 0$ and $A = 1$, giving $B = 2$. Thus

$$\int \frac{3x^2+1}{x^3+x}\,dx = \int\left(\frac{1}{x} + \frac{2x}{x^2+1}\right)dx = \ln|x| + \ln|x^2+1| + K.$$

Using the substitution $w = x^3 + x$, we get $dw = (3x^2+1)dx$, so we have

$$\int \frac{3x^2+1}{x^3+x}\,dx = \int \frac{dw}{w} = \ln|w| + K = \ln|x^3+x| + K.$$

The properties of logarithms show that the two results are the same:

$$\ln|x| + \ln|x^2+1| + K = \ln|x(x^2+1)| + K = \ln|x^3+x| + K.$$

We use K as the constant of integration, since we already used C in the problem.

65. (a) If $a > 0$, then

$$x^2 - a = (x - \sqrt{a})(x + \sqrt{a}).$$

This means that we can use partial fractions:

$$\frac{1}{x^2 - a} = \frac{A}{x - \sqrt{a}} + \frac{B}{x + \sqrt{a}},$$

giving

$$1 = A(x + \sqrt{a}) + B(x - \sqrt{a}),$$

so $A + B = 0$ and $(A - B)\sqrt{a} = 1$. Thus, $A = -B = 1/(2\sqrt{a})$.
So

$$\int \frac{1}{x^2 - a}\,dx = \int \frac{1}{2\sqrt{a}}\left(\frac{1}{x-\sqrt{a}} - \frac{1}{x+\sqrt{a}}\right)dx = \frac{1}{2\sqrt{a}}(\ln|x-\sqrt{a}| - \ln|x+\sqrt{a}|) + C.$$

(b) If $a = 0$, we have

$$\int \frac{1}{x^2}\,dx = -\frac{1}{x} + C.$$

(c) If $a < 0$, then $-a > 0$ so $x^2 - a = x^2 + (-a)$ cannot be factored. Thus

$$\int \frac{1}{x^2 - a}\,dx = \int \frac{1}{x^2 + (-a)}\,dx = \frac{1}{\sqrt{-a}}\arctan\left(\frac{x}{\sqrt{-a}}\right) + C.$$

Solutions for Section 7.5

Exercises

1. (a) The approximation LEFT(2) uses two rectangles, with the height of each rectangle determined by the left-hand endpoint. See Figure 7.2. We see that this approximation is an underestimate.

Figure 7.2

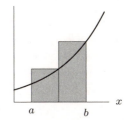

Figure 7.3

(b) The approximation RIGHT(2) uses two rectangles, with the height of each rectangle determined by the right-hand endpoint. See Figure 7.3. We see that this approximation is an overestimate.

(c) The approximation TRAP(2) uses two trapezoids, with the height of each trapezoid given by the secant line connecting the two endpoints. See Figure 7.4. We see that this approximation is an overestimate.

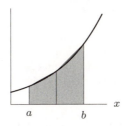

Figure 7.4

(d) The approximation MID(2) uses two rectangles, with the height of each rectangle determined by the height at the midpoint. Alternately, we can view MID(2) as a trapezoid rule where the height is given by the tangent line at the midpoint. Both interpretations are shown in Figure 7.5. We see from the tangent line interpretation that this approximation is an underestimate

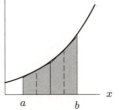

Figure 7.5

5. (a) The approximation LEFT(2) uses two rectangles, with the height of each rectangle determined by the left-hand endpoint. See Figure 7.6. We see that this approximation is an underestimate (that is, it is more negative).

Figure 7.6

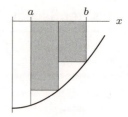

Figure 7.7

(b) The approximation RIGHT(2) uses two rectangles, with the height of each rectangle determined by the right-hand endpoint. See Figure 7.7. We see that this approximation is an overestimate (that is, it is less negative).

(c) The approximation TRAP(2) uses two trapezoids, with the height of each trapezoid given by the secant line connecting the two endpoints. See Figure 7.8. We see that this approximation is an overestimate (that is, it is less negative).

Figure 7.8

(d) The approximation MID(2) uses two rectangles, with the height of each rectangle determined by the height at the midpoint. Alternately, we can view MID(2) as a trapezoid rule where the height is given by the tangent line at the midpoint. Both interpretations are shown in Figure 7.9. We see from the tangent line interpretation that this approximation is an underestimate (that is, it is more negative).

Figure 7.9

9. (a)

$$\text{MID}(2) = 2 \cdot f(1) + 2 \cdot f(3)$$
$$= 2 \cdot 2 + 2 \cdot 10$$
$$= 24$$
$$\text{TRAP}(2) = \frac{\text{LEFT}(2) + \text{RIGHT}(2)}{2}$$
$$= \frac{12 + 44}{2} \quad \text{(see Problem 8)}$$
$$= 28$$

(b)

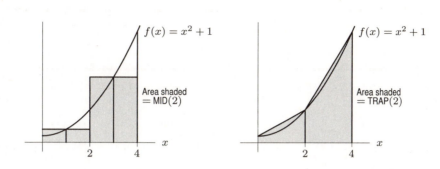

MID(2) is an underestimate, since $f(x) = x^2 + 1$ is concave up and a tangent line will be below the curve. TRAP(2) is an overestimate, since a secant line lies above the curve.

Problems

13. Since the function is decreasing, LEFT is an overestimate and RIGHT is an underestimate. Since the graph is concave down, secant lines lie below the graph so TRAP is an underestimate and tangent lines lie above the graph so MID is an overestimate. We can see that MID and TRAP are closer to the exact value than LEFT and RIGHT. In order smallest to largest, we have:

RIGHT(n) < TRAP(n) < Exact value < MID(n) < LEFT(n).

17. $f(x)$ is decreasing and concave up, so LEFT and TRAP give overestimates and RIGHT and MID give underestimates.

21. (a) $\displaystyle\int_0^{2\pi} \sin\theta \, d\theta = -\cos\theta \Big|_0^{2\pi} = 0.$

(b) See Figure 7.10. MID(1) is 0 since the midpoint of 0 and 2π is π, and $\sin\pi = 0$. Thus MID(1) $= 2\pi(\sin\pi) = 0$. The midpoints we use for MID(2) are $\pi/2$ and $3\pi/2$, and $\sin(\pi/2) = -\sin(3\pi/2)$. Thus MID(2) $= \pi\sin(\pi/2) + \pi\sin(3\pi/2) = 0.$

Figure 7.10

(c) MID(3) $= 0$.

In general, MID$(n) = 0$ for all n, even though your calculator (because of round-off error) might not return it as such. The reason is that $\sin(x) = -\sin(2\pi - x)$. If we use MID$(n)$, we will always take sums where we are adding pairs of the form $\sin(x)$ and $\sin(2\pi - x)$, so the sum will cancel to 0. (If n is odd, we will get a $\sin\pi$ in the sum which does not pair up with anything — but $\sin\pi$ is already 0.)

25.

$$
\begin{aligned}
\text{TRAP}(n) &= \frac{\text{LEFT}(n) + \text{RIGHT}(n)}{2} \\
&= \frac{\text{LEFT}(n) + \text{LEFT}(n) + f(b)\Delta x - f(a)\Delta x}{2} \\
&= \text{LEFT}(n) + \frac{1}{2}(f(b) - f(a))\Delta x
\end{aligned}
$$

Solutions for Section 7.6

Exercises

1. We saw in Problem 7 in Section 7.5 that, for this definite integral, we have LEFT(2) = 27, RIGHT(2) = 135, TRAP(2) = 81, and MID(2) = 67.5. Thus,

$$\text{SIMP}(2) = \frac{2\text{MID}(2) + \text{TRAP}(2)}{3} = \frac{2(67.5) + 81}{3} = 72.$$

Notice that

$$\int_0^6 x^2 \, dx = \frac{x^3}{3} \bigg|_0^6 = \frac{6^3}{3} - \frac{0^3}{3} = 72,$$

and so SIMP(2) gives the exact value of the integral in this case.

Problems

5. (a) $\displaystyle\int_0^4 e^x \, dx = e^x \bigg|_0^4 = e^4 - e^0 \approx 53.598\ldots.$

   (b) Computing the sums directly, since $\Delta x = 2$, we have

   LEFT(2)$= 2 \cdot e^0 + 2 \cdot e^2 \approx 2(1) + 2(7.389) = 16.778;$ error $= 36.820.$
   RIGHT(2)$= 2 \cdot e^2 + 2 \cdot e^4 \approx 2(7.389) + 2(54.598) = 123.974;$ error $= -70.376.$
   TRAP(2)$= \dfrac{16.778 + 123.974}{2} = 70.376;$ error $= 16.778.$
   MID(2)$= 2 \cdot e^1 + 2 \cdot e^3 \approx 2(2.718) + 2(20.086) = 45.608;$ error $= 7.990.$
   SIMP(2)$= \dfrac{2(45.608) + 70.376}{3} = 53.864;$ error $= -0.266.$

   (c) Similarly, since $\Delta x = 1$, we have LEFT(4)$= 31.193;$ error $= 22.405$
   RIGHT(4)$= 84.791;$ error $= -31.193$
   TRAP(4)$= 57.992;$ error $= -4.394$
   MID(4)$= 51.428;$ error $= 2.170$
   SIMP(4)$= 53.616;$ error $= -0.018$

   (d) For LEFT and RIGHT, we expect the error to go down by $1/2$, and this is very roughly what we see. For MID and TRAP, we expect the error to go down by $1/4$, and this is approximately what we see. For SIMP, we expect the error to go down by $1/2^4 = 1/16$, and this is approximately what we see.

9. (a) If $f(x) = 1$, then

$$\int_a^b f(x) \, dx = (b - a).$$

Also,

$$\frac{h}{3} \left(\frac{f(a)}{2} + 2f(m) + \frac{f(b)}{2} \right) = \frac{b - a}{3} \left(\frac{1}{2} + 2 + \frac{1}{2} \right) = (b - a).$$

So the equation holds for $f(x) = 1$.

If $f(x) = x$, then

$$\int_a^b f(x) \, dx = \frac{x^2}{2} \bigg|_a^b = \frac{b^2 - a^2}{2}.$$

Also,

$$\frac{h}{3} \left(\frac{f(a)}{2} + 2f(m) + \frac{f(b)}{2} \right) = \frac{b - a}{3} \left(\frac{a}{2} + 2\frac{a + b}{2} + \frac{b}{2} \right)$$

$$= \frac{b - a}{3} \left(\frac{a}{2} + a + b + \frac{b}{2} \right)$$

$$= \frac{b - a}{3} \left(\frac{3}{2}b + \frac{3}{2}a \right)$$

$$= \frac{(b-a)(b+a)}{2}$$

$$= \frac{b^2 - a^2}{2}.$$

So the equation holds for $f(x) = x$.

If $f(x) = x^2$, then $\int_a^b f(x)\,dx = \left.\frac{x^3}{3}\right|_a^b = \frac{b^3 - a^3}{3}$. Also,

$$\frac{h}{3}\left(\frac{f(a)}{2} + 2f(m) + \frac{f(b)}{2}\right) = \frac{b-a}{3}\left(\frac{a^2}{2} + 2\left(\frac{a+b}{2}\right)^2 + \frac{b^2}{2}\right)$$

$$= \frac{b-a}{3}\left(\frac{a^2}{2} + \frac{a^2 + 2ab + b^2}{2} + \frac{b^2}{2}\right)$$

$$= \frac{b-a}{3}\left(\frac{2a^2 + 2ab + 2b^2}{2}\right)$$

$$= \frac{b-a}{3}\left(a^2 + ab + b^2\right)$$

$$= \frac{b^3 - a^3}{3}.$$

So the equation holds for $f(x) = x^2$.

(b) For any quadratic function, $f(x) = Ax^2 + Bx + C$, the "Facts about Sums and Constant Multiples of Integrands" give us:

$$\int_a^b f(x)\,dx = \int_a^b (Ax^2 + Bx + C)\,dx = A\int_a^b x^2\,dx + B\int_a^b x\,dx + C\int_a^b 1\,dx.$$

Now we use the results of part (a) to get:

$$\int_a^b f(x)\,dx = A\frac{h}{3}\left(\frac{a^2}{2} + 2m^2 + \frac{b^2}{2}\right) + B\frac{h}{3}\left(\frac{a}{2} + 2m + \frac{b}{2}\right) + C\frac{h}{3}\left(\frac{1}{2} + 2\cdot 1 + \frac{1}{2}\right)$$

$$= \frac{h}{3}\left(\frac{Aa^2 + Ba + C}{2} + 2(Am^2 + Bm + C) + \frac{Ab^2 + Bb + C}{2}\right)$$

$$= \frac{h}{3}\left(\frac{f(a)}{2} + 2f(m) + \frac{f(b)}{2}\right).$$

Solutions for Section 7.7

Exercises

1. (a) See Figure 7.11. The area extends out infinitely far along the positive x-axis.

Figure 7.11

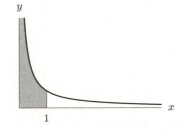

Figure 7.12

(b) See Figure 7.12. The area extends up infinitely far along the positive y-axis.

5. We have

$$\int_1^\infty \frac{1}{5x+2}\,dx = \lim_{b\to\infty}\int_1^b \frac{1}{5x+2}\,dx = \lim_{b\to\infty}\left(\frac{1}{5}\ln(5x+2)\right)\Big|_1^b = \lim_{b\to\infty}\left(\frac{1}{5}\ln(5b+2) - \frac{1}{5}\ln(7)\right).$$

As $b \leftarrow \infty$, we know that $\ln(5b+2) \to \infty$, and so this integral diverges.

9. We have

$$\int_0^\infty xe^{-x^2}\,dx = \lim_{b\to\infty}\int_0^b xe^{-x^2}\,dx = \lim_{b\to\infty}\left(\frac{-1}{2}e^{-x^2}\right)\Big|_0^b = \lim_{b\to\infty}\left(\frac{-1}{2}e^{-b^2} - \frac{-1}{2}\right) = 0 + \frac{1}{2} = \frac{1}{2}.$$

This integral converges to $1/2$.

13.

$$\int_{-\infty}^0 \frac{e^x}{1+e^x}\,dx = \lim_{b\to-\infty}\int_b^0 \frac{e^x}{1+e^x}\,dx$$

$$= \lim_{b\to-\infty}\ln|1+e^x|\Big|_b^0$$

$$= \lim_{b\to-\infty}[\ln|1+e^0| - \ln|1+e^b|]$$

$$= \ln(1+1) - \ln(1+0) = \ln 2.$$

17. This integral is improper because $1/v$ is undefined at $v = 0$. Then

$$\int_0^1 \frac{1}{v}\,dv = \lim_{b\to 0^+}\int_b^1 \frac{1}{v}\,dv = \lim_{b\to 0^+}\left(\ln v\Big|_b^1\right) = -\ln b.$$

As $b \to 0^+$, this goes to infinity and the integral diverges.

21. We use V-26 with $a = 4$ and $b = -4$:

$$\int_0^4 \frac{-1}{u^2-16}\,du = \lim_{b\to 4^-}\int_0^b \frac{-1}{u^2-16}\,du$$

$$= \lim_{b\to 4^-}\int_0^b \frac{-1}{(u-4)(u+4)}\,du$$

$$= \lim_{b\to 4^-}\frac{-(\ln|u-4| - \ln|u+4|)}{8}\Big|_0^b$$

$$= \lim_{b\to 4^-} -\frac{1}{8}\left(\ln|b-4| + \ln 4 - \ln|b+4| - \ln 4\right).$$

As $b \to 4^-$, $\ln|b-4| \to -\infty$, so the limit does not exist and the integral diverges.

25. This is a proper integral; use V-26 in the integral table with $a = 4$ and $b = -4$.

$$\int_{16}^{20} \frac{1}{y^2-16}\,dy = \int_{16}^{20} \frac{1}{(y-4)(y+4)}\,dy$$

$$= \frac{\ln|y-4| - \ln|y+4|}{8}\Big|_{16}^{20}$$

$$= \frac{\ln 16 - \ln 24 - (\ln 12 - \ln 20)}{8}$$

$$= \frac{\ln 320 - \ln 288}{8} = \frac{1}{8}\ln(10/9) = 0.01317.$$

29.

$$\int_0^2 \frac{1}{\sqrt{4 - x^2}}\, dx = \lim_{b \to 2^-} \int_0^b \frac{1}{\sqrt{4 - x^2}}\, dx$$

$$= \lim_{b \to 2^-} \arcsin \frac{x}{2}\bigg|_0^b$$

$$= \lim_{b \to 2^-} \arcsin \frac{b}{2} = \arcsin 1 = \frac{\pi}{2}.$$

33. The integrand is undefined at $y = -3$ and $y = 3$. To consider the limits one at a time, divide the integral at $y = 0$;

$$\int_0^3 \frac{y\, dy}{\sqrt{9 - y^2}} = \lim_{b \to 3^-} \int_0^b \frac{y}{\sqrt{9 - y^2}}\, dy = \lim_{b \to 3^-} \left(-(9 - y^2)^{1/2}\right)\bigg|_0^b$$

$$= \lim_{b \to 3^-} \left(3 - (9 - b^2)^{1/2}\right) = 3.$$

A similar argument shows that

$$\int_{-3}^0 \frac{y\, dy}{\sqrt{9 - y^2}} = \lim_{b \to -3^+} \int_b^0 \frac{y}{\sqrt{9 - y^2}}\, dy = \lim_{b \to -3^+} \left(-(9 - y^2)^{1/2}\right)\bigg|_b^0$$

$$= \lim_{b \to -3^+} \left(-3 + (9 - b^2)^{1/2}\right) = -3.$$

Thus the original integral converges to a value of 0:

$$\int_{-3}^3 \frac{y\, dy}{\sqrt{9 - y^2}} = \int_{-3}^0 \frac{y\, dy}{\sqrt{9 - y^2}} + \int_0^3 \frac{y\, dy}{\sqrt{9 - y^2}} = -3 + 3 = 0.$$

Problems

37. (a) There is no simple antiderivative for this integrand, so we use numerical methods. We find

$$P(1) = \frac{1}{\sqrt{\pi}} \int_0^1 e^{-t^2}\, dt = 0.421.$$

(b) To calculate this improper integral, use numerical methods. If you cannot input infinity into your calculator, increase the upper limit until the value of the integral settles down. We find

$$P(\infty) = \frac{1}{\sqrt{\pi}} \int_0^\infty e^{-t^2}\, dt = 0.500.$$

41. The factor $\ln x$ grows slowly enough (as $x \to 0^+$) not to change the convergence or divergence of the integral, although it will change what it converges or diverges to.

The integral is always improper, because $\ln x$ is not defined for $x = 0$. Integrating by parts (or, alternatively, the integral table) yields

$$\int_0^e x^p \ln x\, dx = \lim_{a \to 0^+} \int_a^e x^p \ln x\, dx$$

$$= \lim_{a \to 0^+} \left(\frac{1}{p + 1} x^{p+1} \ln x - \frac{1}{(p + 1)^2} x^{p+1}\right)\bigg|_a^e$$

$$= \lim_{a \to 0^+} \left[\left(\frac{1}{p + 1} e^{p+1} - \frac{1}{(p + 1)^2} e^{p+1}\right)\right.$$

$$\left. - \left(\frac{1}{p + 1} a^{p+1} \ln a - \frac{1}{(p + 1)^2} a^{p+1}\right)\right].$$

If $p < -1$, then $(p+1)$ is negative, so as $a \to 0^+$, $a^{p+1} \to \infty$ and $\ln a \to -\infty$, and therefore the limit does not exist.

If $p > -1$, then $(p+1)$ is positive and it's easy to see that $a^{p+1} \to 0$ as $a \to 0$. Looking at graphs of $x^{p+1} \ln x$ (for different values of p) shows that $a^{p+1} \ln a \to 0$ as $a \to 0$. This is not so easy to see analytically. It's true because if we let $t = \frac{1}{a}$ then

$$\lim_{a \to 0^+} a^{p+1} \ln a = \lim_{t \to \infty} \left(\frac{1}{t}\right)^{p+1} \ln\left(\frac{1}{t}\right) = \lim_{t \to \infty} -\frac{\ln t}{t^{p+1}}.$$

This last limit is zero because $\ln t$ grows very slowly, much more slowly than t^{p+1}. So if $p > -1$, the integral converges and equals $e^{p+1}[1/(p+1) - 1/(p+1)^2] = pe^{p+1}/(p+1)^2$.

What happens if $p = -1$? Then we get

$$\int_0^e \frac{\ln x}{x} \, dx = \lim_{a \to 0^+} \int_a^e \frac{\ln x}{x} \, dx$$

$$= \lim_{a \to 0^+} \frac{(\ln x)^2}{2} \bigg|_a^e$$

$$= \lim_{a \to 0^+} \left(\frac{1 - (\ln a)^2}{2}\right).$$

Since $\ln a \to -\infty$ as $a \to 0^+$, this limit does not exist.

To summarize, $\int_0^e x^p \ln x$ converges for $p > -1$ to the value $pe^{p+1}/(p+1)^2$.

45. We calculate

$$m_1 = \frac{1}{\sqrt{2\pi}} \int_{-\infty}^{\infty} x e^{-x^2/2} \, dx.$$

Since the integrand is odd, for any b, the integral

$$\int_{-b}^{b} x e^{-x^2/2} \, dx = 0.$$

Thus,

$$m_1 = \frac{1}{\sqrt{2\pi}} \int_{-\infty}^{\infty} x e^{-x^2/2} \, dx = \frac{1}{\sqrt{2\pi}} \lim_{b \to \infty} \int_{-b}^{b} x e^{-x^2/2} \, dx = 0.$$

49. (a)

$$\Gamma(1) = \int_0^{\infty} e^{-t} \, dt$$

$$= \lim_{b \to \infty} \int_0^b e^{-t} \, dt$$

$$= \lim_{b \to \infty} -e^{-t} \bigg|_0^b$$

$$= \lim_{b \to \infty} [1 - e^{-b}] = 1.$$

Using Problem 11,

$$\Gamma(2) = \int_0^{\infty} t e^{-t} \, dt = 1.$$

(b) We integrate by parts. Let $u = t^n$, $v' = e^{-t}$. Then $u' = nt^{n-1}$ and $v = -e^{-t}$, so

$$\int t^n e^{-t} \, dt = -t^n e^{-t} + n \int t^{n-1} e^{-t} \, dt.$$

So

$$\Gamma(n+1) = \int_0^{\infty} t^n e^{-t} \, dt$$

$$= \lim_{b \to \infty} \int_0^b t^n e^{-t} \, dt$$

$$= \lim_{b \to \infty} \left[-t^n e^{-t} \Big|_0^b + n \int_0^b t^{n-1} e^{-t} \, dt \right]$$

$$= \lim_{b \to \infty} -b^n e^{-b} + \lim_{b \to \infty} n \int_0^b t^{n-1} e^{-t} \, dt$$

$$= 0 + n \int_0^\infty t^{n-1} e^{-t} \, dt$$

$$= n\Gamma(n).$$

(c) We already have $\Gamma(1) = 1$ and $\Gamma(2) = 1$. Using $\Gamma(n+1) = n\Gamma(n)$ we can get

$$\Gamma(3) = 2\Gamma(2) = 2$$
$$\Gamma(4) = 3\Gamma(3) = 3 \cdot 2$$
$$\Gamma(5) = 4\Gamma(4) = 4 \cdot 3 \cdot 2.$$

So it appears that $\Gamma(n)$ is just the first $n - 1$ numbers multiplied together, so $\Gamma(n) = (n-1)!$.

Solutions for Section 7.8

Exercises

1. For large x, the integrand behaves like $1/x^2$ because

$$\frac{x^2}{x^4 + 1} \approx \frac{x^2}{x^4} = \frac{1}{x^2}.$$

Since $\int_1^\infty \dfrac{dx}{x^2}$ converges, we expect our integral to converge. More precisely, since $x^4 + 1 > x^4$, we have

$$\frac{x^2}{x^4 + 1} < \frac{x^2}{x^4} = \frac{1}{x^2}.$$

Since $\int_1^\infty \dfrac{dx}{x^2}$ is convergent, the comparison test tells us that $\int_1^\infty \dfrac{x^2}{x^4 + 1} \, dx$ converges also.

5. The integrand is continuous for all $x \geq 1$, so whether the integral converges or diverges depends only on the behavior of the function as $x \to \infty$. As $x \to \infty$, polynomials behave like the highest powered term. Thus, as $x \to \infty$, the integrand $\dfrac{x}{x^2 + 2x + 4}$ behaves like $\dfrac{x}{x^2}$ or $\dfrac{1}{x}$. Since $\int_1^\infty \dfrac{1}{x} \, dx$ diverges, we predict that the given integral will diverge.

9. The integrand is continuous for all $x \geq 1$, so whether the integral converges or diverges depends only on the behavior of the function as $x \to \infty$. As $x \to \infty$, polynomials behave like the highest powered term. Thus, as $x \to \infty$, the integrand $\dfrac{x^2 + 4}{x^4 + 3x^2 + 11}$ behaves like $\dfrac{x^2}{x^4}$ or $\dfrac{1}{x^2}$. Since $\int_1^\infty \dfrac{1}{x^2} \, dx$ converges, we predict that the given integral will converge.

13. The integrand is unbounded as $t \to 5$. We substitute $w = t - 5$, so $dw = dt$. When $t = 5$, $w = 0$ and when $t = 8$, $w = 3$.

$$\int_5^8 \frac{6}{\sqrt{t-5}} \, dt = \int_0^3 \frac{6}{\sqrt{w}} \, dw.$$

Since

$$\int_0^3 \frac{6}{\sqrt{w}} \, dw = \lim_{a \to 0^+} 6 \int_a^3 \frac{1}{\sqrt{w}} \, dw = 6 \lim_{a \to 0^+} 2w^{1/2} \Big|_a^3 = 12 \lim_{a \to 0^+} (\sqrt{3} - \sqrt{a}) = 12\sqrt{3},$$

our integral converges.

17. Since $\dfrac{1}{u+u^2} < \dfrac{1}{u^2}$ for $u \geq 1$, and since $\displaystyle\int_1^\infty \dfrac{du}{u^2}$ converges, $\displaystyle\int_1^\infty \dfrac{du}{u+u^2}$ converges.

21. Since $\dfrac{1}{1+e^y} \leq \dfrac{1}{e^y} = e^{-y}$ and $\displaystyle\int_0^\infty e^{-y}\,dy$ converges, the integral $\displaystyle\int_0^\infty \dfrac{dy}{1+e^y}$ converges.

25. Since $\dfrac{3+\sin\alpha}{\alpha} \geq \dfrac{2}{\alpha}$ for $\alpha \geq 4$, and since $\displaystyle\int_4^\infty \dfrac{2}{\alpha}\,d\alpha$ diverges, then $\displaystyle\int_4^\infty \dfrac{3+\sin\alpha}{\alpha}\,d\alpha$ diverges.

Problems

29. The convergence or divergence of an improper integral depends on the long-term behavior of the integrand, not on its short-term behavior. Figure 7.13 suggests that $g(x) \leq f(x)$ for all values of x beyond $x = k$. Since $\int_k^\infty f(x)\,dx$ converges, we expect $\int_k^\infty g(x)\,dx$ converges also.

However we are interested in $\int_a^\infty g(x)\,dx$. Breaking the integral into two parts enables us to use the fact that $\int_k^\infty g(x)\,dx$ is finite:

$$\int_a^\infty g(x)\,dx = \int_a^k g(x)\,dx + \int_k^\infty g(x)\,dx.$$

The first integral is also finite because the interval from a to k is finite. Therefore, we expect $\int_a^\infty g(x)\,dx$ converges.

Figure 7.13

33. (a) The tangent line to e^t has slope $(e^t)' = e^t$. Thus at $t = 0$, the slope is $e^0 = 1$. The line passes through $(0, e^0) = (0, 1)$. Thus the equation of the tangent line is $y = 1 + t$. Since e^t is everywhere concave up, its graph is always above the graph of any of its tangent lines; in particular, e^t is always above the line $y = 1 + t$. This is tantamount to saying

$$1 + t \leq e^t,$$

with equality holding only at the point of tangency, $t = 0$.

(b) If $t = \dfrac{1}{x}$, then the above inequality becomes

$$1 + \frac{1}{x} \leq e^{1/x}, \text{ or } e^{1/x} - 1 \geq \frac{1}{x}.$$

Since $t = \dfrac{1}{x}$, t is never zero. Therefore, the inequality is strict, and we write

$$e^{1/x} - 1 > \frac{1}{x}.$$

(c) Since $e^{1/x} - 1 > \dfrac{1}{x}$,

$$\frac{1}{x^5\left(e^{1/x}-1\right)} < \frac{1}{x^5\left(\frac{1}{x}\right)} = \frac{1}{x^4}.$$

Since $\displaystyle\int_1^\infty \dfrac{dx}{x^4}$ converges, $\displaystyle\int_1^\infty \dfrac{dx}{x^5\left(e^{1/x}-1\right)}$ converges.

Solutions for Chapter 7 Review

1. $\frac{1}{3}(t+1)^3$

5. Using the power rule gives $\frac{3}{2}w^2 + 7w + C$.

9. Let $5z = w$, then $5dz = dw$, which means $dz = \frac{1}{5}dw$, so

$$\int e^{5z}\, dz = \int e^w \cdot \frac{1}{5}dw = \frac{1}{5}\int e^w\, dw = \frac{1}{5}e^w + C = \frac{1}{5}e^{5z} + C,$$

where C is a constant.

13. The power rule gives $\frac{2}{5}x^{5/2} + \frac{3}{5}x^{5/3} + C$

17. Dividing by x^2 gives

$$\int\left(\frac{x^3 + x + 1}{x^2}\right)dx = \int\left(x + \frac{1}{x} + \frac{1}{x^2}\right)dx = \frac{1}{2}x^2 + \ln|x| - \frac{1}{x} + C.$$

21. Integration by parts twice gives

$$\int x^2 e^{2x}\, dx = \frac{x^2 e^{2x}}{2} - \int 2xe^{2x}\, dx = \frac{x^2}{2}e^{2x} - \frac{x}{2}e^{2x} + \frac{1}{4}e^{2x} + C$$
$$= (\frac{1}{2}x^2 - \frac{1}{2}x + \frac{1}{4})e^{2x} + C.$$

Or use the integral table, III-14 with $p(x) = x^2$ and $a = 1$.

25. We integrate by parts, using $u = (\ln x)^2$ and $v' = 1$. Then $u' = 2\frac{\ln x}{x}$ and $v = x$, so

$$\int (\ln x)^2\, dx = x(\ln x)^2 - 2\int \ln x\, dx.$$

But, integrating by parts or using the integral table, $\int \ln x\, dx = x\ln x - x + C$. Therefore,

$$\int (\ln x)^2\, dx = x(\ln x)^2 - 2x\ln x + 2x + C.$$

Check:
$$\frac{d}{dx}\left[x(\ln x)^2 - 2x\ln x + 2x + C\right] = (\ln x)^2 + x\frac{2\ln x}{x} - 2\ln x - 2x\frac{1}{x} + 2 = (\ln x)^2.$$

29. Substitute $w = 4 - x^2$, $dw = -2x\, dx$:

$$\int x\sqrt{4 - x^2}\, dx = -\frac{1}{2}\int \sqrt{w}\, dw = -\frac{1}{3}w^{3/2} + C = -\frac{1}{3}(4 - x^2)^{3/2} + C.$$

Check
$$\frac{d}{dx}\left[-\frac{1}{3}(4 - x^2)^{3/2} + C\right] = -\frac{1}{3}\left[\frac{3}{2}(4 - x^2)^{1/2}(-2x)\right] = x\sqrt{4 - x^2}.$$

33. Denote $\int \cos^2\theta\, d\theta$ by A. Let $u = \cos\theta$, $v' = \cos\theta$. Then, $v = \sin\theta$ and $u' = -\sin\theta$. Integrating by parts, we get:

$$A = \cos\theta\sin\theta - \int(-\sin\theta)\sin\theta\, d\theta.$$

Employing the identity $\sin^2 \theta = 1 - \cos^2 \theta$, the equation above becomes:

$$A = \cos \theta \sin \theta + \int d\theta - \int \cos^2 \theta \, d\theta$$
$$= \cos \theta \sin \theta + \theta - A + C.$$

Solving this equation for A, and using the identity $\sin 2\theta = 2 \cos \theta \sin \theta$ we get:

$$A = \int \cos^2 \theta \, d\theta = \frac{1}{4} \sin 2\theta + \frac{1}{2}\theta + C.$$

[Note: An alternate solution would have been to use the identity $\cos^2 \theta = \frac{1}{2} \cos 2\theta + \frac{1}{2}$.]

37. Multiplying out, dividing, and then integrating yields

$$\int \frac{(t+2)^2}{t^3} \, dt = \int \frac{t^2 + 4t + 4}{t^3} \, dt = \int \frac{1}{t} \, dt + \int \frac{4}{t^2} \, dt + \int \frac{4}{t^3} \, dt = \ln|t| - \frac{4}{t} - \frac{2}{t^2} + C,$$

where C is a constant.

41. Let $\cos \theta = w$, then $-\sin \theta \, d\theta = dw$, so

$$\int \tan \theta \, d\theta = \int \frac{\sin \theta}{\cos \theta} \, d\theta = \int \frac{-1}{w} \, dw$$
$$= -\ln|w| + C = -\ln|\cos \theta| + C,$$

where C is a constant.

45. Let $w = 2z$, so $dw = 2dz$. Then, since $\dfrac{d}{dw} \arctan w = \dfrac{1}{1 + w^2}$, we have

$$\int \frac{dz}{1 + 4z^2} = \int \frac{\frac{1}{2}dw}{1 + w^2} = \frac{1}{2} \arctan w + C = \frac{1}{2} \arctan 2z + C.$$

49. If $u = t - 10$, $t = u + 10$ and $dt = 1 \, du$, so substituting we get

$$\int (u + 10)u^{10} du = \int (u^{11} + 10u^{10}) \, du = \frac{1}{12}u^{12} + \frac{10}{11}u^{11} + C$$
$$= \frac{1}{12}(t - 10)^{12} + \frac{10}{11}(t - 10)^{11} + C.$$

53. Let $x^2 = w$, then $2x \, dx = dw$, $x = 1 \Rightarrow w = 1$, $x = 3 \Rightarrow w = 9$. Thus,

$$\int_1^3 x(x^2 + 1)^{70} \, dx = \int_1^9 (w + 1)^{70} \frac{1}{2} \, dw$$
$$= \frac{1}{2} \cdot \frac{1}{71}(w + 1)^{71} \Big|_1^9$$
$$= \frac{1}{142}(10^{71} - 2^{71}).$$

57. Let $w = \ln x$, then $dw = (1/x)dx$ which gives

$$\int \frac{1}{x} \tan(\ln x) \, dx = \int \tan w \, dw = \int \frac{\sin w}{\cos w} \, dw = -\ln(|\cos w|) + C = -\ln(|\cos(\ln x)|) + C.$$

61. Dividing and then integrating term by term, we get

$$\int \frac{e^{2y} + 1}{e^{2y}} \, dy = \int \left(\frac{e^{2y}}{e^{2y}} + \frac{1}{e^{2y}} \right) dy = \int (1 + e^{-2y}) \, dy = \int dy + \left(-\frac{1}{2} \right) \int e^{-2y}(-2) \, dy$$
$$= y - \frac{1}{2}e^{-2y} + C.$$

65. Let $w = \sqrt{x^2 + 1}$, then $dw = \dfrac{x\,dx}{\sqrt{x^2 + 1}}$ so that

$$\int \frac{x}{\sqrt{x^2 + 1}}\, \cos\sqrt{x^2 + 1}\, dx = \int \cos w\, dw = \sin w + C = \sin\sqrt{x^2 + 1} + C.$$

69. $\displaystyle\int e^{\sqrt{2x + 3}}\, dx = \frac{1}{\sqrt{2}} \int e^{\sqrt{2x + 3}} \sqrt{2}\, dx.$ If $u = \sqrt{2x + 3}$, $du = \sqrt{2}\, dx$, so

$$\frac{1}{\sqrt{2}} \int e^u\, du = \frac{1}{\sqrt{2}} e^u + C = \frac{1}{\sqrt{2}} e^{\sqrt{2x + 3}} + C.$$

73. The integral table yields

$$\int \frac{5x + 6}{x^2 + 4}\, dx = \frac{5}{2} \ln|x^2 + 4| + \frac{6}{2} \arctan\frac{x}{2} + C$$

$$= \frac{5}{2} \ln|x^2 + 4| + 3\arctan\frac{x}{2} + C.$$

Check:

$$\frac{d}{dx}\left(\frac{5}{2} \ln|x^2 + 4| + \frac{6}{2} \arctan\frac{x}{2} + C\right) = \frac{5}{2}\left(\frac{1}{x^2 + 4}(2x) + 3\frac{1}{1 + (x/2)^2}\frac{1}{2}\right)$$

$$= \frac{5x}{x^2 + 4} + \frac{6}{x^2 + 4} = \frac{5x + 6}{x^2 + 4}.$$

77. Integration by parts will be used twice. First let $u = e^{-ct}$ and $dv = \sin(kt)dt$, then $du = -ce^{-ct}dt$ and $v = (-1/k)\cos kt$. Then

$$\int e^{-ct} \sin kt\, dt = -\frac{1}{k} e^{-ct} \cos kt - \frac{c}{k} \int e^{-ct} \cos kt\, dt$$

$$= -\frac{1}{k} e^{-ct} \cos kt - \frac{c}{k}\left(\frac{1}{k} e^{-ct} \sin kt + \frac{c}{k} \int e^{-ct} \sin kt\, dt\right)$$

$$= -\frac{1}{k} e^{-ct} \cos kt - \frac{c}{k^2} e^{-ct} \sin kt - \frac{c^2}{k^2} \int e^{-ct} \sin kt\, dt.$$

Solving for $\int e^{-ct} \sin kt\, dt$ gives

$$\frac{k^2 + c^2}{k^2} \int e^{-ct} \sin kt\, dt = -\frac{e^{-ct}}{k^2} \left(k\cos kt + c\sin kt\right),$$

so

$$\int e^{-ct} \sin kt\, dt = -\frac{e^{-ct}}{k^2 + c^2} \left(k\cos kt + c\sin kt\right) + C.$$

81. By completing the square, we get

$$x^2 - 3x + 2 = (x^2 - 3x + (-\frac{3}{2})^2) + 2 - \frac{9}{4} = (x - \frac{3}{2})^2 - \frac{1}{4}.$$

Then

$$\int \frac{1}{\sqrt{x^2 - 3x + 2}}\, dx = \int \frac{1}{\sqrt{(x - \frac{3}{2})^2 - \frac{1}{4}}}\, dx.$$

Let $w = (x - (3/2))$, then $dw = dx$ and $a^2 = 1/4$. Then we have

$$\int \frac{1}{\sqrt{x^2 - 3x + 2}}\, dx = \int \frac{1}{\sqrt{w^2 - a^2}}\, dw$$

and from VI-29 of the integral table we have

$$\int \frac{1}{\sqrt{w^2 - a^2}}\, dw = \ln \left| w + \sqrt{w^2 - a^2} \right| + C$$

$$= \ln \left| \left(x - \frac{3}{2} \right) + \sqrt{\left(x - \frac{3}{2} \right)^2 - \frac{1}{4}} \right| + C$$

$$= \ln \left| \left(x - \frac{3}{2} \right) + \sqrt{x^2 - 3x + 2} \right| + C.$$

85. Let $w = ax^2 + 2bx + c$, then $dw = (2ax + 2b)dx$ so that

$$\int \frac{ax + b}{ax^2 + 2bx + c}\, dx = \frac{1}{2} \int \frac{dw}{w} = \frac{1}{2} \ln |w| + C = \frac{1}{2} \ln |ax^2 + 2bx + c| + C.$$

89. Multiplying out and integrating term by term gives

$$\int (x^2 + 5)^3\, dx = \int (x^6 + 15x^4 + 75x^2 + 125)dx = \frac{1}{7}x^7 + 15\frac{x^5}{5} + 75\frac{x^3}{3} + 125x + C$$

$$= \frac{1}{7}x^7 + 3x^5 + 25x^3 + 125x + C.$$

93. If $u = 1 + \cos^2 w$, $du = 2(\cos w)^1 (-\sin w)\, dw$, so

$$\int \frac{\sin w \cos w}{1 + \cos^2 w}\, dw = -\frac{1}{2} \int \frac{-2\sin w \cos w}{1 + \cos^2 w}\, dw = -\frac{1}{2} \int \frac{1}{u}\, du = -\frac{1}{2} \ln |u| + C$$

$$= -\frac{1}{2} \ln |1 + \cos^2 w| + C.$$

97. If $u = \sqrt{x + 1}$, $u^2 = x + 1$ with $x = u^2 - 1$ and $dx = 2u\, du$. Substituting, we get

$$\int \frac{x}{\sqrt{x + 1}}\, dx = \int \frac{(u^2 - 1)2u\, du}{u} = \int (u^2 - 1)2\, du = 2 \int (u^2 - 1)\, du$$

$$= \frac{2u^3}{3} - 2u + C = \frac{2(\sqrt{x + 1})^3}{3} - 2\sqrt{x + 1} + C.$$

101. Letting $u = z - 5$, $z = u + 5$, $dz = du$, and substituting, we have

$$\int \frac{z}{(z - 5)^3}\, dz = \int \frac{u + 5}{u^3}\, du = \int (u^{-2} + 5u^{-3})du = \frac{u^{-1}}{-1} + 5\left(\frac{u^{-2}}{-2} \right) + C$$

$$= \frac{-1}{(z - 5)} + \frac{-5}{2(z - 5)^2} + C.$$

105. Let $w = y^2 - 2y + 1$, so $dw = 2(y - 1)\, dy$. Then

$$\int \sqrt{y^2 - 2y + 1}(y - 1)\, dy = \int w^{1/2} \frac{1}{2}\, dw = \frac{1}{2} \cdot \frac{2}{3} w^{3/2} + C = \frac{1}{3}(y^2 - 2y + 1)^{3/2} + C.$$

Alternatively, notice the integrand can be written $\sqrt{(y - 1)^2}(y - 1) = (y - 1)^2$. This leads to a different-looking but equivalent answer.

109. If $u = 2\sin x$, then $du = 2\cos x\, dx$, so

$$\int \cos(2\sin x) \cos x\, dx = \frac{1}{2} \int \cos(2\sin x)2\cos x\, dx = \frac{1}{2} \int \cos u\, du$$

$$= \frac{1}{2} \sin u + C = \frac{1}{2} \sin(2\sin x) + C.$$

113. We use the substitution $w = x^2 + 2x$ and $dw = (2x + 2)\, dx$ so

$$\int (x + 1) \sinh(x^2 + 2x)\, dx = \frac{1}{2} \int \sinh w\, dw = \frac{1}{2} \cosh w + C = \frac{1}{2} \cosh(x^2 + 2x) + C.$$

Check this answer by taking the derivative: $\dfrac{d}{dx} \left[\dfrac{1}{2} \cosh(x^2 + 2x) + C \right] = \dfrac{1}{2}(2x+2) \sinh(x^2+2x) = (x+1) \sinh(x^2 + 2x)$.

117. Let $w = 1 + 5x^2$. We have $dw = 10x\, dx$, so $\dfrac{dw}{10} = x\, dx$. When $x = 0$, $w = 1$. When $x = 1$, $w = 6$.

$$\frac{x\, dx}{1 + 5x^2} = \int_1^6 \frac{\frac{1}{10}\, dw}{w} = \frac{1}{10} \int_1^6 \frac{dw}{w} = \frac{1}{10} \ln |w| \Big|_1^6$$

$$= \frac{1}{10}(\ln 6 - \ln 1) = \frac{\ln 6}{10}$$

121. Integrating by parts, we take $u = e^{2x}$, $u' = 2e^{2x}$, $v' = \sin 2x$, and $v = -\frac{1}{2} \cos 2x$, so

$$\int e^{2x} \sin 2x\, dx = -\frac{e^{2x}}{2} \cos 2x + \int e^{2x} \cos 2x\, dx.$$

Integrating by parts again, with $u = e^{2x}$, $u' = 2e^{2x}$, $v' = \cos 2x$, and $v = \frac{1}{2} \sin 2x$, we get

$$\int e^{2x} \cos 2x\, dx = \frac{e^{2x}}{2} \sin 2x - \int e^{2x} \sin 2x\, dx.$$

Substituting into the previous equation, we obtain

$$\int e^{2x} \sin 2x\, dx = -\frac{e^{2x}}{2} \cos 2x + \frac{e^{2x}}{2} \sin 2x - \int e^{2x} \sin 2x\, dx.$$

Solving for $\int e^{2x} \sin 2x\, dx$ gives

$$\int e^{2x} \sin 2x\, dx = \frac{1}{4} e^{2x} (\sin 2x - \cos 2x) + C.$$

This result can also be obtained using II-8 in the integral table. Thus

$$\int_{-\pi}^{\pi} e^{2x} \sin 2x = \left[\frac{1}{4} e^{2x} (\sin 2x - \cos 2x) \right] \Big|_{-\pi}^{\pi} = \frac{1}{4}(e^{-2\pi} - e^{2\pi}) \approx -133.8724.$$

We get -133.37 using Simpson's rule with 10 intervals. With 100 intervals, we get -133.8724. Thus our answer matches the approximation of Simpson's rule.

125.

$$\int_0^1 \frac{dx}{x^2 + 1} = \tan^{-1} x \Big|_0^1 = \tan^{-1} 1 - \tan^{-1} 0 = \frac{\pi}{4} - 0 = \frac{\pi}{4}.$$

129. (a) We split $\dfrac{1}{x^2 - x} = \dfrac{1}{x(x - 1)}$ into partial fractions:

$$\frac{1}{x^2 - x} = \frac{A}{x} + \frac{B}{x - 1}.$$

Multiplying by $x(x - 1)$ gives

$$1 = A(x - 1) + Bx = (A + B)x - A,$$

so $-A = 1$ and $A + B = 0$, giving $A = -1$, $B = 1$. Therefore,

$$\int \frac{1}{x^2 - x}\, dx = \int \left(\frac{1}{x - 1} - \frac{1}{x} \right) dx = \ln |x - 1| - \ln |x| + C.$$

(b) $\displaystyle \int \frac{1}{x^2 - x}\, dx = \int \frac{1}{(x - 1)(x)}\, dx$. Using $a = 1$ and $b = 0$ in V-26, we get $\ln |x - 1| - \ln |x| + C$.

133. Splitting the integrand into partial fractions with denominators x and $(x + 5)$, we have

$$\frac{1}{x(x + 5)} = \frac{A}{x} + \frac{B}{x + 5}.$$

Multiplying by $x(x + 5)$ gives the identity

$$1 = A(x + 5) + Bx$$

so

$$1 = (A + B)x + 5A.$$

Since this equation holds for all x, the constant terms on both sides must be equal. Similarly, the coefficient of x on both sides must be equal. So

$$5A = 1$$
$$A + B = 0.$$

Solving these equations gives $A = 1/5$, $B = -1/5$ and the integral becomes

$$\int \frac{1}{x(x + 5)} dx = \frac{1}{5} \int \frac{1}{x} dx - \frac{1}{5} \int \frac{1}{x + 5} dx = \frac{1}{5} \left(\ln |x| - \ln |x + 5| \right) + C.$$

137. Splitting the integrand into partial fractions with denominators $(1 + x)$, $(1 + x)^2$ and x, we have

$$\frac{1 + x^2}{x(1 + x)^2} = \frac{A}{1 + x} + \frac{B}{(1 + x)^2} + \frac{C}{x}.$$

Multiplying by $x(1 + x)^2$ gives the identity

$$1 + x^2 = Ax(1 + x) + Bx + C(1 + x)^2$$

so

$$1 + x^2 = (A + C)x^2 + (A + B + 2C)x + C.$$

Since this equation holds for all x, the constant terms on both sides must be equal. Similarly, the coefficient of x and x^2 on both sides must be equal. So

$$C = 1$$
$$A + B + 2C = 0$$
$$A + C = 1.$$

Solving these equations gives $A = 0$, $B = -2$ and $C = 1$. The integral becomes

$$\int \frac{1 + x^2}{(1 + x)^2 x} dx = -2 \int \frac{1}{(1 + x)^2} dx + \int \frac{1}{x} dx = \frac{2}{1 + x} + \ln |x| + K.$$

We use K as the constant of integration, since we already used C in the problem.

141. Using the substitution $w = \sin x$, we get $dw = \cos x \, dx$, so we have

$$\int \frac{\cos x}{\sin^3 x + \sin x} dx = \int \frac{dw}{w^3 + w}.$$

But

$$\frac{1}{w^3 + w} = \frac{1}{w(w^2 + 1)} = \frac{1}{w} - \frac{w}{w^2 + 1},$$

so

$$\int \frac{\cos x}{\sin^3 x + \sin x} dx = \int \left(\frac{1}{w} - \frac{w}{w^2 + 1} \right) dw$$

$$= \ln |w| - \frac{1}{2} \ln \left| w^2 + 1 \right| + C$$

$$= \ln |\sin x| - \frac{1}{2} \ln \left| \sin^2 x + 1 \right| + C.$$

145. To find $\int we^{-w}\, dw$, integrate by parts, with $u = w$ and $v' = e^{-w}$. Then $u' = 1$ and $v = -e^{-w}$.
Then

$$\int we^{-w}\, dw = -we^{-w} + \int e^{-w}\, dw = -we^{-w} - e^{-w} + C.$$

Thus

$$\int_0^\infty we^{-w}\, dw = \lim_{b \to \infty} \int_0^b we^{-w}\, dw = \lim_{b \to \infty} \left. \left(-we^{-w} - e^{-w}\right) \right|_0^b = 1.$$

149. We find the exact value:

$$\int_{10}^\infty \frac{1}{z^2 - 4}\, dz = \int_{10}^\infty \frac{1}{(z+2)(z-2)}\, dz$$

$$= \lim_{b \to \infty} \int_{10}^b \frac{1}{(z+2)(z-2)}\, dz$$

$$= \lim_{b \to \infty} \left. \frac{1}{4}\left(\ln|z-2| - \ln|z+2|\right) \right|_{10}^b$$

$$= \frac{1}{4} \lim_{b \to \infty} \left[\left(\ln|b-2| - \ln|b+2|\right) - \left(\ln 8 - \ln 12\right)\right]$$

$$= \frac{1}{4} \lim_{b \to \infty} \left[\left(\ln \frac{b-2}{b+2}\right) + \ln \frac{3}{2}\right]$$

$$= \frac{1}{4}\left(\ln 1 + \ln 3/2\right) = \frac{\ln 3/2}{4}.$$

153. The integrand $\dfrac{x}{x+1} \to 1$ as $x \to \infty$, so there's no way $\displaystyle\int_1^\infty \frac{x}{x+1}\, dx$ can converge.

Problems

157. Since $(e^x)^2 = e^{2x}$, we have

$$\text{Area} = \int_0^1 (e^x)^2\, dx = \int_0^1 e^{2x}\, dx = \left. \frac{1}{2}e^{2x} \right|_0^1 = \frac{1}{2}(e^2 - 1).$$

161. In the interval of $0 \le x \le 2$, the equation $y = \sqrt{4 - x^2}$ represents one quadrant of the radius 2 circle centered at the origin. Interpreting the integral as an area gives a value of $\frac{1}{4}\pi 2^2 = \pi$.

165. *First solution*: After the substitution $w = x + 1$, the first integral becomes

$$\int \frac{w - 1}{w}\, dw = w - \int w^{-1}\, dw.$$

With this same substitution, the second integral becomes

$$\int w^{-1}\, dw.$$

Second solution: We note that the sum of the integrands is 1, so the sum of the integrals is x. Thus

$$\int \frac{x}{x+1}\, dx = x - \int \frac{1}{x+1}\, dx.$$

169. (a) (i) Multiplying out gives

$$\int (x^2 + 10x + 25)\, dx = \frac{x^3}{3} + 5x^2 + 25x + C.$$

(ii) Substituting $w = x + 5$, so $dw = dx$, gives

$$\int (x+5)^2\, dx = \int w^2\, dw = \frac{w^3}{3} + C = \frac{(x+5)^3}{3} + C.$$

(b) The results of the two calculations are not the same since

$$\frac{(x+5)^3}{3} + C = \frac{x^3}{3} + \frac{15x^2}{3} + \frac{75x}{3} + \frac{125}{3} + C.$$

However they differ only by a constant, $125/3$, as guaranteed by the Fundamental Theorem of Calculus.

173. (a) i. 0 ii. $\frac{2}{\pi}$ iii. $\frac{1}{2}$

(b) Average value of $f(t) <$ Average value of $k(t) <$ Average value of $g(t)$

We can look at the three functions in the range $-\frac{\pi}{2} \le x \le \frac{3\pi}{2}$, since they all have periods of 2π ($|\cos t|$ and $(\cos t)^2$ also have a period of π, but that does not hurt our calculation). It is clear from the graphs of the three functions below that the average value for $\cos t$ is 0 (since the area above the x-axis is equal to the area below it), while the average values for the other two are positive (since they are everywhere positive, except where they are 0).

It is also fairly clear from the graphs that the average value of $g(t)$ is greater than the average value of $k(t)$; it is also possible to see this algebraically, since

$$(\cos t)^2 = |\cos t|^2 \le |\cos t|$$

because $|\cos t| \le 1$ (and both of these $\le$'s are $<$'s at all the points where the functions are not 0 or 1).

177. (a) $f(x) = 1 + e^{-x}$ is concave up for $0 \le x \le 0.5$, so trapezoids will overestimate $\int_0^{0.5} f(x)dx$, and the midpoint rule will underestimate.

(b) $f(x) = e^{-x^2}$ is concave down for $0 \le x \le 0.5$, so trapezoids will underestimate $\int_0^{0.5} f(x)dx$ and midpoint will overestimate the integral.

(c) Both the trapezoid rule and the midpoint rule will give the exact value of the integral. Note that upper and lower sums will not, unless the line is horizontal.

181. If $I(t)$ is average per capita income t years after 2005, then $I'(t) = r(t)$.

(a) Since $t = 10$ in 2015, by the Fundamental Theorem,

$$I(10) - I(0) = \int_0^{10} r(t)\, dt = \int_0^{10} 1556.37 e^{0.045t}\, dt = \frac{1556.37}{0.045} e^{0.045t} \bigg|_0^{10} = 19{,}655.65 \text{ dollars.}$$

so $I(10) = 34{,}586 + 19{,}655.65 = 54{,}241.65$.

(b) We have

$$I(t) - I(0) = \int_0^t r(t)\, dt = \int_0^t 1556.37 e^{0.045t}\, dt = \frac{1556.37}{0.045} e^{0.045t} \bigg|_0^t = 34{,}586 \left(e^{0.045t} - 1 \right).$$

Thus, since $I(0) = 34{,}586$,

$$I(t) = 34{,}586 + 34{,}586(e^{0.045t} - 1) = 34{,}586 e^{0.045t} \text{ dollars.}$$

185. We want to calculate

$$\int_0^1 C_n \sin(n\pi x) \cdot C_m \sin(m\pi x)\, dx.$$

We use II-11 from the table of integrals with $a = n\pi$, $b = m\pi$. Since $n \neq m$, we see that

$$\int_0^1 \Psi_n(x) \cdot \Psi_m(x)\, dx = C_n C_m \int_0^1 \sin(n\pi x)\sin(m\pi x)\, dx$$

$$= \frac{C_n C_m}{m^2\pi^2 - n^2\pi^2}\left(n\pi\cos(n\pi x)\sin(m\pi x) - m\pi\sin(n\pi x)\cos(m\pi x)\right)\Big|_0^1$$

$$= \frac{C_n C_m}{(m^2 - n^2)\pi^2}\Big(n\pi\cos(n\pi)\sin(m\pi) - m\pi\sin(n\pi)\cos(m\pi)$$

$$-n\pi\cos(0)\sin(0) + m\pi\sin(0)\cos(0)\Big)$$

$$= 0$$

since $\sin(0) = \sin(n\pi) = \sin(m\pi) = 0$.

CAS Challenge Problems

189. (a) A possible answer is

$$\int \sin x \cos x \cos(2x)\, dx = -\frac{\cos(4x)}{16}.$$

Different systems may give the answer in a different form.

(b)

$$\frac{d}{dx}\left(-\frac{\cos(4x)}{16}\right) = \frac{\sin(4x)}{4}.$$

(c) Using the double angle formula $\sin 2A = 2\sin A\cos A$ twice, we get

$$\frac{\sin(4x)}{4} = \frac{2\sin(2x)\cos(2x)}{4} = \frac{2\cdot 2\sin x\cos x\cos(2x)}{4} = \sin x\cos x\cos(2x).$$

CHECK YOUR UNDERSTANDING

1. True. Let $w = f(x)$, so $dw = f'(x)\, dx$, then

$$\int f'(x)\cos(f(x))\, dx = \int \cos w\, dw = \sin w + C = \sin(f(x)) + C.$$

5. False. Completing the square gives

$$\int \frac{dx}{x^2 + 4x + 5} = \int \frac{dx}{(x+2)^2 + 1} = \arctan(x+2) + C.$$

9. True. $y^2 - 1$ is concave up, and the midpoint rule always underestimates for a function that is concave up.

13. False. Let $f(x) = 1/(x+1)$. Then

$$\int_0^\infty \frac{1}{x+1}\, dx = \lim_{b\to\infty} \ln|x+1|\Big|_0^b = \lim_{b\to\infty} \ln(b+1),$$

but $\lim_{b\to\infty} \ln(b+1)$ does not exist.

17. True. By properties of integrals and limits,

$$\lim_{b\to\infty} \int_0^b af(x)\, dx = a \lim_{b\to\infty} \int_0^b f(x)\, dx.$$

Thus, the limit on the left of the equation is finite exactly when the limit on the right side of the equation is finite. Thus $\int_0^\infty af(x)\, dx$ converges if $\int_0^\infty f(x)\, dx$ converges.

21. False. The subdivision size $\Delta x = (1/10)(6-2) = 4/10$.

25. True. We have

$$\text{LEFT}(n) - \text{RIGHT}(n) = (f(x_0) + f(x_1) + \cdots + f(x_{n-1}))\Delta x - (f(x_1) + f(x_2) + \cdots + f(x_n))\Delta x.$$

On the right side of the equation, all terms cancel except the first and last, so:

$$\text{LEFT}(n) - \text{RIGHT}(n) = (f(x_0) - f(x_n))\Delta x = (f(2) - f(6))\Delta x.$$

This is also discussed in Section 5.1.

29. False. This is true if f is an increasing function or if f is a decreasing function, but it is not true in general. For example, suppose that $f(2) = f(6)$. Then $\text{LEFT}(n) = \text{RIGHT}(n)$ for all n, which means that if $\int_2^6 f(x)dx$ lies between $\text{LEFT}(n)$ and $\text{RIGHT}(n)$, then it must equal $\text{LEFT}(n)$, which is not always the case.

For example, if $f(x) = (x-4)^2$ and $n = 1$, then $f(2) = f(6) = 4$, so

$$\text{LEFT}(1) = \text{RIGHT}(1) = 4 \cdot (6-2) = 16.$$

However

$$\int_2^6 (x-4)^2 dx = \left.\frac{(x-4)^3}{3}\right|_2^6 = \frac{2^3}{3} - \left(-\frac{2^3}{3}\right) = \frac{16}{3}.$$

In this example, since $\text{LEFT}(n) = \text{RIGHT}(n)$, we have $\text{TRAP}(n) = \text{LEFT}(n)$. However trapezoids overestimate the area, since the graph of f is concave up. This is also discussed in Section 7.5.

CHAPTER EIGHT

Solutions for Section 8.1

Exercises

1. Each strip is a rectangle of length 3 and width Δx, so

$$\text{Area of strip} = 3\Delta x, \quad \text{so}$$

$$\text{Area of region} = \int_0^5 3\,dx = 3x\Big|_0^5 = 15.$$

Check: This area can also be computed using Length $\times$ Width $= 5 \cdot 3 = 15$.

5. The strip has width Δy, so the variable of integration is y. The length of the strip is x. Since $x^2 + y^2 = 10$ and the region is in the first quadrant, solving for x gives $x = \sqrt{10 - y^2}$. Thus

$$\text{Area of strip} \approx x\Delta y = \sqrt{10 - y^2}\,dy.$$

The region stretches from $y = 0$ to $y = \sqrt{10}$, so

$$\text{Area of region} = \int_0^{\sqrt{10}} \sqrt{10 - y^2}\,dy.$$

Evaluating using VI-30 from the Table of Integrals, we have

$$\text{Area} = \frac{1}{2}\left(y\sqrt{10 - y^2} + 10\arcsin\left(\frac{y}{\sqrt{10}}\right)\right)\Bigg|_0^{\sqrt{10}} = 5(\arcsin 1 - \arcsin 0) = \frac{5}{2}\pi.$$

Check: This area can also be computed using the formula $\frac{1}{4}\pi r^2 = \frac{1}{4}\pi(\sqrt{10})^2 = \frac{5}{2}\pi$.

9. Each slice is a circular disk with radius $r = 2$ cm.

$$\text{Volume of disk} = \pi r^2 \Delta x = 4\pi\Delta x \text{ cm}^3.$$

Summing over all disks, we have

$$\text{Total volume} \approx \sum 4\pi\Delta x \text{ cm}^3.$$

Taking a limit as $\Delta x \to 0$, we get

$$\text{Total volume} = \lim_{\Delta x \to 0}\sum 4\pi\Delta x = \int_0^9 4\pi\,dx \text{ cm}^3.$$

Evaluating gives

$$\text{Total volume} = 4\pi x\Big|_0^9 = 36\pi \text{ cm}^3.$$

Check: The volume of the cylinder can also be calculated using the formula $V = \pi r^2 h = \pi 2^2 \cdot 9 = 36\pi$ cm^3.

13. Each slice is a circular disk. See Figure 8.1. The radius of the sphere is 5 mm, and the radius r at height y is given by the Pythagorean Theorem

$$y^2 + r^2 = 5^2.$$

Solving gives $r = \sqrt{5^2 - y^2}$ mm. Thus,

$$\text{Volume of disk} \approx \pi r^2 \Delta y = \pi(5^2 - y^2)\Delta y \text{ mm}^3.$$

Summing over all disks, we have

$$\text{Total volume} \approx \sum \pi(5^2 - y^2)\Delta y \text{ mm}^3.$$

Taking the limit as $\Delta y \to 0$, we get

$$\text{Total volume} = \lim_{\Delta y \to 0} \sum \pi(5^2 - y^2)\Delta y = \int_0^5 \pi(5^2 - y^2)\, dy \text{ mm}^3.$$

Evaluating gives

$$\text{Total volume} = \pi\left(25y - \frac{y^3}{3}\right)\Bigg|_0^5 = \frac{250}{3}\pi \text{ mm}^3.$$

Check: The volume of a hemisphere can be calculated using the formula $V = \frac{2}{3}\pi r^3 = \frac{2}{3}\pi 5^3 = \frac{250}{3}\pi \text{ mm}^3$.

Figure 8.1

Problems

17. Quarter circle of radius $r = \sqrt{15}$. See Figure 8.2.

Figure 8.2

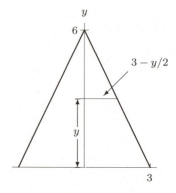

Figure 8.3

21. Cone with height 6 and radius 3. See Figure 8.3.

25.

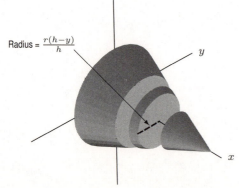

This cone is what you get when you rotate the line $x = r(h - y)/h$ about the y-axis. So slicing perpendicular to the y-axis yields

$$V = \int_{y=0}^{y=h} \pi x^2\, dy = \pi \int_0^h \left(\frac{(h-y)r}{h}\right)^2 dy$$

$$= \pi \frac{r^2}{h^2} \int_0^h (h^2 - 2hy + y^2)\, dy$$

$$= \frac{\pi r^2}{h^2}\left[h^2 y - hy^2 + \frac{y^3}{3}\right]\Bigg|_0^h = \frac{\pi r^2 h}{3}.$$

Solutions for Section 8.2

Exercises

1. The volume is given by

$$V = \int_0^1 \pi y^2 \, dx = \int_0^1 \pi x^4 \, dx = \pi \frac{x^5}{5} \Big|_0^1 = \frac{\pi}{5}.$$

5. The volume is given by

$$V = \int_{-1}^1 \pi y^2 \, dx = \int_{-1}^1 \pi (e^x)^2 \, dx = \int_{-1}^1 \pi e^{2x} \, dx = \frac{\pi}{2} e^{2x} \Big|_{-1}^1 = \frac{\pi}{2}(e^2 - e^{-2}).$$

9. Since the graph of $y = x^2$ is below the graph of $y = x$ for $0 \le x \le 1$, the volume is given by

$$V = \int_0^1 \pi x^2 \, dx - \int_0^1 \pi (x^2)^2 \, dx = \pi \int_0^1 (x^2 - x^4) \, dx = \pi \left(\frac{x^3}{3} - \frac{x^5}{5} \right) \Big|_0^1 = \frac{2\pi}{15}.$$

13. Since $f'(x) = 1/(x + 1)$, we evaluate the integral numerically to get

$$\text{Arc length} = \int_0^2 \sqrt{1 + \left(\frac{1}{x+1} \right)^2} \, dx = 2.302.$$

17. The length is

$$\int_1^2 \sqrt{(x'(t))^2 + (y'(t))^2 + (z'(t))^2} \, dt = \int_1^2 \sqrt{5^2 + 4^2 + (-1)^2} \, dt = \sqrt{42}.$$

This is the length of a straight line from the point $(8, 5, 2)$ to $(13, 9, 1)$.

Problems

21. The two functions intersect at $(0,0)$ and $(8, 2)$. We slice the volume with planes perpendicular to the line $x = 9$. This divides the solid into thin washers with volume

$$\text{Volume of slice} = \pi r_{out}^2 \Delta y - \pi r_{in}^2 \Delta y.$$

The outer radius is the horizontal distance from the line $x = 9$ to the curve $x = y^3$, so $r_{out} = 9 - y^3$. Similarly, the inner radius is the horizontal distance from the line $x = 9$ to the curve $x = 4y$, so $r_{in} = 9 - 4y$. Integrating from $y = 0$ to $y = 2$ we have

$$V = \int_0^2 \left[\pi (9 - y^3)^2 - \pi (9 - 4y)^2 \right] dy.$$

25. One arch of the sine curve lies between $x = -\pi/2$ and $x = \pi/2$. Since $d(\cos x)/dx = -\sin x$, evaluating the integral numerically gives

$$\text{Arc length} = \int_{-\pi/2}^{\pi/2} \sqrt{1 + \sin^2 x} \, dx = 3.820.$$

29.

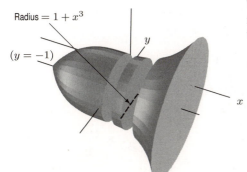

Radius $= 1 + x^3$

$(y = -1)$

We slice the region perpendicular to the x–axis. The Riemann sum we get is $\sum \pi (x^3 + 1)^2 \Delta x$. So the volume V is the integral

$$V = \int_{-1}^1 \pi (x^3 + 1)^2 \, dx$$

$$= \pi \int_{-1}^1 (x^6 + 2x^3 + 1) \, dx$$

$$= \pi \left(\frac{x^7}{7} + \frac{x^4}{2} + x \right) \Big|_{-1}^1$$

$$= (16/7)\pi \approx 7.18.$$

33. Slice the object into rings vertically, as is Figure 8.4. A typical ring has thickness Δx and outer radius $y = 1$ and inner radius $y = x^2$.

$$\text{Volume of slice} \approx \pi 1^2 \Delta x - \pi y^2 \Delta x = \pi(1 - x^4)\,\Delta x.$$

$$\text{Volume of solid} = \lim_{\Delta x \to 0} \sum \pi(1 - x^4)\,\Delta x = \int_0^1 \pi(1 - x^4)\,dx = \pi\left(x - \frac{x^5}{5}\right)\Bigg|_0^1 = \frac{4}{5}\pi.$$

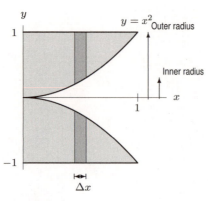

Figure 8.4: Cross-section of solid

Figure 8.5: Base of solid

37. An equilateral triangle of side s has height $\sqrt{3}s/2$ and

$$\text{Area} = \frac{1}{2} \cdot s \cdot \frac{\sqrt{3}s}{2} = \frac{\sqrt{3}}{4}s^2.$$

Slicing perpendicularly to the y-axis gives equilateral triangles whose thickness is Δy and whose side is $x = \sqrt{y}$. See Figure 8.5. Thus

$$\text{Volume of triangular slice} \approx \frac{\sqrt{3}}{4}(\sqrt{y})^2\Delta y = \frac{\sqrt{3}}{4}y\,\Delta y.$$

$$\text{Volume of solid} = \int_0^1 \frac{\sqrt{3}}{4}y\,dy = \frac{\sqrt{3}}{4}\frac{y^2}{2}\Bigg|_0^1 = \frac{\sqrt{3}}{8}.$$

41. We now slice perpendicular to the x-axis. As stated in the problem, the cross-sections obtained thereby will be squares, with base length e^x. The volume of one square slice is $(e^x)^2\,dx$. (See Figure 8.6.) Adding the volumes of the slices yields

$$\text{Volume} = \int_{x=0}^{x=1} y^2\,dx = \int_0^1 e^{2x}\,dx = \frac{e^{2x}}{2}\Bigg|_0^1 = \frac{e^2 - 1}{2} = 3.195.$$

Figure 8.6

45. If $y = e^{-x^2/2}$, then $x = \sqrt{-2 \ln y}$. (Note that since $0 < y \le 1$, $\ln y \le 0$.) A typical slice has thickness Δy and radius x. See Figure 8.7. So

$$\text{Volume of slice} = \pi x^2 \, \Delta y = -2\pi \ln y \, \Delta y.$$

Thus,

$$\text{Total volume} = -2\pi \int_0^1 \ln y \, dy.$$

Since $\ln y$ is not defined at $y = 0$, this is an improper integral:

$$\text{Total Volume} = -2\pi \int_0^1 \ln y \, dy = -2\pi \lim_{a \to 0} \int_a^1 \ln y \, dy$$

$$= -2\pi \lim_{a \to 0} (y \ln y - y) \Big|_a^1 = -2\pi \lim_{a \to 0} (-1 - a \ln a + a).$$

By looking at the graph of $x \ln x$ on a calculator, we see that $\lim_{a \to 0} a \ln a = 0$. Thus,

$$\text{Total volume} = -2\pi(-1) = 2\pi.$$

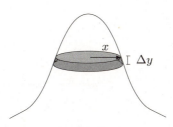

Figure 8.7

49. We can find the volume of the tree by slicing it into a series of thin horizontal cylinders of height dh and circumference C. The volume of each cylindrical disk will then be

$$V = \pi r^2 \, dh = \pi \left(\frac{C}{2\pi} \right)^2 \, dh = \frac{C^2 \, dh}{4\pi}.$$

Summing all such cylinders, we have the total volume of the tree as

$$\text{Total volume} = \frac{1}{4\pi} \int_0^{120} C^2 \, dh.$$

We can estimate this volume using a trapezoidal approximation to the integral with $\Delta h = 20$:

$$\text{LEFT estimate} = \frac{1}{4\pi} [20(31^2 + 28^2 + 21^2 + 17^2 + 12^2 + 8^2)] = \frac{1}{4\pi} (53660).$$

$$\text{RIGHT estimate} = \frac{1}{4\pi} [20(28^2 + 21^2 + 17^2 + 12^2 + 8^2 + 2^2)] = \frac{1}{4\pi} (34520).$$

$$\text{TRAP} = \frac{1}{4\pi} (44090) \approx 3509 \text{ cubic inches.}$$

53. As can be seen in Figure 8.8, the region has three straight sides and one curved one. The lengths of the straight sides are 1, 1, and e. The curved side is given by the equation $y = f(x) = e^x$. We can find its length by the formula

$$\int_0^1 \sqrt{1 + f'(x)^2} \, dx = \int_0^1 \sqrt{1 + (e^x)^2} \, dx = \int_0^1 \sqrt{1 + e^{2x}} \, dx.$$

Evaluating the integral numerically gives 2.0035. The total length, therefore, is about $1 + 1 + e + 2.0035 \approx 6.722$.

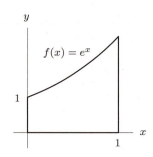

Figure 8.8

57. (a) For $n = 1$, we have

$$|x| + |y| = 1.$$

In the first quadrant, the equation is the line

$$x + y = 1.$$

By symmetry, the graph in the other quadrants gives the square in Figure 8.9.

For $n = 2$, the equation is of a circle of radius 1, centered at the origin:

$$x^2 + y^2 = 1.$$

For $n = 4$, the equation is

$$x^4 + y^4 = 1.$$

The graph is similar to a circle, but bulging out more. See Figure 8.9.

(b) For $n = 1$, the arc length is the perimeter of the square. Each side is the hypotenuse of a right triangle of sides $1, 1, \sqrt{2}$. Thus

$$\text{Arc length } = 4\sqrt{2} = 5.657.$$

For $n = 2$, the arc length is the perimeter of the circle of radius 1. Thus

$$\text{Arc length } = 2\pi \cdot 1 = 2\pi = 6.283.$$

For $n = 4$, we find the arc length using the formula

$$L = \int_a^b \sqrt{1 + (f'(x))^2}\, dx.$$

We find the arc length of the top half of the curve, given by $y = (1 - x^4)^{1/4}$, and double it. Since

$$\frac{dy}{dx} = \frac{1}{4}(1 - x^4)^{-3/4}(-4x^3) = \frac{-x^3}{(1 - x^4)^{3/4}},$$

$$\text{Arc length } = 2\int_{-1}^{1} \sqrt{1 + \left(\frac{-x^3}{(1 - x^4)^{3/4}}\right)^2}\, dx = 2\int_{-1}^{1} \sqrt{1 + \frac{x^6}{(1 - x^4)^{3/2}}}\, dx.$$

The integral is improper because the integral is not defined at $x = \pm 1$. Using numerical methods, we find

$$\text{Arc length } = 2\int_{-1}^{1} \sqrt{1 + \frac{x^6}{(1 - x^4)^{3/2}}}\, dx = 7.018.$$

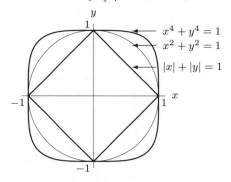

Figure 8.9

Solutions for Section 8.3

Exercises

1. With $r = 1$ and $\theta = 2\pi/3$, we find $x = r\cos\theta = 1 \cdot \cos(2\pi/3) = -1/2$ and $y = r\sin\theta = 1 \cdot \sin(2\pi/3) = \sqrt{3}/2$.
The rectangular coordinates are $(-1/2, \sqrt{3}/2)$.

5. With $x = 1$ and $y = 1$, find r from $r = \sqrt{x^2 + y^2} = \sqrt{1^2 + 1^2} = \sqrt{2}$. Find θ from $\tan\theta = y/x = 1/1 = 1$. Thus,
$\theta = \tan^{-1}(1) = \pi/4$. Since $(1,1)$ is in the first quadrant this is a correct θ. The polar coordinates are $(\sqrt{2}, \pi/4)$.

9. (a) Table 8.1 contains values of $r = 1 - \sin\theta$, both exact and rounded to one decimal.

Table 8.1

θ	0	$\pi/3$	$\pi/2$	$2\pi/3$	π	$4\pi/3$	$3\pi/2$	$5\pi/3$	2π	$7\pi/3$	$5\pi/2$	$8\pi/3$
r	1	$1 - \sqrt{3}/2$	0	$1 - \sqrt{3}/2$	1	$1 + \sqrt{3}/2$	2	$1 + \sqrt{3}/2$	1	$1 - \sqrt{3}/2$	0	$1 - \sqrt{3}/2$
r	1	0.134	0	0.134	1	1.866	2	1.866	1	0.134	0	0.134

(b) See Figure 8.10.

Figure 8.10

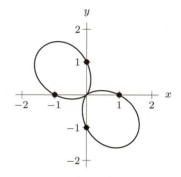

Figure 8.11

(c) The circle has equation $r = 1/2$. The cardioid is $r = 1 - \sin\theta$. Solving these two simultaneously gives

$$1/2 = 1 - \sin\theta,$$

or

$$\sin\theta = 1/2.$$

Thus, $\theta = \pi/6$ or $5\pi/6$. This gives the points $(x, y) = ((1/2)\cos\pi/6, (1/2)\sin\pi/6) = (\sqrt{3}/4, 1/4)$ and $(x, y) = ((1/2)\cos 5\pi/6, (1/2)\sin 5\pi/6) = (-\sqrt{3}/4, 1/4)$ as the location of intersection.

(d) The curve $r = 1 - \sin 2\theta$, pictured in Figure 8.11, has two regions instead of the one region that $r = 1 - \sin\theta$ has.
This is because $1 - \sin 2\theta$ will be 0 twice for every 2π cycle in θ, as opposed to once for every 2π cycle in θ for $1 - \sin\theta$.

13. See Figures 8.12 and 8.13. The first curve will be similar to the second curve, except the cardioid (heart) will be rotated clockwise by $90°$ ($\pi/2$ radians). This makes sense because of the identity $\sin\theta = \cos(\theta - \pi/2)$.

Figure 8.12: $r = 1 - \cos\theta$

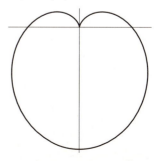

Figure 8.13: $r = 1 - \sin\theta$

17. The region is given by $\sqrt{8} \le r \le \sqrt{18}$ and $\pi/4 \le \theta \le \pi/2$.

21. We have $\cos \theta = x/\sqrt{x^2 + y^2}$ for each point. By definition $\cos^{-1}(x/\sqrt{x^2 + y^2})$ is an angle between 0 and π, in quadrant I or II, which matches points A and B. The formula is valid for A and B.

Problems

25. The region between the spirals is shaded in Figure 8.14.

$$\text{Area} = \frac{1}{2} \int_0^{2\pi} \left((2\theta)^2 - \theta^2 \right) d\theta = \frac{1}{2} \int_0^{2\pi} 3\theta^2 \, d\theta = \frac{1}{2} \theta^3 \Big|_0^{2\pi} = 4\pi^3.$$

Figure 8.14: Region between the inner spiral, $r = \theta$, and the outer spiral, $r = 2\theta$

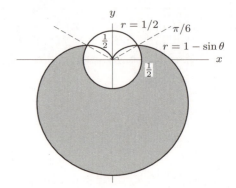

Figure 8.15

29. The two curves intersect where

$$1 - \sin \theta = \frac{1}{2}$$

$$\sin \theta = \frac{1}{2}$$

$$\theta = \frac{\pi}{6}, \frac{5\pi}{6}.$$

See Figure 8.15. We find the area of the right half and multiply that answer by 2 to get the entire area. The integrals can be computed numerically with a calculator or, as we show, using integration by parts or formula IV-17 in the integral tables.

$$\begin{aligned}
\text{Area of right half} &= \frac{1}{2} \int_{-\pi/2}^{\pi/6} \left((1 - \sin \theta)^2 - \left(\frac{1}{2}\right)^2 \right) d\theta \\
&= \frac{1}{2} \int_{-\pi/2}^{\pi/6} \left(1 - 2\sin \theta + \sin^2 \theta - \frac{1}{4} \right) d\theta \\
&= \frac{1}{2} \int_{-\pi/2}^{\pi/6} \left(\frac{3}{4} - 2\sin \theta + \sin^2 \theta \right) d\theta \\
&= \frac{1}{2} \left(\frac{3}{4}\theta + 2\cos \theta - \frac{1}{2}\sin \theta \cos \theta + \frac{1}{2}\theta \right) \Big|_{-\pi/2}^{\pi/6} \\
&= \frac{1}{2} \left(\frac{5\pi}{6} + \frac{7\sqrt{3}}{8} \right).
\end{aligned}$$

Thus,

$$\text{Total area} = \frac{5\pi}{6} + \frac{7\sqrt{3}}{8}.$$

33. (a) See Figure 8.16.

(b) The curves intersect when $r^2 = 2$

$$4 \cos 2\theta = 2$$
$$\cos 2\theta = \frac{1}{2}.$$

In the first quadrant:

$$2\theta = \frac{\pi}{3} \quad \text{so} \quad \theta = \frac{\pi}{6}.$$

Using symmetry, the area in the first quadrant can be multiplied by 4 to find the area of the total bounded region.

$$\begin{aligned}
\text{Area} &= 4\left(\frac{1}{2}\right) \int_0^{\pi/6} (4 \cos 2\theta - 2)\, d\theta \\
&= 2\left(\frac{4 \sin 2\theta}{2} - 2\theta\right)\Big|_0^{\pi/6} \\
&= 4 \sin \frac{\pi}{3} - \frac{2}{3}\pi \\
&= 4\frac{\sqrt{3}}{2} - \frac{2}{3}\pi \\
&= 2\sqrt{3} - \frac{2}{3}\pi = 1.370.
\end{aligned}$$

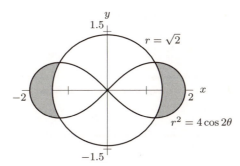

Figure 8.16

37. (a) Expressing x and y parametrically in terms of θ, we have

$$x = r \cos \theta = \frac{\cos \theta}{\theta} \quad \text{and} \quad y = r \sin \theta = \frac{\sin \theta}{\theta}.$$

The slope of the tangent line is given by

$$\frac{dy}{dx} = \frac{dy/d\theta}{dx/d\theta} = \left(\frac{\theta \cos \theta - \sin \theta}{\theta^2}\right) \Big/ \left(\frac{-\theta \sin \theta - \cos \theta}{\theta^2}\right) = \frac{\sin \theta - \theta \cos \theta}{\cos \theta + \theta \sin \theta}.$$

At $\theta = \pi/2$, we have

$$\frac{dy}{dx}\Big|_{\theta=\pi/2} = \frac{1 - (\pi/2)0}{0 + (\pi/2)1} = \frac{2}{\pi}.$$

At $\theta = \pi/2$, we have $x = 0$, $y = 2/\pi$, so the equation of the tangent line is

$$y = \frac{2}{\pi}x + \frac{2}{\pi}.$$

(b) As $\theta \to 0$,

$$x = \frac{\cos \theta}{\theta} \to \infty \quad \text{and} \quad y = \frac{\sin \theta}{\theta} \to 1.$$

Thus, $y = 1$ is a horizontal asymptote. See Figure 8.17.

Figure 8.17

41. Parameterized by θ, the curve $r = f(\theta)$ is given by $x = f(\theta)\cos\theta$ and $y = f(\theta)\sin\theta$. Then

$$
\begin{aligned}
\text{Arc length} &= \int_\alpha^\beta \sqrt{\left(\frac{dx}{d\theta}\right)^2 + \left(\frac{dy}{d\theta}\right)^2}\, d\theta \\
&= \int_\alpha^\beta \sqrt{(f'(\theta)\cos\theta - f(\theta)\sin\theta)^2 + (f'(\theta)\sin\theta + f(\theta)\cos\theta)^2}\, d\theta \\
&= \int_\alpha^\beta \sqrt{(f'(\theta))^2\cos^2\theta - 2f'(\theta)f(\theta)\cos\theta\sin\theta + (f(\theta))^2\sin^2\theta} \\
&\qquad\qquad \overline{+(f'(\theta))^2\sin^2\theta + 2f'(\theta)f(\theta)\sin\theta\cos\theta + (f(\theta))^2\cos^2\theta}\, d\theta \\
&= \int_\alpha^\beta \sqrt{(f'(\theta))^2(\cos^2\theta + \sin^2\theta) + (f(\theta))^2(\sin^2\theta + \cos^2\theta)}\, d\theta \\
&= \int_\alpha^\beta \sqrt{(f'(\theta))^2 + (f(\theta))^2}\, d\theta.
\end{aligned}
$$

Solutions for Section 8.4

Exercises

1. Since density is e^{-x} gm/cm,

$$
\text{Mass} = \int_0^{10} e^{-x}\, dx = -e^{-x}\Big|_0^{10} = 1 - e^{-10} \text{ gm.}
$$

5. (a) Figure 8.18 shows a graph of the density function.

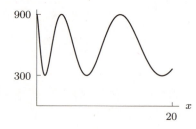

Figure 8.18

(b) Suppose we choose an x, $0 \le x \le 20$. We approximate the density of the number of the cars between x and $x + \Delta x$ miles as $\delta(x)$ cars per mile. Therefore, the number of cars between x and $x + \Delta x$ is approximately $\delta(x)\Delta x$. If we slice the 20 mile strip into N slices, we get that the total number of cars is

$$
C \approx \sum_{i=1}^N \delta(x_i)\Delta x = \sum_{i=1}^N \left[600 + 300\sin(4\sqrt{x_i + 0.15})\right]\Delta x,
$$

where $\Delta x = 20/N$. (This is a right-hand approximation; the corresponding left-hand approximation is $\displaystyle\sum_{i=0}^{N-1} \delta(x_i)\Delta x$.)

(c) As $N \to \infty$, the Riemann sum above approaches the integral

$$C = \int_0^{20} \left(600 + 300 \sin 4\sqrt{x + 0.15}\right) dx.$$

If we calculate the integral numerically, we find $C \approx 11513$. We can also find the integral exactly as follows:

$$C = \int_0^{20} \left(600 + 300 \sin 4\sqrt{x + 0.15}\right) dx$$

$$= \int_0^{20} 600 \, dx + \int_0^{20} 300 \sin 4\sqrt{x + 0.15} \, dx$$

$$= 12000 + 300 \int_0^{20} \sin 4\sqrt{x + 0.15} \, dx.$$

Let $w = \sqrt{x + 0.15}$, so $x = w^2 - 0.15$ and $dx = 2w \, dw$. Then

$$\int_{x=0}^{x=20} \sin 4\sqrt{x + 0.15} \, dx = 2 \int_{w=\sqrt{0.15}}^{w=\sqrt{20.15}} w \sin 4w \, dw, \text{ (using integral table III-15)}$$

$$= 2 \left[-\frac{1}{4} w \cos 4w + \frac{1}{16} \sin 4w \right] \Bigg|_{\sqrt{0.15}}^{\sqrt{20.15}}$$

$$\approx -1.624.$$

Using this, we have $C \approx 12000 + 300(-1.624) \approx 11513$, which matches our numerical approximation.

9. We slice the block horizontally. A slice has area $10 \cdot 3 = 30$ and thickness Δz. On such a slice, the density is approximately constant. Thus

$$\text{Mass of slice} \approx \text{Density} \cdot \text{Volume} \approx (2 - z) \cdot 30\Delta z,$$

so we have

$$\text{Mass of block} \approx \sum (2 - z) 30\Delta z.$$

In the limit as $\Delta z \to 0$, the sum becomes an integral and the approximation becomes exact. Thus

$$\text{Mass of block} = \int_0^1 (2 - z) 30 \, dz = 30 \left(2z - \frac{z^2}{2} \right) \Bigg|_0^1 = 30 \left(2 - \frac{1}{2} \right) = 45.$$

Problems

13. (a) We form a Riemann sum by slicing the region into concentric rings of radius r and width Δr. Then the volume deposited on one ring will be the height $H(r)$ multiplied by the area of the ring. A ring of width Δr will have an area given by

$$\text{Area} = \pi(r + \Delta r)^2 - \pi(r^2)$$
$$= \pi(r^2 + 2r\Delta r + (\Delta r)^2 - r^2)$$
$$= \pi(2r\Delta r + (\Delta r)^2).$$

Since Δr is approaching zero, we can approximate

$$\text{Area of ring} \approx \pi(2r\Delta r + 0) = 2\pi r\Delta r.$$

From this, we have

$$\Delta V \approx H(r) \cdot 2\pi r\Delta r.$$

Thus, summing the contributions from all rings we have

$$V \approx \sum H(r) \cdot 2\pi r\Delta r.$$

Taking the limit as $\Delta r \to 0$, we get

$$V = \int_0^5 2\pi r \left(0.115 e^{-2r} \right) dr.$$

(b) We use integration by parts:

$$V = 0.23\pi \int_0^5 \left(re^{-2r}\right) dr$$

$$= 0.23\pi \left(\frac{re^{-2r}}{-2} - \frac{e^{-2r}}{4}\right)\bigg|_0^5$$

$$\approx 0.181(\text{millimeters}) \cdot (\text{kilometers})^2 = 0.181 \cdot 10^{-3} \cdot 10^6 \text{ meters}^3 = 181 \text{ cubic meters.}$$

17. We slice time into small intervals. Since t is given in seconds, we convert the minute to 60 seconds. We consider water loss over the time interval $0 \le t \le 60$. We also need to convert inches into feet since the velocity is given in ft/sec. Since 1 inch = $1/12$ foot, the square hole has area $1/144$ square feet. For water flowing through a hole with constant velocity v, the amount of water which has passed through in some time, Δt, can be pictured as the rectangular solid in Figure 8.19, which has volume

$$\text{Area} \cdot \text{Height} = \text{Area} \cdot \text{Velocity} \cdot \text{Time.}$$

Figure 8.19: Volume of water passing through hole

Over a small time interval of length Δt, starting at time t, water flows with a nearly constant velocity $v = g(t)$ through a hole $1/144$ square feet in area. In Δt seconds, we know that

$$\text{Water lost} \approx \left(\frac{1}{144} \text{ ft}^2\right)(g(t) \text{ ft/sec})(\Delta t \text{ sec}) = \left(\frac{1}{144}\right)g(t)\,\Delta t \text{ ft}^3.$$

Adding the water from all subintervals gives

$$\text{Total water lost} \approx \sum \frac{1}{144}g(t)\,\Delta t \text{ ft}^3.$$

As $\Delta t \to 0$, the sum tends to the definite integral:

$$\text{Total water lost} = \int_0^{60} \frac{1}{144}g(t)\,dt \text{ ft}^3.$$

21. The center of mass is

$$\bar{x} = \frac{\int_0^\pi x(2 + \sin x)\,dx}{\int_0^\pi (2 + \sin x)\,dx}.$$

The numerator is $\int_0^\pi (2x + x\sin x)\,dx = (x^2 - x\cos x + \sin x)\bigg|_0^\pi = \pi^2 + \pi$.

The denominator is $\int_0^\pi (2 + \sin x)\,dx = (2x - \cos x)\bigg|_0^\pi = 2\pi + 2$. So the center of mass is at

$$\bar{x} = \frac{\pi^2 + \pi}{2\pi + 2} = \frac{\pi(\pi + 1)}{2(\pi + 1)} = \frac{\pi}{2}.$$

25. (a) Since the density is constant, the mass is the product of the area of the plate and its density.

$$\text{Area of the plate } = \int_0^1 x^2 \, dx = \frac{1}{3}x^3 \Big|_0^1 = \frac{1}{3} \text{ cm}^2.$$

Thus the mass of the plate is $2 \cdot 1/3 = 2/3$ gm.

(b) See Figure 8.20. Since the region is "fatter" closer to $x = 1$, $\bar{x}$ is greater than $1/2$.

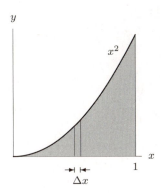

Figure 8.20

(c) To find the center of mass along the x-axis, we slice the region into vertical strips of width Δx. See Figure 8.20. Then

$$\text{Area of strip } = A_x(x)\Delta x \approx x^2 \Delta x$$

Then, since the density is 2 gm/cm^2, we have

$$\bar{x} = \frac{\int_0^1 2x^3 \, dx}{2/3} = \frac{3}{2} \cdot \frac{2x^4}{4} \Big|_0^1 = 3\left(\frac{1}{4}\right) = \frac{3}{4}\text{cm}.$$

This is greater than $1/2$, as predicted in part (b).

29. Since the density is constant, the total mass of the solid is the product of the volume of the solid and its density: $\delta\pi(1 - e^{-2})/2$. By symmetry, $\bar{y} = 0$. To find $\bar{x}$, we slice the solid into disks of width Δx, perpendicular to the x-axis. See Figure 8.21. A disk at x has radius $y = e^{-x}$, so

$$\text{Volume of disk } = A_x(x)\Delta x = \pi y^2 \Delta x = \pi e^{-2x} \Delta x.$$

Since the density is δ, we have

$$\bar{x} = \frac{\int_0^1 x \cdot \delta\pi e^{-2x} dx}{\text{Total mass}} = \frac{\delta\pi \int_0^1 x e^{-2x} dx}{\delta\pi(1 - e^{-2})/2} = \frac{2}{1 - e^{-2}} \int_0^1 x e^{-2x} \, dx.$$

The integral $\int x e^{-2x} \, dx$ can be done by parts: let $u = x$ and $v' = e^{-2x}$. Then $u' = 1$ and $v = e^{-2x}/(-2)$. So

$$\int x e^{-2x} \, dx = \frac{x e^{-2x}}{-2} - \int \frac{e^{-2x}}{-2} \, dx = \frac{x e^{-2x}}{-2} - \frac{e^{-2x}}{4}.$$

and then

$$\int_0^1 x e^{-2x} \, dx = \left(\frac{x e^{-2x}}{-2} - \frac{e^{-2x}}{4}\right)\Big|_0^1 = \left(\frac{e^{-2}}{-2} - \frac{e^{-2}}{4}\right) - \left(0 - \frac{1}{4}\right) = \frac{1 - 3e^{-2}}{4}.$$

The final result is:

$$\bar{x} = \frac{2}{1 - e^{-2}} \cdot \frac{1 - 3e^{-2}}{4} = \frac{1 - 3e^{-2}}{2 - 2e^{-2}} \approx 0.343.$$

Notice that $\bar{x}$ is less that $1/2$, as we would expect from the fact that the solid is wider near the origin.

$y = e^{-x}$

Radius $= y$

Figure 8.21

Solutions for Section 8.5

Exercises

1. The work done is given by

$$W = \int_1^2 3x \, dx = \frac{3}{2}x^2 \bigg|_1^2 = \frac{9}{2} \text{ joules.}$$

5. The force exerted on the satellite by the earth (and vice versa!) is GMm/r^2, where r is the distance from the center of the earth to the center of the satellite, m is the mass of the satellite, M is the mass of the earth, and G is the gravitational constant. So the total work done is

$$\int_{6.4 \cdot 10^6}^{8.4 \cdot 10^6} F \, dr = \int_{6.4 \cdot 10^6}^{8.4 \cdot 10^6} \frac{GMm}{r^2} \, dr = \left(\frac{-GMm}{r}\right) \bigg|_{6.4 \cdot 10^6}^{8.4 \cdot 10^6} \approx 1.489 \cdot 10^{10} \text{ joules.}$$

Problems

9. The bucket moves upward at $40/10 = 4$ meters/minute. If time is in minutes, at time t the bucket is at a height of $x = 4t$ meters above the ground. See Figure 8.22.

The water drips out at a rate of $5/10 = 0.5$ kg/minute. Initially there is 20 kg of water in the bucket, so at time t minutes, the mass of water remaining is

$$m = 20 - 0.5t \text{ kg.}$$

Consider the time interval between t and $t + \Delta t$. During this time the bucket moves a distance $\Delta x = 4\Delta t$ meters. So, during this interval,

$$\text{Work done} \approx mg\Delta x = (20 - 0.5t)g4\Delta t \text{ joules.}$$

$$\text{Total work done} = \lim_{\Delta t \to 0} \sum (20 - 0.5t)g4\Delta t = 4g \int_0^{10} (20 - 0.5t) \, dt$$

$$= 4g(20t - 0.25t^2) \bigg|_0^{10} = 700g = 700(9.8) = 6860 \text{ joules.}$$

Figure 8.22

13. Let x be the distance measured from the bottom the tank. See Figure 8.23. To pump a layer of water of thickness Δx at x feet from the bottom, the work needed is

$$(62.4)\pi 6^2 (20 - x)\Delta x.$$

Therefore, the total work is

$$W = \int_0^{10} 36 \cdot (62.4)\pi(20 - x)dx$$

$$= 36 \cdot (62.4)\pi\left(20x - \frac{1}{2}x^2\right)\Big|_0^{10}$$

$$= 36 \cdot (62.4)\pi(200 - 50)$$

$$\approx 1{,}058{,}591.1 \text{ ft-lb.}$$

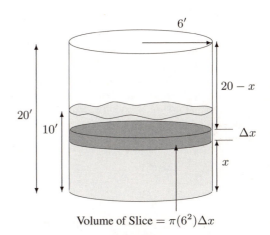

Volume of Slice $= \pi(6^2)\Delta x$

Figure 8.23

17. We slice the water horizontally as in Figure 8.24. We use similar triangles to find the radius r of the slice at height h in terms of h:

$$\frac{r}{h} = \frac{4}{12} \quad \text{so} \quad r = \frac{1}{3}h.$$

At height h,

$$\text{Volume of slice} \approx \pi r^2 \Delta h = \pi\left(\frac{1}{3}h\right)^2 \Delta h \text{ ft}^3.$$

The density of water is δ lb/ft^3, so

$$\text{Weight of slice} \approx \delta\pi\left(\frac{1}{3}h\right)^3 \Delta h \text{ lb.}$$

The water at height h must be lifted a distance of $12 - h$ ft, so

$$\text{Work to move slice} = \delta \cdot \text{Volume} \cdot \text{Distance lifted}$$

$$\approx \delta\left(\pi\left(\frac{1}{3}h\right)^2 \Delta h\right)(12 - h) \text{ ft-lb.}$$

The work done, W, to lift all the water is the sum of the work done on the pieces:

$$W \approx \sum \delta\pi(\frac{1}{3}h)^2\Delta h(12 - h) \text{ ft-lb.}$$

As $\Delta h \to 0$, we obtain a definite integral. Since h varies from $h = 0$ to $h = 9$, and $\delta = 62.4$, we have:

$$W = \int_0^9 \delta\pi(\frac{1}{3}h)^2(12 - h)dh = \frac{62.4\pi}{9}\int_0^9 (12h^2 - h^3)dh = \frac{62.4\pi}{9}\left(4h^3 - \frac{h^4}{4}\right)\Big|_0^9 = 27{,}788 \text{ ft-lb.}$$

The work to pump all the water out is 27,788 ft-lbs.

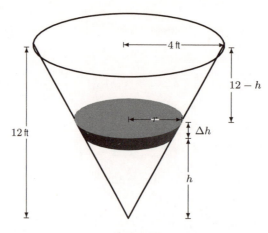

Figure 8.24

21. (a) Divide the wall into N horizontal strips, each of which is of height Δh. See Figure 8.25. The area of each strip is $1000\Delta h$, and the pressure at depth h_i is $62.4h_i$, so we approximate the force on the strip as $1000(62.4h_i)\Delta h$.

Figure 8.25

Therefore,

$$\text{Force on the Dam} \approx \sum_{i=0}^{N-1} 1000(62.4h_i)\Delta h.$$

(b) As $N \to \infty$, the Riemann sum becomes the integral, so the force on the dam is

$$\int_0^{50} (1000)(62.4h)\,dh = 62400\frac{h^2}{2}\bigg|_0^{50} = 78{,}000{,}000 \text{ pounds.}$$

25. (a) Since the density of water is $\delta = 1000 \text{ kg/m}^3$, at the base of the dam, water pressure $\delta gh = 1000 \cdot 9.8 \cdot 180 = 1.76 \cdot 10^6 \text{ nt/m}^2$.

(b) To set up a definite integral giving the force, we divide the dam into horizontal strips. We use horizontal strips because the pressure along each strip is approximately constant, since each part is at approximately the same depth. See Figure 8.26.

$$\text{Area of strip} = 2000\Delta h \text{ m}^2.$$

Pressure at depth of h meters $= \delta gh = 9800h \text{ nt/m}^2$. Thus,

$$\text{Force on strip} \approx \text{Pressure} \times \text{Area} = 9800h \cdot 2000\Delta h = 1.96 \cdot 10^7 h\Delta h \text{ nt.}$$

Summing over all strips and letting $\Delta h \to 0$ gives:

$$\text{Total force} = \lim_{\Delta h \to 0} \sum 1.96 \cdot 10^7 h\Delta h = 1.96 \cdot 10^7 \int_0^{180} h\,dh \text{ newtons.}$$

Evaluating gives

$$\text{Total force} = 1.96 \cdot 10^7 \frac{h^2}{2}\bigg|_0^{180} = 3.2 \cdot 10^{11} \text{ newtons.}$$

```
                    ─────────────── 2000 m ───────────────

        ┌───────────────────────────────────────────────┐    ↑
        │                                                 │    h
 180 m  │═════════════════════════════════════════════════│   ↓  Δh
        │                                                 │
        └───────────────────────────────────────────────┘
```

Figure 8.26

29. We need to divide the disk up into circular rings of charge and integrate their contributions to the potential (at P) from 0 to a. These rings, however, are not uniformly distant from the point P. A ring of radius z is $\sqrt{R^2 + z^2}$ away from point P (see Figure 8.27).

The ring has area $2\pi z\,\Delta z$, and charge $2\pi z\sigma\,\Delta z$. The potential of the ring is then $\dfrac{2\pi z\sigma\,\Delta z}{\sqrt{R^2 + z^2}}$ and the total potential at point P is

$$\int_0^a \frac{2\pi z\sigma\,dz}{\sqrt{R^2 + z^2}} = \pi\sigma \int_0^a \frac{2z\,dz}{\sqrt{R^2 + z^2}}.$$

We make the substitution $u = z^2$. Then $du = 2z\,dz$. We obtain

$$\pi\sigma \int_0^a \frac{2z\,dz}{\sqrt{R^2 + z^2}} = \pi\sigma \int_0^{a^2} \frac{du}{\sqrt{R^2 + u}} = \pi\sigma (2\sqrt{R^2 + u})\Big|_0^{a^2}$$

$$= \pi\sigma(2\sqrt{R^2 + z^2})\Big|_0^a = 2\pi\sigma(\sqrt{R^2 + a^2} - R).$$

(The substitution $u = R^2 + z^2$ or $\sqrt{R^2 + z^2}$ works also.)

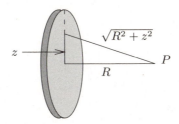

Figure 8.27

33. This time, let's split the second rod into small slices of length dr. See Figure 8.28. Each slice is of mass $\frac{M_2}{l_2}\,dr$, since the density of the second rod is $\frac{M_2}{l_2}$. Since the slice is small, we can treat it as a particle at distance r away from the end of the first rod, as in Problem 32. By that problem, the force of attraction between the first rod and particle is

$$\frac{GM_1 \frac{M_2}{l_2}\,dr}{(r)(r + l_1)}.$$

So the total force of attraction between the rods is

$$\int_a^{a+l_2} \frac{GM_1 \frac{M_2}{l_2}\,dr}{(r)(r + l_1)} = \frac{GM_1 M_2}{l_2} \int_a^{a+l_2} \frac{dr}{(r)(r + l_1)}$$

$$= \frac{GM_1 M_2}{l_2} \int_a^{a+l_2} \frac{1}{l_1}\left(\frac{1}{r} - \frac{1}{r + l_1}\right) dr.$$

$$= \frac{GM_1 M_2}{l_1 l_2} (\ln|r| - \ln|r + l_1|)\Big|_a^{a+l_2}$$

$$= \frac{GM_1 M_2}{l_1 l_2} [\ln|a + l_2| - \ln|a + l_1 + l_2| - \ln|a| + \ln|a + l_1|]$$

$$= \frac{GM_1 M_2}{l_1 l_2} \ln\left[\frac{(a + l_1)(a + l_2)}{a(a + l_1 + l_2)}\right].$$

This result is symmetric: if you switch l_1 and l_2 or M_1 and M_2, you get the same answer. That means it's not important which rod is "first," and which is "second."

Figure 8.28

Solutions for Section 8.6

Exercises

1. At any time t, in a time interval Δt, an amount of $1000\Delta t$ is deposited into the account. This amount earns interest for $(10 - t)$ years giving a future value of $1000e^{(0.08)(10-t)}$. Summing all such deposits, we have

$$\text{Future value} = \int_0^{10} 1000e^{0.08(10-t)}\, dt = \$15,319.30.$$

Problems

5.

$/year

t (years from present)

The graph reaches a peak each summer, and a trough each winter. The graph shows sunscreen sales increasing from cycle to cycle. This gradual increase may be due in part to inflation and to population growth.

9. You should choose the payment which gives you the highest present value. The immediate lump-sum payment of $2800 obviously has a present value of exactly $2800, since you are getting it now. We can calculate the present value of the installment plan as:

$$\text{PV} = 1000e^{-0.06(0)} + 1000e^{-0.06(1)} + 1000e^{-0.06(2)}$$
$$\approx \$2828.68.$$

Since the installment payments offer a (slightly) higher present value, you should accept this option.

13. (a) Suppose the oil extracted over the time period $[0, M]$ is S. (See Figure 8.29.) Since $q(t)$ is the rate of oil extraction, we have:

$$S = \int_0^M q(t)dt = \int_0^M (a - bt)dt = \int_0^M (10 - 0.1t)\, dt.$$

To calculate the time at which the oil is exhausted, set $S = 100$ and try different values of M. We find $M = 10.6$ gives

$$\int_0^{10.6} (10 - 0.1t)\, dt = 100,$$

so the oil is exhausted in 10.6 years.

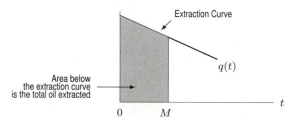

Extraction Curve

$q(t)$

Area below
the extraction curve
is the total oil extracted

0 M t

Figure 8.29

(b) Suppose p is the oil price, C is the extraction cost per barrel, and r is the interest rate. We have the present value of the profit as

$$\text{Present value of profit} = \int_0^M (p - C)q(t)e^{-rt}dt$$

$$= \int_0^{10.6} (20-10)(10-0.1t)e^{-0.1t}\, dt$$
$$= 624.9 \text{ million dollars.}$$

17.

$$\int_0^{q^*} (p^* - S(q))\, dq = \int_0^{q^*} p^*\, dq - \int_0^{q^*} S(q)\, dq$$
$$= p^* q^* - \int_0^{q^*} S(q)\, dq.$$

Using Problem 16, this integral is the extra amount consumers pay (i.e., suppliers earn over and above the minimum they would be willing to accept for supplying the good). It results from charging the equilibrium price.

Solutions for Section 8.7

Exercises

1. The two humps of probability in density (a) correspond to two intervals on which its cumulative distribution function is increasing. Thus (a) and (II) correspond.

A density function increases where its cumulative distribution funciton is concave up, and it decreases where its cumulative distribution function is concave down. Density (b) matches the distribution with both concave up and concave down sections, which is (I). Density (c) matches (III) which has a concave down section but no interval over which it is concave up.

5. Since the function takes on the value of 4, it cannot be a cdf (whose maximum value is 1). In addition, the function decreases for $x > c$, which means that it is not a cdf. Thus, this function is a pdf. The area under a pdf is 1, so $4c = 1$ giving $c = \frac{1}{4}$. The pdf is $p(x) = 4$ for $0 \le x \le \frac{1}{4}$, so the cdf is given in Figure 8.30 by

$$P(x) = \begin{cases} 0 & \text{for} \quad x < 0 \\ 4x & \text{for} \quad 0 \le x \le \dfrac{1}{4} \\ 1 & \text{for} \quad x > \dfrac{1}{4} \end{cases}$$

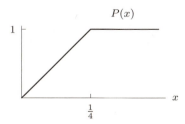

Figure 8.30

9. This function increases and levels off to c. The area under the curve is not finite, so it is not 1. Thus, the function must be a cdf, not a pdf, and $3c = 1$, so $c = 1/3$.

The pdf, $p(x)$ is the derivative, or slope, of the function shown, so, using $c = 1/3$,

$$p(x) = \begin{cases} 0 & \text{for} \quad x < 0 \\ (1/3 - 0)/(2 - 0) = 1/6 & \text{for} \quad 0 \le x \le 2 \\ (1 - 1/3)/(4 - 2) = 1/3 & \text{for} \quad 2 < x \le 4 \\ 0 & \text{for} \quad x > 4. \end{cases}$$

See Figure 8.31.

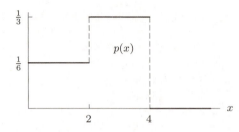

Figure 8.31

Problems

13. (a) $F(7) = 0.6$ tells us that 60% of the trees in the forest have height 7 meters or less.

(b) $F(7) > F(6)$. There are more trees of height less than 7 meters than trees of height less than 6 meters because every tree of height ≤ 6 meters also has height ≤ 7 meters.

17. (a) The area under the graph of the height density function $p(x)$ is concentrated in two humps centered at 0.5 m and 1.1 m. The plants can therefore be separated into two groups, those with heights in the range 0.3 m to 0.7 m, corresponding to the first hump, and those with heights in the range 0.9 m to 1.3 m, corresponding to the second hump. This grouping of the grasses according to height is probably close to the species grouping. Since the second hump contains more area than the first, there are more plants of the tall grass species in the meadow.

(b) As do all cumulative distribution functions, the cumulative distribution function $P(x)$ of grass heights rises from 0 to 1 as x increases. Most of this rise is achieved in two spurts, the first as x goes from 0.3 m to 0.7 m, and the second as x goes from 0.9 m to 1.3 m. The plants can therefore be separated into two groups, those with heights in the range 0.3 m to 0.7 m, corresponding to the first spurt, and those with heights in the range 0.9 m to 1.3 m, corresponding to the second spurt. This grouping of the grasses according to height is the same as the grouping we made in part (a), and is probably close to the species grouping.

(c) The fraction of grasses with height less than 0.7 m equals $P(0.7) = 0.25 = 25\%$. The remaining 75% are the tall grasses.

21. (a) We must have $\int_0^\infty f(t)dt = 1$, for even though it is possible that any given person survives the disease, everyone eventually dies. Therefore,

$$\int_0^\infty cte^{-kt}dt = 1.$$

Integrating by parts gives

$$\int_0^b cte^{-kt}dt = -\frac{c}{k}te^{-kt}\bigg|_0^b + \int_0^b \frac{c}{k}e^{-kt}dt$$

$$= \left(-\frac{c}{k}te^{-kt} - \frac{c}{k^2}e^{-kt}\right)\bigg|_0^b$$

$$= \frac{c}{k^2} - \frac{c}{k}be^{-kb} - \frac{c}{k^2}e^{-kb}.$$

As $b \to \infty$, we see

$$\int_0^\infty cte^{-kt}dt = \frac{c}{k^2} = 1 \quad \text{so} \quad c = k^2.$$

(b) We are told that $\int_0^5 f(t)dt = 0.4$, so using the fact that $c = k^2$ and the antiderivatives from part (a), we have

$$\int_0^5 k^2te^{-kt}dt = \left(-\frac{k^2}{k}te^{-kt} - \frac{k^2}{k^2}e^{-kt}\right)\bigg|_0^5$$

$$= 1 - 5ke^{-5k} - e^{-5k} = 0.4$$

so

$$5ke^{-5k} + e^{-5k} = 0.6.$$

Since this equation cannot be solved exactly, we use a calculator or computer to find $k = 0.275$. Since $c = k^2$, we have $c = (0.275)^2 = 0.076$.

(c) The cumulative death distribution function, $C(t)$, represents the fraction of the population that have died up to time t. Thus,

$$C(t) = \int_0^t k^2 x e^{-kx} dx = \left(-kxe^{-kx} - e^{-kx} \right) \Big|_0^t$$

$$= 1 - kte^{-kt} - e^{-kt}.$$

Solutions for Section 8.8

Exercises

1.

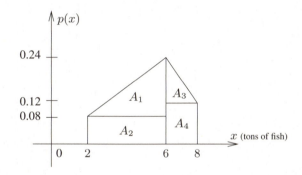

Splitting the figure into four pieces, we see that

$$\text{Area under the curve} = A_1 + A_2 + A_3 + A_4$$
$$= \frac{1}{2}(0.16)4 + 4(0.08) + \frac{1}{2}(0.12)2 + 2(0.12)$$
$$= 1.$$

We expect the area to be 1, since $\int_{-\infty}^{\infty} p(x)\,dx = 1$ for any probability density function, and $p(x)$ is 0 except when $2 \le x \le 8$.

Problems

5. (a) We can find the proportion of students by integrating the density $p(x)$ between $x = 1.5$ and $x = 2$:

$$P(2) - P(1.5) = \int_{1.5}^2 \frac{x^3}{4}\, dx$$

$$= \frac{x^4}{16} \Big|_{1.5}^2$$

$$= \frac{(2)^4}{16} - \frac{(1.5)^4}{16} = 0.684,$$

so that the proportion is $0.684 : 1$ or 68.4%.

(b) We find the mean by integrating x times the density over the relevant range:

$$\text{Mean} = \int_0^2 x \left(\frac{x^3}{4} \right) \, dx$$

$$= \int_0^2 \frac{x^4}{4} \, dx$$

$$= \frac{x^5}{20} \bigg|_0^2$$

$$= \frac{2^5}{20} = 1.6 \text{ hours.}$$

(c) The median will be the time T such that exactly half of the students are finished by time T, or in other words

$$\frac{1}{2} = \int_0^T \frac{x^3}{4} \, dx$$

$$\frac{1}{2} = \frac{x^4}{16} \bigg|_0^T$$

$$\frac{1}{2} = \frac{T^4}{16}$$

$$T = \sqrt[4]{8} = 1.682 \text{ hours.}$$

9. (a) Since $\mu = 100$ and $\sigma = 15$:

$$p(x) = \frac{1}{15\sqrt{2\pi}} e^{-\frac{1}{2}\left(\frac{x-100}{15} \right)^2}.$$

(b) The fraction of the population with IQ scores between 115 and 120 is (integrating numerically)

$$\int_{115}^{120} p(x) \, dx = \int_{115}^{120} \frac{1}{15\sqrt{2\pi}} e^{-\frac{(x-100)^2}{450}} \, dx$$

$$= \frac{1}{15\sqrt{2\pi}} \int_{115}^{120} e^{-\frac{(x-100)^2}{450}} \, dx$$

$$\approx 0.067 = 6.7\% \text{ of the population.}$$

13. It is not (a) since a probability density must be a non-negative function; not (c) since the total integral of a probability density must be 1; (b) and (d) are probability density functions, but (d) is not a good model. According to (d), the probability that the next customer comes after 4 minutes is 0. In real life there should be a positive probability of not having a customer in the next 4 minutes. So (b) is the best answer.

Solutions for Chapter 8 Review

Exercises

1. Vertical slices are circular. Horizontal slices would be similar to ellipses in cross-section, or at least ovals (a word derived from *ovum*, the Latin word for egg).

Figure 8.32

5. We slice the region vertically. Each rotated slice is approximately a cylinder with radius $y = x^2 + 1$ and thickness Δx. See Figure 8.33. The volume of a typical slice is $\pi(x^2 + 1)^2 \Delta x$. The volume, V, of the object is the sum of the volumes of the slices:

$$V \approx \sum \pi(x^2 + 1)^2 \Delta x.$$

As $\Delta x \to 0$ we obtain an integral.

$$V = \int_0^4 \pi(x^2 + 1)^2 dx = \pi \int_0^4 (x^4 + 2x^2 + 1)dx = \pi \left(\frac{x^5}{5} + \frac{2x^3}{3} + x \right) \Big|_0^4 = \frac{3772\pi}{15} = 790.006.$$

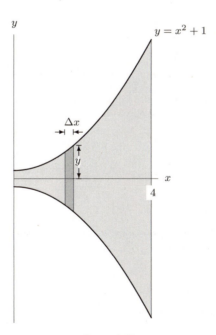

Figure 8.33

9. We divide the region into vertical strips of thickness Δx. As a slice is rotated about the x-axis, it creates a disk of radius r_{out} from which has been removed a smaller circular disk of inside radius r_{in}. We see in Figure 8.34 that $r_{\text{out}} = 2x$ and $r_{\text{in}} = x$. Thus,

$$\text{Volume of a slice } \approx \pi(r_{\text{out}})^2 \Delta x - \pi(r_{\text{in}})^2 \Delta x = \pi(2x)^2 \Delta x - \pi(x)^2 \Delta x.$$

To find the total volume, V, we integrate this quantity between $x = 0$ and $x = 3$:

$$V = \int_0^3 (\pi(2x)^2 - \pi(x)^2)dx = \pi \int_0^3 (4x^2 - x^2)\, dx = \pi \int_0^3 3x^2\, dx = \pi x^3 \Big|_0^3 = 27\pi = 84.823.$$

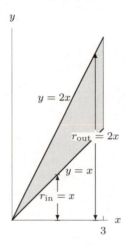

Figure 8.34

13. The region is bounded by $x = 4y$, the x-axis and $x = 8$. The two lines $x = 4y$ and $x = 8$ intersect at $(8, 2)$. We slice the volume with planes that are perpendicular to the line $x = 10$. This divides the solid into thin washers with

$$\text{Volume} \approx \pi r_{out}^2 dy - \pi r_{in}^2 dy.$$

The inner radius is the distance from the line $x = 10$ to the line $x = 8$ and the outer radius is the distance from the line $x = 10$ to the line $x = 4y$. Integrating from $y = 0$ to $y = 2$ we have

$$V = \int_0^2 \left[\pi(10 - 4y)^2 - \pi(2)^2\right]\, dy.$$

17. We slice the cone horizontally into cylindrical disks with radius r and thickness Δh. See Figure 8.35. The volume of each disk is $\pi r^2 \Delta h$. We use the similar triangles in Figure 8.36 to write r as a function of h:

$$\frac{r}{h} = \frac{3}{12} \quad \text{so} \quad r = \frac{1}{4}h.$$

The volume of the disk at height h is $\pi(\frac{1}{4}h)^2 \Delta h$. To find the total volume, we integrate this quantity from $h = 0$ to $h = 12$.

$$V = \int_0^{12} \pi \left(\frac{1}{4}h\right)^2 dh = \frac{\pi}{16} \frac{h^3}{3} \Big|_0^{12} = 36\pi = 113.097 \text{ m}^3.$$

Figure 8.35

Figure 8.36

21. We'll find the arc length of the top half of the ellipse, and multiply that by 2. In the top half of the ellipse, the equation $(x^2/a^2) + (y^2/b^2) = 1$ implies

$$y = +b\sqrt{1 - \frac{x^2}{a^2}}.$$

Differentiating $(x^2/a^2) + (y^2/b^2) = 1$ implicitly with respect to x gives us

$$\frac{2x}{a^2} + \frac{2y}{b^2}\frac{dy}{dx} = 0,$$

so

$$\frac{dy}{dx} = \frac{\frac{-2x}{a^2}}{\frac{2y}{b^2}} = -\frac{b^2 x}{a^2 y}.$$

Substituting this into the arc length formula, we get

$$\text{Arc Length} = \int_{-a}^{a} \sqrt{1 + \left(-\frac{b^2 x}{a^2 y}\right)^2}\, dx$$

$$= \int_{-a}^{a} \sqrt{1 + \left(\frac{b^4 x^2}{a^4 (b^2)(1 - \frac{x^2}{a^2})}\right)}\, dx$$

$$= \int_{-a}^{a} \sqrt{1 + \left(\frac{b^2 x^2}{a^2 (a^2 - x^2)}\right)}\, dx.$$

Hence the arc length of the entire ellipse is

$$2\int_{-a}^{a} \sqrt{1 + \left(\frac{b^2 x^2}{a^2 (a^2 - x^2)}\right)}\, dx.$$

25. The arc length is given by

$$L = \int_{1}^{2} \sqrt{1 + e^{2x}}\, dx \approx 4.785.$$

Note that $\sqrt{1 + e^{2x}}$ does not have an obvious elementary antiderivative, so we use an approximation method to find an approximate value for L.

Problems

29. (a) The points of intersection are $x = 0$ to $x = 2$, so we have

$$\text{Area} = \int_{0}^{2} (2x - x^2)dx = x^2 - \frac{x^3}{3}\bigg|_{0}^{2} = \frac{4}{3} = 1.333.$$

(b) The outside radius is $2x$ and the inside radius is x^2, so we have

$$\text{Volume} = \int_{0}^{2} (\pi(2x)^2 - \pi(x^2)^2)dx = \pi\int_{0}^{2} (4x^2 - x^4)dx = \frac{\pi}{15}(20x^3 - 3x^5)\bigg|_{0}^{2} = \frac{64\pi}{15} = 13.404.$$

(c) The length of the perimeter is equal to the length of the top plus the length of the bottom. Using the arclength formula, and the fact that the derivative of $2x$ is 2 and the derivative of x^2 is $2x$, we have

$$L = \int_{0}^{2} \sqrt{1 + 2^2}dx + \int_{0}^{2} \sqrt{1 + (2x)^2}dx = 4.4721 + 4.6468 = 9.119.$$

33. (a) Since $y = e^{-bx}$ is non-negative, we integrate to find the area:

$$\text{Area} = \int_0^1 (e^{-bx})dx = \frac{-1}{b}e^{-bx}\Big|_0^1 = \frac{1}{b}(1 - e^{-b}).$$

(b) Each slice of the object is approximately a cylinder with radius e^{-bx} and thickness Δx. We have

$$\text{Volume} = \int_0^1 \pi(e^{-bx})^2 dx = \pi \int_0^1 e^{-2bx} dx = \frac{-\pi}{2b}e^{-2bx}\Big|_0^1 = \frac{\pi}{2b}(1 - e^{-2b}).$$

37. Slice the object into disks vertically, as in Figure 8.37. A typical disk has thickness Δx and radius $y = \sqrt{1 - x^2}$. Thus

$$\text{Volume of disk} \approx \pi y^2 \Delta x = \pi(1 - x^2)\,\Delta x.$$

$$\text{Volume of solid} = \lim_{\Delta x \to 0} \sum \pi(1 - x^2)\,\Delta x = \int_0^1 \pi(1 - x^2)\,dx = \pi\left(x - \frac{x^3}{3}\right)\Big|_0^1 = \frac{2\pi}{3}.$$

Note: As we expect, this is the volume of a half sphere.

Figure 8.37

41. Slicing perpendicularly to the y-axis gives semicircles whose thickness is Δy and whose diameter is $x = \sqrt{1 - y^2}$. See Figure 8.38. Thus

$$\text{Volume of semicircular slice} \approx \frac{\pi}{2}\left(\frac{\sqrt{1 - y^2}}{2}\right)^2 \Delta y = \frac{\pi}{8}(1 - y^2)\,\Delta y.$$

$$\text{Volume of solid} = \int_0^1 \frac{\pi}{8}(1 - y^2)\,dy = \frac{\pi}{8}\left(y - \frac{y^3}{3}\right)\Big|_0^1 = \frac{\pi}{8} \cdot \frac{2}{3} = \frac{\pi}{24}.$$

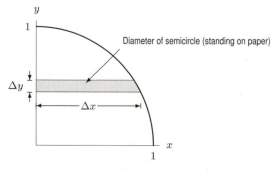

Figure 8.38: Base of Solid

45. (a) The line $y = ax$ must pass through (l, b). Hence $b = al$, so $a = b/l$.

(b) Cut the cone into N slices, slicing perpendicular to the x–axis. Each piece is almost a cylinder. The radius of the ith cylinder is $r(x_i) = \dfrac{bx_i}{l}$, so the volume

$$V \approx \sum_{i=1}^{N} \pi \left(\frac{bx_i}{l} \right)^2 \Delta x.$$

Therefore, as $N \to \infty$, we get

$$V = \int_0^l \pi b^2 l^{-2} x^2 \, dx$$

$$= \pi \frac{b^2}{l^2} \left[\frac{x^3}{3} \right]_0^l = \left(\pi \frac{b^2}{l^2} \right) \left(\frac{l^3}{3} \right) = \frac{1}{3} \pi b^2 l.$$

49. Multiplying $r = 2a \cos \theta$ by r, converting to Cartesian coordinates, and completing the square gives

$$r^2 = 2ar \cos \theta$$
$$x^2 + y^2 = 2ax$$
$$x^2 - 2ax + a^2 + y^2 = a^2$$
$$(x - a)^2 + y^2 = a^2.$$

This is the standard form of the equation of a circle with radius a and center $(x, y) = (a, 0)$.

To check the limits on θ note that the circle is in the right half plane, where $-\pi/2 \le \theta \le \pi/2$. Rays from the origin at all these angles meet the circle because the circle is tangent to the y-axis at the origin.

53. Writing C in parametric form gives

$$x = 2a \cos^2 \theta \quad \text{and} \quad y = 2a \cos \theta \sin \theta,$$

so

$$\text{Arc length} = \int_{-\pi/2}^{\pi/2} \sqrt{(-4a \cos \theta \sin \theta)^2 + (-2a \sin^2 \theta + 2a \cos^2 \theta)^2} \, d\theta$$

$$= 2a \int_{-\pi/2}^{\pi/2} \sqrt{4 \cos^2 \theta \sin^2 \theta + \sin^4 \theta - 2 \sin^2 \theta \cos^2 \theta + \cos^4 \theta} \, d\theta$$

$$= 2a \int_{-\pi/2}^{\pi/2} \sqrt{\sin^4 \theta + 2 \sin^2 \theta \cos^2 \theta + \cos^4 \theta} \, d\theta$$

$$= 2a \int_{-\pi/2}^{\pi/2} \sqrt{(\sin^2 \theta + \cos^2 \theta)^2} \, d\theta$$

$$= 2a \int_{-\pi/2}^{\pi/2} d\theta = 2\pi a.$$

57. Let x be the distance from the bucket to the surface of the water. It follows that $0 \le x \le 40$. At x feet, the bucket weighs $(30 - \frac{1}{4}x)$, where the $\frac{1}{4}x$ term is due to the leak. When the bucket is x feet from the surface of the water, the work done by raising it Δx feet is $(30 - \frac{1}{4}x) \Delta x$. So the total work required to raise the bucket to the top is

$$W = \int_0^{40} (30 - \frac{1}{4}x) dx$$

$$= \left(30x - \frac{1}{8}x^2 \right) \Big|_0^{40}$$

$$= 30(40) - \frac{1}{8}40^2 = 1000 \text{ ft-lb.}$$

61. Let h be height above the bottom of the dam. Then

$$\text{Water force } = \int_0^{25} (62.4)(25 - h)(60)\, dh$$

$$= (62.4)(60) \left(25h - \frac{h^2}{2} \right) \Big|_0^{25}$$

$$= (62.4)(60)(625 - 312.5)$$

$$= (62.4)(60)(312.5)$$

$$= 1{,}170{,}000 \text{ lbs.}$$

65. We divide up time between 1971 and 1992 into intervals of length Δt, and calculate how much of the strontium-90 produced during that time interval is still around.

Strontium-90 decays exponentially, so if a quantity S_0 was produced t years ago, and S is the quantity around today, $S = S_0 e^{-kt}$. Since the half-life is 28 years, $\frac{1}{2} = e^{-k(28)}$, giving $k = -\ln(1/2)/28 \approx 0.025$.

We measure t in years from 1971, so that 1992 is $t = 21$.

$$
\begin{array}{ccc}
1971\ (t = 0) & \Delta t & 1992\ (t = 21)
\end{array}
$$

$$
\vdash\!\!\!\!\!\dashv\!\!\dashv \qquad \qquad \dashv
$$

$$
t \qquad \qquad (21 - t)
$$

Since strontium-90 is produced at a rate of 3 kg/year, during the interval Δt, a quantity $3\Delta t$ kg was produced. Since this was $(21 - t)$ years ago, the quantity remaining now is $(3\Delta t)e^{-0.025(21-t)}$. Summing over all such intervals gives

$$\text{Strontium remaining} \atop \text{in 1992} \approx \int_0^{21} 3e^{-0.025(21-t)}\, dt = \frac{3e^{-0.025(21-t)}}{0.025} \Big|_0^{21} = 49 \text{ kg.}$$

[Note: This is like a future value problem from economics, but with a negative interest rate.]

69. Graph B is more spread out to the right, and so it represents a gas in which more of the molecules are moving at faster velocities. Thus the average velocity in gas B is larger.

73. (a) Slicing horizontally, as shown in Figure 8.39, we see that the volume of one disk-shaped slab is

$$\Delta V \approx \pi x^2 \Delta y = \frac{\pi y}{a} \Delta y.$$

Thus, the volume of the water is given by

$$V = \int_0^h \frac{\pi}{a} y\, dy = \frac{\pi}{a} \frac{y^2}{2} \Big|_0^h = \frac{\pi h^2}{2a}.$$

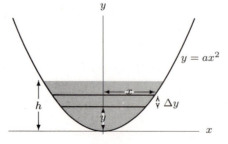

Figure 8.39

(b) The surface of the water is a circle of radius x. Since at the surface, $y = h$, we have $h = ax^2$. Thus, at the surface, $x = \sqrt{(h/a)}$. Therefore the area of the surface of water is given by

$$A = \pi x^2 = \frac{\pi h}{a}.$$

(c) If the rate at which water is evaporating is proportional to the surface area, we have

$$\frac{dV}{dt} = -kA.$$

(The negative sign is included because the volume is decreasing.) By the chain rule, $\frac{dV}{dt} = \frac{dV}{dh} \cdot \frac{dh}{dt}$. We know $\frac{dV}{dh} = \frac{\pi h}{a}$ and $A = \frac{\pi h}{a}$ so

$$\frac{\pi h}{a}\frac{dh}{dt} = -k\frac{\pi h}{a} \quad \text{giving} \quad \frac{dh}{dt} = -k.$$

(d) Integrating gives

$$h = -kt + h_0.$$

Solving for t when $h = 0$ gives

$$t = \frac{h_0}{k}.$$

CAS Challenge Problems

77. (a) We need to check that the point with the given coordinates is on the curve, i.e., that

$$x = a\sin^2 t, \quad y = \frac{a\sin^3 t}{\cos t}$$

satisfies the equation

$$y = \sqrt{\frac{x^3}{a-x}}.$$

This can be done by substituting into the computer algebra system and asking it to simplify the difference between the two sides, or by hand calculation:

$$\text{Right-hand side} = \sqrt{\frac{(a\sin^2 t)^3}{a - a\sin^2 t}} = \sqrt{\frac{a^3 \sin^6 t}{a(1 - \sin^2 t)}}$$

$$= \sqrt{\frac{a^3 \sin^6 t}{a\cos^2 t}} = \sqrt{\frac{a^2 \sin^6 t}{\cos^2 t}}$$

$$= \frac{a\sin^3 t}{\cos t} = y = \text{Left-hand side.}$$

We chose the positive square root because both $\sin t$ and $\cos t$ are nonnegative for $0 \le t \le \pi/2$. Thus the point always lies on the curve. In addition, when $t = 0$, $x = 0$ and $y = 0$, so the point starts at $x = 0$. As t approaches $\pi/2$, the value of $x = a\sin^2 t$ approaches a and the value of $y = a\sin^3 t/\cos t$ increases without bound (or approaches ∞), so the point on the curve approaches the vertical asymptote at $x = a$.

(b) We calculate the volume using horizontal slices. See the graph of $y = \sqrt{x^3/(a-x)}$ in Figure 8.40.

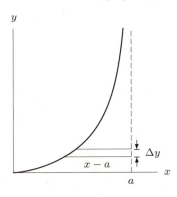

Figure 8.40

The slice at y is a disk of thickness Δy and radius $x - a$, hence it has volume $\pi(x-a)^2\Delta y$. So the volume is given by the improper integral

$$\text{Volume} = \int_0^\infty \pi(x-a)^2 \, dy.$$

(c) We substitute

$$x = a \sin^2 t, \quad y = \frac{a \sin^3 t}{\cos t}$$

and

$$dy = \frac{d}{dt}\left(\frac{a \sin^3 t}{\cos t}\right) dt = a\left(3 \sin^2 t + \frac{\sin^4 t}{\cos^2 t}\right) dt.$$

Since $t = 0$ where $y = 0$ and $t = \pi/2$ at the asymptote where $y \to \infty$, we get

$$\text{Volume} = \int_0^{\pi/2} \pi(a \sin^2 t - a)^2 a\left(3 \sin^2 t + \frac{\sin^4 t}{\cos^2 t}\right) dt$$

$$= \pi a^3 \int_0^{\pi/2} (3 \sin^2 t \cos^4 t + \sin^4 t \cos^2 t)\, dt = \frac{\pi^2 a^3}{8}.$$

You can use a CAS to calculate this integral; it can also be done using trigonometric identities.

CHECK YOUR UNDERSTANDING

1. True. Since $y = \pm\sqrt{9 - x^2}$ represent the top and bottom halves of the sphere, slicing disks perpendicular to the x-axis gives

$$\text{Volume of slice} \approx \pi y^2 \Delta x = \pi(9 - x^2)\Delta x$$

$$\text{Volume} = \int_{-3}^3 \pi(9 - x^2)\, dx.$$

5. False. Volume is always positive, like area.

9. True. One way to look at it is that the center of mass should not change if you change the units by which you measure the masses. If you double the masses, that is no different than using as a new unit of mass half the old unit. Alternatively, let the masses be m_1, m_2, and m_3 located at x_1, x_2, and x_3. Then the center of mass is given by:

$$\bar{x} = \frac{x_1 m_1 + x_2 m_2 + x_3 m_3}{m_1 + m_2 + m_3}.$$

Doubling the masses does not change the center of mass, since it doubles both the numerator and the denominator.

13. False. Work is the product of force and distance moved, so the work done in either case is 200 ft-lb.

17. False. The pressure is positive and when integrated gives a positive force.

21. False. It is true that $p(x) \geq 0$ for all x, but we also need $\int_{-\infty}^\infty p(x)dx = 1$. Since $p(x) = 0$ for $x \leq 0$, we need only check the integral from 0 to ∞. We have

$$\int_0^\infty xe^{-x^2}\, dx = \lim_{b \to \infty}\left(-\frac{1}{2}e^{-x^2}\right)\Big|_0^b = \frac{1}{2}.$$

25. False. Since f is concave down, this means that $f'(x)$ is decreasing, so $f'(x) \leq f'(0) = 3/4$ on the interval $[0, 4]$. However, it could be that $f'(x)$ becomes negative so that $(f'(x))^2$ becomes large, making the integral for the arc length large also. For example, $f(x) = (3/4)x - x^2$ is concave down and $f'(0) = 3/4$, but $f(0) = 0$ and $f(4) = -13$, so the graph of f on the interval $[0, 4]$ has arc length at least 13.

29. False. Note that p is the density function for the population, not the cumulative density function. Thus $p(10) = p(20)$ means that x values near 10 are as likely as x values near 20.

CHAPTER NINE

Solutions for Section 9.1

Exercises

1. The terms look like powers of 2 so we guess $s_n = 2^n$. This makes the first term $2^1 = 2$ rather than 4. We try instead $s_n = 2^{n+1}$. If we now check, we get the terms $4, 8, 16, 32, 64, \ldots$, which is right.

5. The numerator is n. The denominator is then $2n + 1$, so $s_n = n/(2n + 1)$.

9. The first term is $2 \cdot 1/(2 \cdot 1 + 1) = 2/3$. The second term is $2 \cdot 2/(2 \cdot 2 + 1) = 4/5$. The first five terms are

$$2/3, 4/5, 6/7, 8/9, 10/11.$$

Problems

13. Since 2^n increases without bound as n increases, the sequence diverges.

17. We have:
$$\lim_{n \to \infty} \left(\frac{n}{10} + \frac{10}{n} \right) = \lim_{n \to \infty} \frac{n}{10} + \lim_{n \to \infty} 10n.$$
Since $n/10$ gets arbitrarily large and $10/n$ approaches 0 as $n \to \infty$, the sequence diverges.

21. Since the exponential function 2^n dominates the power function n^3 as $n \to \infty$, the series diverges.

25. (a) matches (IV), since the sequence increases toward 1.
   (b) matches (III), since the odd terms increase toward 1 and the even terms decrease toward 1.
   (c) matches (II), since the sequence decreases toward 0.
   (d) matches (I), since the sequence decreases toward 1.

29. We have $s_2 = s_1 + 2 = 3$ and $s_3 = s_2 + 3 = 6$. Continuing, we get

$$1, \ 3, \ 6, \ 10, \ 15, \ 21.$$

33. The first 6 terms of the sequence for the sampling is

$$\cos 0.5, \ \cos 1.0, \ \cos 1.5, \ \cos 2.0, \ \cos 2.5, \ \cos 3.0$$
$$= 0.878, \ 0.540, \ 0.071, \ -0.416, \ -0.801, \ -0.990.$$

37. The first smoothing gives
$$1.5, \ 2, \ 3, \ 4, \ 5, \ 6, \ 7 \ldots$$
The second smoothing gives
$$1.75, \ 2.17, \ 3, \ 4, \ 5, \ 6 \ldots$$
Terms which are already the same as their average with their neighbors are not changed.

41. The differences between consecutive terms are $4, 9, 16, 25$, so, for example, $s_2 = s_1 + 4$ and $s_3 = s_2 + 9$. Thus, a possible recursive definition is $s_n = s_{n-1} + n^2$ for $n > 1$ and $s_1 = 1$.

45. For $n > 1$, if $s_n = n(n + 1)/2$, then $s_{n-1} = (n - 1)(n - 1 + 1)/2 = n(n - 1)/2$. Since

$$s_n = \frac{1}{2}(n^2 + n) = \frac{n^2}{2} + \frac{n}{2} \quad \text{and} \quad s_{n-1} = \frac{1}{2}(n^2 - n) = \frac{n^2}{2} - \frac{n}{2},$$

we have
$$s_n - s_{n-1} = \frac{n}{2} + \frac{n}{2} = n,$$
so
$$s_n = s_{n-1} + n.$$

In addition, $s_1 = 1(2)/2 = 1$.

49. The sequence appears to be decreasing toward 0, but at a slower and slower rate. Even after 100 terms, it is hard to guess what the series will eventually do. It can be shown that it converges to 0.

53. In year 1, the payment is

$$p_1 = 10{,}000 + 0.05(100{,}000) = 15{,}000.$$

The balance in year 2 is $100{,}000 - 10{,}000 = 90{,}000$, so

$$p_2 = 10{,}000 + 0.05(90{,}000) = 14{,}500.$$

The balance in year 3 is 80,000, so

$$p_3 = 10{,}000 + 0.05(80{,}000) = 14{,}000.$$

Thus,

$$p_n = 10{,}000 + 0.05(100{,}000 - (n-1) \cdot 10{,}000)$$
$$= 15{,}500 - 500n.$$

57. (a) The first 12 terms are

$$1, 1, 2, 3, 5, 8, 13, 21, 34, 55, 89, 144.$$

(b) The sequence of ratios is

$$1, 2, \frac{3}{2}, \frac{5}{3}, \frac{8}{5}, \frac{13}{8}, \frac{21}{13}, \frac{34}{21}, \frac{55}{34}, \frac{89}{55} \cdots$$

To three decimal places, the first ten ratios are

$$1, 2, 1.500, 1.667, 1.600, 1.625, 1.615, 1.619, 1.618, 1.618.$$

It appears that the sequence of ratios is converging to $r = 1.618$. We find $(1.618)^2 = 2.618 = 1.618 + 1$ so r seems to satisfy $r^2 = r + 1$. Alternatively, by the quadratic formula, the positive root of $x^2 - x - 1 = 0$ is $(1 + \sqrt{5})/2 = 1.618$.

(c) If we multiply both sides of the equation $r^2 = r + 1$ by Ar^{n-2}, we obtain

$$Ar^n = Ar^{n-1} + Ar^{n-2}.$$

Thus, if $s_n = Ar^n$, then $s_{n-1} = Ar^{n-1}$ and $s_{n-2} = Ar^{n-2}$, so the sequence satisfies $s_n = s_{n-1} + s_{n-2}$.

Solutions for Section 9.2

Exercises

1. Yes, $a = 5$, ratio $= -2$.

5. No. Ratio between successive terms is not constant: $\dfrac{2x^2}{x} = 2x$, while $\dfrac{3x^3}{2x^2} = \dfrac{3}{2}x$.

9. Yes, $a = 1$, ratio $= -y^2$.

13. The series has 9 terms. The first term is $a = 0.00002$ and the constant ratio is $x = 0.1$, so

$$\text{Sum} = \frac{0.00002(1 - x^9)}{(1-x)} = \frac{0.00002(1 - (0.1)^9)}{0.9} = 0.0000222.$$

17. Using the formula for the sum of a finite geometric series,

$$\sum_{n=4}^{20} \left(\frac{1}{3}\right)^n = \left(\frac{1}{3}\right)^4 + \left(\frac{1}{3}\right)^5 + \cdots + \left(\frac{1}{3}\right)^{20} = \left(\frac{1}{3}\right)^4 \left(1 + \frac{1}{3} + \left(\frac{1}{3}\right)^2 + \cdots \left(\frac{1}{3}\right)^{16}\right) = \frac{(1/3)^4(1 - (1/3)^{17})}{1 - (1/3)} = \frac{3^{17} - 1}{2 \cdot 3^{20}}.$$

21. This is a geometric series with first term 2 and ratio $-2z$,

$$2 - 4z + 8z^2 - 16z^3 + \cdots = \frac{2}{1 - (-2z)} = \frac{2}{1 + 2z}.$$

This series converges for $|-2z| < 1$, that is for $-1/2 < z < 1/2$.

Problems

25. Since the amount of ampicillin excreted during the time interval between tablets is 250 mg, we have

$$\text{Amount of ampicillin excreted} = \text{Original quantity} - \text{Final quantity}$$
$$250 = Q - (0.04)Q.$$

Solving for Q gives, as before,

$$Q = \frac{250}{1 - 0.04} \approx 260.42.$$

29. If a payment M in the future has present value P, then

$$M = P(1 + r)^t,$$

where t is the number of periods in the future and r is the interest rate. Thus

$$P = \frac{M}{(1 + r)^t}.$$

The monthly interest rate here is $0.09/12 = 0.0075$, so the present value of first payment, made at the end of the first month, is $M/(1.0075)$. The present value of the second payment is $M/(1.0075)^2$. Continuing in this way, the sum of the present value of all of the payments is

$$\frac{M}{(1.0075)} + \frac{M}{(1.0075)^2} + \cdots + \frac{M}{(1.0075)^{240}}.$$

This is a finite geometric series with 240 terms, with sum

$$\frac{M}{1.0075}\left(\frac{1 - 1.0075^{-240}}{1 - 1.0075^{-1}}\right) = 111.145M.$$

Setting this equal to the loan amount of 200,000 gives a monthly payment of $M = \$200,000/111.145 = \$1799.45$.

33. If the half-life is T hours, then the exponential decay formula $Q = Q_0 e^{-kt}$ gives $k = \ln 2/T$. If we start with $Q_0 = 1$ tablet, then the amount of drug present in the body after $5T$ hours is

$$Q = e^{-5kT} = e^{-5\ln 2} = 0.03125,$$

so 3.125% of a tablet remains. Thus, immediately after taking the first tablet, there is one tablet in the body. Five half-lives later, this has reduced to $1 \cdot 0.03125 = 0.03125$ tablets, and immediately after the second tablet there are $1 + 0.03125$ tablets in the body. Continuing this forever leads to

$$\text{Number of tablets in body} = 1 + 0.03125 + (0.03125)^2 + \cdots + (0.03125)^n + \cdots.$$

This is an infinite geometric series, with common ratio $x = 0.03125$, and sum $1/(1 - x)$. Thus

$$\text{Number of tablets in body} = \frac{1}{1 - 0.03125} = 1.0323.$$

37. (a) The acceleration of gravity is 32 ft/sec^2 so acceleration $= 32$ and velocity $v = 32t + C$. Since the ball is dropped, its initial velocity is 0 so $v = 32t$. Thus the position is $s = 16t^2 + C$. Calling the initial position $s = 0$, we have $s = 6t$. The distance traveled is h so $h = 16t$. Solving for t we get $t = \frac{1}{4}\sqrt{h}$.

(b) The first drop from 10 feet takes $\frac{1}{4}\sqrt{10}$ seconds. The first full bounce (to $10 \cdot \left(\frac{3}{4}\right)$ feet) takes $\frac{1}{4}\sqrt{10 \cdot \left(\frac{3}{4}\right)}$ seconds to rise, therefore the same time to come down. Thus, the full bounce, up and down, takes $2\left(\frac{1}{4}\right)\sqrt{10 \cdot \left(\frac{3}{4}\right)}$ seconds. The next full bounce takes $2\left(\frac{1}{4}\right)10 \cdot \left(\frac{3}{4}\right)^2 = 2\left(\frac{1}{4}\right)\sqrt{10}\left(\sqrt{\frac{3}{4}}\right)^2$ seconds. The nth bounce takes $2\left(\frac{1}{4}\right)\sqrt{10}\left(\sqrt{\frac{3}{4}}\right)^n$ seconds. Therefore the

Total amount of time

$$= \frac{1}{4}\sqrt{10} + \underbrace{\frac{2}{4}\sqrt{10}\sqrt{\frac{3}{4}} + \frac{2}{4}\sqrt{10}\left(\sqrt{\frac{3}{4}}\right)^2 + \frac{2}{4}\sqrt{10}\left(\sqrt{\frac{3}{4}}\right)^3}_{\text{Geometric series with } a = \frac{2}{4}\sqrt{10}\sqrt{\frac{3}{4}} = \frac{1}{2}\sqrt{10}\sqrt{\frac{3}{4}} \text{ and } x = \sqrt{\frac{3}{4}}} + \cdots$$

$$= \frac{1}{4}\sqrt{10} + \frac{1}{2}\sqrt{10}\sqrt{\frac{3}{4}}\left(\frac{1}{1 - \sqrt{3/4}}\right) \text{ seconds.}$$

Solutions for Section 9.3

Exercises

1. The series is $1 + 2 + 3 + 4 + 5 + \cdots$. The sequence of partial sums is

$$S_1 = 1, \quad S_2 = 1 + 2, \quad S_3 = 1 + 2 + 3, \quad S_4 = 1 + 2 + 3 + 4, \quad S_5 = 1 + 2 + 3 + 4 + 5, \ldots$$

which is

$$1, \quad 3, \quad 6, \quad 10, \quad 15 \ldots.$$

5. We use the integral test with $f(x) = x/(x^2 + 1)$ to determine whether this series converges or diverges. We determine whether the corresponding improper integral $\displaystyle\int_1^\infty \frac{x}{x^2 + 1}\,dx$ converges or diverges:

$$\int_1^\infty \frac{x}{x^2+1}\,dx = \lim_{b\to\infty}\int_1^b \frac{x}{x^2+1}\,dx = \lim_{b\to\infty}\frac{1}{2}\ln(x^2+1)\Big|_1^b = \lim_{b\to\infty}\left(\frac{1}{2}\ln(b^2+1) - \frac{1}{2}\ln 2\right) = \infty.$$

Since the integral $\displaystyle\int_1^\infty \frac{x}{x^2+1}\,dx$ diverges, we conclude from the integral test that the series $\displaystyle\sum_{n=1}^\infty \frac{n}{n^2+1}$ diverges.

9. The improper integral $\displaystyle\int_0^\infty \frac{1}{x^2+1}\,dx$ converges to $\dfrac{\pi}{2}$, since

$$\int_0^b \frac{1}{x^2+1}\,dx = \arctan x\Big|_0^b = \arctan b - \arctan 0 = \arctan b,$$

and $\displaystyle\lim_{b\to\infty}\arctan b = \frac{\pi}{2}$. The terms of the series $\displaystyle\sum_{n=1}^\infty \frac{1}{n^2+1}$ form a right hand sum for the improper integral; each term represents the area of a rectangle of width 1 fitting completely under the graph of the function $\dfrac{1}{x^2+1}$. (See Figure 9.1.) Thus the sequence of partial sums is bounded above by $\dfrac{\pi}{2}$. Since the partial sums are increasing (every new term added is positive), the series is guaranteed to converge to some number less than or equal to $\pi/2$ by Theorem 9.1.

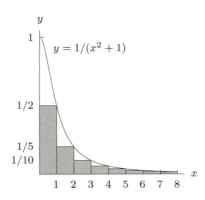

Figure 9.1

Problems

13. Using the integral test, we compare the series with

$$\int_0^\infty \frac{3}{x+2}\,dx = \lim_{b\to\infty}\int_0^b \frac{3}{x+2}\,dx = 3\ln|x+2|\Big|_0^b.$$

Since $\ln(b+2)$ is unbounded as $b\to\infty$, the integral diverges and therefore so does the series.

17. Since the terms in the series are positive and decreasing, we can use the integral test. We calculate the corresponding improper integral using the substitution $w = 1+x^2$:

$$\int_0^\infty \frac{2x}{(1+x^2)^2}\,dx = \lim_{b\to\infty}\int_0^b \frac{2x}{(1+x^2)^2}\,dx = \lim_{b\to\infty}\frac{-1}{(1+x^2)}\Big|_0^b = \lim_{b\to\infty}\left(\frac{-1}{1+b^2}+1\right) = 1.$$

Since the limit exists, the integral converges, so the series $\displaystyle\sum_{n=0}^{\infty}\frac{2n}{(1+n^2)^2}$ converges.

21. Using the integral test, we compare the series with

$$\int_0^\infty \frac{3}{x^2+4}\,dx = \lim_{b\to\infty}\int_0^b \frac{3}{x^2+4}\,dx = \frac{3}{2}\lim_{b\to\infty}\arctan\left(\frac{x}{2}\right)\Big|_0^b = \frac{3}{2}\lim_{b\to\infty}\arctan\left(\frac{b}{2}\right) = \frac{3\pi}{4},$$

by integral table V-24. Since the integral converges so does the series.

25. Both $\displaystyle\sum_{n=1}^{\infty}\left(\frac{1}{2}\right)^n$ and $\displaystyle\sum_{n=1}^{\infty}\left(\frac{2}{3}\right)^n$ are convergent geometric series. Therefore, by Property 1 of Theorem 9.2, the series $\displaystyle\sum_{n=1}^{\infty}\left(\frac{1}{2}\right)^n + \left(\frac{2}{3}\right)^n$ converges.

29. Since the terms in the series are positive and decreasing, we can use the integral test. We calculate the corresponding improper integral using the substitution $w = 1+\ln x$:

$$\int_1^\infty \frac{1}{x(1+\ln x)}\,dx = \lim_{b\to\infty}\int_1^b \frac{1}{x(1+\ln x)}\,dx = \lim_{b\to\infty}\ln(1+\ln x)\Big|_1^b = \lim_{b\to\infty}\ln(1+\ln b).$$

Since the limit does not exist, the integral diverges, so the series $\displaystyle\sum_{n=1}^{\infty}\frac{1}{n(1+\ln n)}$ diverges.

33. Using $\ln(2^n) = n\ln 2$, we see that

$$\sum \frac{1}{\ln(2^n)} = \sum \frac{1}{(\ln 2)n}.$$

The series on the right is the harmonic series multiplied by $1/\ln 2$. Since the harmonic series diverges, $\sum_{n=1}^{\infty} 1/\ln(2^n)$ diverges.

37. (a) A common denominator is $k(k+1)$ so

$$\frac{1}{k} - \frac{1}{k+1} = \frac{k+1}{k(k+1)} - \frac{k}{k(k+1)} = \frac{k+1-k}{k(k+1)} = \frac{1}{k(k+1)}.$$

(b) Using the result of part (a), the partial sum can be written as

$$S_3 = \frac{1}{1\cdot 2} + \frac{1}{2\cdot 3} + \frac{1}{3\cdot 4} = \frac{1}{1} - \frac{1}{2} + \frac{1}{2} - \frac{1}{3} + \frac{1}{3} - \frac{1}{4} = 1 - \frac{1}{4}.$$

All of the intermediate terms cancel out, leaving only the first and last terms. Thus $S_{10} = 1 - \dfrac{1}{11}$ and $S_n = 1 - \dfrac{1}{n+1}$.

(c) The limit of S_n as $n \to \infty$ is $\displaystyle\lim_{n\to\infty}\left(1 - \frac{1}{n+1}\right) = 1 - 0 = 1$. Thus the series $\displaystyle\sum_{k=1}^{\infty} \frac{1}{k(k+1)}$ converges to 1.

41. We have $a_n = S_n - S_{n-1}$. If $\sum a_n$ converges, then $S = \lim_{n\to\infty} S_n$ exists. Hence $\lim_{n\to\infty} S_{n-1}$ exists and is equal to S also. Thus

$$\lim_{n\to\infty} a_n = \lim_{n\to\infty}(S_n - S_{n-1}) = \lim_{n\to\infty} S_n - \lim_{n\to\infty} S_{n-1} = S - S = 0.$$

45. (a) Let N an integer with $N \geq c$. Consider the series $\displaystyle\sum_{i=N+1}^{\infty} a_i$. The partial sums of this series are increasing because all the terms in the series are positive. We show the partial sums are bounded using the right-hand sum in Figure 9.2. We see that for each positive integer k

$$f(N+1) + f(N+2) + \cdots + f(N+k) \leq \int_{N}^{N+k} f(x)\,dx.$$

Since $f(n) = a_n$ for all n, and $c \leq N$, we have

$$a_{N+1} + a_{N+2} + \cdots + a_{N+k} \leq \int_{c}^{N+k} f(x)\,dx.$$

Since $f(x)$ is a positive function, $\int_{c}^{N+k} f(x)\,dx \leq \int_{c}^{b} f(x)\,dx$ for all $b \geq N+k$. Since f is positive and $\int_{c}^{\infty} f(x)\,dx$ is convergent, $\int_{c}^{N+k} f(x)\,dx < \int_{c}^{\infty} f(x)\,dx$, so we have

$$a_{N+1} + a_{N+2} + \cdots + a_{N+k} \leq \int_{c}^{\infty} f(x)\,dx \quad \text{for all } k.$$

Thus, the partial sums of the series $\displaystyle\sum_{i=N+1}^{\infty} a_i$ are bounded and increasing, so this series converges by Theorem 9.1.

Now use Theorem 9.2, property 2, to conclude that $\displaystyle\sum_{i=1}^{\infty} a_i$ converges.

Figure 9.2

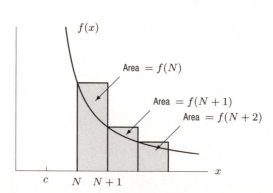

Figure 9.3

(b) We now suppose $\int_c^\infty f(x)\,dx$ diverges. In Figure 9.3 we see that for each positive integer k

$$\int_N^{N+k+1} f(x)\,dx \le f(N) + f(N+1) + \cdots + f(N+k).$$

Since $f(n) = a_n$ for all n, we have

$$\int_N^{N+k+1} f(x)\,dx \le a_N + a_{N+1} + \cdots + a_{N+k}.$$

Since $f(x)$ is defined for all $x \ge c$, if $\int_c^\infty f(x)\,dx$ is divergent, then $\int_N^\infty f(x)\,dx$ is divergent. So as $k \to \infty$, the the integral $\int_N^{N+k+1} f(x)\,dx$ diverges, so the partial sums of the series $\sum_{i=N}^\infty a_i$ diverge. Thus, the series $\sum_{i=1}^\infty a_i$ diverges.

More precisely, suppose the series converged. Then the partial sums would be bounded. (The partial sums would be less than the sum of the series, since all the terms in the series are positive.) But that would imply that the integral converged, by Theorem 9.1 on Convergence of Monotone Bounded Sequences. This contradicts the assumption that $\int_N^\infty f(x)\,dx$ is divergent.

49. (a) A calculator or computer gives

$$\sum_1^{20} \frac{1}{n^2} = \frac{1}{1^2} + \frac{1}{2^2} + \cdots + \frac{1}{20^2} = 1.596.$$

(b) Since $\sum_1^\infty \frac{1}{n^2} = \frac{\pi^2}{6}$, the answer to part (a) gives

$$\frac{\pi^2}{6} \approx 1.596$$

$$\pi \approx \sqrt{6 \cdot 1.596} = 3.09$$

(c) A calculator or computer gives

$$\sum_1^{100} \frac{1}{n^2} = \frac{1}{1^2} + \frac{1}{2^2} + \cdots + \frac{1}{100^2} = 1.635,$$

so

$$\frac{\pi^2}{6} \approx 1.635$$

$$\pi \approx \sqrt{6 \cdot 1.635} = 3.13.$$

(d) The error in approximating $\pi^2/6$ by $\sum_1^{20} 1/n^2$ is the tail of the series $\sum_{21}^\infty 1/n^2$. From Figure 9.4, we see that

$$\sum_{21}^\infty \frac{1}{n^2} < \int_{20}^\infty \frac{dx}{x^2} = -\frac{1}{x}\Big|_{20}^\infty = \frac{1}{20} = 0.05.$$

A similar argument leads to a bound for the error in approximating $\pi^2/6$ by $\sum_1^{100} 1/n^2$ as

$$\sum_{101}^\infty \frac{1}{n^2} < \int_{100}^\infty \frac{dx}{x^2} = -\frac{1}{x}\Big|_{100}^\infty = \frac{1}{100} = 0.01.$$

Figure 9.4

Solutions for Section 9.4

Exercises

1. Let $a_n = 1/(n-3)$, for $n \geq 4$. Since $n - 3 < n$, we have $1/(n-3) > 1/n$, so

$$a_n > \frac{1}{n}.$$

The harmonic series $\sum_{n=4}^{\infty} \frac{1}{n}$ diverges, so the comparison test tells us that the series $\sum_{n=4}^{\infty} \frac{1}{n-3}$ also diverges.

5. As n gets large, polynomials behave like the leading term, so for large n,

$$\frac{n+4}{n^3 + 5n - 3} \quad \text{behaves like} \quad \frac{n}{n^3} = \frac{1}{n^2}.$$

Since the series $\sum_{n=1}^{\infty} 1/n^2$ converges, we predict that the given series will converge.

9. Let $a_n = 1/(n^4 + e^n)$. Since $n^4 + e^n > n^4$, we have

$$\frac{1}{n^4 + e^n} < \frac{1}{n^4},$$

so

$$0 < a_n < \frac{1}{n^4}.$$

Since the p-series $\sum_{n=1}^{\infty} \frac{1}{n^4}$ converges, the comparison test tells us that the series $\sum_{n=1}^{\infty} \frac{1}{n^4 + e^n}$ also converges.

13. Let $a_n = (2^n + 1)/(n2^n - 1)$. Since $n2^n - 1 < n2^n + n = n(2^n + 1)$, we have

$$\frac{2^n + 1}{n2^n - 1} > \frac{2^n + 1}{n(2^n + 1)} = \frac{1}{n}.$$

Therefore, we can compare the series $\sum_{n=1}^{\infty} \frac{2^n + 1}{n2^n - 1}$ with the divergent harmonic series $\sum_{n=1}^{\infty} \frac{1}{n}$. The comparison test tells us that $\sum_{n=1}^{\infty} \frac{2^n + 1}{n2^n - 1}$ also diverges.

17. Since $a_n = 1/(r^n n!)$, replacing n by $n+1$ gives $a_{n+1} = 1/(r^{n+1}(n+1)!)$. Thus

$$\frac{|a_{n+1}|}{|a_n|} = \frac{\dfrac{1}{r^{n+1}(n+1)!}}{\dfrac{1}{r^n n!}} = \frac{r^n n!}{r^{n+1}(n+1)!} = \frac{1}{r(n+1)},$$

so

$$L = \lim_{n \to \infty} \frac{|a_{n+1}|}{|a_n|} = \frac{1}{r} \lim_{n \to \infty} \frac{1}{n+1} = 0.$$

Since $L = 0$, the ratio test tells us that $\sum_{n=1}^{\infty} \frac{1}{r^n n!}$ converges for all $r > 0$.

21. Since $\cos(n\pi) = (-1)^n$, this is an alternating series.

25. Let $a_n = 1/(2n + 1)$. Then replacing n by $n+1$ gives $a_{n+1} = 1/(2n + 3)$. Since $2n + 3 > 2n + 1$, we have

$$0 < a_{n+1} = \frac{1}{2n+3} < \frac{1}{2n+1} = a_n.$$

We also have $\lim_{n \to \infty} a_n = 0$. Therefore, the alternating series test tells us that the series $\sum_{n=1}^{\infty} \frac{(-1)^{n-1}}{2n+1}$ converges.

29. The series $\sum \dfrac{(-1)^n}{2n}$ converges by the alternating series test. However $\sum \dfrac{1}{2n}$ diverges because it is a multiple of the harmonic series. Thus $\sum \dfrac{(-1)^n}{2n}$ is conditionally convergent.

33. We first check absolute convergence by deciding whether $\sum \arcsin(1/n)$ converges. Since $\sin\theta \approx \theta$ for small angles θ, writing $x = \sin\theta$ we see that
$$\arcsin x \approx x.$$
Since $\arcsin x$ "behaves like" x for small x, we expect $\arcsin(1/n)$ to "behave like" $1/n$ for large n. To confirm this we calculate
$$\lim_{n\to\infty} \frac{\arcsin(1/n)}{1/n} = \lim_{x\to 0} \frac{\arcsin x}{x} = \lim_{\theta\to 0} \frac{\theta}{\sin\theta} = 1.$$
Thus $\sum \arcsin(1/n)$ diverges by limit comparison with the harmonic series $\sum 1/n$.

We now check conditional convergence. The original series is alternating, so we check whether $a_{n+1} < a_n$. Consider $a_n = f(n)$, where $f(x) = \arcsin(1/x)$. Since
$$\frac{d}{dx}\arcsin\frac{1}{x} = \frac{1}{\sqrt{1-(1/x)^2}}\left(-\frac{1}{x^2}\right)$$
is negative for $x > 1$, we know that a_n is decreasing for $n > 1$. Therefore, for $n > 1$
$$a_{n+1} = \arcsin(1/(n+1)) < \arcsin(1/n) = a_n.$$
Since $\arcsin(1/n) \to 0$ as $n \to \infty$, we see that $\sum (-1)^{n-1}\arcsin(1/n)$ is conditionally convergent.

37. The nth term is $a_n = 1 - \cos(1/n)$ and we are taking $b_n = 1/n^2$. We have
$$\lim_{n\to\infty} \frac{a_n}{b_n} = \lim_{n\to\infty} \frac{1-\cos(1/n)}{1/n^2}.$$
This limit is of the indeterminate form $0/0$ so we evaluate it using l'Hopital's rule. We have
$$\lim_{n\to\infty} \frac{1-\cos(1/n)}{1/n^2} = \lim_{n\to\infty} \frac{\sin(1/n)(-1/n^2)}{-2/n^3} = \lim_{n\to\infty} \frac{1}{2}\frac{\sin(1/n)}{1/n} = \lim_{x\to 0} \frac{1}{2}\frac{\sin x}{x} = \frac{1}{2}.$$
The limit comparison test applies with $c = 1/2$. The p-series $\sum 1/n^2$ converges because $p = 2 > 1$. Therefore $\sum(1-\cos(1/n))$ also converges.

41. The nth term $a_n = 2^n/(3^n - 1)$ behaves like $2^n/3^n$ for large n, so we take $b_n = 2^n/3^n$. We have
$$\lim_{n\to\infty} \frac{a_n}{b_n} = \lim_{n\to\infty} \frac{2^n/(3^n-1)}{2^n/3^n} = \lim_{n\to\infty} \frac{3^n}{3^n-1} = \lim_{n\to\infty} \frac{1}{1-3^{-n}} = 1.$$
The limit comparison test applies with $c = 1$. The geometric series $\sum 2^n/3^n = \sum(2/3)^n$ converges. Therefore $\sum 2^n/(3^n-1)$ also converges.

Problems

45. The comparison test requires that $a_n = \sin n$ be positive for all n. It is not.

49. The alternating series test requires $a_n = \sin n$ be positive, which it is not. This is not an alternating series.

53. The partial sums look like: $S_1 = 1$, $S_2 = 0$, $S_3 = 0.5$, $S_4 = 0.3333$, $S_5 = 0.375$, $S_{10} = 0.3679$, $S_{20} = 0.3679$, and higher partial sums agree with these first 4 decimal places. The series appears to be converging to about 0.3679.

Since $a_n = 1/n!$ is positive and decreasing and $\lim_{n\to\infty} 1/n! = 0$, the alternating series test confirms the convergence of this series.

57. We use the ratio test and calculate
$$\lim_{n\to\infty} \frac{|a_{n+1}|}{|a_n|} = \lim_{n\to\infty} \frac{n!/(n+1)^2}{(n-1)!/n^2} = \lim_{n\to\infty}\left(\frac{n!}{(n-1)!}\cdot\frac{n^2}{(n+1)^2}\right) = \lim_{n\to\infty}\left(n\cdot\frac{n^2}{(n+1)^2}\right).$$
Since the limit does not exist (it is ∞), the series diverges.

61. We compare the series with the convergent series $\sum 1/n^2$. From the graph of $\tan x$, we see that $\tan x < 2$ for $0 \le x \le 1$, so $\tan(1/n) < 2$ for all n. Thus

$$\frac{1}{n^2} \tan\left(\frac{1}{n}\right) < \frac{1}{n^2} 2,$$

so the series converges, since $2 \sum 1/n^2$ converges. Alternatively, we try the integral test. Since the terms in the series are positive and decreasing, we can use the integral test. We calculate the corresponding integral using the substitution $w = 1/x$:

$$\int_1^\infty \frac{1}{x^2} \tan\left(\frac{1}{x}\right) dx = \lim_{b\to\infty} \int_1^b \frac{1}{x^2} \tan\left(\frac{1}{x}\right) dx = \lim_{b\to\infty} \ln\left(\cos\frac{1}{x}\right)\Big|_1^b = \lim_{b\to\infty} \left(\ln\left(\cos\left(\frac{1}{b}\right)\right) - \ln(\cos 1)\right) = -\ln(\cos 1).$$

Since the limit exists, the integral converges, so the series $\sum_{n=1}^\infty \frac{1}{n^2} \tan(1/n)$ converges.

65. Since the exponential, 2^n, grows faster than the power, n^2, the terms are growing in size. Thus, $\lim_{n\to\infty} a_n \ne 0$. We conclude that this series diverges.

69. As $n \to \infty$, we see that

$$\frac{n+2}{n^2-1} \to \frac{n}{n^2} = \frac{1}{n}.$$

Since $\sum(1/n)$ diverges, we expect our series to have the same behavior.
    More precisely, for all $n \ge 2$, we have

$$0 \le \frac{1}{n} = \frac{n}{n^2} \le \frac{n+2}{n^2-1},$$

so $\sum_{n=2}^\infty \frac{n+2}{n^2-1}$ diverges by comparison with the divergent series $\sum \frac{1}{n}$.

73. Factoring gives $\sum_{n=1}^\infty \left(\frac{2}{n}\right)^a = 2^a \sum_{n=1}^\infty \frac{1}{n^a}$. This is a constant times a p-series that converges if $a > 1$ and diverges if $a \le 1$.

77. To use the alternating series test, consider $a_n = f(n)$, where $f(x) = \arctan(a/x)$. We need to show that $f(x)$ is decreasing. Since

$$f'(x) = \frac{1}{1+(a/x)^2}\left(-\frac{a}{x^2}\right),$$

we have $f'(x) < 0$ for $a > 0$, so $f(x)$ is decreasing for all x. Thus $a_{n+1} < a_n$ for all n, and as $\lim_{n\to\infty} \arctan(a/n) = 0$ for all a, by the alternating series test,

$$\sum_{n=1}^\infty (-1)^n \arctan(a/n)$$

converges.

81. Since $0 \le c_n \le 2^{-n}$ for all n, and since $\sum 2^{-n}$ is a convergent geometric series, $\sum c_n$ converges by the Comparison Test. Similarly, since $2^n \le a_n$, and since $\sum 2^n$ is a divergent geometric series, $\sum a_n$ diverges by the Comparison Test. We do not have enough information to determine whether or not $\sum b_n$ and $\sum d_n$ converge.

85. Each term in $\sum b_n$ is greater than or equal to a_1 times a term in the harmonic series:

$$b_1 = a_1 \cdot 1$$
$$b_2 = \frac{a_1+a_2}{2} > a_1 \cdot \frac{1}{2}$$
$$b_3 = \frac{a_1+a_2+a_3}{3} > a_1 \cdot \frac{1}{3}$$
$$\vdots$$
$$b_n = \frac{a_1+a_2+\cdots+a_n}{n} > a_1 \cdot \frac{1}{n}$$

Adding these inequalities gives

$$\sum b_n > a_1 \sum \frac{1}{n}.$$

Since the harmonic series $\sum 1/n$ diverges, a_1 times the harmonic series also diverges. Then, by the comparison test, the series $\sum b_n$ diverges.

89. The limit

$$\lim_{n \to \infty} \sqrt[n]{a_n} = \lim_{n \to \infty} \frac{5n+1}{3n^2} = 0 < 1,$$

so the series converges.

Solutions for Section 9.5

Exercises

1. Yes.

5. The general term can be written as $\dfrac{1 \cdot 3 \cdot 5 \cdots (2n-1)}{2^n \cdot n!} x^n$ for $n \geq 1$. Other answers are possible.

9. The general term can be written as $\dfrac{(x-a)^n}{2^{n-1} \cdot n!}$ for $n \geq 1$. Other answers are possible.

13. Since $C_n = (n+1)/(2^n + n)$, replacing n by $n+1$ gives $C_{n+1} = (n+2)/(2^{n+1} + n + 1)$. Using the ratio test, we have

$$\frac{|a_{n+1}|}{|a_n|} = |x| \frac{|C_{n+1}|}{|C_n|} = |x| \frac{(n+2)/(2^{n+1}+n+1)}{(n+1)/(2^n+n)} = |x| \frac{n+2}{2^{n+1}+n+1} \cdot \frac{2^n+n}{n+1} = |x| \frac{n+2}{n+1} \cdot \frac{2^n+n}{2^{n+1}+n+1}.$$

Since

$$\lim_{n \to \infty} \frac{n+2}{n+1} = 1$$

and

$$\lim_{n \to \infty} \left(\frac{2^n+n}{2^{n+1}+n+1} \right) = \frac{1}{2} \lim_{n \to \infty} \left(\frac{2^n+n}{2^n+(n+1)/2} \right) = \frac{1}{2},$$

because 2^n dominates n as $n \to \infty$, we have

$$\lim_{n \to \infty} \frac{|a_{n+1}|}{|a_n|} = \frac{1}{2}|x|.$$

Thus the radius of convergence is $R = 2$.

17. The coefficient of the nth term is $C_n = (-1)^{n+1}/n^2$. Now consider the ratio

$$\left| \frac{a_{n+1}}{a_n} \right| = \left| \frac{n^2 x^{n+1}}{(n+1)^2 x^n} \right| \to |x| \quad \text{as} \quad n \to \infty.$$

Thus, the radius of convergence is $R = 1$.

21. We write the series as

$$x - \frac{x^3}{3} + \frac{x^5}{5} - \frac{x^7}{7} + \cdots + (-1)^{n-1} \frac{x^{2n-1}}{2n-1} + \cdots,$$

so

$$a_n = (-1)^{n-1} \frac{x^{2n-1}}{2n-1}.$$

Replacing n by $n+1$, we have

$$a_{n+1} = (-1)^{n+1-1} \frac{x^{2(n+1)-1}}{2(n+1)-1} = (-1)^n \frac{x^{2n+1}}{2n+1}.$$

Thus

$$\frac{|a_{n+1}|}{|a_n|} = \left| \frac{(-1)^n x^{2n+1}}{2n+1} \right| \cdot \left| \frac{2n-1}{(-1)^{n-1} x^{2n-1}} \right| = \frac{2n-1}{2n+1} x^2,$$

so

$$L = \lim_{n \to \infty} \frac{|a_{n+1}|}{|a_n|} = \lim_{n \to \infty} \frac{2n-1}{2n+1} x^2 = x^2.$$

By the ratio test, this series converges if $L < 1$, that is, if $x^2 < 1$, so $R = 1$.

Problems

25. We use the ratio test:

$$\left| \frac{a_{n+1}}{a_n} \right| = \left| \frac{(x-3)^{n+1}}{n+1} \cdot \frac{n}{(x-3)^n} \right| = \frac{n}{n+1} \cdot |x-3|.$$

Since $n/(n+1) \to 1$ as $n \to \infty$, we have

$$\lim_{n\to\infty} \left| \frac{a_{n+1}}{a_n} \right| = |x-3|.$$

The series converges for $|x-3| < 1$. The radius of convergence is 1 and the series converges for $2 < x < 4$.
   We check the endpoints. For $x = 2$, we have

$$\sum_{n=2}^{\infty} \frac{(x-3)^n}{n} = \sum_{n=2}^{\infty} \frac{(2-3)^n}{n} = \sum_{n=2}^{\infty} \frac{(-1)^n}{n}.$$

This is the alternating harmonic series and converges. For $x = 4$, we have

$$\sum_{n=2}^{\infty} \frac{(x-3)^n}{n} = \sum_{n=2}^{\infty} \frac{(4-3)^n}{n} = \sum_{n=2}^{\infty} \frac{1}{n}.$$

This is the harmonic series and diverges. The series converges at $x = 2$ and diverges at $x = 4$. Therefore, the interval of convergence is $2 \le x < 4$.

29. We use the ratio test to find the radius of convergence

$$\left| \frac{a_{n+1}}{a_n} \right| = \left| \frac{(n+1)! x^{n+1}}{n! x^n} \right| = |x|(n+1).$$

Since $\lim_{n\to\infty} |x|(n+1) = \infty$ for all $x \ne 0$, the radius of convergence is $R = 0$. There are no endpoints to check. The series converges only for $x = 0$, so the interval of convergence is the single point $x = 0$.

33. To compare $8/(4+y)$ with $a/(1-x)$, divide the numerator and denominator by 4. This gives $8/(4+y) = 2/(1+y/4)$, which is the sum of a geometric series with $a = 2$, $x = -y/4$. The power series is

$$2 + 2(-y/4) + 2(-y/4)^2 + 2(-y/4)^3 + \cdots = \sum_{n=0}^{\infty} 2(-y/4)^n.$$

This series converges for $|y/4| < 1$, that is, for $-4 < y < 4$.

37. The series is centered at $x = -7$. Since the series converges at $x = 0$, which is a distance of 7 from $x = -7$, the radius of convergence, R, is at least 7. Since the series diverges at $x = -17$, which is a distance of 10 from $x = -7$, the radius of convergence is no more than 10. That is, $7 \le R \le 10$.

41. (a) We have

$$f(x) = 1 + x + \frac{x^2}{2} + \cdots,$$

so

$$f(0) = 1 + 0 + 0 + \cdots = 1.$$

(b) To find the domain of f, we find the interval of convergence.

$$\lim_{n\to\infty} \frac{|a_{n+1}|}{|a_n|} = \lim_{n\to\infty} \frac{|x^{n+1}/(n+1)!|}{|x^n/n!|} = \lim_{n\to\infty} \left(\frac{|x|^{n+1} n!}{|x|^n (n+1)!} \right) = |x| \lim_{n\to\infty} \frac{1}{n+1} = 0.$$

Thus the series converges for all x, so the domain of f is all real numbers.

(c) Differentiating term-by-term gives

$$f'(x) = \frac{d}{dx}\left(\sum_{n=0}^{\infty} \frac{x^n}{n!} \right) = \frac{d}{dx}\left(1 + x + \frac{x^2}{2!} + \frac{x^3}{3!} + \frac{x^4}{4!} + \cdots \right)$$

$$= 0 + 1 + 2\frac{x}{2!} + 3\frac{x^2}{3!} + 4\frac{x^3}{4!} + \cdots$$

$$= 1 + x + \frac{x^2}{2!} + \frac{x^3}{3!} + \cdots.$$

Thus, the series for f and f' are the same, so

$$f(x) = f'(x).$$

(d) We guess $f(x) = e^x$.

Solutions for Chapter 9 Review

Exercises

1. If $b = 1$, then the sum is 6. If $b \neq 1$, we use the formula for the sum of a finite geometric series. This is a six-term geometric series ($n = 6$) with initial term $a = b^5$ and constant ratio $x = b$:

$$\text{Sum} = \frac{a(1 - x^n)}{1 - x} = \frac{b^5(1 - b^6)}{1 - b}.$$

5. We have:

$$\lim_{n \to \infty} \left(\frac{n+1}{n} \right) = 1.$$

The terms of the sequence do not approach 0, and oscillate between values that are getting closer to $+1$ and -1. Thus the sequence diverges.

9. We use the integral test to determine whether this series converges or diverges. To do so we determine whether the corresponding improper integral $\int_1^{\infty} \frac{3x^2 + 2x}{x^3 + x^2 + 1} dx$ converges or diverges. The integral can be calculated using the substitution $w = x^3 + x^2 + 1$, $dw = (3x^2 + 2x)\,dx$.

$$\int_1^{\infty} \frac{3x^2 + 2x}{x^3 + x^2 + 1} dx = \lim_{b \to \infty} \int_1^b \frac{3x^2 + 2x}{x^3 + x^2 + 1} dx$$

$$= \lim_{b \to \infty} \ln |x^3 + x^2 + 1| \Big|_1^b$$

$$= \lim_{b \to \infty} \left(\ln |b^3 + b^2 + 1| - \ln 3 \right) = \infty.$$

Since the integral $\int_1^{\infty} \frac{3x^2 + 2x}{x^3 + x^2 + 1} dx$ diverges, we conclude from the integral test that the series $\sum_{n=1}^{\infty} \frac{3n^2 + 2n}{n^3 + n^2 + 1}$ diverges.

13. Since $a_n = 1/(2^n n!)$, replacing n by $n+1$ gives $a_{n+1} = 1/(2^{n+1}(n+1)!)$. Thus

$$\frac{|a_{n+1}|}{|a_n|} = \frac{\dfrac{1}{2^{n+1}(n+1)!}}{\dfrac{1}{2^n n!}} = \frac{2^n n!}{2^{n+1}(n+1)!} = \frac{1}{2(n+1)},$$

so

$$L = \lim_{n \to \infty} \frac{|a_{n+1}|}{|a_n|} = \lim_{n \to \infty} \frac{1}{2n + 2} = 0.$$

Since $L < 1$, the ratio test tells us that $\sum_{n=1}^{\infty} \frac{1}{2^n n!}$ converges.

17. Let $a_n = 1/\sqrt{n^2 + 1}$. Then replacing n by $n+1$ we have $a_{n+1} = 1/\sqrt{(n+1)^2 + 1}$. Since $\sqrt{(n+1)^2 + 1} > \sqrt{n^2 + 1}$, we have

$$\frac{1}{\sqrt{(n+1)^2 + 1}} < \frac{1}{\sqrt{n^2 + 1}},$$

so

$$0 < a_{n+1} < a_n.$$

In addition, $\lim_{n \to \infty} a_n = 0$ so $\sum_{n=0}^{\infty} \frac{(-1)^n}{\sqrt{n^2 + 1}}$ converges by the alternating series test.

21. Since

$$\lim_{n\to\infty} a_n = \lim_{n\to\infty} \frac{1}{\arctan n} = \frac{2}{\pi} \neq 0$$

we know that $\sum \frac{(-1)^{n-1}}{\arctan n}$ diverges by Property 3 of Theorem 9.2.

25. The nth term $a_n = (n^3 - 2n^2 + n + 1)/(n^5 - 2)$ behaves like $n^3/n^5 = 1/n^2$ for large n, so we take $b_n = 1/n^2$. We have

$$\lim_{n\to\infty} \frac{a_n}{b_n} = \lim_{n\to\infty} \frac{(n^3 - 2n^2 + n + 1)/(n^5 - 2)}{1/n^2} = \lim_{n\to\infty} \frac{n^5 - 2n^4 + n^3 + n^2}{n^5 - 2} = 1.$$

The limit comparison test applies with $c = 1$. The p-series $\sum 1/n^2$ converges because $p = 2 > 1$. Therefore the series $\sum \left(n^3 - 2n^2 + n + 1\right) / \left(n^5 - 2\right)$ also converges.

29. This is a p-series with $p > 1$, so it converges.

33. We use the integral test to determine whether this series converges or diverges. To do so we determine whether the corresponding improper integral $\int_1^\infty \frac{x^2}{x^3 + 1} dx$ converges or diverges:

$$\int_1^\infty \frac{x^2}{x^3 + 1} dx = \lim_{b\to\infty} \int_1^b \frac{x^2}{x^3 + 1} dx = \lim_{b\to\infty} \frac{1}{3} \ln|x^3 + 1|\Big|_1^b = \lim_{b\to\infty} \left(\frac{1}{3} \ln(b^3 + 1) - \frac{1}{3} \ln 2\right).$$

Since the limit does not exist, the integral $\int_1^\infty \frac{x^2}{x^3 + 1} dx$ diverges an so we conclude from the integral test that the series $\sum_{n=1}^\infty \frac{n^2}{n^3 + 1}$ diverges. The limit comparison test with $b_n = 1/n$ can also be used.

37. Let $a_n = 2^{-n} \frac{(n+1)}{(n+2)} = \left(\frac{n+1}{n+2}\right)\left(\frac{1}{2^n}\right)$. Since $\frac{(n+1)}{(n+2)} < 1$ and $\frac{1}{2^n} = \left(\frac{1}{2}\right)^n$, we have

$$0 < a_n < \left(\frac{1}{2}\right)^n,$$

so that we can compare the series $\sum_{n=1}^\infty 2^{-n} \frac{(n+1)}{(n+2)}$ with the convergent geometric series $\sum_{n=1}^\infty \left(\frac{1}{2}\right)^n$. The comparison test tells us that

$$\sum_{n=1}^\infty 2^{-n} \frac{(n+1)}{(n+2)}$$

also converges.

41. Writing $a_n = 1/(2 + \sin n)$, we have $\lim_{n\to\infty} a_n$ does not exist, so the series diverges by Property 3 of Theorem 9.2.

45. Since $\ln(1 + 1/k) = \ln((k+1)/k) = \ln(k+1) - \ln k$, the nth partial sum of this series is

$$S_n = \sum_{k=1}^n \ln\left(1 + \frac{1}{k}\right)$$
$$= \sum_{k=1}^n \ln(k+1) - \sum_{k=1}^n \ln k$$
$$= (\ln 2 + \ln 3 + \cdots + \ln(n+1)) - (\ln 1 + \ln 2 + \cdots + \ln n)$$
$$= \ln(n+1) - \ln 1$$
$$= \ln(n+1).$$

Thus, the partial sums, S_n, grow without bound as $n \to \infty$, so the series diverges by the definition.

49. Let $C_n = \frac{(2n)!}{(n!)^2}$. Then replacing n by $n+1$, we have $C_{n+1} = \frac{(2n+2)!}{((n+1)!)^2}$. Thus, with $a_n = (2n)!x^n/(n!)^2$, we have

$$\frac{|a_{n+1}|}{|a_n|} = |x|\frac{|C_{n+1}|}{|C_n|} = |x|\frac{(2n+2)!/((n+1)!)^2}{(2n)!/(n!)^2} = |x|\frac{(2n+2)!}{(2n)!} \cdot \frac{(n!)^2}{((n+1)!)^2}.$$

Since $(2n+2)! = (2n+2)(2n+1)(2n)!$ and $(n+1)! = (n+1)n!$ we have

$$\frac{|C_{n+1}|}{|C_n|} = \frac{(2n+2)(2n+1)}{(n+1)(n+1)},$$

so

$$\lim_{n\to\infty} \frac{|a_{n+1}|}{|a_n|} = |x| \lim_{n\to\infty} \frac{|C_{n+1}|}{|C_n|} = |x| \lim_{n\to\infty} \frac{(2n+2)(2n+1)}{(n+1)(n+1)} = |x| \lim_{n\to\infty} \frac{4n+2}{n+1} = 4|x|,$$

so the radius of convergence of this series is $R = 1/4$.

53. Here the coefficient of the nth term is $C_n = n/(2n+1)$. Now we have

$$\left|\frac{a_{n+1}}{a_n}\right| = \left|\frac{((n+1)/(2n+3))x^{n+1}}{(n/(2n+1))x^n}\right| = \frac{(n+1)(2n+1)}{n(2n+3)}|x| \to |x| \text{ as } n \to \infty.$$

Thus, by the ratio test, the radius of convergence is $R = 1$.

57. We use the ratio test to find the radius of convergence;

$$\left|\frac{a_{n+1}}{a_n}\right| = \left|\frac{x^{2(n+1)+1}}{(n+1)!} \cdot \frac{n!}{x^{2n+1}}\right| = \left|\frac{x^2}{n+1}\right|$$

Since $\lim_{n\to\infty} |x^2|/(n+1) = 0$ for all x, the radius of convergence is $R = \infty$. There are no endpoints to check. The interval of convergence is all real numbers $-\infty < x < \infty$.

Problems

61. To get $\$20,000$ the day he retires he needs to invest a present value P such that $P(1+5/100)^{20} = \$20,000$. Solving for P gives the present value $P = \$20,000 \cdot 1.05^{-20}$. To fund the second payment he needs to invest $\$20,000 \cdot 1.05^{-21}$, and so on. To fund the payment 10 years after his retirement he needs to invest $\$20,000 \cdot 1.05^{-30}$. There are 11 payments in all, so

$$\text{Total investment} = 20,000 \cdot 1.05^{-20} + 20,000 \cdot 1.05^{-21} + \ldots + 20,000 \cdot 1.05^{-30}$$
$$= 20,000 \cdot 1.05^{-20} \left(1 + 1.05^{-1} + \ldots + 1.05^{-10}\right)$$
$$= 20,000 \cdot 1.05^{-20} \left(\frac{1 - 1.05^{-11}}{1 - 1.05^{-1}}\right)$$
$$= 20,000 \cdot 1.05^{-20} \cdot 8.7217$$
$$= \$65,742.60.$$

65.

$$\text{Present value of first coupon} = \frac{50}{1.04}$$
$$\text{Present value of second coupon} = \frac{50}{(1.04)^2}, \text{etc.}$$

$$\text{Total present value} = \underbrace{\frac{50}{1.04} + \frac{50}{(1.04)^2} + \cdots + \frac{50}{(1.04)^{10}}}_{\text{coupons}} + \underbrace{\frac{1000}{(1.04)^{10}}}_{\text{principal}}$$
$$= \frac{50}{1.04}\left(1 + \frac{1}{1.04} + \cdots + \frac{1}{(1.04)^9}\right) + \frac{1000}{(1.04)^{10}}$$
$$= \frac{50}{1.04}\left(\frac{1 - \left(\frac{1}{1.04}\right)^{10}}{1 - \frac{1}{1.04}}\right) + \frac{1000}{(1.04)^{10}}$$
$$= 405.545 + 675.564$$
$$= \$1081.11$$

69. A person should expect to pay the present value of the bond on the day it is bought.

$$\text{Present value of first payment} = \frac{10}{1.04}$$

$$\text{Present value of second payment} = \frac{10}{(1.04)^2}, \text{ etc.}$$

Therefore,

$$\text{Total present value} = \frac{10}{1.04} + \frac{10}{(1.04)^2} + \frac{10}{(1.04)^3} + \cdots.$$

This is a geometric series with $a = \dfrac{10}{1.04}$ and $x = \dfrac{1}{1.04}$, so

$$\text{Total present value} = \frac{\frac{10}{1.04}}{1 - \frac{1}{1.04}} = £250.$$

73. No. If the series $\displaystyle\sum_{n=1}^{\infty}(-1)^{n-1}a_n$ converges then, using Theorem 9.2, part 3, we have $\displaystyle\lim_{n\to\infty}(-1)^{n-1}a_n = 0$, which cannot happen if $\displaystyle\lim_{n\to\infty}a_n \neq 0$.

77. Since $\sum a_n$ converge, we know that $\lim_{n\to\infty} a_n = 0$. Thus $\lim_{n\to\infty}(1/a_n)$ does not exist, and it follows that $\sum(1/a_n)$ diverges by Property 3 of Theorem 9.2.

81. The series

$$\sum_{n=1}^{\infty}\left(\frac{1}{n} + \frac{1}{n}\right) = \sum_{n=1}^{\infty}\frac{2}{n}$$

diverges by Theorem 9.2 and the fact that $\displaystyle\sum_{n=1}^{\infty}\frac{1}{n}$ diverges.

The series

$$\sum_{n=1}^{\infty}\left(\frac{1}{n} - \frac{1}{n}\right) = \sum_{n=1}^{\infty}0 = 0$$

converges. But $\displaystyle\sum_{n=1}^{\infty}-\frac{1}{n}$ diverges by Theorem 9.2 and the fact that $\displaystyle\sum_{n=1}^{\infty}\frac{1}{n}$ diverges.

Thus, if $a_n = 1/n$ and $b_n = 1/n$, so that $\sum a_n$ and $\sum b_n$ both diverge, we see that $\sum(a_n + b_n)$ may diverge.

If, on the other hand, $a_n = 1/n$ and $b_n = -1/n$, so that $\sum a_n$ and $\sum b_n$ both diverge, we see that $\sum(a_n + b_n)$ may converge.

Therefore, if $\sum a_n$ and $\sum b_n$ both diverge, we cannot tell whether $\sum(a_n + b_n)$ converges or diverges. Thus the statement is true.

CHECK YOUR UNDERSTANDING

1. False. The first 1000 terms could be the same for two different sequences and yet one sequence converges and the other diverges. For example, $s_n = 0$ for all n is a convergent sequence, but

$$t_n = \begin{cases} 0 & \text{if } n \leq 1000 \\ n & \text{if } n > 1000 \end{cases}$$

is a divergent sequence.

5. False. The terms s_n tend to the limit of the sequence which may not be zero. For example, $s_n = 1 + 1/n$ is a convergent sequence and s_n tends to 1 as n increases.

9. False. The sequence $-1, 1, -1, 1, \ldots$ given by $s_n = (-1)^n$ alternates in sign but does not converge.

13. False. This power series has an interval of convergence about $x = 0$. Knowing the power series converges for $x = 1$ does not tell us whether the series converges for $x = 2$. Since the series converges at $x = 1$, we know the radius of convergence is at least 1. However, we do not know whether the interval of convergence extends as far as $x = 2$, so we cannot say whether the series converges at $x = 2$.

For example, $\sum \dfrac{x^n}{2^n}$ converges for $x = 1$ (it is a geometric series with ratio of $1/2$), but does not converge for $x = 2$ (the terms do not go to 0).

Since this statement is not true for all C_n, the statement is false.

17. True. Consider the series $\sum(-b_n)$ and $\sum(-a_n)$. The series $\sum(-b_n)$ converges, since $\sum b_n$ converges, and

$$0 \le -a_n \le -b_n.$$

By the comparison test, $\sum(-a_n)$ converges, so $\sum a_n$ converges.

21. False, since if we write out the terms of the series, using the fact that $\cos 0 = 1, \cos \pi = -1, \cos(2\pi) = 1, \cos(3\pi) = -1$, and so on, we have

$$(-1)^0 \cos 0 + (-1)^1 \cos \pi + (-1)^2 \cos 2\pi + (-1)^3 \cos 3\pi + \cdots$$
$$= (1)(1) + (-1)(-1) + (1)(1) + (-1)(-1) + \cdots$$
$$= 1 + 1 + 1 + 1 + \cdots.$$

This is not an alternating series.

25. False. For example, if $a_n = (-1)^{n-1}/n$, then $\sum a_n$ converges by the alternating series test. But $(-1)^n a_n = (-1)^n (-1)^{n-1}/n = (-1)^{2n-1}/n = -1/n$. Thus, $\sum (-1)^n a_n$ is the negative of the harmonic series and does not converge.

29. True. Since the series is alternating, Theorem 9.9 gives the error bound. Summing the first 100 terms gives S_{100}, and if the true sum is S,

$$|S - S_{100}| < a_{101} = \frac{1}{101} < 0.01.$$

33. False. Consider the series $\displaystyle\sum_{n=1}^{\infty} 1/n$. This series does not converge, but $1/n \to 0$ as $n \to \infty$.

37. False. The alternating harmonic series $\sum \dfrac{(-1)^n}{n}$ is conditionally convergent because it converges by the Alternating Series test, but the harmonic series $\sum \left| \dfrac{(-1)^n}{n} \right| = \sum \dfrac{1}{n}$ is divergent. The alternating harmonic series is not absolutely convergent.

41. True. Since the power series converges at $x = 10$, the radius of convergence is at least 10. Thus, $x = -9$ must be within the interval of convergence.

45. True. The interval of convergence is centered on $x = a$, so $a = (-11 + 1)/2 = -5$.

CHAPTER TEN

Solutions for Section 10.1

Exercises

1. Let $f(x) = \dfrac{1}{1-x} = (1-x)^{-1}$. Then $f(0) = 1$.

$$f'(x) = 1!(1-x)^{-2} \quad f'(0) = 1!,$$
$$f''(x) = 2!(1-x)^{-3} \quad f''(0) = 2!,$$
$$f'''(x) = 3!(1-x)^{-4} \quad f'''(0) = 3!,$$
$$f^{(4)}(x) = 4!(1-x)^{-5} \quad f^{(4)}(0) = 4!,$$
$$f^{(5)}(x) = 5!(1-x)^{-6} \quad f^{(5)}(0) = 5!,$$
$$f^{(6)}(x) = 6!(1-x)^{-7} \quad f^{(6)}(0) = 6!,$$
$$f^{(7)}(x) = 7!(1-x)^{-8} \quad f^{(7)}(0) = 7!.$$

$$P_3(x) = 1 + x + x^2 + x^3,$$
$$P_5(x) = 1 + x + x^2 + x^3 + x^4 + x^5,$$
$$P_7(x) = 1 + x + x^2 + x^3 + x^4 + x^5 + x^6 + x^7.$$

5. Let $f(x) = \cos x$. Then $f(0) = \cos(0) = 1$, and

$$f'(x) = -\sin x \quad f'(0) = 0,$$
$$f''(x) = -\cos x \quad f''(0) = -1,$$
$$f'''(x) = \sin x \quad f'''(0) = 0,$$
$$f^{(4)}(x) = \cos x \quad f^{(4)}(0) = 1,$$
$$f^{(5)}(x) = -\sin x \quad f^{(5)}(0) = 0,$$
$$f^{(6)}(x) = -\cos x \quad f^{(6)}(0) = -1.$$

Thus,

$$P_2(x) = 1 - \frac{x^2}{2!},$$
$$P_4(x) = 1 - \frac{x^2}{2!} + \frac{x^4}{4!},$$
$$P_6(x) = 1 - \frac{x^2}{2!} + \frac{x^4}{4!} - \frac{x^6}{6!}.$$

9. Let $f(x) = \dfrac{1}{\sqrt{1+x}} = (1+x)^{-1/2}$. Then $f(0) = 1$.

$$f'(x) = -\tfrac{1}{2}(1+x)^{-3/2} \quad f'(0) = -\tfrac{1}{2},$$
$$f''(x) = \tfrac{3}{2^2}(1+x)^{-5/2} \quad f''(0) = \tfrac{3}{2^2},$$
$$f'''(x) = -\tfrac{3\cdot5}{2^3}(1+x)^{-7/2} \quad f'''(0) = -\tfrac{3\cdot5}{2^3},$$
$$f^{(4)}(x) = \tfrac{3\cdot5\cdot7}{2^4}(1+x)^{-9/2} \quad f^{(4)}(0) = \tfrac{3\cdot5\cdot7}{2^4}$$

Then,

$$P_2(x) = 1 - \frac{1}{2}x + \frac{1}{2!}\frac{3}{2^2}x^2 = 1 - \frac{1}{2}x + \frac{3}{8}x^2,$$

$$P_3(x) = P_2(x) - \frac{1}{3!}\frac{3 \cdot 5}{2^3}x^3 = 1 - \frac{1}{2}x + \frac{3}{8}x^2 - \frac{5}{16}x^3,$$

$$P_4(x) = P_3(x) + \frac{1}{4!}\frac{3 \cdot 5 \cdot 7}{2^4}x^4 = 1 - \frac{1}{2}x + \frac{3}{8}x^2 - \frac{5}{16}x^3 + \frac{35}{128}x^4.$$

13. Let $f(x) = \sqrt{1+x} = (1+x)^{1/2}$.

Then $f'(x) = \frac{1}{2}(1+x)^{-1/2}$, $f''(x) = -\frac{1}{4}(1+x)^{-3/2}$, and $f'''(x) = \frac{3}{8}(1+x)^{-5/2}$. The Taylor polynomial of degree three about $x = 1$ is thus

$$P_3(x) = (1+1)^{1/2} + \frac{1}{2}(1+1)^{-1/2}(x-1) + \frac{-\frac{1}{4}(1+1)^{-3/2}}{2!}(x-1)^2$$

$$+ \frac{\frac{3}{8}(1+1)^{-5/2}}{3!}(x-1)^3$$

$$= \sqrt{2}\left(1 + \frac{x-1}{4} - \frac{(x-1)^2}{32} + \frac{(x-1)^3}{128}\right).$$

Problems

17. Using the fact that

$$f(x) \approx P_3(x) = f(0) + f'(0)x + \frac{f''(0)}{2!}x^2 + \frac{f'''(0)}{3!}x^3$$

and identifying coefficients with those given for $P_3(x)$, we obtain the following:

(a) $f(0) = $ constant term which equals 2,   so $f(0) = 2$.
(b) $f'(0) = $ coefficient of x which equals -1,   so $f'(0) = -1$.
(c) $\frac{f''(0)}{2!} = $ coefficient of x^2 which equals $-1/3$,   so $f''(0) = -2/3$.
(d) $\frac{f'''(0)}{3!} = $ coefficient of x^3 which equals 2,   so $f'''(0) = 12$.

21. Since $P_2(x)$ is the second degree Taylor polynomial for $f(x)$ about $x = 0$, $P_2(0) = f(0)$, which says $a = f(0)$. Since

$$\frac{d}{dx}P_2(x)\Big|_{x=0} = f'(0),$$

$b = f'(0)$; and since

$$\frac{d^2}{dx^2}P_2(x)\Big|_{x=0} = f''(0),$$

$2c = f''(0)$. In other words, a is the y-intercept of $f(x)$, b is the slope of the tangent line to $f(x)$ at $x = 0$ and c tells us the concavity of $f(x)$ near $x = 0$. So $c < 0$ since f is concave down; $b > 0$ since f is increasing; $a > 0$ since $f(0) > 0$.

25.

$$\lim_{x \to 0}\frac{\sin x}{x} = \lim_{x \to 0}\frac{x - \frac{x^3}{3!}}{x} = \lim_{x \to 0}\left(1 - \frac{x^2}{3!}\right) = 1.$$

29. (a) Since the coefficient of the x-term of each f is 1, we know $f_1'(0) = f_2'(0) = f_3'(0) = 1$. Thus, each of the fs slopes upward near 0, and are in the second figure.

The coefficient of the x-term in g_1 and in g_2 is 1, so $g_1'(0) = g_2'(0) = 1$. For g_3 however, $g_3'(0) = -1$. Thus, g_1 and g_2 slope up near 0, but g_3 slopes down. The gs are in the first figure.

(b) Since $g_1(0) = g_2(0) = g_3(0) = 1$, the point A is $(0, 1)$.
Since $f_1(0) = f_2(0) = f_3(0) = 2$, the point B is $(0, 2)$.

(c) Since g_3 slopes down, g_3 is I. Since the coefficient of x^2 for g_1 is 2, we know

$$\frac{g_1''(0)}{2!} = 2 \qquad \text{so} \qquad g_1''(0) = 4.$$

By similar reasoning $g_2''(0) = 2$. Since g_1 and g_2 are concave up, and g_1 has a larger second derivative, g_1 is III and g_2 is II.

Calculating the second derivatives of the fs from the coefficients x^2, we find

$$f_1''(0) = 4 \qquad f_2''(0) = -2 \qquad f_3''(0) = 2.$$

Thus, f_1 and f_3 are concave up, with f_1 having the larger second derivative, so f_1 is III and f_3 is II. Then f_2 is concave down and is I.

33. (a) The equation $\sin x = 0.2$ has one solution near $x = 0$ and infinitely many others, one near each multiple of π. See Figure 10.1. The equation $x - \dfrac{x^3}{3!} = 0.2$ has three solutions, one near $x = 0$ and two others. See Figure 10.2.

Figure 10.1: Graph of $y = \sin x$ and $y = 0.2$ **Figure 10.2**: Graph of $y = x - \frac{x^3}{3!}$ and $y = 0.2$

(b) Near $x = 0$, the cubic Taylor polynomial $x - x^3/3! \approx \sin x$. Thus, the solutions to the two equations near $x = 0$ are approximately equal. The other solutions are not close. The reason is that $x - x^3/3!$ only approximates $\sin x$ near $x = 0$ but not further away. See Figure 10.3.

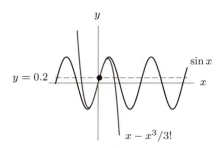

Figure 10.3

Solutions for Section 10.2

Exercises

1. Differentiating $(1 + x)^{3/2}$:

$$
\begin{aligned}
f(x) &= (1+x)^{3/2} & f(0) &= 1, \\
f'(x) &= (3/2)(1+x)^{1/2} & f'(0) &= \tfrac{3}{2}, \\
f''(x) &= (1/2)(3/2)(1+x)^{-1/2} = (3/4)(1+x)^{-1/2} & f''(0) &= \tfrac{3}{4}, \\
f'''(x) &= (-1/2)(3/4)(1+x)^{-3/2} = (-3/8)(1+x)^{-3/2} & f'''(0) &= -\tfrac{3}{8}.
\end{aligned}
$$

$$
\begin{aligned}
f(x) = (1+x)^{3/2} &= 1 + \frac{3}{2} \cdot x + \frac{(3/4)x^2}{2!} + \frac{(-3/8)x^3}{3!} + \cdots \\
&= 1 + \frac{3x}{2} + \frac{3x^2}{8} - \frac{x^3}{16} + \cdots
\end{aligned}
$$

5.

$$f(x) = \tfrac{1}{1-x} = (1-x)^{-1} \qquad\qquad f(0) = 1,$$
$$f'(x) = -(1-x)^{-2}(-1) = (1-x)^{-2} \qquad f'(0) = 1,$$
$$f''(x) = -2(1-x)^{-3}(-1) = 2(1-x)^{-3} \quad f''(0) = 2,$$
$$f'''(x) = -6(1-x)^{-4}(-1) = 6(1-x)^{-4} \quad f'''(0) = 6.$$

$$f(x) = \frac{1}{1-x} = 1 + 1 \cdot x + \frac{2x^2}{2!} + \frac{6x^3}{3!} + \cdots$$
$$= 1 + x + x^2 + x^3 + \cdots$$

9.

$$f(x) = \sin x \qquad f(\tfrac{\pi}{4}) = \tfrac{\sqrt{2}}{2},$$
$$f'(x) = \cos x \qquad f'(\tfrac{\pi}{4}) = \tfrac{\sqrt{2}}{2},$$
$$f''(x) = -\sin x \quad f''(\tfrac{\pi}{4}) = -\tfrac{\sqrt{2}}{2},$$
$$f'''(x) = -\cos x \quad f'''(\tfrac{\pi}{4}) = -\tfrac{\sqrt{2}}{2}.$$

$$\sin x = \frac{\sqrt{2}}{2} + \frac{\sqrt{2}}{2}\left(x - \frac{\pi}{4}\right) - \frac{\sqrt{2}}{2}\frac{(x-\frac{\pi}{4})^2}{2!} - \frac{\sqrt{2}}{2}\frac{(x-\frac{\pi}{4})^3}{3!} - \cdots$$
$$= \frac{\sqrt{2}}{2} + \frac{\sqrt{2}}{2}\left(x - \frac{\pi}{4}\right) - \frac{\sqrt{2}}{4}\left(x - \frac{\pi}{4}\right)^2 - \frac{\sqrt{2}}{12}\left(x - \frac{\pi}{4}\right)^3 - \cdots$$

13.

$$f(x) = \tan x \qquad\qquad\qquad f(\tfrac{\pi}{4}) = 1,$$
$$f'(x) = \frac{1}{\cos^2 x} \qquad\qquad\qquad f'(\tfrac{\pi}{4}) = 2,$$
$$f''(x) = \frac{-2(-\sin x)}{\cos^3 x} = \frac{2\sin x}{\cos^3 x} \qquad f''(\tfrac{\pi}{4}) = 4,$$
$$f'''(x) = \frac{-6\sin x(-\sin x)}{\cos^4 x} + \frac{2}{\cos^2 x} \quad f'''(\tfrac{\pi}{4}) = 16.$$

$$\tan x = 1 + 2\left(x - \frac{\pi}{4}\right) + 4\frac{(x-\frac{\pi}{4})^2}{2!} + 16\frac{(x-\frac{\pi}{4})^3}{3!} + \cdots$$
$$= 1 + 2\left(x - \frac{\pi}{4}\right) + 2\left(x - \frac{\pi}{4}\right)^2 + \frac{8}{3}\left(x - \frac{\pi}{4}\right)^3 + \cdots$$

17. The general term can be written as x^n for $n \geq 0$.

21. The general term can be written as $(-1)^k x^{2k+1}/(2k+1)!$ for $k \geq 0$.

Problems

25. By looking at Figure 10.4 we can that the Taylor polynomials are reasonable approximations for the function $f(x) = \frac{1}{\sqrt{1+x}}$ between $x = -1$ and $x = 1$. Thus a good guess is that the interval of convergence is $-1 < x < 1$.

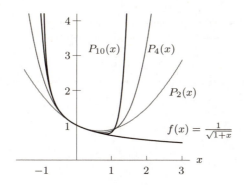

Figure 10.4

29. (a) We have shown that the series is

$$1 + px + \frac{p(p-1)}{2!}x^2 + \frac{p(p-1)(p-2)}{3!}x^3 + \cdots$$

so the general term is

$$\frac{p(p-1)\ldots(p-(n-1))}{n!}x^n.$$

(b) We use the ratio test

$$\lim_{n\to\infty}\frac{|a_{n+1}|}{|a_n|} = |x|\lim_{n\to\infty}\left|\frac{p(p-1)\ldots(p-(n-1))(p-n)\cdot n!}{(n+1)!p(p-1)\ldots(p-(n-1))}\right| = |x|\lim_{n\to\infty}\left|\frac{p-n}{n+1}\right|.$$

Since p is fixed, we have

$$\lim_{n\to\infty}\left|\frac{p-n}{n+1}\right| = 1, \quad \text{so} \quad R = 1.$$

33. We define $e^{i\theta}$ to be

$$e^{i\theta} = 1 + i\theta + \frac{(i\theta)^2}{2!} + \frac{(i\theta)^3}{3!} + \frac{(i\theta)^4}{4!} + \frac{(i\theta)^5}{5!} + \frac{(i\theta)^6}{6!} + \cdots$$

Suppose we consider the expression $\cos\theta + i\sin\theta$, with $\cos\theta$ and $\sin\theta$ replaced by their Taylor series:

$$\cos\theta + i\sin\theta = \left(1 - \frac{\theta^2}{2!} + \frac{\theta^4}{4!} - \frac{\theta^6}{6!} + \cdots\right) + i\left(\theta - \frac{\theta^3}{3!} + \frac{\theta^5}{5!} - \cdots\right)$$

Reordering terms, we have

$$\cos\theta + i\sin\theta = 1 + i\theta - \frac{\theta^2}{2!} - \frac{i\theta^3}{3!} + \frac{\theta^4}{4!} + \frac{i\theta^5}{5!} - \frac{\theta^6}{6!} - \cdots$$

Using the fact that $i^2 = -1$, $i^3 = -i$, $i^4 = 1$, $i^5 = i, \cdots$, we can rewrite the series as

$$\cos\theta + i\sin\theta = 1 + i\theta + \frac{(i\theta)^2}{2!} + \frac{(i\theta)^3}{3!} + \frac{(i\theta)^4}{4!} + \frac{(i\theta)^5}{5!} + \frac{(i\theta)^6}{6!} + \cdots$$

Amazingly enough, this series is the Taylor series for e^x with $i\theta$ substituted for x. Therefore, we have shown that

$$\cos\theta + i\sin\theta = e^{i\theta}.$$

37. This is the series for $\cos x$ with x replaced by 10, so the series converges to $\cos 10$.

41. This is the series for $\cos x$ with $x = 1$ substituted. Thus

$$1 - \frac{1}{2!} + \frac{1}{4!} - \frac{1}{6!} + \cdots = \cos 1.$$

Solutions for Section 10.3

Exercises

1. Substitute $y = -x$ into $e^y = 1 + y + \frac{y^2}{2!} + \frac{y^3}{3!} + \cdots$. We get

$$e^{-x} = 1 + (-x) + \frac{(-x)^2}{2!} + \frac{(-x)^3}{3!} + \cdots$$
$$= 1 - x + \frac{x^2}{2!} - \frac{x^3}{3!} + \cdots.$$

5. Since $\frac{d}{dx}(\arcsin x) = \frac{1}{\sqrt{1-x^2}} = 1 + \frac{1}{2}x^2 + \frac{3}{8}x^4 + \frac{5}{16}x^6 + \cdots$, integrating gives

$$\arcsin x = c + x + \frac{1}{6}x^3 + \frac{3}{40}x^5 + \frac{5}{112}x^7 + \cdots.$$

Since $\arcsin 0 = 0$, $c = 0$.

9.

$$\phi^3 \cos(\phi^2) = \phi^3 \left(1 - \frac{(\phi^2)^2}{2!} + \frac{(\phi^2)^4}{4!} - \frac{(\phi^2)^6}{6!} + \cdots \right)$$
$$= \phi^3 - \frac{\phi^7}{2!} + \frac{\phi^{11}}{4!} - \frac{\phi^{15}}{6!} + \cdots$$

13. Multiplying out gives $(1+x)^3 = 1 + 3x + 3x^2 + x^3$. Since this polynomial equals the original function for all x, it must be the Taylor series. The general term is $0 \cdot x^n$ for $n \geq 4$.

17. Using the binomial expansion for $(1+x)^{1/2}$ with $x = h/T$:

$$\sqrt{T+h} = \left(T + \frac{T}{T}h \right)^{1/2} = \left(T \left(1 + \frac{h}{T} \right) \right)^{1/2} = \sqrt{T} \left(1 + \frac{h}{T} \right)^{1/2}$$
$$= \sqrt{T} \left(1 + (1/2) \left(\frac{h}{T} \right) + \frac{(1/2)(-1/2)}{2!} \left(\frac{h}{T} \right)^2 + \frac{(1/2)(-1/2)(-3/2)}{3!} \left(\frac{h}{T} \right)^3 \cdots \right)$$
$$= \sqrt{T} \left(1 + \frac{1}{2} \left(\frac{h}{T} \right) - \frac{1}{8} \left(\frac{h}{T} \right)^2 + \frac{1}{16} \left(\frac{h}{T} \right)^3 \cdots \right).$$

21.

$$\frac{a}{\sqrt{a^2 + x^2}} = \frac{a}{a(1 + \frac{x^2}{a^2})^{\frac{1}{2}}} = \left(1 + \frac{x^2}{a^2} \right)^{-\frac{1}{2}}$$
$$= 1 + \left(-\frac{1}{2} \right) \frac{x^2}{a^2} + \frac{1}{2!} \left(-\frac{1}{2} \right) \left(-\frac{3}{2} \right) \left(\frac{x^2}{a^2} \right)^2$$
$$+ \frac{1}{3!} \left(-\frac{1}{2} \right) \left(-\frac{3}{2} \right) \left(-\frac{5}{2} \right) \left(\frac{x^2}{a^2} \right)^3 + \cdots$$
$$= 1 - \frac{1}{2} \left(\frac{x}{a} \right)^2 + \frac{3}{8} \left(\frac{x}{a} \right)^4 - \frac{5}{16} \left(\frac{x}{a} \right)^6 + \cdots$$

Problems

25. From the series for $\ln(1 + y)$,

$$\ln(1 + y) = y - \frac{y^2}{2} + \frac{y^3}{3} - \frac{y^4}{4} + \cdots,$$

we get

$$\ln(1 + y^2) = y^2 - \frac{y^4}{2} + \frac{y^6}{3} - \frac{y^8}{4} + \cdots$$

The Taylor series for $\sin y$ is

$$\sin y = y - \frac{y^3}{3!} + \frac{y^5}{5!} - \frac{y^7}{7!} + \cdots$$

So

$$\sin y^2 = y^2 - \frac{y^6}{3!} + \frac{y^{10}}{5!} - \frac{y^{14}}{7!} + \cdots$$

The Taylor series for $\cos y$ is

$$\cos y = 1 - \frac{y^2}{2!} + \frac{y^4}{4!} - \frac{y^6}{6!} + \cdots$$

So

$$1 - \cos y = \frac{y^2}{2!} - \frac{y^4}{4!} + \frac{y^6}{6!} + \cdots$$

Near $y = 0$, we can drop terms beyond the fourth degree in each expression:

$$\ln(1 + y^2) \approx y^2 - \frac{y^4}{2}$$

$$\sin y^2 \approx y^2$$

$$1 - \cos y \approx \frac{y^2}{2!} - \frac{y^4}{4!}.$$

(Note: These functions are all even, so what holds for negative y will hold for positive y.)
Clearly $1 - \cos y$ is smallest, because the y^2 term has a factor of $\frac{1}{2}$. Thus, for small y,

$$\frac{y^2}{2!} - \frac{y^4}{4!} < y^2 - \frac{y^4}{2} < y^2$$

so

$$1 - \cos y < \ln(1 + y^2) < \sin(y^2).$$

29. Since $e^x = \sum_{n=0}^{\infty} \frac{x^n}{n!}$ and $\sinh 2x = (e^{2x} - e^{-2x})/2$, the Taylor expansion for $\sinh 2x$ is

$$\sinh 2x = \frac{1}{2}\left(\sum_{n=0}^{\infty} \frac{(2x)^n}{n!} - \sum_{n=0}^{\infty} \frac{(-2x)^n}{n!}\right) = \frac{1}{2}\left(\sum_{n=0}^{\infty}(1 - (-1)^n)\frac{(2x)^n}{n!}\right)$$

$$= \sum_{m=0}^{\infty} \frac{(2x)^{2m+1}}{(2m+1)!}.$$

Since $\cosh 2x = (e^{2x} + e^{-2x})/2$, we have

$$\cosh 2x = \frac{1}{2}\left(\sum_{n=0}^{\infty} \frac{(2x)^n}{n!} + \sum_{n=0}^{\infty} \frac{(-2x)^n}{n!}\right) = \frac{1}{2}\left(\sum_{n=0}^{\infty}(1 + (-1)^n)\frac{(2x)^n}{n!}\right)$$

$$= \sum_{m=0}^{\infty} \frac{(2x)^{2m}}{(2m)!}.$$

33. (a) $\mu = \frac{mM}{m + M}.$

If $M \gg m$, then the denominator $m + M \approx M$, so $\mu \approx \frac{mM}{M} = m.$

(b)

$$\mu = m\left(\frac{M}{m + M}\right) = m\left(\frac{\frac{1}{M}M}{\frac{m}{M} + \frac{M}{M}}\right) = m\left(\frac{1}{1 + \frac{m}{M}}\right)$$

We can use the binomial expansion since $\frac{m}{M} < 1$.

$$\mu = m\left[1 - \frac{m}{M} + \left(\frac{m}{M}\right)^2 - \left(\frac{m}{M}\right)^3 + \cdots\right]$$

(c) If $m \approx \dfrac{1}{1836}M$, then $\dfrac{m}{M} \approx \dfrac{1}{1836} \approx 0.000545$.

So a first order approximation to μ would give $\mu = m(1 - 0.000545)$. The percentage difference from $\mu = m$ is -0.0545%.

37. (a) To find when V takes on its minimum values, set $\frac{dV}{dr} = 0$. So

$$-V_0 \frac{d}{dr}\left(2\left(\frac{r_0}{r}\right)^6 - \left(\frac{r_0}{r}\right)^{12}\right) = 0$$

$$-V_0\left(-12r_0^6 r^{-7} + 12r_0^{12} r^{-13}\right) = 0$$

$$12r_0^6 r^{-7} = 12r_0^{12} r^{-13}$$

$$r_0^6 = r^6$$

$$r = r_0.$$

Rewriting $V'(r)$ as $\dfrac{12r_0^6 V_0}{r^7}\left(1 - \left(\dfrac{r_0}{r}\right)^6\right)$, we see that $V'(r) > 0$ for $r > r_0$ and $V'(r) < 0$ for $r < r_0$. Thus, $V = -V_0(2(1)^6 - (1)^{12}) = -V_0$ is a minimum.

(Note: We discard the negative root $-r_0$ since the distance r must be positive.)

(b)

$$V(r) = -V_0\left(2\left(\frac{r_0}{r}\right)^6 - \left(\frac{r_0}{r}\right)^{12}\right)$$

$$V'(r) = -V_0(-12r_0^6 r^{-7} + 12r_0^{12} r^{-13})$$

$$V''(r) = -V_0(84r_0^6 r^{-8} - 156r_0^{12} r^{-14})$$

$$V(r_0) = -V_0$$
$$V'(r_0) = 0$$
$$V''(r_0) = 72V_0 r_0^{-2}$$

The Taylor series is thus:

$$V(r) = -V_0 + 72V_0 r_0^{-2} \cdot (r - r_0)^2 \cdot \frac{1}{2} + \cdots$$

(c) The difference between V and its minimum value $-V_0$ is

$$V - (-V_0) = 36V_0 \frac{(r - r_0)^2}{r_0^2} + \cdots$$

which is approximately proportional to $(r - r_0)^2$ since terms containing higher powers of $(r - r_0)$ have relatively small values for r near r_0.

(d) From part (a) we know that $dV/dr = 0$ when $r = r_0$, hence $F = 0$ when $r = r_0$. Since, if we discard powers of $(r - r_0)$ higher than the second,

$$V(r) \approx -V_0\left(1 - 36\frac{(r - r_0)^2}{r_0^2}\right)$$

giving

$$F = -\frac{dV}{dr} \approx 72 \cdot \frac{r - r_0}{r_0^2}(-V_0) = -72V_0 \frac{r - r_0}{r_0^2}.$$

So F is approximately proportional to $(r - r_0)$.

Solutions for Section 10.4

Exercises

1. The error bound in approximating e^{01} using the Taylor polynomial of degree 3 for $f(x) = e^x$ about $x = 0$ is:

$$|E_3| = |f(0.1) - P_3(0.1)| \leq \frac{M \cdot |0.1 - 0|^4}{4!} = \frac{M(0.1)^4}{24},$$

where $|f^{(4)}(x)| \leq M$ for $0 \leq x \leq 0.1$. Now, $f^{(4)}(x) = e^x$. Since e^x is increasing for all x, we see that $|f^{(4)}(x)|$ is maximized for x between 0 and 0.1 when $x = 0.1$. Thus,

$$|f^{(4)}| \leq e^{0.1},$$

so

$$|E_3| \leq \frac{e^{0.1} \cdot (0.1)^4}{24} = 0.00000460.$$

The Taylor polynomial of degree 3 is

$$P_3(x) = 1 + x + \frac{1}{2!}x^2 + \frac{1}{3!}x^3.$$

The approximation is $P_3(0.1)$, so the actual error is

$$E_3 = e^{0.1} - P_3(0.1) = 1.10517092 - 1.10516667 = 0.00000425,$$

which is slightly less than the bound.

5. The error bound in approximating $\ln(1.5)$ using the Taylor polynomial of degree 3 for $f(x) = \ln(1+x)$ about $x = 0$ is:

$$|E_4| = |f(0.5) - P_3(0.5)| \leq \frac{M \cdot |0.5 - 0|^4}{4!} = \frac{M(0.5)^4}{24},$$

where $|f^{(4)}(x)| \leq M$ for $0 \leq x \leq 0.5$. Since

$$f^{(4)}(x) = \frac{-3!}{(1+x)^4}$$

and the denominator attains its minimum when $x = 0$, we have $|f^{(4)}(x)| \leq 3!$, so

$$|E_4| \leq \frac{3! \, (0.5)^4}{24} = 0.0156.$$

The Taylor polynomial of degree 3 is

$$P_3(x) = 0 + x + (-1)\frac{x^2}{2!} + (-1)(-2)\frac{x^3}{3!}$$
$$= x - \frac{1}{2}x^2 + \frac{1}{3}x^3.$$

The approximation is $P_3(0.5)$, so the actual error is

$$E_3 = \ln(1.5) - P_3(0.5) = 0.4055 - 0.4167 = -0.0112$$

which is slightly less, in absolute value, than the bound.

Problems

9. Let $f(x) = \sqrt{1+x}$. We use a Taylor polynomial with $x = 1$ to approximate $\sqrt{2}$. The error bound for the Taylor approximation of degree three for $f(x) = \sqrt{2}$ about $x = 0$ is:

$$|E_3| = |f(1) - P_3(1)| \leq \frac{M \cdot |1 - 0|^4}{4!} = \frac{M}{24},$$

where $|f^{(4)}(x)| \leq M$ for $0 \leq x \leq 1$.

Now,

$$f^{(4)}(x) = -\frac{15}{16}(1+x)^{-7/2} = \frac{-15}{16(1+x)^{7/2}}.$$

Since $1 \leq (1+x)^{7/2}$ for x between 0 and 1, we see that

$$|f^{(4)}(x)| = \frac{15}{16(1+x)^{\frac{7}{2}}} \leq \frac{15}{16}$$

for x between 0 and 1. Thus,

$$|E_3| \leq \frac{15}{16 \cdot 24} < 0.039$$

13. (a) The second-degree Taylor polynomial for $f(t) = e^t$ is $P_2(t) = 1 + t + t^2/2$. Since the full expansion of $e^t = 1 + t + t^2/2 + t^3/6 + t^4/24 + \cdots$ is clearly larger than $P_2(t)$ for $t > 0$, $P_2(t)$ is an underestimate on $[0, 0.5]$.

(b) Using the second-degree error bound, if $|f^{(3)}(t)| \leq M$ for $0 \leq t \leq 0.5$, then

$$|E_2| \leq \frac{M}{3!} \cdot |t|^3 \leq \frac{M(0.5)^3}{6}.$$

Since $|f^{(3)}(t)| = e^t$, and e^t is increasing on $[0, 0.5]$,

$$f^{(3)}(t) \leq e^{0.5} < \sqrt{4} = 2.$$

So

$$|E_2| \leq \frac{(2)(0.5)^3}{6} < 0.047.$$

17. (a)

Table 10.1		
$E_1 = \sin x - x$		
x	$\sin x$	E
-0.5	-0.4794	0.0206
-0.4	-0.3894	0.0106
-0.3	-0.2955	0.0045
-0.2	-0.1987	0.0013
-0.1	-0.0998	0.0002

Table 10.2		
$E_1 = \sin x - x$		
x	$\sin x$	E
0	0	0
0.1	0.0998	-0.0002
0.2	0.1987	-0.0013
0.3	0.2955	-0.0045
0.4	0.3894	-0.0106
0.5	0.4794	-0.0206

(b) See answer to part (a) above.

(c)

The fact that the graph of E_1 lies between the horizontal lines at ± 0.03 shows that $|E_1| < 0.03$ for $-0.5 \leq x \leq 0.5$.

21.

$$\sin x = x - \frac{x^3}{3!} + \frac{x^5}{5!} - \cdots$$

Write the error in approximating $\sin x$ by the Taylor polynomial of degree $n = 2k + 1$ as E_n so that

$$\sin x = x - \frac{x^3}{3!} + \frac{x^5}{5!} - \cdots (-1)^k \frac{x^{2k+1}}{(2k+1)!} + E_n.$$

(Notice that $(-1)^k = 1$ if k is even and $(-1)^k = -1$ if k is odd.) We want to show that if x is fixed, $E_n \to 0$ as $k \to \infty$. Since $f(x) = \sin x$, all the derivatives of $f(x)$ are $\pm \sin x$ or $\pm \cos x$, so we have for all n and all x

$$|f^{(n+1)}(x)| \leq 1.$$

Using the bound on the error given in the text on page 528, we see that

$$|E_n| \leq \frac{1}{(2k+2)!}|x|^{2k+2}.$$

By the argument in the text on page 528, we know that for all x,

$$\frac{|x|^{2k+2}}{(2k+2)!} = \frac{|x|^{n+1}}{(n+1)!} \to 0 \quad \text{as} \quad n = 2k+1 \to \infty.$$

Thus the Taylor series for $\sin x$ does converge to $\sin x$ for every x.

Solutions for Section 10.5

Exercises

1. No, a Fourier series has terms of the form $\cos nx$, not $\cos^n x$.

5.

$$a_0 = \frac{1}{2\pi} \int_{-\pi}^{\pi} f(x)\, dx = \frac{1}{2\pi}\left[\int_{-\pi}^{0} -1\, dx + \int_{0}^{\pi} 1\, dx \right] = 0$$

$$a_1 = \frac{1}{\pi} \int_{-\pi}^{\pi} f(x) \cos x\, dx = \frac{1}{\pi}\left[\int_{-\pi}^{0} -\cos x\, dx + \int_{0}^{\pi} \cos x\, dx \right]$$

$$= \frac{1}{\pi}\left[-\sin x \Big|_{-\pi}^{0} + \sin x \Big|_{0}^{\pi} \right] = 0.$$

Similarly, a_2 and a_3 are both 0.
(In fact, notice $f(x) \cos nx$ is an odd function, so $\int_{-\pi}^{\pi} f(x) \cos nx = 0$.)

$$b_1 = \frac{1}{\pi} \int_{-\pi}^{\pi} f(x) \sin x\, dx = \frac{1}{\pi}\left[\int_{-\pi}^{0} -\sin x\, dx + \int_{0}^{\pi} \sin x\, dx \right]$$

$$= \frac{1}{\pi}\left[\cos x \Big|_{-\pi}^{0} + (-\cos x) \Big|_{0}^{\pi} \right] = \frac{4}{\pi}$$

$$b_2 = \frac{1}{\pi} \int_{-\pi}^{\pi} f(x) \sin 2x\, dx = \frac{1}{\pi}\left[\int_{-\pi}^{0} -\sin 2x\, dx + \int_{0}^{\pi} \sin 2x\, dx \right]$$

$$= \frac{1}{\pi}\left[\frac{1}{2}\cos 2x \Big|_{-\pi}^{0} + (-\frac{1}{2}\cos 2x) \Big|_{0}^{\pi} \right] = 0.$$

$$b_3 = \frac{1}{\pi} \int_{-\pi}^{\pi} f(x) \sin 3x\, dx = \frac{1}{\pi}\left[\int_{-\pi}^{0} -\sin 3x\, dx + \int_{0}^{\pi} \sin 3x\, dx \right]$$

$$= \frac{1}{\pi}\left[\frac{1}{3}\cos 3x \Big|_{-\pi}^{0} + (-\frac{1}{3}\cos 3x) \Big|_{0}^{\pi} \right] = \frac{4}{3\pi}.$$

Thus, $F_1(x) = F_2(x) = \frac{4}{\pi} \sin x$ and $F_3(x) = \frac{4}{\pi} \sin x + \frac{4}{3\pi} \sin 3x$.

$F_1(x) = F_2(x) = \frac{4}{\pi} \sin x$

$F_3(x) = \frac{4}{\pi} \sin x + \frac{4}{3\pi} \sin 3x$

9.

$$a_0 = \frac{1}{2\pi} \int_{-\pi}^{\pi} h(x)\, dx = \frac{1}{2\pi} \int_{0}^{\pi} x\, dx = \frac{\pi}{4}$$

As in Problem 10, we use the integral table (III-15 and III-16) to find formulas for a_n and b_n.

$$a_n = \frac{1}{\pi} \int_{-\pi}^{\pi} h(x) \cos(nx)\, dx = \frac{1}{\pi} \int_{0}^{\pi} x \cos nx\, dx = \frac{1}{\pi}\left(\frac{x}{n}\sin(nx) + \frac{1}{n^2}\cos(nx) \right)\Big|_{0}^{\pi}$$

$$= \frac{1}{\pi}\left(\frac{1}{n^2}\cos(n\pi) - \frac{1}{n^2} \right)$$

$$= \frac{1}{n^2\pi}\left(\cos(n\pi) - 1 \right).$$

Note that since $\cos(n\pi) = (-1)^n$, $a_n = 0$ if n is even and $a_n = -\frac{2}{n^2\pi}$ if n is odd.

$$b_n = \frac{1}{\pi}\int_{-\pi}^{\pi} h(x)\cos(nx)\,dx = \frac{1}{\pi}\int_0^{\pi} x\sin x\,dx$$

$$= \frac{1}{\pi}\left(-\frac{x}{n}\cos(nx) + \frac{1}{n^2}\sin(nx) \right)\Big|_0^{\pi}$$

$$= \frac{1}{\pi}\left(-\frac{\pi}{n}\cos(n\pi) \right)$$

$$= -\frac{1}{n}\cos(n\pi)$$

$$= \frac{1}{n}(-1)^{n+1} \quad \text{if } n \geq 1$$

We have that the nth Fourier polynomial for h (for $n \geq 1$) is

$$H_n(x) = \frac{\pi}{4} + \sum_{i=1}^{n}\left(\frac{1}{i^2\pi}\Big(\cos(i\pi) - 1 \Big)\cdot\cos(ix) + \frac{(-1)^{i+1}\sin(ix)}{i} \right).$$

This can also be written as

$$H_n(x) = \frac{\pi}{4} + \sum_{i=1}^{n}\frac{(-1)^{i+1}\sin(ix)}{i} + \sum_{i=1}^{\left[\frac{n}{2}\right]}\frac{-2}{(2i-1)^2\pi}\cos((2i-1)x)$$

where $\left[\frac{n}{2}\right]$ denotes the biggest integer smaller than or equal to $\frac{n}{2}$. In particular, we have the graphs in Figure 10.5.

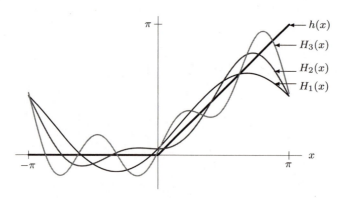

Figure 10.5

Problems

13. We have $f(x) = x$, $0 \leq x < 1$. Let $t = 2\pi x - \pi$. Notice that as x varies from 0 to 1, t varies from $-\pi$ to π. Thus if we rewrite the function in terms of t, we can find the Fourier series in terms of t in the usual way. To do this, let $g(t) = f(x) = x = \frac{t+\pi}{2\pi}$ on $-\pi \leq t < \pi$. We now find the fourth degree Fourier polynomial for g.

$$a_o = \frac{1}{2\pi}\int_{-\pi}^{\pi} g(t)\,dt = \frac{1}{2\pi}\int_{-\pi}^{\pi}\frac{t+\pi}{2\pi}\,dt = \frac{1}{(2\pi)^2}\left(\frac{t^2}{2} + \pi t \right)\Big|_{-\pi}^{\pi} = \frac{1}{2}$$

Notice, a_0 is the average value of both f and g. For $n \geq 1$,

$$a_n = \frac{1}{\pi} \int_{-\pi}^{\pi} \frac{t+\pi}{2\pi} \cos(nt)dt = \frac{1}{2\pi^2} \int_{-\pi}^{\pi} (t\cos(nt) + \pi \cos(nt))dt$$

$$= \frac{1}{2\pi^2} \left[\frac{t}{n} \sin(nt) + \frac{1}{n^2} \cos(nt) + \frac{\pi}{n} \sin(nt) \right] \Bigg|_{-\pi}^{\pi}$$

$$= 0.$$

$$b_n = \frac{1}{\pi} \int_{-\pi}^{\pi} \frac{t+\pi}{2\pi} \sin(nt)\, dt = \frac{1}{2\pi^2} \int_{-\pi}^{\pi} (t\sin(nt) + \pi \sin(nt))\, dt$$

$$= \frac{1}{2\pi^2} \left[-\frac{t}{n} \cos(nt) + \frac{1}{n^2} \sin(nt) - \frac{\pi}{n} \cos(nt) \right] \Bigg|_{-\pi}^{\pi}$$

$$= \frac{1}{2\pi^2} \left(-\frac{4\pi}{n} \cos(\pi n) \right) = -\frac{2}{\pi n} \cos(\pi n) = \frac{2}{\pi n} (-1)^{n+1}.$$

We get the integrals for a_n and b_n using the integral table (formulas III-15 and III-16).

Thus, the Fourier polynomial of degree 4 for g is:

$$G_4(t) = \frac{1}{2} + \frac{2}{\pi} \sin t - \frac{1}{\pi} \sin 2t + \frac{2}{3\pi} \sin 3t - \frac{1}{2\pi} \sin 4t.$$

Now, since $g(t) = f(x)$, the Fourier polynomial of degree 4 for f can be found by replacing t in terms of x again. Thus,

$$F_4(x) = \frac{1}{2} + \frac{2}{\pi} \sin(2\pi x - \pi) - \frac{1}{\pi} \sin(4\pi x - 2\pi) + \frac{2}{3\pi} \sin(6\pi x - 3\pi) - \frac{1}{2\pi} \sin(8\pi x - 4\pi).$$

Now, using the fact that $\sin(x - \pi) = -\sin x$ and $\sin(x - 2\pi) = \sin x$, etc., we have:

$$F_4(x) = \frac{1}{2} - \frac{2}{\pi} \sin(2\pi x) - \frac{1}{\pi} \sin(4\pi x) - \frac{2}{3\pi} \sin(6\pi x) - \frac{1}{2\pi} \sin(8\pi x).$$

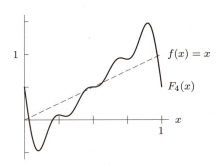

17. Let $f(x) = a_k \cos kx + b_k \sin kx$. Then the energy of f is given by

$$\frac{1}{\pi} \int_{-\pi}^{\pi} (f(x))^2 \, dx = \frac{1}{\pi} \int_{-\pi}^{\pi} (a_k \cos kx + b_k \sin kx)^2 \, dx$$

$$= \frac{1}{\pi} \int_{-\pi}^{\pi} (a_k^2 \cos^2 kx - 2a_k b_k \cos kx \sin kx + b_k^2 \sin^2 kx) \, dx$$

$$= \frac{1}{\pi} \left[a_k^2 \int_{-\pi}^{\pi} \cos^2 kx \, dx - 2a_k b_k \int_{-\pi}^{\pi} \cos kx \sin kx \, dx + b_k^2 \int_{-\pi}^{\pi} \sin^2 kx \, dx \right]$$

$$= \frac{1}{\pi} \left[a_k^2 \pi - 2a_k b_k \cdot 0 + b_k^2 \pi \right] = a_k^2 + b_k^2.$$

21. (a)

The energy of the pulse train f is

$$E = \frac{1}{\pi} \int_{-\pi}^{\pi} (f(x))^2 \, dx = \frac{1}{\pi} \int_{-1/5}^{1/5} 1^2 \, dx = \frac{1}{\pi} \left(\frac{1}{5} - \left(-\frac{1}{5} \right) \right) = \frac{2}{5\pi}.$$

Next, find the Fourier coefficients:

$$a_0 = \text{average value of } f \text{ on } [-\pi, \pi] = \frac{1}{2\pi} (\text{ Area}) = \frac{1}{2\pi} \left(\frac{2}{5} \right) = \frac{1}{5\pi},$$

$$a_k = \frac{1}{\pi} \int_{-\pi}^{\pi} f(x) \cos kx \, dx = \frac{1}{\pi} \int_{-1/5}^{1/5} \cos kx \, dx = \frac{1}{k\pi} \sin kx \Big|_{-1/5}^{1/5}$$

$$= \frac{1}{k\pi} \left(\sin \left(\frac{k}{5} \right) - \sin \left(-\frac{k}{5} \right) \right) = \frac{1}{k\pi} \left(2 \sin \left(\frac{k}{5} \right) \right),$$

$$b_k = \frac{1}{\pi} \int_{-\pi}^{\pi} f(x) \sin kx \, dx = \frac{1}{\pi} \int_{-1/5}^{1/5} \sin kx \, dx = -\frac{1}{k\pi} \cos kx \Big|_{-1/5}^{1/5}$$

$$= -\frac{1}{k\pi} \left(\cos \left(\frac{k}{5} \right) - \cos \left(-\frac{k}{5} \right) \right) = \frac{1}{k\pi} (0) = 0.$$

The energy of f contained in the constant term is

$$A_0^2 = 2a_0^2 = 2 \left(\frac{1}{5\pi} \right)^2 = \frac{2}{25\pi^2}$$

which is

$$\frac{A_0^2}{E} = \frac{2/25\pi^2}{2/5\pi} = \frac{1}{5\pi} \approx 0.063662 = 6.3662\% \quad \text{of the total.}$$

The fraction of energy contained in the first harmonic is

$$\frac{A_1^2}{E} = \frac{a_1^2}{E} = \frac{\left(\frac{2 \sin \frac{1}{5}}{\pi} \right)^2}{\frac{2}{5\pi}} \approx 0.12563.$$

The fraction of energy contained in both the constant term and the first harmonic together is

$$\frac{A_0^2}{E} + \frac{A_1^2}{E} \approx 0.06366 + 0.12563 = 0.18929 = 18.929\%.$$

(b) The formula for the energy of the kth harmonic is

$$A_k^2 = a_k^2 + b_k^2 = \left(\frac{2 \sin \frac{k}{5}}{k\pi} \right)^2 + 0^2 = \frac{4 \sin^2 \frac{k}{5}}{k^2 \pi^2}.$$

By graphing this formula as a continuous function for $k \geq 1$, we see its overall behavior as k gets larger in Figure 10.6. The energy spectrum for the first five terms is shown in Figure 10.7.

Figure 10.6   Figure 10.7

(c) The constant term and the first five harmonics contain

$$\frac{A_0^2}{E} + \frac{A_1^2}{E} + \frac{A_2^2}{E} + \frac{A_3^2}{E} + \frac{A_4^2}{E} + \frac{A_5^2}{E} \approx 61.5255\%$$

of the total energy of f.

(d) The fifth Fourier approximation to f is

$$F_5(x) = \tfrac{1}{5\pi} + \tfrac{2\sin(\frac{1}{5})}{\pi}\cos x + \tfrac{\sin(\frac{2}{5})}{\pi}\cos 2x + \tfrac{2\sin(\frac{3}{5})}{3\pi}\cos 3x + \tfrac{\sin(\frac{4}{5})}{2\pi}\cos 4x + \tfrac{2\sin 1}{5\pi}\cos 5x.$$

For comparison, below is the thirteenth Fourier approximation to f.

25. We make the substitution $u = mx$, $dx = \frac{1}{m}du$. Then

$$\int_{-\pi}^{\pi} \cos^2 mx\, dx = \frac{1}{m}\int_{u=-m\pi}^{u=m\pi} \cos^2 u\, du.$$

By Formula IV-18 of the integral table, this equals

$$\frac{1}{m}\left[\frac{1}{2}\cos u \sin u\right]\Big|_{-m\pi}^{m\pi} + \frac{1}{m}\frac{1}{2}\int_{-m\pi}^{m\pi} 1\, du = 0 + \frac{1}{2m}u\Big|_{-m\pi}^{m\pi} = \frac{1}{2m}u\Big|_{-m\pi}^{m\pi}$$

$$= \frac{1}{2m}(2m\pi) = \pi.$$

29. (a) To show that $g(t)$ is periodic with period 2π, we calculate

$$g(t + 2\pi) = f\left(\frac{b(t+2\pi)}{2\pi}\right) = f\left(\frac{bt}{2\pi} + b\right) = f\left(\frac{bt}{2\pi}\right) = g(t).$$

Since $g(t + 2\pi) = g(t)$ for all t, we know that $g(t)$ is periodic with period 2π. In addition

$$g\left(\frac{2\pi x}{b}\right) = f\left(\frac{b(2\pi x/b)}{2\pi}\right) = f(x).$$

(b) We make the change of variable $t = 2\pi x/b$, $dt = (2\pi/b)dx$ in the usual formulas for the Fourier coefficients of $g(t)$, as follows:

$$a_0 = \frac{1}{2\pi} \int_{t=-\pi}^{\pi} g(t)dt = \frac{1}{2\pi} \int_{x=-b/2}^{b/2} g\left(\frac{2\pi x}{b}\right) \frac{2\pi}{b} dx = \frac{1}{b} \int_{-\frac{b}{2}}^{\frac{b}{2}} f(x)\, dx$$

$$a_k = \frac{1}{\pi} \int_{t=-\pi}^{\pi} g(t)\cos(kt)\, dt = \frac{1}{\pi} \int_{x=-b/2}^{b/2} g\left(\frac{2\pi x}{b}\right) \cos\left(\frac{2\pi kx}{b}\right) \frac{2\pi}{b}\, dx$$

$$= \frac{2}{b} \int_{-b/2}^{b/2} f(x)\cos\left(\frac{2\pi kx}{b}\right)\, dx$$

$$b_k = \frac{1}{\pi} \int_{t=-\pi}^{\pi} g(t)\sin(kt)\, dt = \frac{1}{\pi} \int_{x=-b/2}^{b/2} g\left(\frac{2\pi x}{b}\right) \sin\left(\frac{2\pi kx}{b}\right) \frac{2\pi}{b}\, dx$$

$$= \frac{2}{b} \int_{-b/2}^{b/2} f(x)\sin\left(\frac{2\pi kx}{b}\right)\, dx$$

(c) By part (a), the Fourier series for $f(x)$ can be obtained by substituting $t = 2\pi x/b$ into the Fourier series for $g(t)$ which was found in part (b).

Solutions for Chapter 10 Review

Exercises

1. $e^x \approx 1 + e(x-1) + \dfrac{e}{2}(x-1)^2$

5. $f'(x) = 3x^2 + 14x - 5$, $f''(x) = 6x + 14$, $f'''(x) = 6$. The Taylor polynomial about $x = 1$ is

$$P_3(x) = 4 + \frac{12}{1!}(x-1) + \frac{20}{2!}(x-1)^2 + \frac{6}{3!}(x-1)^3$$
$$= 4 + 12(x-1) + 10(x-1)^2 + (x-1)^3.$$

Notice that if you multiply out and collect terms in $P_3(x)$, you will get $f(x)$ back.

9. Substituting $y = t^2$ in $\sin y = y - \dfrac{y^3}{3!} + \dfrac{y^5}{5!} - \dfrac{y^7}{7!} + \cdots$ gives

$$\sin t^2 = t^2 - \frac{t^6}{3!} + \frac{t^{10}}{5!} - \frac{t^{14}}{7!} + \cdots$$

13. We use the binomial series to expand $1/\sqrt{1-z^2}$ and multiply by z^2. Since

$$\frac{1}{\sqrt{1+x}} = (1+x)^{-1/2} = 1 - \frac{1}{2}x + \frac{(-1/2)(-3/2)}{2!}x^2 + \frac{(-1/2)(-3/2)(-5/2)}{3!}x^3 + \cdots$$

$$= 1 - \frac{1}{2}x + \frac{3}{8}x^2 - \frac{5}{16}x^3 + \cdots.$$

Substituting $x = -z^2$ gives

$$\frac{z^2}{\sqrt{1-z^2}} = 1 - \frac{1}{2}(-z^2) + \frac{3}{8}(-z^2)^2 - \frac{5}{16}(-z^2)^3 + \cdots$$

$$= 1 + \frac{1}{2}z^2 + \frac{3}{8}z^4 + \frac{5}{16}z^6 + \cdots.$$

Multiplying by z^2, we have

$$\frac{z^2}{\sqrt{1-z^2}} = z^2 + \frac{1}{2}z^4 + \frac{3}{8}z^6 + \frac{5}{16}z^8 + \cdots.$$

17.

$$\frac{a}{a+b} = \frac{a}{a(1+\frac{b}{a})} = \left(1+\frac{b}{a}\right)^{-1} = 1 - \frac{b}{a} + \left(\frac{b}{a}\right)^2 - \left(\frac{b}{a}\right)^3 + \cdots$$

Problems

21. Factoring out a 3, we see

$$3\left(1 + 1 + \frac{1}{2!} + \frac{1}{3!} + \frac{1}{4!} + \frac{1}{5!} + \cdots\right) = 3e^1 = 3e.$$

25. This is the series for $\sin x$ with $x = 2$ substituted. Thus

$$2 - \frac{8}{3!} + \frac{32}{5!} - \frac{128}{7!} + \cdots = 2 - \frac{2^3}{3!} + \frac{2^5}{5!} - \frac{2^7}{7!} + \cdots = \sin 2.$$

29. The second degree Taylor polynomial for $f(x)$ around $x = 3$ is

$$f(x) \approx f(3) + f'(3)(x-3) + \frac{f''(3)}{2!}(x-3)^2$$

$$= 1 + 5(x-3) - \frac{10}{2!}(x-3)^2 = 1 + 5(x-3) - 5(x-3)^2.$$

Substituting $x = 3.1$, we get

$$f(3.1) \approx 1 + 5(3.1 - 3) - 5(3.1 - 3)^2 = 1 + 5(0.1) - 5(0.01) = 1.45.$$

33. (a) The series for $\frac{\sin 2\theta}{\theta}$ is

$$\frac{\sin 2\theta}{\theta} = \frac{1}{\theta}\left(2\theta - \frac{(2\theta)^3}{3!} + \frac{(2\theta)^5}{5!} - \cdots\right) = 2 - \frac{4\theta^2}{3} + \frac{4\theta^4}{15} - \cdots$$

so $\lim_{\theta \to 0} \dfrac{\sin 2\theta}{\theta} = 2$.

(b) Near $\theta = 0$, we make the approximation

$$\frac{\sin 2\theta}{\theta} \approx 2 - \frac{4}{3}\theta^2$$

so the parabola is $y = 2 - \frac{4}{3}\theta^2$.

37.

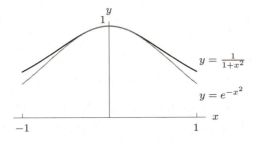

(a)

$$e^{-x^2} = 1 - x^2 + \frac{x^4}{2!} - \frac{x^6}{3!} + \cdots$$

$$\frac{1}{1+x^2} = 1 - x^2 + x^4 - x^6 + \cdots$$

Notice that the first two terms are the same in both series.

(b) $\dfrac{1}{1+x^2}$ is greater.

(c) Even, because the only terms involved are of even degree.

(d) The coefficients for e^{-x^2} become extremely small for higher powers of x, and we can "counteract" the effect of these powers for large values of x. The series for $\frac{1}{1+x^2}$ has no such coefficients.

41. (a) If h is much smaller than R, we can say that $(R + h) \approx R$, giving the approximation

$$F = \frac{mgR^2}{(R+h)^2} \approx \frac{mgR^2}{R^2} = mg.$$

(b)

$$F = \frac{mgR^2}{(R+h)^2} = \frac{mg}{(1+h/R)^2} = mg(1+h/R)^{-2}$$

$$= mg\left(1 + \frac{(-2)}{1!}\left(\frac{h}{R}\right) + \frac{(-2)(-3)}{2!}\left(\frac{h}{R}\right)^2 + \frac{(-2)(-3)(-4)}{3!}\left(\frac{h}{R}\right)^3 + \cdots\right)$$

$$= mg\left(1 - \frac{2h}{R} + \frac{3h^2}{R^2} - \frac{4h^3}{R^3} + \cdots\right)$$

(c) The first order correction comes from term $-2h/R$. The approximation for F is then given by

$$F \approx mg\left(1 - \frac{2h}{R}\right).$$

If the first order correction alters the estimate for F by 10%, we have

$$\frac{2h}{R} = 0.10 \qquad \text{so} \quad h = 0.05R \approx 0.05(6400) = 320 \text{ km.}$$

The approximation $F \approx mg$ is good to within 10% — that is, up to about 300 km.

45. The situation is more complicated. Let's first consider the case when $g'''(0) \neq 0$. To be specific let $g'''(0) > 0$. Then

$$g(x) \approx P_3(x) = g(0) + \frac{g'''(0)}{3!}x^3.$$

So, $g(x) - g(0) \approx \frac{g'''(0)}{3!}x^3$. (Notice that $\frac{g'''(0)}{3!} > 0$ is a constant.) Now, no matter how small an open interval I around $x = 0$ is, there are always some x_1 and x_2 in I such that $x_1 < 0$ and $x_2 > 0$, which means that $\frac{g'''(0)}{3!}x_1^3 < 0$ and $\frac{g'''(0)}{3!}x_2^3 > 0$, i.e. $g(x_1) - g(0) < 0$ and $g(x_2) - g(0) > 0$. Thus, $g(0)$ is neither a local minimum nor a local maximum. (If $g'''(0) < 0$, the same conclusion still holds. Try it! The reasoning is similar.)

Now let's consider the case when $g'''(0) = 0$. If $g^{(4)}(0) > 0$, then by the fourth degree Taylor polynomial approximation to g at $x = 0$, we have

$$g(x) - g(0) \approx \frac{g^{(4)}(0)}{4!}x^4 > 0$$

for x in a small open interval around $x = 0$. So $g(0)$ is a local minimum. (If $g^{(4)}(0) < 0$, then $g(0)$ is a local maximum.)

In general, suppose that $g^{(k)}(0) \neq 0$, $k \geq 2$, and all the derivatives of g with order less than k are 0. In this case g looks like cx^k near $x = 0$, which determines its behavior there. Then $g(0)$ is neither a local minimum nor a local maximum if k is odd. For k even, $g(0)$ is a local minimum if $g^{(k)}(0) > 0$, and $g(0)$ is a local maximum if $g^{(k)}(0) < 0$.

49. Since $g(x) = f(x+c)$, we have that $[g(x)]^2 = [f(x+c)]^2$, so g^2 is f^2 shifted horizontally by c. Since f has period 2π, so does f^2 and g^2. If you think of the definite integral as an area, then because of the periodicity, integrals of f^2 over any interval of length 2π have the same value. So

$$\text{Energy of } f = \int_{-\pi}^{\pi} (f(x))^2 \, dx = \int_{-\pi+c}^{\pi+c} (f(x))^2 \, dx.$$

Now we know that

$$\text{Energy of } g = \frac{1}{\pi}\int_{-\pi}^{\pi} (g(x))^2 \, dx$$

$$= \frac{1}{\pi}\int_{-\pi}^{\pi} (f(x+c))^2 \, dx.$$

Using the substitution $t = x + c$, we see that the two energies are equal.

CAS Challenge Problems

53. (a) The Taylor polynomial is

$$P_{10}(x) = 1 + \frac{x^2}{12} - \frac{x^4}{720} + \frac{x^6}{30240} - \frac{x^8}{1209600} + \frac{x^{10}}{47900160}$$

(b) All the terms have even degree. A polynomial with only terms of even degree is an even function. This suggests that f might be an even function.

(c) To show that f is even, we must show that $f(-x) = f(x)$.

$$f(-x) = \frac{-x}{e^{-x} - 1} + \frac{-x}{2} = \frac{x}{1 - \frac{1}{e^x}} - \frac{x}{2} = \frac{xe^x}{e^x - 1} - \frac{x}{2}$$

$$= \frac{xe^x - \frac{1}{2}x(e^x - 1)}{e^x - 1}$$

$$= \frac{xe^x - \frac{1}{2}xe^x + \frac{1}{2}x}{e^x - 1} = \frac{\frac{1}{2}xe^x + \frac{1}{2}x}{e^x - 1} = \frac{\frac{1}{2}x(e^x - 1) + x}{e^x - 1}$$

$$= \frac{1}{2}x + \frac{x}{e^x - 1} = \frac{x}{e^x - 1} + \frac{x}{2} = f(x)$$

CHECK YOUR UNDERSTANDING

1. False. For example, both $f(x) = x^2$ and $g(x) = x^2 + x^3$ have $P_2(x) = x^2$.

5. False. The Taylor series for $\sin x$ about $x = \pi$ is calculated by taking derivatives and using the formula

$$f(a) + f'(a)(x - a) + \frac{f''(a)}{2!}(x - a)^2 + \cdots .$$

The series for $\sin x$ about $x = \pi$ turns out to be

$$-(x - \pi) + \frac{(x - \pi)^3}{3!} - \frac{(x - \pi)^5}{5!} + \cdots .$$

9. False. The derivative of $f(x)g(x)$ is not $f'(x)g'(x)$. If this statement were true, the Taylor series for $(\cos x)(\sin x)$ would have all zero terms.

13. True. For large x, the graph of $P_{10}(x)$ looks like the graph of its highest powered term, $x^{10}/10!$. But e^x grows faster than any power, so e^x gets further and further away from $x^{10}/10! \approx P_{10}(x)$.

17. True. Since f is even, $f(x)\sin(mx)$ is odd for any m, so

$$b_m = \frac{1}{\pi}\int_{-\pi}^{\pi} f(x)\sin x(mx)\,dx = 0.$$

21. False. The quadratic approximation to $f_1(x)f_2(x)$ near $x = 0$ is

$$f_1(0)f_2(0) + (f_1'(0)f_2(0) + f_1(0)f_2'(0))x + \frac{f_1''(0)f_2(0) + 2f_1'(0)f_2'(0) + f_1(0)f_2''(0)}{2}x^2.$$

On the other hand, we have
$$L_1(x) = f_1(0) + f_1'(0)x, \quad L_2(x) = f_2(0) + f_2'(0)x,$$

so

$$L_1(x)L_2(x) = (f_1(0) + f_1'(0)x)(f_2(0) + f_2'(0)x) = f_1(0)f_2(0) + (f_1'(0)f_2(0) + f_2'(0)f_1(0))x + f_1'(0)f_2'(0)x^2.$$

The first two terms of the right side agree with the quadratic approximation to $f_1(x)f_2(x)$ near $x = 0$, but the term of degree 2 does not.

For example, the linear approximation to e^x is $1 + x$, but the quadratic approximation to $(e^x)^2 = e^{2x}$ is $1 + 2x + 2x^2$, not $(1 + x)^2 = 1 + 2x + x^2$.

CHAPTER ELEVEN

Solutions for Section 11.1

Exercises

1. (a) = (III), (b) = (IV), (c) = (I), (d) = (II).

5. In order to prove that $y = A + Ce^{kt}$ is a solution to the differential equation

$$\frac{dy}{dt} = k(y - A),$$

we must show that the derivative of y with respect to t is in fact equal to $k(y - A)$:

$$y = A + Ce^{kt}$$
$$\frac{dy}{dt} = 0 + (Ce^{kt})(k)$$
$$= kCe^{kt}$$
$$= k(Ce^{kt} + A - A)$$
$$= k\left((Ce^{kt} + A) - A\right)$$
$$= k(y - A).$$

9. Differentiating and using the fact that

$$\frac{d}{dt}(\cosh t) = \sinh t \quad \text{and} \quad \frac{d}{dt}(\sinh t) = \cosh t,$$

we see that

$$\frac{dx}{dt} = \omega C_1 \sinh \omega t + \omega C_2 \cosh \omega t$$
$$\frac{d^2 x}{dt^2} = \omega^2 C_1 \cosh \omega t + \omega^2 C_2 \sinh \omega t$$
$$= \omega^2 \left(C_1 \cosh \omega t + C_2 \sinh \omega t\right).$$

Therefore, we see that

$$\frac{d^2 x}{dt^2} = \omega^2 x.$$

Problems

13. Since $y = x^2 + k$, we know that $y' = 2x$. Substituting $y = x^2 + k$ and $y' = 2x$ into the differential equation, we get

$$10 = 2y - xy'$$
$$= 2(x^2 + k) - x(2x)$$
$$= 2x^2 + 2k - 2x^2$$
$$= 2k.$$

Thus, $k = 5$ is the only solution.

17. (a) If $y = Cx^n$ is a solution to the given differential equation, then we must have

$$x \frac{d\left(Cx^n\right)}{dx} - 3(Cx^n) = 0$$
$$x(Cnx^{n-1}) - 3(Cx^n) = 0$$
$$Cnx^n - 3Cx^n = 0$$
$$C(n-3)x^n = 0.$$

Thus, if $C = 0$, we get $y = 0$ is a solution, for every n. If $C \neq 0$, then $n = 3$, and so $y = Cx^3$ is a solution.

(b) Because $y = 40$ for $x = 2$, we cannot have $C = 0$. Thus, by part (a), we get $n = 3$. The solution to the differential equation is

$$y = Cx^3.$$

To determine C if $y = 40$ when $x = 2$, we substitute these values into the equation.

$$40 = C \cdot 2^3$$
$$40 = C \cdot 8$$
$$C = 5.$$

So, now both C and n are fixed at specific values.

21.

(I) $y = e^x$, $y' = e^x$, $y'' = e^x$

(II) $y = x^3$, $y' = 3x^2$, $y'' = 6x$

(III) $y = e^{-x}$, $y' = -e^{-x}$, $y'' = e^{-x}$

(IV) $y = x^{-2}$, $y' = -2x^{-3}$, $y'' = 6x^{-4}$

and so:

(a) (I),(III) because $y'' = y$ in each case.

(b) (IV) because $x^2 y'' + 2xy' - 2y = x^2(6x^{-4}) + 2x(-2x^{-3}) - 2x^{-2} = 6x^{-2} - 4x^{-2} - 2x^{-2} = 0$.

(c) (II),(IV) because $x^2 y'' = 6y$ in each case.

Solutions for Section 11.2

Exercises

1. See Figure 11.1. Other choices of solution curves are, of course, possible.

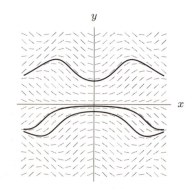

Figure 11.1

5. (a) See Figure 11.2.

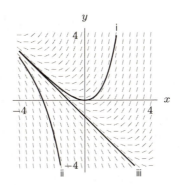

Figure 11.2

(b) The solution through $(-1, 0)$ appears to be linear, so its equation is $y = -x - 1$.
(c) If $y = -x - 1$, then $y' = -1$ and $x + y = x + (-x - 1) = -1$, so this checks as a solution.

Problems

9. (a) and (b) See Figure 11.3

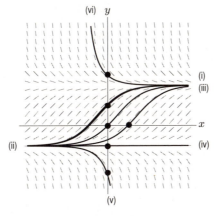

Figure 11.3

(c) Figure 11.3 shows that a solution will be increasing if its y-values fall in the range $-1 < y < 2$. This makes sense since if we examine the equation $y' = 0.5(1 + y)(2 - y)$, we will find that $y' > 0$ if $-1 < y < 2$. Notice that if the y-value ever gets to 2, then $y' = 0$ and the function becomes constant, following the line $y = 2$. (The same is true if ever $y = -1$.)

From the graph, the solution is decreasing if $y > 2$ or $y < -1$. Again, this also follows from the equation, since in either case $y' < 0$.

The curve has a horizontal tangent if $y' = 0$, which only happens if $y = 2$ or $y = -1$. This also can be seen on the graph in Figure 11.3.

13. (a) II   (b) VI   (c) IV   (d) I   (e) III   (f) V

Solutions for Section 11.3

Exercises

1. (a) In Table 11.1, we see that $y(0.4) \approx 1.5282$.

(b) In Table 11.2, we see that $y(0.4) = -1.4$. (This answer is exact.)

Table 11.1   *Euler's method for*
$y' = x + y$ *with* $y(0) = 1$

x	y	$\Delta y =$(slope)Δx
0	1	$0.1 = (1)(0.1)$
0.1	1.1	$0.12 = (1.2)(0.1)$
0.2	1.22	$0.142 = (1.42)(0.1)$
0.3	1.362	$0.1662 = (1.662)(0.1)$
0.4	1.5282	

Table 11.2   *Euler's method for*
$y' = x + y$ *with* $y(-1) = 0$

x	y	$\Delta y =$(slope)Δx
-1	0	$-0.1 = (-1)(0.1)$
-0.9	-0.1	$-0.1 = (-1)(0.1)$
-0.8	-0.2	$-0.1 = (-1)(0.1)$
-0.7	-0.3	
$\vdots$	$\vdots$	Notice that y
0	-1	decreases by 0.1
$\vdots$	$\vdots$	for every step
0.4	-1.4	

Problems

5. (a) $\Delta x = 0.5$

Table 11.3   *Euler's method for*
$y' = 2x$, *with* $y(0) = 1$

x	y	$\Delta y =$(slope)Δx
0	1	$0 = (2 \cdot 0)(0.5)$
0.5	1	$0.5 = (2 \cdot 0.5)(0.5)$
1	1.5	

$\Delta x = 0.25$

Table 11.4   *Euler's method for* $y' = 2x$,
with $y(0) = 1$

x	y	$\Delta y =$(slope)Δx
0	1	$0 = (2 \cdot 0)(0.25)$
0.25	1	$0.125 = (2 \cdot 0.25)(0.25)$
0.50	1.125	$0.25 = (2 \cdot 0.5)(0.25)$
0.75	1.375	$0.375 = (2 \cdot 0.75)(0.25)$
1	1.75	

(b) General solution is $y = x^2 + C$, and $y(0) = 1$ gives $C = 1$. Thus, the solution is $y = x^2 + 1$. So the true value of y when $x = 1$ is $y = 1^2 + 1 = 2$.

(c) When $\Delta x = 0.5$, error $= 0.5$.
When $\Delta x = 0.25$, error $= 0.25$.
Thus, decreasing Δx by a factor of 2 has decreased the error by a factor of 2, as expected.

9. By looking at the slope fields, or by computing the second derivative

$$\frac{d^2y}{dx^2} = 2x - 2y\frac{dy}{dx} = 2x - 2x^2y + 2y^3,$$

we see that the solution curve is concave up, so Euler's method gives an underestimate.

13. Assume that $x > 0$ and that we use n steps in Euler's method. Label the x-coordinates we use in the process $x_0, x_1, \ldots, x_n$, where $x_0 = 0$ and $x_n = x$. Then using Euler's method to find $y(x)$, we get

Table 11.5

	x	y	$\Delta y = (\text{slope})\Delta x$
P_0	$0 = x_0$	0	$f(x_0)\Delta x$
P_1	x_1	$f(x_0)\Delta x$	$f(x_1)\Delta x$
P_2	x_2	$f(x_0)\Delta x + f(x_1)\Delta x$	$f(x_2)\Delta x$
$\vdots$	$\vdots$	$\vdots$	$\vdots$
P_n	$x = x_n$	$\sum_{i=0}^{n-1} f(x_i)\Delta x$	

Thus the result from Euler's method is $\sum_{i=0}^{n-1} f(x_i)\Delta x$. We recognize this as the left-hand Riemann sum that approximates $\int_0^x f(t)\, dt$.

Solutions for Section 11.4

Exercises

1. Separating variables gives

$$\int \frac{1}{P}\, dP = -\int 2dt,$$

so

$$\ln|P| = -2t + C.$$

Therefore

$$P = \pm e^{-2t+C} = Ae^{-2t}.$$

The initial value $P(0) = 1$ gives $1 = A$, so

$$P = e^{-2t}.$$

5. Separating variables gives

$$\int \frac{dy}{y} = -\int \frac{1}{3}\, dx$$

$$\ln|y| = -\frac{1}{3}x + C.$$

Solving for y, we have

$$y = Ae^{-\frac{1}{3}x}, \text{ where } A = \pm e^C.$$

Since $y(0) = A = 10$, we have

$$y = 10e^{-\frac{1}{3}x}.$$

9. Separating variables gives

$$\int \frac{dz}{z} = \int 5\, dt$$

$$\ln|z| = 5t + C.$$

Solving for z, we have

$$z = Ae^{5t}, \text{ where } A = \pm e^C.$$

Using the fact that $z(1) = 5$, we have $z(1) = Ae^5 = 5$, so $A = 5/e^5$. Therefore,

$$z = \frac{5}{e^5}e^{5t} = 5e^{5t-5}.$$

13. Separating variables gives

$$\int \frac{dy}{y - 200} = \int 0.5dt$$
$$\ln|y - 200| = 0.5t + C$$
$$y = 200 + Ae^{0.5t}, \quad \text{where } A = \pm e^C.$$

The initial condition, $y(0) = 50$, gives

$$50 = 200 + A, \quad \text{so} \quad A = -150.$$

Thus,

$$y = 200 - 150e^{0.5t}.$$

17. Factoring out the 0.1 gives

$$\frac{dm}{dt} = 0.1m + 200 = 0.1(m + 2000)$$
$$\int \frac{dm}{m + 2000} = \int 0.1\,dt,$$

so

$$\ln|m + 2000| = 0.1t + C,$$

and

$$m = Ae^{0.1t} - 2000, \quad \text{where } A = \pm e^C.$$

Using the initial condition, $m(0) = Ae^{(0.1)\cdot 0} - 2000 = 1000$, gives $A = 3000$. Thus

$$m = 3000e^{0.1t} - 2000.$$

21. Separating variables gives

$$\frac{dz}{dt} = te^z$$
$$e^{-z}dz = tdt$$
$$\int e^{-z}\,dz = \int t\,dt,$$

so

$$-e^{-z} = \frac{t^2}{2} + C.$$

Since the solution passes through the origin, $z = 0$ when $t = 0$, we must have

$$-e^{-0} = \frac{0}{2} + C, \text{ so } C = -1.$$

Thus

$$-e^{-z} = \frac{t^2}{2} - 1,$$

or

$$z = -\ln\left(1 - \frac{t^2}{2}\right).$$

25. Separating variables gives

$$\frac{dw}{d\theta} = \theta w^2 \sin \theta^2$$

$$\int \frac{dw}{w^2} = \int \theta \sin \theta^2 \, d\theta,$$

so

$$-\frac{1}{w} = -\frac{1}{2} \cos \theta^2 + C.$$

According to the initial conditions, $w(0) = 1$, so $-1 = -\frac{1}{2} + C$ and $C = -\frac{1}{2}$. Thus,

$$-\frac{1}{w} = -\frac{1}{2} \cos \theta^2 - \frac{1}{2}$$

$$\frac{1}{w} = \frac{\cos \theta^2 + 1}{2}$$

$$w = \frac{2}{\cos \theta^2 + 1}.$$

Problems

29. Separating variables gives

$$\int \frac{dR}{R} = \int k \, dt.$$

Integrating gives

$$\ln |R| = kt + C,$$

so

$$|R| = e^{kt+C} = e^{kt} e^C$$
$$R = A e^{kt}, \quad \text{where } A = \pm e^C \quad \text{or} \quad A = 0.$$

33. Separating variables gives

$$\int \frac{dP}{P - a} = \int k \, dt.$$

Integrating yields

$$\ln |P - a| = kt + C,$$

so

$$P = a + A e^{kt} \quad \text{where } A = \pm e^C \quad \text{or } A = 0.$$

37. Separating variables and integrating gives

$$\int \frac{1}{R^2 + 1} \, dR = \int a \, dx$$

or

$$\arctan R = ax + C$$

so that

$$R = \tan(ax + C).$$

41. Separating variables gives

$$\frac{dx}{dt} = \frac{x \ln x}{t},$$

so

$$\int \frac{dx}{x \ln x} = \int \frac{dt}{t},$$

and thus

$$\ln|\ln x| = \ln t + C,$$

so

$$|\ln x| = e^C e^{\ln t} = e^C t.$$

Therefore

$$\ln x = At, \quad \text{where } A = \pm e^C \quad \text{or} \quad A = 0, \quad \text{so} \quad x = e^{At}.$$

45. (a), (b) See Figure 11.4.

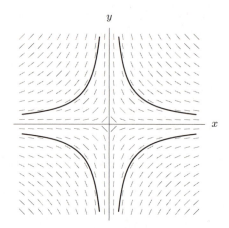

Figure 11.4

(c) Since

$$\frac{dy}{dx} = -\frac{y}{x},$$

we have

$$\int \frac{dy}{y} = -\int \frac{dx}{x},$$

so

$$\ln|y| = -\ln|x| + C,$$

giving

$$|y| = e^{-\ln|x|+C} = (|x|)^{-1}e^C.$$

Thus,

$$y = \frac{A}{x}, \quad \text{where } A = \pm e^C \quad \text{or} \quad A = 0.$$

Solutions for Section 11.5

Exercises

1. (a) (I)
    (b) (IV)
    (c) (II) and (IV)
    (d) (II) and (III)

5. (a) The equilibrium solutions occur where the slope $y' = 0$, which occurs on the slope field where the lines are horizontal, or (looking at the equation) at $y = 2$ and $y = -1$. Looking at the slope field, we can see that $y = 2$ is stable, since the slopes at nearby values of y point toward it, whereas $y = -1$ is unstable.

(b) Draw solution curves passing through the given points by starting at these points and following the flow of the slopes, as shown in Figure 11.5.

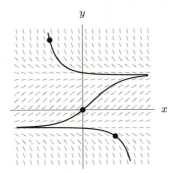

Figure 11.5

Problems

9. (a) Separating variables

$$\int \frac{dy}{100 - y} = \int dt$$
$$-\ln|100 - y| = t + C$$
$$|100 - y| = e^{-t-C}$$
$$100 - y = (\pm e^{-C})e^{-t} = Ae^{-t} \quad \text{where } A = \pm e^{-C}$$
$$y = 100 - Ae^{-t}.$$

(b) See Figure 11.6.

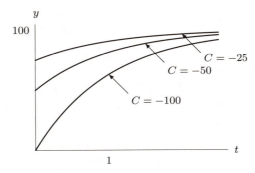

Figure 11.6

(c) Substituting $y = 0$ when $t = 0$ gives

$$0 = 100 + Ce^{-0}$$

so $C = -100$. Thus solution is

$$y = 100 - 100e^{-t}.$$

13. (a) Since we are told that the rate at which the quantity of the drug decreases is proportional to the amount of the drug left in the body, we know the differential equation modeling this situation is

$$\frac{dQ}{dt} = kQ.$$

Since we are told that the quantity of the drug is decreasing, we know that $k < 0$.

(b) We know that the general solution to the differential equation

$$\frac{dQ}{dt} = kQ$$

is

$$Q = Ce^{kt}.$$

(c) We are told that the half life of the drug is 3.8 hours. This means that at $t = 3.8$, the amount of the drug in the body is half the amount that was in the body at $t = 0$, or, in other words,

$$0.5Q(0) = Q(3.8).$$

Solving this equation gives

$$0.5Q(0) = Q(3.8)$$
$$0.5Ce^{k(0)} = Ce^{k(3.8)}$$
$$0.5C = Ce^{k(3.8)}$$
$$0.5 = e^{k(3.8)}$$
$$\ln(0.5) = k(3.8)$$
$$\frac{\ln(0.5)}{3.8} = k$$
$$k \approx -0.182.$$

(d) From part (c) we know that the formula for Q is

$$Q = Ce^{-0.182t}.$$

We are told that initially there are 10 mg of the drug in the body. Thus at $t = 0$, we get

$$10 = Ce^{-0.182(0)}$$

so

$$C = 10.$$

Thus our equation becomes

$$Q(t) = 10e^{-0.182t}.$$

Substituting $t = 12$, we get

$$Q(t) = 10e^{-0.182t}$$
$$Q(12) = 10e^{-0.182(12)}$$
$$= 10e^{-2.184}$$
$$Q(12) \approx 1.126 \text{ mg}.$$

17. According to Newton's Law of Cooling, the temperature, T, of the roast as a function of time, t, satisfies

$$T'(t) = k(350 - T)$$
$$T(0) = 40.$$

Solving this differential equation, we get that $T = 350 - 310e^{-kt}$ for some $k > 0$. To find k, we note that at $t = 1$ we have $T = 90$, so

$$90 = 350 - 310e^{-k(1)}$$
$$\frac{260}{310} = e^{-k}$$
$$k = -\ln\left(\frac{260}{310}\right)$$
$$\approx 0.17589.$$

Thus, $T = 350 - 310e^{-0.17589t}$. Solving for t when $T = 140$, we have

$$140 = 350 - 310e^{-0.17589t}$$

$$\frac{210}{310} = e^{-0.17589t}$$

$$t = \frac{\ln(210/310)}{-0.17589}$$

$$t \approx 2.21 \text{ hours.}$$

21. Lake Superior will take the longest, because the lake is largest (V is largest) and water is moving through it most slowly (r is smallest). Lake Erie looks as though it will take the least time because V is smallest and r is close to the largest. For Erie, $k = r/V = 175/460 \approx 0.38$. The lake with the largest value of r is Ontario, where $k = r/V = 209/1600 \approx 0.13$. Since e^{-kt} decreases faster for larger k, Lake Erie will take the shortest time for any fixed fraction of the pollution to be removed.

For Lake Superior,

$$\frac{dQ}{dt} = -\frac{r}{V}Q = -\frac{65.2}{12{,}200}Q \approx -0.0053Q$$

so

$$Q = Q_0 e^{-0.0053t}.$$

When 80% of the pollution has been removed, 20% remains so $Q = 0.2Q_0$. Substituting gives us

$$0.2Q_0 = Q_0 e^{-0.0053t}$$

so

$$t = -\frac{\ln(0.2)}{0.0053} \approx 301 \text{ years.}$$

(Note: The 301 is obtained by using the exact value of $\frac{r}{V} = \frac{65.2}{12{,}200}$, rather than 0.0053. Using 0.0053 gives 304 years.)
For Lake Erie, as in the text

$$\frac{dQ}{dt} = -\frac{r}{V}Q = -\frac{175}{460}Q \approx -0.38Q$$

so

$$Q = Q_0 e^{-0.38t}.$$

When 80% of the pollution has been removed

$$0.2Q_0 = Q_0 e^{-0.38t}$$

$$t = -\frac{\ln(0.2)}{0.38} \approx 4 \text{ years.}$$

So the ratio is

$$\frac{\text{Time for Lake Superior}}{\text{Time for Lake Erie}} \approx \frac{301}{4} \approx 75.$$

In other words it will take about 75 times as long to clean Lake Superior as Lake Erie.

25. (a) The differential equation is

$$\frac{dT}{dt} = -k(T - A),$$

where $A = 10°$F is the outside temperature.

(b) Integrating both sides yields

$$\int \frac{dT}{T - A} = -\int k \, dt.$$

Then $\ln|T - A| = -kt + C$, so $T = A + Be^{-kt}$. Thus

$$T = 10 + 58e^{-kt}.$$

Since 10:00 pm corresponds to $t = 9$,

$$57 = 10 + 58e^{-9k}$$

$$\frac{47}{58} = e^{-9k}$$

$$\ln \frac{47}{58} = -9k$$

$$k = -\frac{1}{9}\ln\frac{47}{58} \approx 0.0234.$$

At 7:00 the next morning ($t = 18$) we have

$$T \approx 10 + 58e^{18(-0.0234)}$$
$$= 10 + 58(0.66)$$
$$\approx 48°\text{F},$$

so the pipes won't freeze.

(c) We assumed that the temperature outside the house stayed constant at $10°$F. This is probably incorrect because the temperature was most likely warmer during the day (between 1 pm and 10 pm) and colder after (between 10 pm and 7 am). Thus, when the temperature in the house dropped from $68°$F to $57°$F between 1 pm and 10 pm, the outside temperature was probably higher than $10°$F, which changes our calculation of the value of the constant k. The house temperature will most certainly be lower than $48°$F at 7 am, but not by much—not enough to freeze.

Solutions for Section 11.6

Exercises

1. (a) If $B = f(t)$, where t is in years,

$$\frac{dB}{dt} = \text{Rate of money earned from interest} + \text{Rate of money deposited}$$
$$\frac{dB}{dt} = 0.10B + 1000.$$

(b) We use separation of variables to solve the differential equation

$$\frac{dB}{dt} = 0.1B + 1000.$$

$$\int \frac{1}{0.1B + 1000} dB = \int dt$$
$$\frac{1}{0.1} \ln|0.1B + 1000| = t + C_1$$
$$0.1B + 1000 = C_2 e^{0.1t}$$
$$B = Ce^{0.1t} - 10,000$$

For $t = 0$, $B = 0$, hence $C = 10,000$. Therefore, $B = 10,000e^{0.1t} - 10,000$.

5. Using (Rate balance changes) = (Rate interest is added)− (Rate payments are made), when the interest rate is i, we have

$$\frac{dB}{dt} = iB - 100.$$

Solving this equation, we find:

$$\frac{dB}{dt} = i\left(B - \frac{100}{i}\right)$$
$$\int \frac{dB}{B - \frac{100}{i}} = \int i\,dt$$
$$\ln\left|B - \frac{100}{i}\right| = it + C$$
$$B - \frac{100}{i} = Ae^{it}, \text{ where } A = \pm e^{C}.$$

At time $t = 0$ we start with a balance of \$1000. Thus
$1000 - \frac{100}{i} = Ae^0$, so $A = 1000 - \frac{100}{i}$.

Thus $B = \frac{100}{i} + (1000 - \frac{100}{i})e^{it}$.
When $i = 0.05$, $B = 2000 - 1000e^{0.05t}$.
When $i = 0.1$, $B = 1000$.
When $i = 0.15$, $B = 666.67 + 333.33e^{0.15t}$.
We now look at the graph in Figure 11.7 when $i = 0.05$, $i = 0.1$, and $i = 0.15$.

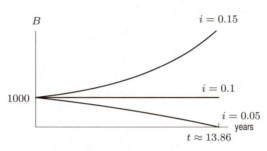

Figure 11.7

Problems

9. Let $C(t)$ be the current flowing in the circuit at time t, then

$$\frac{dC}{dt} = -\alpha C$$

where $\alpha > 0$ is the constant of proportionality between the rate at which the current decays and the current itself.

The general solution of this differential equation is $C(t) = Ae^{-\alpha t}$ but since $C(0) = 30$, we have that $A = 30$, and so we get the particular solution $C(t) = 30e^{-\alpha t}$.

When $t = 0.01$, the current has decayed to 11 amps so that $11 = 30e^{-\alpha 0.01}$ which gives $\alpha = -100\ln(11/30) = 100.33$ so that,

$$C(t) = 30e^{-100.33t}.$$

13. Let the depth of the water at time t be y. Then $\frac{dy}{dt} = -k\sqrt{y}$, where k is a positive constant. Separating variables,

$$\int \frac{dy}{\sqrt{y}} = -\int k\,dt,$$

so

$$2\sqrt{y} = -kt + C.$$

When $t = 0$, $y = 36$; $2\sqrt{36} = -k \cdot 0 + C$, so $C = 12$.
When $t = 1$, $y = 35$; $2\sqrt{35} = -k + 12$, so $k \approx 0.17$.
Thus, $2\sqrt{y} \approx -0.17t + 12$. We are looking for t such that $y = 0$; this happens when $t \approx \frac{12}{0.17} \approx 71$ hours, or about 3 days.

17. Let $V(t)$ be the volume of water in the tank at time t, then

$$\frac{dV}{dt} = k\sqrt{V}$$

This is a separable equation which has the solution

$$V(t) = (\frac{kt}{2} + C)^2$$

Since $V(0) = 200$ this gives $200 = C^2$ so

$$V(t) = (\frac{kt}{2} + \sqrt{200})^2.$$

However, $V(1) = 180$ therefore

$$180 = (\frac{k}{2} + \sqrt{200})^2,$$

so that $k = 2\left(\sqrt{180} - \sqrt{200}\right) = -1.45146$. Therefore,

$$V(t) = (-0.726t + \sqrt{200})^2.$$

The tank will be half-empty when $V(t) = 100$, so we solve

$$100 = (-0.726t + \sqrt{200})^2$$

to obtain $t = 5.7$ days. The tank will be half empty in 5.7 days.

The volume after 4 days is $V(4)$ which is approximately 126.32 liters.

21. (a) The balance in the account at the beginning of the month is given by the following sum

$$\begin{pmatrix} \text{balance in} \\ \text{account} \end{pmatrix} = \begin{pmatrix} \text{previous month's} \\ \text{balance} \end{pmatrix} + \begin{pmatrix} \text{interest on} \\ \text{previous month's balance} \end{pmatrix} + \begin{pmatrix} \text{monthly deposit} \\ \text{of \$100} \end{pmatrix}$$

Denote month i's balance by B_i. Assuming the interest is compounded continuously, we have

$$\begin{pmatrix} \text{previous month's} \\ \text{balance} \end{pmatrix} + \begin{pmatrix} \text{interest on previous} \\ \text{month's balance} \end{pmatrix} = B_{i-1}e^{0.1/12}.$$

Since the interest rate is $10\% = 0.1$ per year, interest is $\frac{0.1}{12}$ per month. So at month i, the balance is

$$B_i = B_{i-1}e^{\frac{0.1}{12}} + 100$$

Explicitly, we have for the five years (60 months) the equations:

$$B_0 = 0$$
$$B_1 = B_0 e^{\frac{0.1}{12}} + 100$$
$$B_2 = B_1 e^{\frac{0.1}{12}} + 100$$
$$B_3 = B_2 e^{\frac{0.1}{12}} + 100$$
$$\vdots \quad \vdots$$
$$B_{60} = B_{59} e^{\frac{0.1}{12}} + 100$$

In other words,

$$B_1 = 100$$
$$B_2 = 100e^{\frac{0.1}{12}} + 100$$
$$B_3 = (100e^{\frac{0.1}{12}} + 100)e^{\frac{0.1}{12}} + 100$$
$$= 100e^{\frac{(0.1)2}{12}} + 100e^{\frac{0.1}{12}} + 100$$
$$B_4 = 100e^{\frac{(0.1)3}{12}} + 100e^{\frac{(0.1)2}{12}} + 100e^{\frac{(0.1)}{12}} + 100$$
$$\vdots \quad \vdots$$
$$B_{60} = 100e^{\frac{(0.1)59}{12}} + 100e^{\frac{(0.1)58}{12}} + \cdots + 100e^{\frac{(0.1)1}{12}} + 100$$
$$B_{60} = \sum_{k=0}^{59} 100e^{\frac{(0.1)k}{12}}$$

(b) The sum $B_{60} = \sum_{k=0}^{59} 100e^{\frac{(0.1)k}{12}}$ can be written as $B_{60} = \sum_{k=0}^{59} 1200e^{\frac{(0.1)k}{12}}(\frac{1}{12})$ which is the left Riemann sum for

$\int_0^5 1200e^{0.1t}dt$, with $\Delta t = \frac{1}{12}$ and $N = 60$. Evaluating the sum on a calculator gives $B_{60} = 7752.26$.

(c) The situation described by this problem is almost the same as that in Problem 20, except that here the money is being deposited once a month rather than continuously; however the nominal yearly rates are the same. Thus we would expect the balance after 5 years to be approximately the same in each case. This means that the answer to part (b) of this problem should be approximately the same as the answer to part (c) to Problem 20. Since the deposits in this problem start at the end of the first month, as opposed to right away, we would expect the balance after 5 years to be slightly smaller than in Problem 20, as is the case.

Alternatively, we can use the Fundamental Theorem of Calculus to show that the integral can be computed exactly

$$\int_0^5 1200e^{0.1t}\,dt = 12000(e^{(0.1)5} - 1) = 7784.66$$

Thus $\int_0^5 1200e^{0.1t}\,dt$ represents the exact solution to Problem 20. Since $1200e^{0.1t}$ is an increasing function, the left hand sum we calculated in part (b) of this problem underestimates the integral. Thus the answer to part (b) of this problem should be less than the answer to part (c) of Problem 20.

25. (a)

$$\frac{dQ}{dt} = r - \alpha Q = -\alpha\left(Q - \frac{r}{\alpha}\right)$$

$$\int \frac{dQ}{Q - r/\alpha} = = -\alpha \int dt$$

$$\ln\left|Q - \frac{r}{\alpha}\right| = -\alpha t + C$$

$$Q - \frac{r}{\alpha} = Ae^{-\alpha t}$$

When $t = 0$, $Q = 0$, so $A = -\frac{r}{\alpha}$ and

$$Q = \frac{r}{\alpha}(1 - e^{-\alpha t})$$

So,

$$Q_\infty = \lim_{t \to \infty} Q = \frac{r}{\alpha}.$$

(b) Doubling r doubles Q_∞. Since $Q_\infty = r/\alpha$, the time to reach $\frac{1}{2}Q_\infty$ is obtained by solving

$$\frac{r}{2\alpha} = \frac{r}{\alpha}(1 - e^{-\alpha t})$$

$$\frac{1}{2} = 1 - e^{-\alpha t}$$

$$e^{-\alpha t} = \frac{1}{2}$$

$$t = -\frac{\ln(1/2)}{\alpha} = \frac{\ln 2}{\alpha}.$$

So altering r does not alter the time it takes to reach $\frac{1}{2}Q_\infty$. See Figure 11.8.

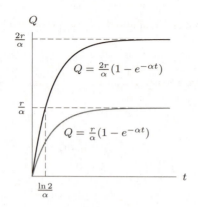

Figure 11.8

(c) Q_∞ is halved by doubling α, and so is the time, $t = \frac{\ln 2}{\alpha}$, to reach $\frac{1}{2}Q_\infty$.

29. (a) Newton's Law of Motion says that

$$\text{Force} = (\text{mass}) \times (\text{acceleration}).$$

Since acceleration, dv/dt, is measured upward and the force due to gravity acts downward,

$$-\frac{mgR^2}{(R+h)^2} = m\frac{dv}{dt}$$

so

$$\frac{dv}{dt} = -\frac{gR^2}{(R+h)^2}.$$

(b) Since $v = \frac{dh}{dt}$, the chain rule gives

$$\frac{dv}{dt} = \frac{dv}{dh} \cdot \frac{dh}{dt} = \frac{dv}{dh} \cdot v.$$

Substituting into the differential equation in part (a) gives

$$v\frac{dv}{dh} = -\frac{gR^2}{(R+h)^2}.$$

(c) Separating variables gives

$$\int v\, dv = -\int \frac{gR^2}{(R+h)^2}\, dh$$

$$\frac{v^2}{2} = \frac{gR^2}{(R+h)} + C$$

Since $v = v_0$ when $h = 0$,

$$\frac{v_0{}^2}{2} = \frac{gR^2}{(R+0)} + C \quad \text{gives} \quad C = \frac{v_0{}^2}{2} - gR,$$

so the solution is

$$\frac{v^2}{2} = \frac{gR^2}{(R+h)} + \frac{v_0{}^2}{2} - gR$$

$$v^2 = v_0{}^2 + \frac{2gR^2}{(R+h)} - 2gR$$

(d) The escape velocity v_0 ensures that $v^2 \geq 0$ for all $h \geq 0$. Since the positive quantity $\dfrac{2gR^2}{(R+h)} \to 0$ as $h \to \infty$, to ensure that $v^2 \geq 0$ for all h, we must have

$$v_0{}^2 \geq 2gR.$$

When $v_0{}^2 = 2gR$ so $v_0 = \sqrt{2gR}$, we say that v_0 is the escape velocity.

Solutions for Section 11.7

Exercises

1. We see from the differential equation that $k = 0.05$ and $L = 2800$, so the general solution is

$$P = \frac{2800}{1 + Ae^{-0.05t}}.$$

5. We rewrite

$$10P - 5P^2 = 10P\left(1 - \frac{P}{2}\right),$$

so $k = 10$ and $L = 2$. Since $P_0 = L/4$, we have $A = (L - P_0)/P_0 = 3$. Thus

$$P = \frac{2}{1 + 3e^{-10t}}.$$

The time to peak dP/dt is

$$t = \frac{1}{k}\ln A = \ln(3)/10.$$

9. (a) We see that $k = 0.035$ which tells us that the quantity P grows by about 3.5% per unit time when P is very small relative to L. We also see that $L = 6000$ which tells us the upper limit on the value of P if P is initially below 6000.

   (b) The largest rate of change occurs when $P = L/2 = 3000$.

Problems

13. A continuous growth rate of 0.2% means that

$$\frac{1}{P}\frac{dP}{dt} = 0.2\% = 0.002.$$

Separating variables and integrating gives

$$\int \frac{dP}{P} = \int 0.002\, dt$$

$$P = P_0 e^{0.002t} = (6.6 \times 10^6)e^{0.002t}.$$

17. (a) For 1800 we have

$$\frac{1}{P}\frac{dP}{dt} = \frac{1}{P(1800)}\frac{P(1810) - P(1790)}{20} = \frac{1}{5.3}\frac{7.2 - 3.9}{20} = 0.0311.$$

For 1930 we have

$$\frac{1}{P}\frac{dP}{dt} = \frac{1}{P(1930)}\frac{P(1940) - P(1920)}{20} = \frac{1}{122.8}\frac{131.7 - 105.7}{20} = 0.0106.$$

   (b) The slope between these points is

$$\text{Slope} = \frac{0.0311 - 0.0106}{5.3 - 122.8} = -0.000174.$$

Thus the fitted line has an equation of the form

$$\frac{1}{P}\frac{dP}{dt} = k - 0.000174P.$$

Using the point $P = 5.3$, $(1/P)dP/dt = 0.0311$, we solve for the vertical intercept k:

$$0.0311 = k - 0.000174(5.3)$$

$$k = 0.0311 + 0.000174(5.3) = 0.032.$$

Thus $-k/L = -0.000174$, so $L = 0.0320/.000174 = 184$.

   (c) These are quite close to the values given in the text.

21. By rewriting the equation, we see that it is logistic:

$$\frac{1}{P}\frac{dP}{dt} = \frac{(100 - P)}{1000}.$$

Before looking at its solution, we explain why there must always be at least 100 individuals. Since the population begins at 200, the quantity dP/dt is initially negative, so the population initially decreases. It continues to do so while $P > 100$. If the population ever reached 100, then dP/dt would be 0. This would mean the population stopped changing—so if the population ever decreased to 100, that's where it would stay. The fact that dP/dt is always negative for $P > 100$ also shows that the population is always under 200, as shown in Figure 11.9.

Figure 11.9

The solution, as given by the formula derived in the chapter, is

$$P = \frac{100}{1 - 0.5e^{-0.1t}}$$

25. (a) Let I be the number of informed people at time t, and I_0 the number who know initially. Then this model predicts that $dI/dt = k(M - I)$ for some positive constant k. Solving this, we find the solution is

$$I = M - (M - I_0)e^{-kt}.$$

We sketch the solution with $I_0 = 0$. Notice that dI/dt is largest when I is smallest, so the information spreads fastest in the beginning, at $t = 0$. In addition, Figure 11.10 shows that $I \to M$ as $t \to \infty$, meaning that everyone gets the information eventually.

(b) In this case, the model suggests that $dI/dt = kI(M - I)$ for some positive constant k. This is a logistic model with carrying capacity M. We sketch the solutions for three different values of I_0 in Figure 11.11.

Figure 11.10

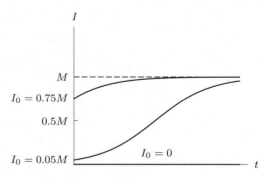

Figure 11.11

(i) If $I_0 = 0$ then $I = 0$ for all t. In other words, if nobody knows something, it does not spread by word of mouth!

(ii) If $I_0 = 0.05M$, then dI/dt is increasing up to $I = M/2$. Thus, the information is spreading fastest at $I = M/2$.

(iii) If $I_0 = 0.75M$, then dI/dt is always decreasing for $I > M/2$, so dI/dt is largest when $t = 0$.

29. (a) See Figure 11.12.

(b) See Figure 11.13.

Figure 11.12

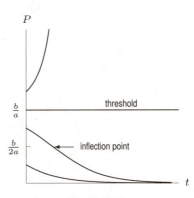

Figure 11.13

Figure 11.12 shows that dP/dt is negative for $P < \frac{b}{a}$, making P a decreasing function when $P(0) < \frac{b}{a}$. When $P > \frac{b}{a}$, the sign of dP/dt is positive, so P is an increasing function. Thus solution curves starting above $\frac{b}{a}$ are increasing, and those starting below $\frac{b}{a}$ are decreasing. See Figure 11.13.

For $P > \frac{b}{a}$, the slope, $\frac{dP}{dt}$, increases with P, so the graph of P against t is concave up. For $0 < P < \frac{b}{a}$, the value of P decreases with time. As P decreases, the slope $\frac{dP}{dt}$ decreases for $\frac{b}{2a} < P < \frac{b}{a}$, and increases toward 0 for $0 < P < \frac{b}{2a}$. Thus solution curves starting just below the threshold value of $\frac{b}{a}$ are concave down for $\frac{b}{2a} < P < \frac{b}{a}$ and concave up and asymptotic to the t-axis for $0 < P < \frac{b}{2a}$. See Figure 11.13.

(c) $P = \frac{b}{a}$ is called the threshold population because for populations greater than $\frac{b}{a}$, the population will increase without bound. For populations less than $\frac{b}{a}$, the population will go to zero, i.e. to extinction.

33. The population dies out if H is large enough that $dP/dt < 0$ for all P. The largest value for $dP/dt = kP(1 - P/L)$ occurs when $P = L/2$; then

$$\frac{dP}{dt} = kP\left(1 - \frac{P}{L}\right) = k\frac{L}{2}\left(1 - \frac{L/2}{L}\right) = \frac{kL}{4}.$$

Thus if $H > kL/4$, we have $dP/dt < 0$ for all P and the population dies out if the quota is met.

Solutions for Section 11.8

Exercises

1. Since

$$\frac{dS}{dt} = -aSI,$$

$$\frac{dI}{dt} = aSI - bI,$$

$$\frac{dR}{dt} = bI$$

we have

$$\frac{dS}{dt} + \frac{dI}{dt} + \frac{dR}{dt} = -aSI + aSI - bI + bI = 0.$$

Thus $\frac{d}{dt}(S + I + R) = 0$, so $S + I + R = $ constant.

5. If $w = 2$ and $r = 2$, then $\frac{dw}{dt} = -2$ and $\frac{dr}{dt} = 2$, so initially the number of worms decreases and the number of robins increases. In the long run, however, the populations will oscillate; they will even go back to $w = 2$ and $r = 2$. See Figure 11.14.

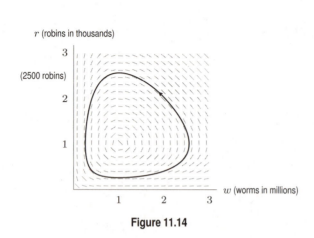

r (robins in thousands)

(2500 robins)

w (worms in millions)

Figure 11.14

Figure 11.15

9. The numbers of robins begins to increase while the number of worms remains approximately constant. See Figure 11.15. The numbers of robins and worms oscillate periodically between 0.2 and 3, with the robin population lagging behind the worm population.

13. x decreases quickly while y increases more slowly.

Problems

17. (a) The x population is unaffected by the y population—it grows exponentially no matter what the y population is, even if $y = 0$. If alone, the y population decreases to zero exponentially, because its equation becomes $dy/dt = -0.1y$.
   (b) Here, interaction between the two populations helps the y population but does not effect the x population. This is not a predator-prey relationship; instead, this is a one-way relationship, where the y population is helped by the existence of x's. These equations could, for instance, model the interaction of rhinoceroses (x) and dung beetles (y).

21. (a) Thinking of y as a function of x and x as a function of t, then by the chain rule: $\dfrac{dy}{dt} = \dfrac{dy}{dx}\dfrac{dx}{dt}$, so:

$$\frac{dy}{dx} = \frac{dy/dt}{dx/dt} = \frac{-0.01x}{-0.05y} = \frac{x}{5y}$$

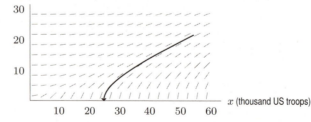

y (thousand Japanese troops)

x (thousand US troops)

   (b) The figure above shows the slope field for this differential equation and the trajectory starting at $x_0 = 54$, $y_0 = 21.5$. The trajectory goes to the x-axis, where $y = 0$, meaning that the Japanese troops were all killed or wounded before the US troops were, and thus predicts the US victory (which did occur). Since the trajectory meets the x-axis at $x \approx 25$, the differential equation predicts that about 25,000 US troops would survive the battle.
   (c) The fact that the US got reinforcements, while the Japanese did not, does not alter the predicted outcome (a US victory). The US reinforcements have the effect of changing the trajectory, altering the number of troops surviving the battle. See the graph below.

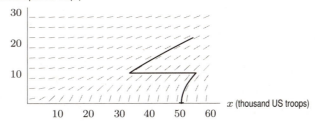

y (thousand Japanese troops)

x (thousand US troops)

25. (a) Taking the constants of proportionality to be a and b, with $a > 0$ and $b > 0$, the equations are

$$\frac{dx}{dt} = -axy$$

$$\frac{dy}{dt} = -bxy$$

(b) $\dfrac{dy}{dx} = \dfrac{dy/dt}{dx/dt} = \dfrac{-bxy}{-axy} = \dfrac{b}{a}$. Solving the differential equation gives $y = \dfrac{b}{a}x + C$, where C depends on the initial sizes of the two armies.

(c) The sign of C determines which side wins the battle. Looking at the general solution $y = \dfrac{b}{a}x + C$, we see that if $C > 0$ the y-intercept is at C, so y wins the battle by virtue of the fact that it still has troops when $x = 0$. If $C < 0$ then the curve intersects the axes at $x = -\frac{a}{b}C$, so x wins the battle because it has troops when $y = 0$. If $C = 0$, then the solution goes to the point $(0, 0)$, which represents the case of mutual annihilation.

(d) We assume that an army wins if the opposing force goes to 0 first.

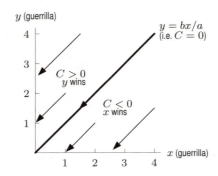

Solutions for Section 11.9

Exercises

1. (a) To find the equilibrium points we set

$$20x - 10xy = 0$$

$$25y - 5xy = 0.$$

So, $x = 0$, $y = 0$ is an equilibrium point. Another one is given by

$$10y = 20$$

$$5x = 25.$$

Therefore, $x = 5$, $y = 2$ is the other equilibrium point.

(b) At $x = 2$, $y = 4$,

$$\frac{dx}{dt} = 20x - 10xy = 40 - 80 = -40$$

$$\frac{dy}{dt} = 25y - 5xy = 100 - 40 = 60.$$

Since these are not both zero, this point is not an equilibrium point.

Problems

5. We first find the nullclines. Vertical nullclines occur where $\frac{dx}{dt} = 0$, which happens when $x = 0$ or $y = \frac{1}{3}(2 - x)$. Horizontal nullclines occur where $\frac{dy}{dt} = y(1 - 2x) = 0$, which happens when $y = 0$ or $x = \frac{1}{2}$. These nullclines are shown in Figure 11.16.

Equilibrium points (also shown in Figure 11.16) occur at the intersections of vertical and horizontal nullclines. There are three such points for this system of equations; $(0, 0)$, $(\frac{1}{2}, \frac{1}{2})$ and $(2, 0)$.

The nullclines divide the positive quadrant into four regions as shown in Figure 11.16. Trajectory directions for these regions are shown in Figure 11.17.

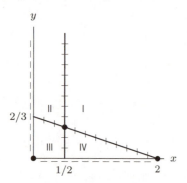

Figure 11.16: Nullclines and equilibrium points (dots)

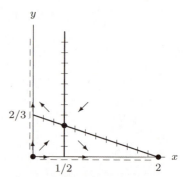

Figure 11.17: General directions of trajectories and equilibrium points (dots)

9. We assume that x, $y \geq 0$ and then find the nullclines. $\frac{dx}{dt} = x(1 - \frac{x}{2} - y) = 0$ when $x = 0$ or $y + \frac{x}{2} = 1$. $\frac{dy}{dt} = y(1 - \frac{y}{3} - x) = 0$ when $y = 0$ or $x + \frac{y}{3} = 1$.
We find the equilibrium points. They are $(2, 0)$, $(0, 3)$, $(0, 0)$, and $(\frac{4}{5}, \frac{3}{5})$. The nullclines and equilibrium points are shown in Figure 11.18.

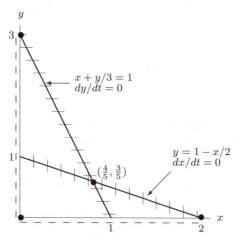

Figure 11.18: Nullclines and equilibrium points (dots)

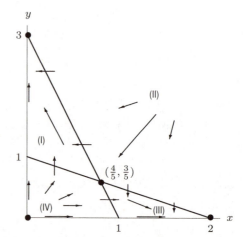

Figure 11.19: General directions of trajectories and equilibrium points (dots)

Figure 11.19 shows that if the initial point is in sector (I), the trajectory heads toward the equilibrium point $(0, 3)$. Similarly, if the trajectory begins in sector (III), then it heads toward the equilibrium $(2, 0)$ over time. If the trajectory begins in sector (II) or (IV), it can go to any of the three equilibrium points $(2, 0)$, $(0, 3)$, or $(\frac{4}{5}, \frac{3}{5})$.

13. (a) See Figure 11.20.

$$\frac{dx}{dt} = 0 \text{ when } x = \frac{10.5}{0.45} = 23.3$$

$$\frac{dy}{dt} = 0 \text{ when } 8.2x - 0.8y - 142 = 0$$

There is an equilibrium point where the trajectories cross at $x = 23.3$, $y = 61.7$

In region I, $\dfrac{dx}{dt} > 0$, $\dfrac{dy}{dt} < 0$.

In region II, $\dfrac{dx}{dt} < 0$, $\dfrac{dy}{dt} < 0$.

In region III, $\dfrac{dx}{dt} < 0$, $\dfrac{dy}{dt} > 0$.

In region IV, $\dfrac{dx}{dt} > 0$, $\dfrac{dy}{dt} > 0$.

(b) See Figure 11.21.

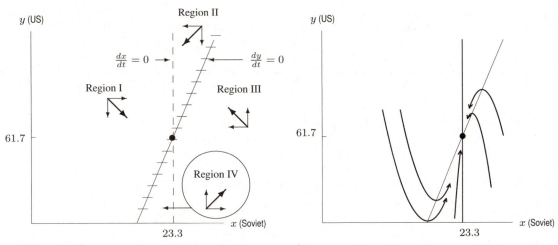

Figure 11.20: Nullclines and equilibrium point (dot) for US-Soviet arms race

Figure 11.21: Trajectories for US-Soviet arms race.

(c) All the trajectories tend toward the equilibrium point $x = 23.3$, $y = 61.7$. Thus the model predicts that in the long run the arms race will level off with the Soviet Union spending 23.3 billion dollars a year on arms and the US 61.7 billion dollars.

(d) As the model predicts, yearly arms expenditure did tend toward 23 billion for the Soviet Union and 62 billion for the US.

Solutions for Section 11.10

Exercises

1. If $y = 2\cos t + 3\sin t$, then $y' = -2\sin t + 3\cos t$ and $y'' = -2\cos t - 3\sin t$. Thus, $y'' + y = 0$.

5. If $y(t) = A\sin(\omega t) + B\cos(\omega t)$ then
$$y' = \omega A\cos(\omega t) - \omega B\sin(\omega t)$$
$$y'' = -\omega^2 A\sin(\omega t) - \omega^2 B\cos(\omega t)$$
therefore
$$y'' + \omega^2 y = -\omega^2 A\sin(\omega t) - \omega^2 B\cos(2t) + \omega^2(A\sin(\omega t) + B\cos(\omega t)) = 0$$
for all values of A and B, so the given function is a solution.

9. The amplitude is $\sqrt{3^2 + 7^2} = \sqrt{58}$.

Problems

13. First, we note that the solutions of:

(a) $x'' + x = 0$ are $x = A\cos t + B\sin t$;

(b) $x'' + 4x = 0$ are $x = A\cos 2t + B\sin 2t$;

(c) $x'' + 16x = 0$ are $x = A\cos 4t + B\sin 4t$.

This follows from what we know about the general solution to $x'' + \omega^2 x = 0$.

The period of the solutions to (a) is 2π, the period of the solutions to (b) is π, and the period of the solutions of (c) is $\frac{\pi}{2}$. Since the t-scales are the same on all of the graphs, we see that graphs (I) and (IV) have the same period, which is twice the period of graph (III). Graph (II) has twice the period of graphs (I) and (IV). Since each graph represents a solution, we have the following:

- equation (a) goes with graph (II)
  equation (b) goes with graphs (I) and (IV)
  equation (c) goes with graph (III)
- The graph of (I) passes through $(0,0)$, so $0 = A\cos 0 + B\sin 0 = A$. Thus, the equation is $x = B\sin 2t$. Since the amplitude is 2, we see that $x = 2\sin 2t$ is the equation of the graph. Similarly, the equation for (IV) is $x = -3\sin 2t$. The graph of (II) also passes through $(0,0)$, so, similarly, the equation must be $x = B\sin t$. In this case, we see that $B = -1$, so $x = -\sin t$.

  Finally, the graph of (III) passes through $(0,1)$, and 1 is the maximum value. Thus, $1 = A\cos 0 + B\sin 0$, so $A = 1$. Since it reaches a local maximum at $(0,1)$, $x'(0) = 0 = -4A\sin 0 + 4B\cos 0$, so $B = 0$. Thus, the solution is $x = \cos 4t$.

17. At $t = 0$, we find that $y = -1$, which is clearly the lowest point on the path. Since $y' = 3\sin 3t$, we see that $y' = 0$ when $t = 0$. Thus, at $t = 0$ the object is at rest, although it will move up after $t = 0$.

21. (a) Since a mass of 3 kg stretches the spring by 2 cm, the spring constant k is given by

$$3g = 2k \quad \text{so} \quad k = \frac{3g}{2}.$$

See Figure 11.22.

equilibrium

x cm

3 kg

Figure 11.22

Suppose we measure the displacement x from the equilibrium; then, using

$$\text{Mass} \cdot \text{Acceleration} = \text{Force}$$

gives

$$3x'' = -kx = -\frac{3gx}{2}$$
$$x'' + \frac{g}{2}x = 0$$

Since at time $t = 0$, the brick is 5 cm below the equilibrium and not moving, the initial conditions are $x(0) = 5$ and $x'(0) = 0$.

(b) The solution to the differential equation is

$$x = A\cos\left(\sqrt{\frac{g}{2}}t\right) + B\sin\left(\sqrt{\frac{g}{2}}t\right).$$

Since $x(0) = 5$, we have

$$x = A\cos(0) + B\sin(0) = 5 \quad \text{so} \quad A = 5.$$

In addition,

$$x'(t) = -5\sqrt{\frac{g}{2}}\sin\left(\sqrt{\frac{g}{2}}t\right) + B\sqrt{\frac{g}{2}}\cos\left(\sqrt{\frac{g}{2}}t\right)$$

so

$$x'(0) = -5\sqrt{\frac{g}{2}}\sin(0) + B\sqrt{\frac{g}{2}}\cos(0) = 0 \quad \text{so} \quad B = 0.$$

Thus,

$$x = 5\cos\sqrt{\frac{g}{2}}t.$$

25. (a) $36\dfrac{d^2Q}{dt^2} + \dfrac{Q}{9} = 0$ so $\dfrac{d^2Q}{dt^2} = -\dfrac{Q}{324}$.
Thus,

$$Q = C_1 \cos\frac{1}{18}t + C_2 \sin\frac{1}{18}t.$$
$$Q(0) = 0 = C_1 \cos 0 + C_2 \sin 0 = C_1,$$
$$\text{so} \quad C_1 = 0.$$

So, $Q = C_2 \sin\dfrac{1}{18}t$, and

$$Q' = I = \frac{1}{18}C_2 \cos\frac{1}{18}t.$$
$$Q'(0) = I(0) = 2 = \frac{1}{18}C_2 \cos\left(\frac{1}{18}\cdot 0\right) = \frac{1}{18}C_2,$$
$$\text{so} \quad C_2 = 36.$$

Therefore, $Q = 36\sin\dfrac{1}{18}t$.

(b) As in part (a), $Q = C_1 \cos\dfrac{1}{18}t + C_2 \sin\dfrac{1}{18}t$.
According to the initial conditions:

$$Q(0) = 6 = C_1 \cos 0 + C_2 \sin 0 = C_1,$$
$$\text{so} \quad C_1 = 6.$$

So $Q = 6\cos\dfrac{1}{18}t + C_2 \sin\dfrac{1}{18}t$.
Thus,

$$Q' = I = -\frac{1}{3}\sin\frac{1}{18}t + \frac{1}{18}C_2 \cos\frac{1}{18}t.$$
$$Q'(0) = I(0) = 0 = -\frac{1}{3}\sin\left(\frac{1}{18}\cdot 0\right) + \frac{1}{18}C_2 \cos\left(\frac{1}{18}\cdot 0\right) = \frac{1}{18}C_2,$$
$$\text{so} \quad C_2 = 0.$$

Therefore, $Q = 6\cos\dfrac{1}{18}t$.

Solutions for Section 11.11

Exercises

1. The characteristic equation is $r^2 + 4r + 3 = 0$, so $r = -1$ or -3.
Therefore $y(t) = C_1 e^{-t} + C_2 e^{-3t}$.

5. The characteristic equation is $r^2 + 7 = 0$, so $r = \pm\sqrt{7}i$.
Therefore $s(t) = C_1 \cos\sqrt{7}t + C_2 \sin\sqrt{7}t$.

9. The characteristic equation is $r^2 + r + 1 = 0$, so $r = -\frac{1}{2} \pm \frac{\sqrt{3}}{2}i$.
Therefore $p(t) = C_1 e^{-t/2}\cos\frac{\sqrt{3}}{2}t + C_2 e^{-t/2}\sin\frac{\sqrt{3}}{2}t$.

13. The characteristic equation is

$$r^2 + 5r + 6 = 0$$

which has the solutions $r = -2$ and $r = -3$ so that

$$y(t) = Ae^{-3t} + Be^{-2t}$$

The initial condition $y(0) = 1$ gives

$$A + B = 1$$

and $y'(0) = 0$ gives

$$-3A - 2B = 0$$

so that $A = -2$ and $B = 3$ and

$$y(t) = -2e^{-3t} + 3e^{-2t}$$

17. The characteristic equation is $r^2 + 6r + 5 = 0$, so $r = -1$ or -5.
Therefore $y(t) = C_1 e^{-t} + C_2 e^{-5t}$.
$y'(t) = -C_1 e^{-t} - 5C_2 e^{-5t}$
$y'(0) = 0 = -C_1 - 5C_2$
$y(0) = 1 = C_1 + C_2$
    Therefore $C_2 = -1/4$, $C_1 = 5/4$ and $y(t) = \frac{5}{4}e^{-t} - \frac{1}{4}e^{-5t}$.

21. The characteristic equation is

$$r^2 + 5r + 6 = 0$$

which has the solutions $r = -2$ and $r = -3$ so that

$$y(t) = Ae^{-2t} + Be^{-3t}$$

The initial condition $y(0) = 1$ gives

$$A + B = 1$$

and $y(1) = 0$ gives

$$Ae^{-2} + Be^{-3} = 0$$

so that $A = \dfrac{1}{1-e}$ and $B = -\dfrac{e}{1-e}$ and

$$y(t) = \frac{1}{1-e}e^{-2t} + \frac{-e}{1-e}e^{-3t}$$

Problems

25. (a) $x'' + 4x = 0$ represents an undamped oscillator, and so goes with (IV).
(b) $x'' - 4x = 0$ has characteristic equation $r^2 - 4 = 0$ and so $r = \pm 2$. The solution is $C_1 e^{-2t} + C_2 e^{2t}$. This represents non-oscillating motion, so it goes with (II).
(c) $x'' - 0.2x' + 1.01x = 0$ has characteristic equation $r^2 - 0.2 + 1.01 = 0$ so $b^2 - 4ac = 0.04 - 4.04 = -4$, and $r = 0.1 \pm i$. So the solution is

$$C_1 e^{(0.1+i)t} + C_2 e^{(0.1-i)t} = e^{0.1t}(A \sin t + B \cos t).$$

The negative coefficient in the x' term represents an amplifying force. This is reflected in the solution by $e^{0.1t}$, which increases as t increases, so this goes with (I).
(d) $x'' + 0.2x' + 1.01x$ has characteristic equation $r^2 + 0.2r + 1.01 = 0$ so $b^2 - 4ac = -4$. This represents a damped oscillator. We have $r = -0.1 \pm i$ and so the solution is $x = e^{-0.1t}(A \sin t + B \cos t)$, which goes with (III).

29. Recall that $s'' + bs' + cs = 0$ is overdamped if the discriminant $b^2 - 4c > 0$, critically damped if $b^2 - 4c = 0$, and underdamped if $b^2 - 4c < 0$. Since $b^2 - 4c = b^2 - 20$, the solution is overdamped if $b > 2\sqrt{5}$ or $b < -2\sqrt{5}$, critically damped if $b = \pm 2\sqrt{5}$, and underdamped if $-2\sqrt{5} < b < 2\sqrt{5}$.

33. The frictional force is $F_{\text{drag}} = -c\frac{ds}{dt}$. Thus spring (iv) has the smallest frictional force.

37. (a) If $r_1 = \frac{-b-\sqrt{b^2-4c}}{2}$ then $r_1 < 0$ since both b and $\sqrt{b^2-4c}$ are positive.

If $r_2 = \frac{-b+\sqrt{b^2-4c}}{2}$, then $r_2 < 0$ because

$$b = \sqrt{b^2} > \sqrt{b^2 - 4c}.$$

(b) The general solution to the differential equation is of the form

$$y = C_1 e^{r_1 t} + C_2 e^{r_2 t}$$

and since r_1 and r_2 are both negative, y must go to 0 as $t \to \infty$.

41. In this case, the differential equation describing the charge is $Q'' + Q' + \frac{1}{4}Q = 0$, so the characteristic equation is $r^2 + r + \frac{1}{4} = 0$. This equation has one root, $r = -\frac{1}{2}$, so the equation for charge is

$$Q(t) = (C_1 + C_2 t)e^{-\frac{1}{2}t},$$
$$Q'(t) = -\frac{1}{2}(C_1 + C_2 t)e^{-\frac{1}{2}t} + C_2 e^{-\frac{1}{2}t}$$
$$= \left(C_2 - \frac{C_1}{2} - \frac{C_2 t}{2}\right)e^{-\frac{1}{2}t}.$$

(a) We have

$$Q(0) = C_1 = 0,$$
$$Q'(0) = C_2 - \frac{C_1}{2} = 2.$$

Thus, $C_1 = 0$, $C_2 = 2$, and

$$Q(t) = 2te^{-\frac{1}{2}t}.$$

(b) We have

$$Q(0) = C_1 = 2,$$
$$Q'(0) = C_2 - \frac{C_1}{2} = 0.$$

Thus, $C_1 = 2$, $C_2 = 1$, and

$$Q(t) = (2 + t)e^{-\frac{1}{2}t}.$$

(c) The resistance was decreased by exactly the amount to switch the circuit from the overdamped case to the critically damped case. Comparing the solutions of parts (a) and (b) in Problems 40, we find that in the critically damped case the net charge goes to 0 much faster as $t \to \infty$.

Solutions for Chapter 11 Review

Exercises

1. The slope fields in (I) and (II) appear periodic. (I) has zero slope at $x = 0$, so (I) matches $y' = \sin x$, whereas (II) matches $y' = \cos x$. The slope in (V) tends to zero as $x \to \pm\infty$, so this must match $y' = e^{-x^2}$. Of the remaining slope fields, only (III) shows negative slopes, matching $y' = xe^{-x}$. The slope in (IV) is zero at $x = 0$, so it matches $y' = x^2 e^{-x}$. This leaves field (VI) to match $y' = e^{-x}$.

5. This equation is separable, so we integrate, giving

$$\int dP = \int t\,dt$$

so

$$P(t) = \frac{t^2}{2} + C.$$

9. This equation is separable, so we integrate, using the table of integrals or partial fractions, to get

$$\int \frac{1}{R - 3R^2}\, dR = 2\int dt$$

$$\int \frac{1}{R}\, dR + \int \frac{3}{1 - 3R}\, dR = 2\int dt$$

so

$$\ln |R| - \ln |1 - 3R| = 2t + C$$

$$\ln \left| \frac{R}{1 - 3R} \right| = 2t + C$$

$$\frac{R}{1 - 3R} = Ae^{2t}$$

$$R = \frac{Ae^{2t}}{1 + 3Ae^{2t}}.$$

13. $1 + y^2 - \frac{dy}{dx} = 0$ gives $\frac{dy}{dx} = y^2 + 1$, so $\int \frac{dy}{1+y^2} = \int dx$ and $\arctan y = x + C$. Since $y(0) = 0$ we have $C = 0$, giving $y = \tan x$.

17. $\frac{dy}{dx} = \frac{0.2y(18 + 0.1x)}{x(100 + 0.5y)}$ giving $\int \frac{(100 + 0.5y)}{0.2y}\, dy = \int \frac{18 + 0.1x}{x}\, dx$, so

$$\int \left(\frac{500}{y} + \frac{5}{2} \right) dy = \int \left(\frac{18}{x} + \frac{1}{10} \right) dx.$$

Therefore, $500 \ln |y| + \frac{5}{2}y = 18 \ln |x| + \frac{1}{10}x + C$. Since the curve passes through (10,10), $500 \ln 10 + 25 = 18 \ln 10 + 1 + C$, so $C = 482 \ln 10 + 24$. Thus, the solution is

$$500 \ln |y| + \frac{5}{2}y = 18 \ln |x| + \frac{1}{10}x + 482 \ln 10 + 24.$$

We cannot solve for y in terms of x, so we leave the answer in this form.

21. $\frac{df}{dx} = \sqrt{x f(x)}$ gives $\int \frac{df}{\sqrt{f(x)}} = \int \sqrt{x}\, dx$, so $2\sqrt{f(x)} = \frac{2}{3}x^{\frac{3}{2}} + C$. Since $f(1) = 1$, we have $2 = \frac{2}{3} + C$ so $C = \frac{4}{3}$.
Thus, $2\sqrt{f(x)} = \frac{2}{3}x^{\frac{3}{2}} + \frac{4}{3}$, so $f(x) = (\frac{1}{3}x^{\frac{3}{2}} + \frac{2}{3})^2$.
(Note: this is only defined for $x \geq 0$.)

25. $(1 + t^2)y\frac{dy}{dt} = 1 - y$ implies that $\int \frac{y\, dy}{1-y} = \int \frac{dt}{1+t^2}$ implies that $\int \left(-1 + \frac{1}{1-y} \right) dy = \int \frac{dt}{1+t^2}$. Therefore $-y - \ln |1-y| = \arctan t + C$. $y(1) = 0$, so $0 = \arctan 1 + C$, and $C = -\frac{\pi}{4}$, so $-y - \ln |1 - y| = \arctan t - \frac{\pi}{4}$. We cannot solve for y in terms of t.

29. The characteristic equation of $9z'' + z = 0$ is

$$9r^2 + 1 = 0$$

If we write this in the form $r^2 + br + c = 0$, we have that $r^2 + 1/9 = 0$ and

$$b^2 - 4c = 0 - (4)(1/9) = -4/9 < 0$$

This indicates underdamped motion and since the roots of the characteristic equation are $r = \pm\frac{1}{3}i$, the general equation is

$$y(t) = C_1 \cos \left(\frac{1}{3}t \right) + C_2 \sin \left(\frac{1}{3}t \right)$$

Problems

33. (a) The slope field for $dy/dx = xy$ is in Figure 11.23.

Figure 11.23

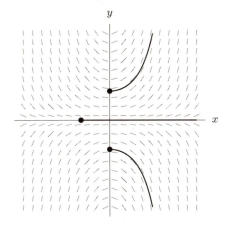

Figure 11.24

(b) Some solution curves are shown in Figure 11.24.

(c) Separating variables gives

$$\int \frac{1}{y} dy = \int x dx$$

or

$$\ln |y| = \frac{1}{2} x^2 + C.$$

Solving for y gives

$$y(x) = Ae^{x^2/2}$$

where $A = \pm e^C$. In addition, $y(x) = 0$ is a solution. So $y(x) = Ae^{x^2/2}$ is a solution for any A.

37. (a) $\Delta x = \frac{1}{5} = 0.2$.

At $x = 0$:

$y_0 = 1, y' = 4$; so $\Delta y = 4(0.2) = 0.8$. Thus, $y_1 = 1 + 0.8 = 1.8$.

At $x = 0.2$:

$y_1 = 1.8, y' = 3.2$; so $\Delta y = 3.2(0.2) = 0.64$. Thus, $y_2 = 1.8 + 0.64 = 2.44$.

At $x = 0.4$:

$y_2 = 2.44, y' = 2.56$; so $\Delta y = 2.56(0.2) = 0.512$. Thus, $y_3 = 2.44 + 0.512 = 2.952$.

At $x = 0.6$:

$y_3 = 2.952, y' = 2.048$; so $\Delta y = 2.048(0.2) = 0.4096$. Thus, $y_4 = 3.3616$.

At $x = 0.8$:

$y_4 = 3.3616, y' = 1.6384$; so $\Delta y = 1.6384(0.2) = 0.32768$. Thus, $y_5 = 3.68928$. So $y(1) \approx 3.689$.

(b)

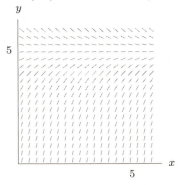

Since solution curves are concave down for $0 \leq y \leq 5$, and $y(0) = 1 < 5$, the estimate from Euler's method will be an overestimate.

(c) Solving by separation:

$$\int \frac{dy}{5 - y} = \int dx, \quad \text{so} \quad -\ln |5 - y| = x + C.$$

Then $5 - y = Ae^{-x}$ where $A = \pm e^{-C}$. Since $y(0) = 1$, we have $5 - 1 = Ae^0$, so $A = 4$.
Therefore, $y = 5 - 4e^{-x}$, and $y(1) = 5 - 4e^{-1} \approx 3.528$.
(Note: as predicted, the estimate in (a) is too large.)

(d) Doubling the value of n will probably halve the error and, therefore, give a value half way between 3.528 and 3.689, which is approximately 3.61.

41. (a) If A is surface area, we know that for some constant K

$$\frac{dV}{dt} = -KA.$$

If r is the radius of the sphere, $V = 4\pi r^3/3$ and $A = 4\pi r^2$. Solving for r in terms of V gives $r = (3V/4\pi)^{1/3}$, so

$$\frac{dV}{dt} = -K(4\pi r^2) = -K4\pi \left(\frac{3V}{4\pi}\right)^{2/3} \quad \text{so} \quad \frac{dV}{dt} = -kV^{2/3}$$

where k is another constant, $k = K(4\pi)^{1/3} 3^{2/3}$.

(b) Separating variables gives

$$\int \frac{dV}{V^{2/3}} = -\int k\, dt$$

$$3V^{1/3} = -kt + C.$$

Since $V = V_0$ when $t = 0$, we have $3V_0^{1/3} = C$, so

$$3V^{1/3} = -kt + 3V_0^{1/3}.$$

Solving for V gives

$$V = \left(-\frac{k}{3}t + V_0^{1/3}\right)^3.$$

This function is graphed in Figure 11.25.

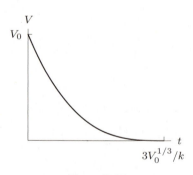

Figure 11.25

(c) The snowball disappears when $V = 0$, that is when

$$-\frac{k}{3}t + V_0^{1/3} = 0$$

giving

$$t = \frac{3V_0^{1/3}}{k}.$$

45. Recall that $s'' + bs' + c = 0$ is overdamped if the discriminant $b^2 - 4c > 0$, critically damped if $b^2 - 4c = 0$, and underdamped if $b^2 - 4c < 0$. Since $b^2 - 4c = 16 - 4c$, the circuit is overdamped if $c < 4$, critically damped if $c = 4$, and underdamped if $c > 4$.

49. Let I be the number of infected people. Then, the number of healthy people in the population is $M - I$. The rate of infection is

$$\text{Infection rate} = \frac{0.01}{M}(M - I)I.$$

and the rate of recovery is

$$\text{Recovery rate} = 0.009I.$$

Therefore,

$$\frac{dI}{dt} = \frac{0.01}{M}(M - I)I - 0.009I$$

or

$$\frac{dI}{dt} = 0.001I(1 - 10\frac{I}{M}).$$

This is a logistic differential equation, and so the solution will look like the following graph:

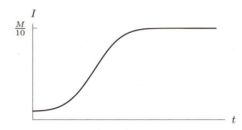

The limiting value for I is $\frac{1}{10}M$, so $1/10$ of the population is infected in the long run.

CAS Challenge Problems

53. (a) Using the integral equation with $n + 1$ replaced by n, we have

$$y_n(a) = b + \int_a^a (y_{n-1}(t)^2 + t^2)\, dt = b + 0 = b.$$

(b) We have $a = 1$ and $b = 0$, so the integral equation tells us that

$$y_{n+1}(s) = \int_1^s (y_n(t)^2 + t^2)\, dt.$$

With $n = 0$, since $y_0(s) = 0$, the CAS gives

$$y_1(s) = \int_1^s 0 + t^2\, dt = -\frac{1}{3} + \frac{s^3}{3}.$$

Then

$$y_2(s) = \int_1^s (y_1(t)^2 + t^2)\, dt = -\frac{17}{42} + \frac{s}{9} + \frac{s^3}{3} - \frac{s^4}{18} + \frac{s^7}{63},$$

and

$$y_3(s) = \int_1^s (y_2(t)^2 + t^2)\, dt$$

$$= -\frac{157847}{374220} + \frac{289\,s}{1764} - \frac{17\,s^2}{378} + \frac{82\,s^3}{243} - \frac{17\,s^4}{252} + \frac{s^5}{42} - \frac{s^6}{486} + \frac{s^7}{63} - \frac{11\,s^8}{1764} +$$
$$\frac{5\,s^9}{6804} + \frac{2\,s^{11}}{2079} - \frac{s^{12}}{6804} + \frac{s^{15}}{59535}.$$

(c) The solution y, and the approximations y_1, y_2, y_3 are graphed in Figure 11.26. The approximations appear to be accurate on the range $0.5 \le s \le 1.5$.

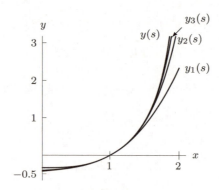

Figure 11.26

CHECK YOUR UNDERSTANDING

1. True. The general solution to $y' = -ky$ is $y = Ce^{-kt}$.

5. True. No matter what initial value you pick, the solution curve has the x-axis as an asymptote.

9. True. Rewrite the equation as $dy/dx = xy + x = x(y + 1)$. Since the equation now has the form $dy/dx = f(x)g(y)$, it can be solved by separation of variables.

13. False. This is a logistic equation with equilibrium values $P = 0$ and $P = 2$. Solution curves do not cross the line $P = 2$ and do not go from $(0, 1)$ to $(1, 3)$.

17. True. Since $f'(x) = g(x)$, we have $f''(x) = g'(x)$. Since $g(x)$ is increasing, $g'(x) > 0$ for all x, so $f''(x) > 0$ for all x. Thus the graph of f is concave up for all x.

21. False. Let $g(x) = 0$ for all x and let $f(x) = 17$. Then $f'(x) = g(x)$ and $\lim_{x \to \infty} g(x) = 0$, but $\lim_{x \to \infty} f(x) = 17$.

25. True. The slope of the graph of f is $dy/dx = 2x - y$. Thus when $x = a$ and $y = b$, the slope is $2a - b$.

29. False. Since $f'(1) = 2(1) - 5 = -3$, the point $(1, 5)$ could not be a critical point of f.

33. True. We will use the hint. Let $w = g(x) - f(x)$. Then:

$$\frac{dw}{dx} = g'(x) - f'(x) = (2x - g(x)) - (2x - f(x)) = f(x) - g(x) = -w.$$

Thus $dw/dx = -w$. This equation is the equation for exponential decay and has the general solution $w = Ce^{-x}$. Thus,

$$\lim_{x \to \infty} (g(x) - f(x)) = \lim_{x \to \infty} Ce^{-x} = 0.$$

37. If we differentiate implicitly the equation for the family, we get $2x - 2y\,dy/dx = 0$. When we solve, we get the differential equation we want $dy/dx = x/y$.

CHAPTER TWELVE

Solutions for Section 12.1

Exercises

1. The distance of a point $P = (x, y, z)$ from the yz-plane is $|x|$, from the xz-plane is $|y|$, and from the xy-plane is $|z|$. So A is closest to the yz-plane, since it has the smallest x-coordinate in absolute value. B lies on the xz-plane, since its y-coordinate is 0. C is farthest from the xy-plane, since it has the largest z-coordinate in absolute value.

5. Your final position is $(1, -1, 1)$. This places you in front of the yz-plane, to the left of the xz-plane, and above the xy-plane.

9. The graph is all points with $y = 4$ and $z = 2$, i.e., a line parallel to the x-axis and passing through the points $(0, 4, 2); (2, 4, 2); (4, 4, 2)$ etc. See Figure 12.1.

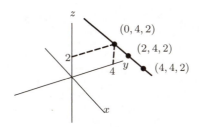

Figure 12.1

13. (a) 80-90°F
(b) 60-72°F
(c) 60-100°F

17. Table 12.1 gives the amount M spent on beef per household per week. Thus, the amount the household spent on beef in a year is $52M$. Since the household's annual income is I thousand dollars, the proportion of income spent on beef is

$$P = \frac{52M}{1000I} = 0.052 \frac{M}{I}.$$

Thus, we need to take each entry in Table 12.1, divide it by the income at the left, and multiply by 0.052. Table 12.2 shows the results.

Table 12.1 *Money spent on beef ($/household/week)*

	\multicolumn{4}{c}{Price of Beef ($)}			
	3.00	3.50	4.00	4.50
20	7.95	9.07	10.04	10.94
40	12.42	14.18	15.76	17.46
60	15.33	17.50	19.88	21.78
80	16.05	18.52	20.76	22.82
100	17.37	20.20	22.40	24.89

Income ($1,000)

Table 12.2 *Proportion of annual income spent on beef*

	\multicolumn{4}{c}{Price of Beef ($)}			
	3.00	3.50	4.00	4.50
20	0.021	0.024	0.026	0.028
40	0.016	0.018	0.020	0.023
60	0.013	0.015	0.017	0.019
80	0.010	0.012	0.013	0.015
100	0.009	0.011	0.012	0.013

Income ($1,000)

Problems

21.

Table 12.3 *Temperature adjusted for wind-chill at 5 mph*

Temperature (°F)	35	30	25	20	15	10	5	0
Adjusted temperature (°F)	31	25	19	13	7	1	−5	−11

Table 12.4 *Temperature adjusted for wind-chill at 20 mph*

Temperature (°F)	35	30	25	20	15	10	5	0
Adjusted temperature (°F)	24	17	11	4	−2	−9	−15	−22

25. The gravitational force on a 100 kg object which is $7,000,000$ meters from the center of the earth (or about 600 km above the earth's surface) is about 820 newtons.

29. The distance of any point with coordinates (x, y, z) from the x-axis is $\sqrt{y^2 + z^2}$. The distance of the point from the xy-plane is $|x|$. Since the condition states that these distances are equal, the equation for the condition is

$$\sqrt{y^2 + z^2} = |x| \qquad \text{i.e.} \qquad y^2 + z^2 = x^2.$$

This is the equation of a cone whose tip is at the origin and which opens along the x-axis with a slope of 1 as shown in Figure 12.2.

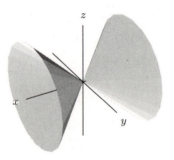

Figure 12.2

33. Using the distance formula, we find that

$$\text{Distance from } P_1 \text{ to } P = \sqrt{206}$$
$$\text{Distance from } P_2 \text{ to } P = \sqrt{152}$$
$$\text{Distance from } P_3 \text{ to } P = \sqrt{170}$$
$$\text{Distance from } P_4 \text{ to } P = \sqrt{113}$$

So $P_4 = (-4, 2, 7)$ is closest to $P = (6, 0, 4)$.

Solutions for Section 12.2

Exercises

1. (a) is (IV), since $z = 2 + x^2 + y^2$ is a paraboloid opening upward with a positive z-intercept.
   (b) is (II), since $z = 2 - x^2 - y^2$ is a paraboloid opening downward.
   (c) is (I), since $z = 2(x^2 + y^2)$ is a paraboloid opening upward and going through the origin.
   (d) is (V), since $z = 2 + 2x - y$ is a slanted plane.
   (e) is (III), since $z = 2$ is a horizontal plane.

5. The graph is a bowl opening up, with vertex at the point $(0, 0, 4)$. See Figure 12.3.

Figure 12.3

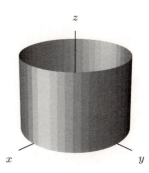

Figure 12.4

9. In the xy-plane, the graph is a circle of radius 2. Since there are no restrictions on z, we extend this circle along the z-axis. The graph is a circular cylinder extended in the z-direction. See Figure 12.4.

Problems

13. The one-variable function $f(a, t)$ represents the effect of an injection of a mg at time t. Figure 12.5 shows the graphs of the four functions $f(1, t) = te^{-4t}$, $f(2, t) = te^{-3t}$, $f(3, t) = te^{-2t}$, and $f(4, t) = te^{-t}$ corresponding to injections of 1, 2, 3, and 4 mg of the drug. The general shape of the graph is the same in every case: The concentration in the blood is zero at the time of injection $t = 0$, then increases to a maximum value, and then decreases toward zero again. We see that if a larger dose of the drug is administered, the peak of the graph is later and higher. This makes sense, since a larger dose will take longer to diffuse fully into the bloodstream and will produce a higher concentration when it does.

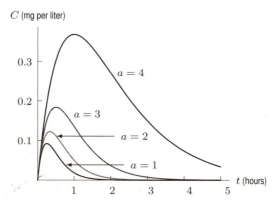

Figure 12.5: Concentration $C = f(a, t)$ of the drug resulting from an a mg injection

17. (a) This is a bowl; z increases as the distance from the origin increases, from a minimum of 0 at $x = y = 0$.
   (b) Neither. This is an upside-down bowl. This function will decrease from 1, at $x = y = 0$, to arbitrarily large negative values as x and y increase due to the negative squared terms of x and y. It will look like the bowl in part (a) except flipped over and raised up slightly.
   (c) This is a plate. Solving the equation for z gives $z = 1 - x - y$ which describes a plane whose x and y slopes are -1. It is perfectly flat, but not horizontal.
   (d) Within its domain, this function is a bowl. It is undefined at points at which $x^2 + y^2 > 5$, but within those limits it describes the bottom half of a sphere of radius $\sqrt{5}$ centered at the origin.
   (e) This function is a plate. It is perfectly flat and horizontal.

21. One possible equation: $z = (x - y)^2$. See Figure 12.6.

Figure 12.6

25. (a) If we have iron stomachs and can consume cola and pizza endlessly without ill effects, then we expect our happiness to increase without bound as we get more cola and pizza. Graph (IV) shows this since it increases along both the pizza and cola axes throughout.

(b) If we get sick upon eating too many pizzas or drinking too much cola, then we expect our happiness to decrease once either or both of those quantities grows past some optimum value. This is depicted in graph (I) which increases along both axes until a peak is reached, and then decreases along both axes.

(c) If we do get sick after too much cola, but are always able to eat more pizza, then we expect our happiness to decrease after we drink some optimum amount of cola, but continue to increase as we get more pizza. This is shown by graph (III) which increases continuously along the pizza axis but, after reaching a maximum, begins to decrease along the cola axis.

29. The paraboloid is $z = x^2 + y^2 + 5$, so it is represented by

$$z = f(x, y) = x^2 + y^2 + 5$$

and

$$g(x, y, z) = x^2 + y^2 + 5 - z = 0.$$

Other answers are possible.

Solutions for Section 12.3

Exercises

1. We'll set $z = 4$ at the peak. See Figure 12.7.

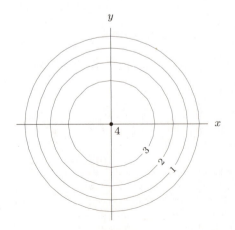

Figure 12.7

5. The contour where $f(x, y) = x + y = c$, or $y = -x + c$, is the graph of the straight line with slope -1 as shown in Figure 12.8. Note that we have plotted the contours for $c = -3, -2, -1, 0, 1, 2, 3$. The contours are evenly spaced.

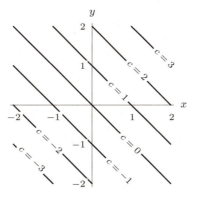

Figure 12.8

9. The contour where $f(x, y) = xy = c$, is the graph of the hyperbola $y = c/x$ if $c \neq 0$ and the coordinate axes if $c = 0$, as shown in Figure 12.9. Note that we have plotted contours for $c = -5, -4, -3, -2, -1, 0, 1, 2, 3, 4, 5$. The contours become more closely packed as we move further from the origin.

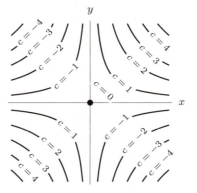

Figure 12.9

13. The contour where $f(x, y) = \cos(\sqrt{x^2 + y^2}) = c$, where $-1 \leq c \leq 1$, is a set of circles centered at $(0, 0)$, with radius $\cos^{-1} c + 2k\pi$ with $k = 0, 1, 2, ..$ and $-\cos^{-1} c + 2k\pi$, with $k = 1, 2, 3, ...$ as shown in Figure 12.10. Note that we have plotted contours for $c = 0, 0.2, 0.4, 0.6, 0.8, 1$.

Figure 12.10

17. The values in Table 12.5 are not constant along rows or columns and therefore cannot be the lines shown in (I) or (IV). Also observe that as you move away from the origin, whose contour value is 0, the z-values on the contours increase. Thus, this table corresponds to diagram (II).

   The values in Table 12.6 are also not constant along rows or columns. Since the contour values are decreasing as you move away from the origin, this table corresponds to diagram (III).

   Table 12.7 shows that for each fixed value of x, we have constant contour value, suggesting a straight vertical line at each x-value, as in diagram (IV).

   Table 12.8 also shows lines, however these are horizontal since for each fixed value of y we have constant contour values. Thus, this table matches diagram (I).

Problems

21. (a) The contour lines are much closer together on path A, so path A is steeper.
   (b) If you are on path A and turn around to look at the countryside, you find hills to your left and right, obscuring the view. But the ground falls away on either side of path B, so you are likely to get a much better view of the countryside from path B.
   (c) There is more likely to be a stream alongside path A, because water follows the direction of steepest descent.

25. Figure 12.11 shows an east-west cross-section along the line $N = 50$ kilometers.
   Figure 12.12 shows an east-west cross-section along the line $N = 100$ kilometers.

Figure 12.11

Figure 12.12

Figure 12.13 shows a north-south cross-section along the line $E = 60$ kilometers.
Figure 12.14 shows a north-south cross-section along the line $E = 120$ kilometers.

Figure 12.13

Figure 12.14

29. (a) Multiply the values on each contour of the original contour diagram by 3. See Figure 12.15.

Figure 12.15: $3f(x, y)$

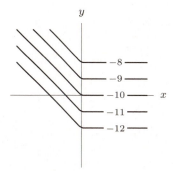

Figure 12.16: $f(x, y) - 10$

(b) Subtract 10 from the values on each contour. See Figure 12.16.
(c) Shift the diagram 2 units to the right and 2 units up. See Figure 12.17.

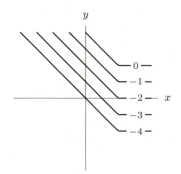

Figure 12.17: $f(x - 2, y - 2)$

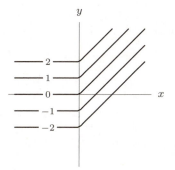

Figure 12.18: $f(-x, y)$

(d) Reflect the diagram about the y-axis. See Figure 12.18.

33. Since $f(x, y) = x^2 - y^2 = (x - y)(x + y) = 0$ gives $x - y = 0$ or $x + y = 0$, the contours $f(x, y) = 0$ are the lines $y = x$ or $y = -x$. In the regions between them, $f(x, y) > 0$ or $f(x, y) < 0$ as shown in Figure 12.19. The surface $z = f(x, y)$ is above the xy-plane where $f > 0$ (that is on the shaded regions containing the x-axis) and is below the xy-plane where $f < 0$. This means that a person could sit on the surface facing along the positive or negative x-axis, and with his/her legs hanging down the sides below the y-axis. Thus, the graph of the function is saddle-shaped at the origin.

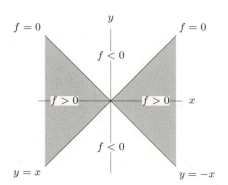

Figure 12.19

Solutions for Section 12.4

Exercises

1.

Table 12.5

$x \backslash y$	0.0	1.0
0.0	−1.0	1.0
2.0	3.0	5.0

5. A table of values is linear if the rows are all linear and have the same slope and the columns are all linear and have the same slope. The table might represent a linear function since the slope in each row is 5 and the slope in each column is 2.

9. Since

$$0 \;= c + m \cdot 0 + n \cdot 0 \qquad c = 0$$
$$-1 = c + m \cdot 0 + n \cdot 2 \qquad c + 2n = -1$$
$$-4 = c + m \cdot (-3) + n \cdot 0 \quad c - 3m = -4$$

we get:

$$c = 0, m = \frac{4}{3}, n = -\frac{1}{2}.$$

Thus, $z = \frac{4}{3}x - \frac{1}{2}y$.

13. (a) Since z is a linear function of x and y with slope 2 in the x-direction, and slope 3 in the y-direction, we have:

$$z = 2x + 3y + c$$

We can write an equation for changes in z in terms of changes in x and y:

$$\Delta z = (2(x + \Delta x) + 3(y + \Delta y) + c) - (2x + 3y + c)$$
$$= 2\Delta x + 3\Delta y$$

Since $\Delta x = 0.5$ and $\Delta y = -0.2$, we have

$$\Delta z = 2(0.5) + 3(-0.2) = 0.4$$

So a 0.5 change in x and a −0.2 change in y produces a 0.4 change in z.

(b) As we know that $z = 2$ when $x = 5$ and $y = 7$, the value of z when $x = 4.9$ and $y = 7.2$ will be

$$z = 2 + \Delta z = 2 + 2\Delta x + 3\Delta y$$

where Δz is the change in z when x changes from 4.9 to 5 and y changes from 7.2 to 7. We have $\Delta x = 4.9 - 5 = -0.1$ and $\Delta y = 7.2 - 7 = 0.2$. Therefore, when $x = 4.9$ and $y = 7.2$, we have

$$z = 2 + 2 \cdot (-0.1) + 5 \cdot 0.2 = 2.4$$

Problems

17. (a) Expenditure, E, is given by the equation:

$$E = (\text{price of raw material 1})m_1 + (\text{price of raw material 2})m_2 + C$$

where C denotes all the other expenses (assumed to be constant). Since the prices of the raw materials are constant, but m_1 and m_2 are variables, we have a linear function.

(b) Revenue, R, is given by the equation:

$$R = (p_1)q_1 + (p_2)q_2.$$

Since p_1 and p_2 are constant, while q_1 and q_2 are variables, we again have a linear function.

(c) Revenue is again given by the equation,

$$R = (p_1)q_1 + (p_2)q_2.$$

Since p_2 and q_2 are now constant, the term $(p_2)q_2$ is also constant. However, since p_1 and q_1 are variables, the $(p_1)q_1$ term means that the function is not linear.

21. The function, g, has a slope of 3 in the x direction and a slope of 1 in the y direction, so $g(x, y) = c + 3x + y$. Since $g(0, 0) = 0$, the formula is $g(x, y) = 3x + y$.

25. See Figure 12.20.

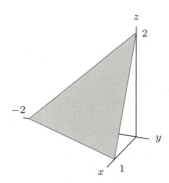

Figure 12.20

29. (a) The contours of f have equation

$$k = c + mx + ny, \quad \text{where } k \text{ is a constant.}$$

Solving for y gives:

$$y = -\frac{m}{n}x + \frac{k - c}{n}$$

Since c, m, n and k are constants, this is the equation of a line. The coefficient of x is the slope and is equal to $-m/n$.

(b) Substituting $x + n$ for x and $y - m$ for y into $f(x, y)$ gives

$$f(x + n, y - m) = c + m(x + n) + n(y - m)$$

Multiplying out and simplifying gives

$$f(x + n, y - m) = c + mx + mn + ny - nm$$

$$f(x + n, y - m) = c + mx + ny = f(x, y)$$

(c) Part (b) tells us that if we move n units in the x direction and $-m$ units in the y direction, the value of the function $f(x, y)$ remains constant. Since contours are lines where the function has a constant value, this implies that we remain on the same contour. This agrees with part (a) which tells us that the slope of any contour line will be $-m/n$. Since the slope is $\Delta y / \Delta x$, it follows that changing y by $-m$ and x by n will keep us on the same contour.

Solutions for Section 12.5

Exercises

1. (a) Observe that setting $f(x, y, z) = c$ gives a cylinder about the x-axis, with radius $\sqrt{c}$. These surfaces are in graph (I).
  (b) By the same reasoning the level curves for $h(x, y, z)$ are cylinders about the y-axis, so they are represented in graph (II).

5. If we solve for z, we get $z = (1 - x^2 - y)^2$, so the level surface is the graph of $f(x, y) = (1 - x^2 - y)^2$.

9. A hyperboloid of two sheets.

13. Yes,

$$z = f(x, y) = \frac{2}{5}x + \frac{3}{5}y - 2.$$

Problems

17. The top half of the sphere is represented by

$$z = f(x,y) = \sqrt{10 - x^2 - y^2}$$

and

$$g(x,y,z) = x^2 + y^2 + z^2 = 10, \quad z \geq 0.$$

Other answers are possible.

21. The equation of any plane parallel to the plane $z = 2x+3y-5$ has x-slope 2 and y-slope 3, so has equation $z = 2x+3y-c$ for any constant c, or $2x + 3y - z = c$. Thus we could take $g(x,y,z) = 2x + 3y - z$. Other answers are possible.

25. In the xz-plane, the equation $x^2/4 + z^2 = 1$ is an ellipse, with widest points at $x = \pm 2$ on the x-axis and crossing the z-axis at $z = \pm 1$. Since the equation has no y term, the level surface is a cylinder of elliptical cross-section, centered along the y-axis.

29. Let's consider the function $y = 2 + \sin z$ drawn in the yz-plane in Figure 12.21.

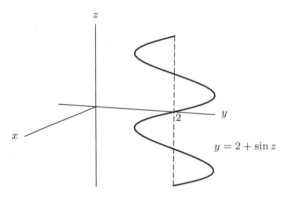

Figure 12.21

Now rotate this graph around the z-axis. Then, a point (x,y,z) is on the surface if and only if $x^2 + y^2 = (2 + \sin z)^2$. Thus, the surface generated is a surface of rotation with the profile shown in Figure 12.21.

Similarly, the surface with equation $x^2 + y^2 = (f(z))^2$ is the surface obtained rotating the graph of $y = f(z)$ around the z-axis.

33. For values of $f < 4$, the level surfaces are spheres, with larger f giving smaller radii. See Figure 12.22.

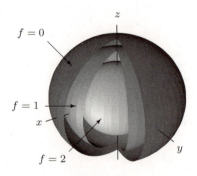

Figure 12.22

Solutions for Section 12.6

Exercises

1. No, $1/(x^2 + y^2)$ is not defined at the origin, so is not continuous at all points in the square $-1 \leq x \leq 1, -1 \leq y \leq 1$.

5. The function $\tan(\theta)$ is undefined when $\theta = \pi/2 \approx 1.57$. Since there are points in the square $-2 \leq x \leq 2, -2 \leq y \leq 2$ with $x \cdot y = \pi/2$ (e.g. $x = 1, y = \pi/2$) the function $\tan(xy)$ is not defined inside the square, hence not continuous.

9. Since f does not depend on y we have:

$$\lim_{(x,y)\to(0,0)} f(x,y) = \lim_{x\to 0} \frac{x}{x^2 + 1} = \frac{0}{0+1} = 0.$$

Problems

13. We want to show that f does not have a limit as (x, y) approaches $(0, 0)$. Let us suppose that (x, y) tends to $(0, 0)$ along the line $y = mx$. Then

$$f(x, y) = f(x, mx) = \frac{x^2 - m^2 x^2}{x^2 + m^2 x^2} = \frac{1 - m^2}{1 + m^2}.$$

Therefore

$$\lim_{x\to 0} f(x, mx) = \frac{1 - m^2}{1 + m^2}$$

and so for $m = 1$ we get

$$\lim_{\substack{(x,y)\to(0,0) \\ y=x}} f(x, y) = \frac{1 - 1}{1 + 1} = \frac{0}{2} = 0$$

and for $m = 0$

$$\lim_{\substack{(x,y)\to(0,0) \\ y=0}} f(x, y) = \frac{1 - 0}{1 + 0} = 1.$$

Thus no matter how close they are to the origin, there will be points (x, y) such that $f(x, y)$ is close to 0 and points (x, y) where $f(x, y)$ is close to 1. So the limit:

$$\lim_{(x,y)\to(0,0)} f(x, y) \text{ does not exist.}$$

17. We will study the continuity of f at $(a, 0)$. Now $f(a, 0) = 1 - a$. In addition:

$$\lim_{\substack{(x,y)\to(a,0) \\ y>0}} f(x, y) = \lim_{x\to a} (1 - x) = 1 - a$$

$$\lim_{\substack{(x,y)\to(a,0) \\ y<0}} f(x, y) = \lim_{x\to a} -2 = -2.$$

If $a = 3$, then

$$\lim_{\substack{(x,y)\to(3,0) \\ y>0}} f(x, y) = 1 - 3 = -2 = \lim_{\substack{(x,y)\to(3,0) \\ y<0}} f(x, y)$$

and so $\lim_{(x,y)\to(3,0)} f(x, y) = -2 = f(3, 0)$. Therefore f is continuous at $(3, 0)$.

On the other hand, if $a \neq 3$, then

$$\lim_{\substack{(x,y)\to(a,0) \\ y>0}} f(x, y) = 1 - a \neq -2 = \lim_{\substack{(x,y)\to(a,0) \\ y<0}} f(x, y)$$

so $\lim_{(x,y)\to(a,0)} f(x, y)$ does not exist. Thus f is not continuous at $(a, 0)$ if $a \neq 3$.

Thus, f is not continuous along the line $y = 0$. (In fact the only point on this line where f is continuous is the point $(3, 0)$.)

21. The function $f(x, y) = x^2 + y^2 + 1$ gets closer and closer to 1 as (x, y) gets closer to the origin. To make f continuous at the origin, we need to have $f(0, 0) = 1$. Thus $c = 1$ will make the function continuous at the origin.

Solutions for Chapter 12 Review

Exercises

1. An example is the line $z = -x$ in the xz-plane. See Figure 12.23.

Figure 12.23

5. Given (x, y) we can solve uniquely for z, namely $z = 2 + \dfrac{x}{5} + \dfrac{y}{5} - \dfrac{3x^2}{5} + y^2$. Thus, z is a function of x and y:

$$z = f(x, y) = 2 + \frac{x}{5} + \frac{y}{5} - \frac{3x^2}{5} + y^2.$$

9. Contours are lines of the form $3x - 5y + 1 = c$ as shown in Figure 12.24. Note that for the regions of x and y given, the c values range from $-12 < c < 12$ and are evenly spaced.

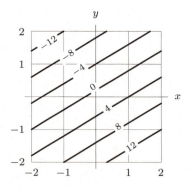

Figure 12.24

13. These conditions describe a line parallel to the z-axis which passes through the xy-plane at $(2, 1, 0)$.

17. A contour diagram is linear if the contours are parallel straight lines, equally spaced for equally spaced values of z. This contour diagram does not represent a linear function.

21. The level surfaces appear to be circular cylinders centered on the z-axis. Since they don't change with z, there is no z in the formula, and we can use the formula for a circle in the xy-plane, $x^2 + y^2 = r^2$. Thus the level surfaces are of the form $f(x, y, z) = x^2 + y^2 = c$ for $c > 0$.

25. (a) The value of z decreases as x increases. See Figure 12.25.
  (b) The value of z increases as y increases. See Figure 12.26.

Figure 12.25

Figure 12.26

Problems

29. Points along the positive x-axis are of the form $(x, 0)$; at these points the function looks like $2x/2x = 1$ everywhere (except at the origin, where it is undefined). On the other hand, along the y-axis, the function looks like $-y^2/y^2 = -1$. Since approaching the origin along two different paths yields numbers that are not the same, the limit does not exist.

33. When h is fixed, say $h = 1$, then
$$V = f(r, 1) = \pi r^2 1 = \pi r^2$$
Similarly,
$$f(r, \tfrac{2}{3}) = \frac{4}{9}\pi r^2 \quad \text{and} \quad f(r, \tfrac{1}{3}) = \frac{\pi}{9}r^2$$
When r is fixed, say $r = 1$, then
$$f(1, h) = \pi(1)^2 h = \pi h$$
Similarly,
$$f(2, h) = 4\pi \quad \text{and} \quad f(3, h) = 9\pi h.$$

Figure 12.27

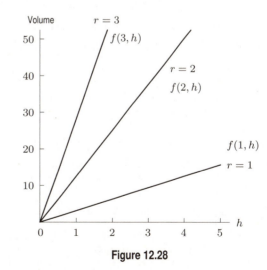

Figure 12.28

37. The function $f(0, t) = \cos t \sin 0 = 0$ gives the displacement of the left end of the string as time varies. Since that point remains stationary, the displacement is zero. The function $f(1, t) = \cos t \sin 1 = 0.84 \cos t$ gives the displacement of the point at $x = 1$ as time varies. Since $\cos t$ oscillates back and forth between 1 and -1, this point moves back and forth with maximum displacement of 0.84 in either direction. Notice the maximum displacements are greatest at $x = \pi/2$ where $\sin x = 1$.

CAS Challenge Problems

41. (a)

$$f(x, f(x, y)) = 3 + x + 2(3 + x + 2y) = (3 + 2 \cdot 3) + (1 + 2)x + 2^2 y = 9 + 3x + 4y$$
$$f(x, f(x, f(x, y))) = 3 + x + 2(3 + x + 2(3 + x + 2y))$$
$$= (3 + 2 \cdot 3 + 2^2 \cdot 3) + (1 + 2 + 2^2)x + 2^3 y = 21 + 7x + 8y$$

(b) From part (a) we guess that the general pattern for k nested fs is

$$(3 + 2 \cdot 3 + 2^2 \cdot 3 + \cdots + 2^{k-1} \cdot 3) + (1 + 2 + 2^2 + \cdots + 2^{k-1})x + 2^k y$$

Thus

$$f(x, f(x, f(x, f(x, f(x, f(x, y)))))) =$$
$$(3 + 2 \cdot 3 + 2^2 \cdot 3 + \cdots + 2^5 \cdot 3) + (1 + 2 + 2^2 + \cdots + 2^5)x + 2^6 y = 189 + 63x + 64y.$$

CHECK YOUR UNDERSTANDING

1. Could not be true. If the origin is on the level curve $z = 1$, then $z = f(0,0) = 1 \neq -1$. So $(0,0)$ cannot be on both $z = 1$ and $z = -1$.

5. True. For every point (x, y), compute the value $z = e^{-(x^2+y^2)}$ at that point. The level curve obtained by getting z equal to that value goes through the point (x, y).

9. False. If, for example, $d = 2$ meters and $H = 57°C$, there could be many times t at which the water temperature is $57\,°C$ at 2 meters depth.

13. True. If there were such an intersection point, that point would have two different temperatures simultaneously.

17. False. The point $(0, 0, 0)$ does not satisfy the equation.

21. True. The x-axis is where $y = z = 0$.

25. False. If $x = 10$, substituting gives $10^2 + y^2 + z^2 = 10$, so $y^2 + z^2 = -90$. Since $y^2 + z^2$ cannot be negative, a point with $x = 10$ cannot satisfy the equation.

29. True. The cross-sections with $y = c$ are of the form $z = 1 - c^2$, which are horizontal lines.

33. False. Wherever $f(x, y) = 0$ the graphs of $f(x, y)$ and $-f(x, y)$ will intersect.

37. True. If $f = c$ then the contours are of the form $c = y^2 + (x - 2)^2$, which are circles centered at $(2, 0)$ if $c > 0$. But if $c = 0$ the contour is the single point $(2, 0)$.

41. False. As a counterexample, consider any function with one variable missing, e.g. $f(x, y) = x^2$. The graph of this is not a plane (it is a *parabolic cylinder*) but has contours which are lines of the form $x = c$.

45. True. The graph of g is the same as the graph of f translated down by 5 units, so the horizontal slice of f at height 5 is the same as the horizontal slice of g at height 0.

49. True. $f(0, 0) = 0$, $f(0, 1) = 4$ give a y slope of 4, but $f(0, 0) = 0$, $f(0, 3) = 5$ give a y slope of $5/3$. Since linearity means the y slope must be the same between any two points, this function cannot be linear.

53. True. Functions can have only one value for a given input, so their graphs can intersect a vertical line at most once. A vertical plane would not satisfy this property, so cannot be the graph of a function.

57. True. Both are the set of all points (x, y, z) in 3-space satisfying $z = x^2 + y^2$.

61. True. The level surfaces are of the form $x + 2y + z = k$, or $z = k - x - 2y$. These are the graphs of the linear functions $f(x, y) = k - x - 2y$, each of which has x-slope of -1 and y-slope equal to -2. Thus they form parallel planes.

65. False. For example, the function $g(x, y, z) = \sin(x + y + z)$ has level surfaces of the form $x + y + z = k$, where $k = \arcsin(c) + n\pi$, for $n = 0, \pm1, \pm2, \ldots$. These surfaces are planes (for $-1 \leq c \leq 1$).

CHAPTER THIRTEEN

Solutions for Section 13.1

Exercises

1. The vectors are $\vec{a} = \vec{i} + 3\vec{j}$, $\vec{b} = 3\vec{i} + 2\vec{j}$, $\vec{v} = -2\vec{i} - 2\vec{j}$, and $\vec{w} = -\vec{i} + 2\vec{j}$.

5. $\vec{a} = \vec{b} = \vec{c} = 3\vec{k}$, $\quad \vec{d} = 2\vec{i} + 3\vec{k}$, $\quad \vec{e} = \vec{j}$, $\quad \vec{f} = -2\vec{i}$

9. $-4\vec{i} + 8\vec{j} - 0.5\vec{i} + 0.5\vec{k} = -4.5\vec{i} + 8\vec{j} + 0.5\vec{k}$

13. $0.6\vec{i} + 0.2\vec{j} - \vec{k} + 0.3\vec{i} + 0.3\vec{k} = 0.9\vec{i} + 0.2\vec{j} - 0.7\vec{k}$

17. $\|\vec{v}\| = \sqrt{1^2 + (-1)^2 + 3^2} = \sqrt{11}$.

21.

$$
\begin{aligned}
5\vec{a} + 2\vec{b} &= 5(2\vec{j} + \vec{k}) + 2(-3\vec{i} + 5\vec{j} + 4\vec{k}) \\
&= (5(2)\vec{j} + 5(1)\vec{k}) + (2(-3)\vec{i} + 2(5)\vec{j} + 2(4)\vec{k}) \\
&= (10\vec{j} + 5\vec{k}) + (-6\vec{i} + 10\vec{j} + 8\vec{k}) = (0-6)\vec{i} + (10+10)\vec{j} + (5+8)\vec{k} \\
&= -6\vec{i} + 20\vec{j} + 13\vec{k}.
\end{aligned}
$$

25.

$$
\begin{aligned}
\|\vec{y} - \vec{x}\| &= \|(4\vec{i} - 7\vec{j}) - (-2\vec{i} + 9\vec{j})\| = \|(4 - (-2))\vec{i} + (-7 - 9)\vec{j}\| = \|6\vec{i} - 16\vec{j}\| \\
&= \sqrt{6^2 + (-16)^2} = \sqrt{36 + 256} = \sqrt{292} = 2\sqrt{73}.
\end{aligned}
$$

Problems

29. If two vectors are parallel, they are scalar multiples of one another. Thus

$$
\frac{a^2}{5a} = \frac{6}{-3}.
$$

Solving for a gives

$$
a^2 = -2 \cdot 5a \quad \text{so} \quad a = 0, -10.
$$

33. (a) The components are $v_1 = 2 \cos \pi/4 = \sqrt{2}$, $v_2 = 2 \sin \pi/4 = \sqrt{2}$. See Figure 13.1. Thus $\vec{v} = \sqrt{2}\vec{i} + \sqrt{2}\vec{j}$.

(b) Since the vector lies in the xz-plane, its y-component is 0. Its x-component is $1 \cos(\frac{\pi}{6})\vec{i} = \frac{\sqrt{3}}{2}\vec{i}$ and its z-component is $1 \sin(\frac{\pi}{6})\vec{j} = \frac{1}{2}\vec{j}$. See Figure 13.2. So the vector is $\frac{\sqrt{3}}{2}\vec{i} + \frac{1}{2}\vec{k}$.

Figure 13.1

Figure 13.2

37.

Figure 13.3

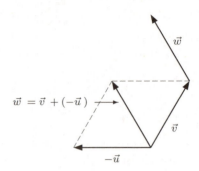

Figure 13.4

Break the hexagon up into 6 equilateral triangles, as shown in Figure 13.3.
Then $\vec{u} - \vec{v} + \vec{w} = \vec{0}$, so $\vec{w} = \vec{v} - \vec{u}$
Similarly, $\vec{x} = -\vec{u}$, $\vec{y} = -\vec{v}$, $\vec{z} = -\vec{w} = \vec{u} - \vec{v}$.

41.

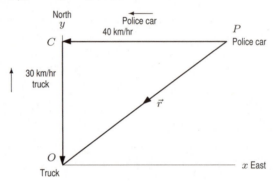

Figure 13.5

Since both vehicles reach the crossroad in exactly one hour, at the present the truck is at O in Figure 13.5; the police car is at P and the crossroads is at C. If $\vec{r}$ is the vector representing the line of sight of the truck with respect to the police car.

$$\vec{r} = -40\vec{i} - 30\vec{j}$$

Solutions for Section 13.2

Exercises

1. Scalar

5. Writing $\vec{P} = (P_1, P_2, \cdots, P_{50})$ where P_i is the population of the i-th state, shows that $\vec{P}$ can be thought of as a vector with 50 components.

9. We need to calculate the length of each vector.

$$\|21\vec{i} + 35\vec{j}\| = \sqrt{21^2 + 35^2} = \sqrt{1666} \approx 40.8,$$
$$\|40\vec{i}\| = \sqrt{40^2} = 40.$$

So the first car is faster.

Problems

13. The velocity vector of the plane with respect to the air has the form

$$\vec{v} = a\vec{i} + 80\vec{k} \text{ where } \|\vec{v}\| = 480.$$

(See Figure 13.6.) Therefore $\sqrt{a^2 + 80^2} = 480$ so $a = \sqrt{480^2 - 80^2} \approx 473.3$ km/hr. We conclude that $\vec{v} \approx 473.3\vec{i} + 80\vec{k}$.

The wind vector is

$$\vec{w} = 100(\cos 45°)\vec{i} + 100(\sin 45°)\vec{j}$$
$$\approx 70.7\vec{i} + 70.7\vec{j}$$

The velocity vector of the plane with respect to the ground is then

$$\vec{v} + \vec{w} = (473.3\vec{i} + 80\vec{k}) + (70.7\vec{i} + 70.7\vec{j})$$
$$= 544\vec{i} + 70.7\vec{j} + 80\vec{k}$$

From Figure 13.7, we see that the velocity relative to the ground is

$$544\vec{i} + 70.7\vec{j}.$$

The ground speed is therefore $\sqrt{544^2 + 70.7^2} \approx 548.6$ km/hr.

Figure 13.6: Side view

Figure 13.7: Top view

17. Let $\vec{R}$ be the resultant force, and let $\vec{F}_1$ and $\vec{F}_2$ be the forces exerted by the larger and smaller tugs. See Figure 13.8. Then $\|\vec{F}_1\| = \frac{5}{4}\|\vec{F}_2\|$. The y components of the vectors $\vec{F}_1$ and $\vec{F}_2$ must cancel each other in order to ensure that the ship travels due east, hence

$$\|\vec{F}_1\| \sin 30° = \|\vec{F}_2\| \sin\theta,$$

so

$$\frac{5}{4}\|\vec{F}_2\| \sin 30° = \|\vec{F}_2\| \sin\theta,$$

giving $\sin\theta = \frac{5}{8}$, and hence $\theta = \sin^{-1}\frac{5}{8} = 38.7°$.

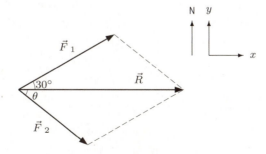

Figure 13.8

21. We want the total force on the object to be zero. We must choose the third force $\vec{F}_3$ so that $\vec{F}_1 + \vec{F}_2 + \vec{F}_3 = 0$. Since $\vec{F}_1 + \vec{F}_2 = 11\vec{i} - 4\vec{j}$, we need $\vec{F}_3 = -11\vec{i} + 4\vec{j}$.

25. (a) Let x-axis be the East direction and y-axis be the North direction. From Figure 13.9,

$$\theta = \sin^{-1}(4/5) = 53.1°.$$

That is, he should steer at $53.1°$ east of south.

Figure 13.9

(b)

Figure 13.10

Let $\vec{R}$ be the resultant of the wind and river velocities, that is

$$\vec{R} = -4\vec{i} + (10\cos(\frac{\pi}{4})\vec{i} + 10\cos(\frac{\pi}{4})\vec{j})$$
$$= (-4 + 5\sqrt{2})\vec{i} + 5\sqrt{2}\vec{j}.$$

From Figure 13.10, we see that to get the the x-component of his rowing velocity and the x-component of $\vec{R}$ to cancel each other, we must have

$$5\sin\theta = -4 + 5\sqrt{2}$$
$$\theta = \sin^{-1}\left(\frac{-4 + 5\sqrt{2}}{5}\right) = 37.9°.$$

However for this value of θ, the y-component of the velocity is

$$5\sqrt{2} - 5\cos(37.9°) = 3.1.$$

Since the y-component is positive, the man will not move across the river in a southward direction.

29.

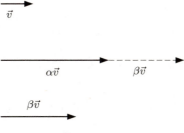

Figure 13.11

The vectors $\vec{v}$, $\alpha\vec{v}$ and $\beta\vec{v}$ are all parallel. Figure 13.11 shows them with $\alpha, \beta > 0$, so all the vectors are in the same direction. Notice that $\alpha\vec{v}$ is a vector α times as long as $\vec{v}$ and $\beta\vec{v}$ is β times as long as $\vec{v}$. Therefore $\alpha\vec{v} + \beta\vec{v}$ is a vector $(\alpha + \beta)$ times as long as $\vec{v}$, and in the same direction. Thus,

$$\alpha\vec{v} + \beta\vec{v} = (\alpha + \beta)\vec{v}.$$

33. According to the definition of scalar multiplication, $1 \cdot \vec{v}$ has the same direction and magnitude as $\vec{v}$, so it is the same as $\vec{v}$.

Solutions for Section 13.3

Exercises

1. $\vec{a} \cdot \vec{y} = (2\vec{j} + \vec{k}) \cdot (4\vec{i} - 7\vec{j}) = -14.$

5. $\vec{c} \cdot \vec{a} + \vec{a} \cdot \vec{y} = (\vec{i} + 6\vec{j}) \cdot (2\vec{j} + \vec{k}) + (2\vec{j} + \vec{k}) \cdot (4\vec{i} - 7\vec{j}) = 12 - 14 = -2.$

9. Since $\vec{c} \cdot \vec{c}$ is a scalar and $(\vec{c} \cdot \vec{c})\vec{a}$ is a vector, the answer to this equation is another scalar. We could calculate $\vec{c} \cdot \vec{c}$, then $(\vec{c} \cdot \vec{c})\vec{a}$, and then take the dot product $((\vec{c} \cdot \vec{c})\vec{a}) \cdot \vec{a}$. Alternatively, we can use the fact that

$$((\vec{c} \cdot \vec{c})\vec{a}) \cdot \vec{a} = (\vec{c} \cdot \vec{c})(\vec{a} \cdot \vec{a}).$$

Since

$$\vec{c} \cdot \vec{c} = (\vec{i} + 6\vec{j}) \cdot (\vec{i} + 6\vec{j}) = 1^2 + 6^2 = 37$$
$$\vec{a} \cdot \vec{a} = (2\vec{j} + \vec{k}) \cdot (2\vec{j} + \vec{k}) = 2^2 + 1^2 = 5,$$

we have,

$$(\vec{c} \cdot \vec{c})(\vec{a} \cdot \vec{a}) = 37(5) = 185$$

13. The equation can be rewritten as

$$z - 5x + 10 = 15 - 3y$$
$$-5x + 3y + z = 5$$

so $\vec{n} = -5\vec{i} + 3\vec{j} + \vec{k}$.

17. (a) Writing the plane in the form $2x + 3y - z = 0$ shows that a normal vector is

$$\vec{n} = 2\vec{i} + 3\vec{j} - \vec{k}.$$

Any multiple of this vector is also a correct answer.

(b) Any vector perpendicular to $\vec{n}$ is parallel to the plane, so one possible answer is

$$\vec{v} = 3\vec{i} - 2\vec{j}.$$

Many other answers are possible.

21. (a) We first find the unit vector in direction $\vec{v}$. Since $||\vec{v}|| = \sqrt{3^2 + 4^2} = 5$, the unit vector in direction of $\vec{v}$ is $\vec{u} = 0.6\vec{i} + 0.8\vec{j}$. Then

$$\vec{F}_{\text{parallel}} = (\vec{F} \cdot \vec{u})\vec{u}$$
$$= (-0.4 \cdot 0.6 + 0.3 \cdot 0.8)\vec{u}$$
$$= \vec{0}.$$

Notice that the component of $\vec{F}$ in direction $\vec{v}$ is equal to $\vec{0}$. This makes sense (and could have been predicted) since $\vec{F}$ is perpendicular to $\vec{v}$.

(b) We have

$$\vec{F}_{\text{perp}} = \vec{F} - \vec{F}_{\text{parallel}} = \vec{F}.$$

(c) Since work is the dot product of the force and displacement vectors, we have

$$W = \vec{F} \cdot \vec{v} = -0.4 \cdot 3 + 0.3 \cdot 4 = 0.$$

Notice that since the force is perpendicular to the displacement, the work done is zero.

25. (a) We first find the unit vector in direction $\vec{v}$. Since $||\vec{v}|| = \sqrt{5^2 + (-1)^2} = \sqrt{26}$, the unit vector in direction of $\vec{v}$ is $\vec{u} = \vec{v}/\sqrt{26}$. Then

$$\vec{F}_{\text{parallel}} = (\vec{F} \cdot \vec{u})\vec{u}$$
$$= (20/\sqrt{26})\vec{u}$$
$$= \frac{100}{26}\vec{i} + \frac{-20}{26}\vec{j}$$
$$= 3.846\vec{i} - 0.769\vec{j}.$$

(b) We have

$$\vec{F}_{\text{perp}} = \vec{F} - \vec{F}_{\text{parallel}} = (-20\vec{j}) - (3.846\vec{i} - 0.769\vec{j}) = -3.846\vec{i} - 19.231\vec{j}.$$

(c) Since work is the dot product of the force and displacement vectors, we have

$$W = \vec{F} \cdot \vec{v} = 20.$$

Problems

29. (a) Perpendicular vectors have a dot product of 0. Since $\vec{a} \cdot \vec{c} = 1(-2) - 3(-1) - 1 \cdot 1 = 0$, and $\vec{b} \cdot \vec{d} = 1(-1) + 1(-1) + 2 \cdot 1 = 0$, the pairs we want are $\vec{a}, \vec{c}$ and $\vec{b}, \vec{d}$.

(b) Parallel vectors are multiples of one another, so there are no parallel vectors in this set.

(c) Since $\vec{v} \cdot \vec{w} = ||\vec{v}|| ||\vec{w}|| \cos\theta$, the dot product of the vectors we want is positive. We have

$$\vec{a} \cdot \vec{b} = 1 \cdot 1 - 3 \cdot 1 - 1 \cdot 2 = -4$$
$$\vec{a} \cdot \vec{d} = 1(-1) - 3(-1) - 1 \cdot 1 = 1$$
$$\vec{b} \cdot \vec{c} = 1(-2) + 3(-1) + 2 \cdot 1 = -1$$
$$\vec{c} \cdot \vec{d} = -2(-1) - 1(-1) + 1 \cdot 1 = 4,$$

and we already know $\vec{a} \cdot \vec{c} = \vec{b} \cdot \vec{d} = 0$. Thus, the pairs of vectors with an angle of less than $\pi/2$ between them are $\vec{a}, \vec{d}$ and $\vec{c}, \vec{d}$.

(d) Vectors with an angle of more than $\pi/2$ between them have a negative dot product, so pairs are $\vec{a}, \vec{b}$ and $\vec{b}, \vec{c}$.

33. If the vectors are perpendicular, we need

$$\vec{v} \cdot \vec{w} = (2a\vec{i} - a\vec{j} + 16\vec{k}) \cdot (5\vec{i} + a\vec{j} - \vec{k}) = 10a - a^2 - 16 = 0.$$

Solving $10a - a^2 - 16 = -(a - 2)(a - 8) = 0$ gives $a = 2, 8$.

37. The plane is

$$3x - y + 4z = 3 \cdot 1 - 1 \cdot 5 + 4 \cdot 2$$
$$3x - y + 4z = 6.$$

41. Two planes are parallel if their normal vectors are parallel. Since the plane $3x + y + z = 4$ has normal vector $\vec{n} = 3\vec{i} + \vec{j} + \vec{k}$, the plane we are looking for has the same normal vector and passes through the point $(-2, 3, 2)$. Thus, it has the equation

$$3x + y + z = 3 \cdot (-2) + 3 + 2 = -1.$$

45. (a) The points A, B and C are shown in Figure 13.12.

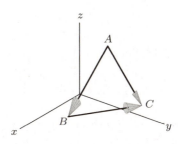

Figure 13.12

First, we calculate the vectors which form the sides of this triangle:
$$\vec{AB} = (4\vec{i} + 2\vec{j} + \vec{k}) - (2\vec{i} + 2\vec{j} + 2\vec{k}) = 2\vec{i} - \vec{k}$$
$$\vec{BC} = (2\vec{i} + 3\vec{j} + \vec{k}) - (4\vec{i} + 2\vec{j} + \vec{k}) = -2\vec{i} + \vec{j}$$
$$\vec{AC} = (2\vec{i} + 3\vec{j} + \vec{k}) - (2\vec{i} + 2\vec{j} + 2\vec{k}) = \vec{j} - \vec{k}$$

Now we calculate the lengths of each of the sides of the triangles:
$$\|\vec{AB}\| = \sqrt{2^2 + (-1)^2} = \sqrt{5}$$
$$\|\vec{BC}\| = \sqrt{(-2)^2 + 1^2} = \sqrt{5}$$
$$\|\vec{AC}\| = \sqrt{1^2 + (-1)^2} = \sqrt{2}$$
Thus the length of the shortest side of S is $\sqrt{2}$.

(b) $\cos \angle BAC = \dfrac{\vec{AB} \cdot \vec{AC}}{\|\vec{AB}\| \cdot \|\vec{AC}\|} = \dfrac{2 \cdot 0 + 0 \cdot 1 + (-1) \cdot (-1)}{\sqrt{5} \cdot \sqrt{2}} \approx 0.32$

49. (a) The speed of the current is $\|\vec{c}\| = \sqrt{5} = 2.24$ m/sec.

(b) The speed of the current in the direction of the canoe's motion is the component of $\vec{c}$ in the direction of $\vec{v}$. This is given by:

$$\text{Speed of current in direction of canoe's motion} = \frac{\vec{c} \cdot \vec{v}}{\|\vec{v}\|} = \frac{(1)(5) + (2)(3)}{\sqrt{5^2 + 3^2}}$$
$$= \frac{11}{\sqrt{34}}$$
$$= 1.89 \text{ m/sec.}$$

Notice that the speed of the current in the direction of the canoe is less than the speed of the current in the direction in which the current is moving.

53. Let $\vec{u} = 3\vec{i} + 4\vec{j}$ and $\vec{v} = 5\vec{i} - 12\vec{j}$. We seek a vector $\vec{w} = x\vec{i} + y\vec{j}$ such that the cosine of the angle between $\vec{u}$ and $\vec{w}$ equals the cosine of the angle between $\vec{v}$ and $\vec{w}$. Thus

$$\frac{\vec{u} \cdot \vec{w}}{\|\vec{u}\| \|\vec{w}\|} = \frac{\vec{v} \cdot \vec{w}}{\|\vec{v}\| \|\vec{w}\|}$$

or

$$\frac{3x + 4y}{5\sqrt{x^2 + y^2}} = \frac{5x - 12y}{13\sqrt{x^2 + y^2}}.$$

Simplifying, we have $x = -8y$. The vector we want is of the form $\vec{w} = -8y\vec{i} + y\vec{j}$, but should we take $y > 0$ or $y < 0$? The smaller of the two angles formed by $\vec{u}$ and $\vec{v}$ is between $0°$ and $180°$, and so $\vec{w}$ must make an acute angle with $\vec{u}$ and $\vec{v}$. If $y > 0$ then $\vec{u} \cdot \vec{w} = -20y < 0$ indicating an obtuse angle and if $y < 0$ then $\vec{u} \cdot \vec{w} = -20y > 0$ indicating an acute angle. We have $\vec{w} = -8y\vec{i} + y\vec{j}$ with $y < 0$. Thus $\vec{w}$ can be any positive multiple of the vector $8\vec{i} - \vec{j}$.

57. Since $\vec{u} \cdot \vec{w} = \vec{v} \cdot \vec{w}$, $(\vec{u} - \vec{v}) \cdot \vec{w} = 0$. This equality holds for any $\vec{w}$, so we can take $\vec{w} = \vec{u} - \vec{v}$. This gives

$$\|\vec{u} - \vec{v}\|^2 = (\vec{u} - \vec{v}) \cdot (\vec{u} - \vec{v}) = 0,$$

that is,

$$\|\vec{u} - \vec{v}\| = 0.$$

This implies $\vec{u} - \vec{v} = 0$, that is, $\vec{u} = \vec{v}$.

61. Let $\vec{u}$ and $\vec{v}$ be the displacement vectors from C to the other two vertices. Then

$$
\begin{aligned}
c^2 &= \|\vec{u} - \vec{v}\|^2 \\
&= (\vec{u} - \vec{v}) \cdot (\vec{u} - \vec{v}) \\
&= \vec{u} \cdot \vec{u} - \vec{v} \cdot \vec{u} - \vec{u} \cdot \vec{v} + \vec{v} \cdot \vec{v} \\
&= \|\vec{u}\|^2 - 2\|u\|\|v\|\cos C + \|\vec{v}\|^2 \\
&= a^2 - 2ab\cos C + b^2
\end{aligned}
$$

Solutions for Section 13.4

Exercises

1. $\vec{v} \times \vec{w} = \vec{k} \times \vec{j} = -\vec{i}$ (remember $\vec{i}, \vec{j}, \vec{k}$ are unit vectors along the axes, and you must use the right hand rule.)

5. $\vec{v} = 2\vec{i} - 3\vec{j} + \vec{k}$, and $\vec{w} = \vec{i} + 2\vec{j} - \vec{k}$

$$\vec{v} \times \vec{w} = \begin{vmatrix} \vec{i} & \vec{j} & \vec{k} \\ 2 & -3 & 1 \\ 1 & 2 & -1 \end{vmatrix} = \vec{i} + 3\vec{j} + 7\vec{k}$$

9.

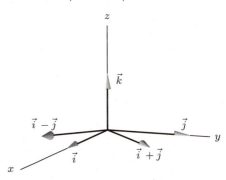

Figure 13.13

By definition, $(\vec{i} + \vec{j}) \times (\vec{i} - \vec{j})$ is in the direction of $-\vec{k}$. The magnitude is

$$\|\vec{i} + \vec{j}\| \cdot \|\vec{i} - \vec{j}\| \sin\frac{\pi}{2} = \sqrt{2} \cdot \sqrt{2} \cdot 1 = 2.$$

So $(\vec{i} + \vec{j}) \times (\vec{i} - \vec{j}) = -2\vec{k}$. See Figure 13.13.

13. The displacement vector from $(3, 4, 2)$ to $(-2, 1, 0)$ is:

$$\vec{a} = -5\vec{i} - 3\vec{j} - 2\vec{k}.$$

The displacement vector from $(3, 4, 2)$ to $(0, 2, 1)$ is:

$$\vec{b} = -3\vec{i} - 2\vec{j} - \vec{k}.$$

Therefore the vector normal to the plane is:

$$\vec{n} = \vec{a} \times \vec{b} = -\vec{i} + \vec{j} + \vec{k}.$$

Using the first point, the equation of the plane can be written as:

$$-(x - 3) + (y - 4) + (z - 2) = 0.$$

The equation of the plane is thus:

$$-x + y + z = 3.$$

Problems

17. We use the same normal $\vec{n} = 4\vec{i} + 26\vec{j} + 14\vec{k}$ and the point $(0,0,0)$ to get $4(x-0) + 26(y-0) + 14(z-0) = 0$, or $4x + 26y + 14z = 0$.

21. The normal vectors to the planes are $\vec{n_1} = 2\vec{i} - 3\vec{j} + 5\vec{k}$ and $\vec{n_2} = 4\vec{i} + \vec{j} - 3\vec{k}$. The line of intersection is perpendicular to both normal vectors (picture the pages in a partially open book). Hence the vector we need is $\vec{n_1} \times \vec{n_2} = 4\vec{i} + 26\vec{j} + 14\vec{k}$.

25. (a) If we let $\overrightarrow{PQ}$ in Figure 13.14 be the vector from point P to point Q and $\overrightarrow{PR}$ be the vector from P to R, then

$$\overrightarrow{PQ} = -\vec{i} + 2\vec{k}$$

$$\overrightarrow{PR} = 2\vec{i} - \vec{k},$$

then the area of the parallelogram determined by $\overrightarrow{PQ}$ and $\overrightarrow{PR}$ is:

$$\text{Area of parallelogram} = \|\overrightarrow{PQ} \times \overrightarrow{PR}\| = \left\| \begin{vmatrix} \vec{i} & \vec{j} & \vec{k} \\ -1 & 0 & 2 \\ 2 & 0 & -1 \end{vmatrix} \right\| = \|3\vec{j}\| = 3.$$

Thus, the area of the triangle PQR is

$$\begin{pmatrix} \text{Area of} \\ \text{triangle} \end{pmatrix} = \frac{1}{2} \begin{pmatrix} \text{Area of} \\ \text{parallelogram} \end{pmatrix} = \frac{3}{2} = 1.5.$$

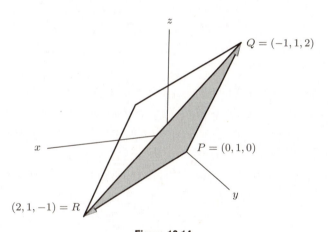

Figure 13.14

(b) Since $\vec{n} = \overrightarrow{PQ} \times \overrightarrow{PR}$ is perpendicular to the plane PQR, and from above, we have $\vec{n} = 3\vec{j}$, the equation of the plane has the form $3y = C$. At the point $(0,1,0)$ we get $3 = C$, therefore $3y = 3$, i.e., $y = 1$.

29. Since

$$\|\vec{v} \times \vec{w}\| = \|\vec{v}\| \cdot \|\vec{w}\| \sin\theta,$$

and

$$\vec{v} \cdot \vec{w} = \|\vec{v}\| \cdot \|\vec{w}\| \cos\theta,$$

so

$$\frac{\|\vec{v} \times \vec{w}\|}{\vec{v} \cdot \vec{w}} = \frac{\|\vec{v}\| \cdot \|\vec{w}\| \sin\theta}{\|\vec{v}\| \cdot \|\vec{w}\| \cos\theta} = \tan\theta,$$

so

$$\tan\theta = \frac{\|2\vec{i} - 3\vec{j} + 5\vec{k}\|}{3} = \frac{\sqrt{38}}{3} = 2.055.$$

33. The quantities $\left|\vec{a} \cdot (\vec{b} \times \vec{c})\right|$ and $\left|(\vec{a} \times \vec{b}) \cdot \vec{c}\right|$ both represent the volume of the same parallelepiped, namely that defined by the three vectors $\vec{a}$, $\vec{b}$, and $\vec{c}$, and therefore must be equal. Thus, the two triple products $\vec{a} \cdot (\vec{b} \times \vec{c})$ and $(\vec{a} \times \vec{b}) \cdot \vec{c}$ must be equal except perhaps for their sign. In fact, both are positive if $\vec{a}$, $\vec{b}$, $\vec{c}$ are right-handed and negative if $\vec{a}$, $\vec{b}$, $\vec{c}$ are left-handed. This can be shown by drawing a picture:

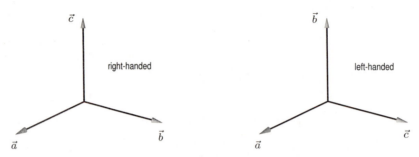

Figure 13.15

37. Write $\vec{v}$ and $\vec{w}$ in components and expand using the distributive property of the cross product.

$$\vec{v} \times \vec{w} = (v_1\vec{i} + v_2\vec{j} + v_3\vec{k}) \times (w_1\vec{i} + w_2\vec{j} + w_3\vec{k})$$
$$= v_1w_1\vec{i} \times \vec{i} + v_1w_2\vec{i} \times \vec{j} + v_1w_3\vec{i} \times \vec{k}$$
$$+ v_2w_1\vec{j} \times \vec{i} + v_2w_2\vec{j} \times \vec{j} + v_2w_3\vec{j} \times \vec{k}$$
$$+ v_3w_1\vec{k} \times \vec{i} + v_3w_2\vec{k} \times \vec{j} + v_3w_3\vec{k} \times \vec{k}$$

Now we use the fact that $\vec{i} \times \vec{i} = \vec{0}, \vec{i} \times \vec{j} = \vec{k}, \vec{i} \times \vec{k} = -\vec{j}, \vec{j} \times \vec{i} = -\vec{k}, \vec{j} \times \vec{j} = \vec{0}, \vec{j} \times \vec{k} = \vec{i}, \vec{k} \times \vec{i} = \vec{j}, \vec{k} \times \vec{j} = -\vec{i}, \vec{k} \times \vec{k} = \vec{0}$. Thus we have

$$\vec{v} \times \vec{w} = \vec{0} + v_1w_2\vec{k} + v_1w_3(-\vec{j}) + v_2w_1(-\vec{k}) + \vec{0} + v_2w_3\vec{i} + v_3w_1\vec{j} + v_3w_2(-\vec{i}) + \vec{0}$$
$$= (v_2w_3 - v_3w_2)\vec{i} + (v_3w_1 - v_1w_3)\vec{j} + (v_1w_2 - v_2w_1)\vec{k}.$$

41. The area vector for face $OAB = \frac{1}{2}\vec{b} \times \vec{a}$.
The area vector for face $OBC = \frac{1}{2}\vec{a} \times \vec{c}$.
The area vector for face $OAC = \frac{1}{2}\vec{b} \times \vec{c}$.
The area vector for face $ABC = \frac{1}{2}(\vec{b} - \vec{a}) \times (\vec{c} - \vec{a})$.

$$\frac{1}{2}\vec{b} \times \vec{a} + \frac{1}{2}\vec{c} \times \vec{b} + \frac{1}{2}\vec{a} \times \vec{c} + \frac{1}{2}(\vec{b} - \vec{a}) \times (\vec{c} - \vec{a}) =$$
$$\frac{1}{2}\vec{b} \times \vec{a} + \frac{1}{2}\vec{c} \times \vec{b} + \frac{1}{2}\vec{a} \times \vec{c} + \frac{1}{2}(\vec{b} \times \vec{c} - \vec{b} \times \vec{a} - \vec{a} \times \vec{c} - \vec{a} \times \vec{a}) = 0.$$

45. (a) Since
$$\vec{u} \times \vec{v} = (u_2v_3 - u_3v_2)\vec{i} + (u_3v_1 - u_1v_3)\vec{j} + (u_1v_2 - u_2v_1)\vec{k},$$

we have
$$\text{Area of } S = \|\vec{u} \times \vec{v}\| = \left((u_2v_3 - u_3v_2)^2 + (u_3v_1 - u_1v_3)^2 + (u_1v_2 - u_2v_1)^2\right)^{1/2}.$$

(b) The two edges of R are given by the projections of $\vec{u}$ and $\vec{v}$ onto the xy-plane. These are the vectors $\vec{U}$ and $\vec{V}$, obtained by omitting the $\vec{k}$-components of $\vec{u}$ and $\vec{v}$: we have $\vec{U} = u_1\vec{i} + u_2\vec{j}$ and $\vec{V} = v_1\vec{i} + v_2\vec{j}$, Thus

$$\text{Area of } R = \|\vec{U} \times \vec{V}\| = \|(u_1v_2 - u_2v_1)\vec{k}\| = |u_1v_2 - u_2v_1|.$$

(c) The vector $m\vec{i} + n\vec{j} - \vec{k}$ is normal to the plane $z = mx + ny + c$. Since the vectors $\vec{u}$ and $\vec{v}$ are in the plane (they're the sides of $\vec{S}$), the vector $\vec{u} \times \vec{v}$ is also normal to the plane. Thus, these two vectors are scalar multiples of one another. Suppose

$$\vec{u} \times \vec{v} = \lambda(m\vec{i} + n\vec{j} - \vec{k})$$

Since the $\vec{k}$ component of $\vec{u} \times \vec{v}$ is $(u_1 v_2 - u_2 v_1)\vec{k}$, comparing the $\vec{k}$-components tells us that

$$\lambda = -(u_1 v_2 - u_2 v_1).$$

Thus,

$$-(u_1 v_2 - u_2 v_1)(m\vec{i} + n\vec{j} - \vec{k}) = \vec{u} \times \vec{v} = (u_2 v_3 - u_3 v_2)\vec{i} + (u_3 v_1 - u_1 v_3)\vec{j} + (u_1 v_2 - u_2 v_1)\vec{k},$$

so

$$m = \frac{u_2 v_3 - u_3 v_2}{u_2 v_1 - u_1 v_2}$$

$$n = \frac{u_3 v_1 - u_1 v_3}{u_2 v_1 - u_1 v_2}.$$

(d) We have

$$(1 + m^2 + n^2) \cdot (\text{Area of } R)^2 = \left(1 + \left(\frac{u_2 v_3 - u_3 v_2}{u_2 v_1 - u_1 v_2}\right)^2 + \left(\frac{u_3 v_1 - u_1 v_3}{u_2 v_1 - u_1 v_2}\right)^2\right)(u_1 v_2 - u_2 v_1)^2$$

$$= (u_2 v_3 - u_3 v_2)^2 + (u_3 v_1 - u_1 v_3)^2 + (u_1 v_2 - u_2 v_1)^2$$

$$= (\text{Area of } S)^2.$$

Solutions for Chapter 13 Review

Exercises

1. Scalar. $\vec{u} \cdot \vec{v} = (2\vec{i} - 3\vec{j} - 4\vec{k}) \cdot (\vec{k} - \vec{j}) = 2 \cdot 0 - 3(-1) - 4 \cdot 1 = -1.$

5. $5\vec{c} = 5\vec{i} + 30\vec{j}$

9. $3\vec{v} - \vec{w} - \vec{v} = 2\vec{v} - \vec{w} = 2(2\vec{i} + 3\vec{j} - \vec{k}) - (\vec{i} - \vec{j} + 2\vec{k}) = 3\vec{i} + 7\vec{j} - 4\vec{k}.$

13. For any vector $\vec{v}$, we have $\vec{v} \times \vec{v} = \vec{0}$.

17. The cross product of two parallel vectors is $\vec{0}$, so the cross product of any vector with itself is $\vec{0}$.

21. (a) We have $\vec{v} \cdot \vec{w} = 3 \cdot 4 + 2 \cdot (-3) + (-2) \cdot 1 = 4.$

(b) We have $\vec{v} \times \vec{w} = -4\vec{i} - 11\vec{j} - 17\vec{k}.$

(c) A vector of length 5 parallel to $\vec{v}$ is

$$\frac{5}{\|\vec{v}\|}\vec{v} = \frac{5}{\sqrt{17}}(3\vec{i} + 2\vec{j} - 2\vec{k}) = 3.64\vec{i} + 2.43\vec{j} - 2.43\vec{k}.$$

(d) The angle between vectors $\vec{v}$ and $\vec{w}$ is found using

$$\cos\theta = \frac{\vec{v} \cdot \vec{w}}{\|\vec{v}\|\|\vec{w}\|} = \frac{4}{\sqrt{17}\sqrt{26}} = 0.190,$$

so $\theta = 79.0°$.

(e) The component of vector $\vec{v}$ in the direction of vector $\vec{w}$ is

$$\frac{\vec{v} \cdot \vec{w}}{\|\vec{w}\|} = \frac{4}{\sqrt{26}} = 0.784.$$

(f) The answer is any vector $\vec{a}$ such that $\vec{a} \cdot \vec{v} = 0$. One possible answer is $2\vec{i} - 2\vec{j} + \vec{k}$.

(g) A vector perpendicular to both is the cross product:

$$\vec{v} \times \vec{w} = -4\vec{i} - 11\vec{j} - 17\vec{k}.$$

25. $\vec{n} = 4\vec{i} + 6\vec{k}$ (the coefficients of x, y, z are the same as the coefficients of $\vec{i}, \vec{j}$, and $\vec{k}$.)

29. Since $\vec{F} = 2\vec{d}$, the two vectors are parallel in the same direction, so

$$\vec{F}_{\text{parallel}} = \vec{F} \text{ and } \vec{F}_{\text{perp}} = \vec{0}.$$

The work done is

$$W = \vec{F} \cdot \vec{d} = 2 + 8 = 10.$$

Notice that this is the same as the magnitude of the force, $\|\vec{F}\| = \sqrt{20}$, times the distance traveled, $\|\vec{d}\| = \sqrt{5}$, since the force is the same direction as the displacement.

33. The unit vector in the direction of $\vec{d}$ is $\vec{u} = (1/\sqrt{2})(\vec{i} + \vec{j})$. Thus

$$\vec{F}_{\text{parallel}} = \left(\vec{F} \cdot \vec{u}\right)\vec{u} = \frac{2}{\sqrt{2}}\vec{u} = \vec{i} + \vec{j},$$

$$\vec{F}_{\text{perp}} = \vec{F} - \vec{F}_{\text{parallel}} = \vec{i} - \vec{j}.$$

Notice that $\vec{F}_{\text{perp}} \cdot \vec{u} = 0$, as we expect. The work done is

$$W = \vec{F} \cdot \vec{d} = 2 - 0 = 2.$$

Problems

37. (a) The velocity vector for the boat is $\vec{b} = 25\vec{i}$ and the velocity vector for the current is

$$\vec{c} = -10\cos(45°)\vec{i} - 10\sin(45°)\vec{j} = -7.07\vec{i} - 7.07\vec{j}.$$

The actual velocity of the boat is

$$\vec{b} + \vec{c} = 17.93\vec{i} - 7.07\vec{j}.$$

(b) $\|\vec{b} + \vec{c}\| = 19.27$ km/hr.

(c) We see in Figure 13.16 that $\tan\theta = \dfrac{7.07}{17.93}$, so $\theta = 21.52°$ south of east.

Figure 13.16

41. Vectors $\vec{v}_1$, $\vec{v}_4$, and $\vec{v}_8$ are all parallel to each other. Vectors $\vec{v}_3$, $\vec{v}_5$, and $\vec{v}_7$ are all parallel to each other, and are all perpendicular to the vectors in the previous sentence. Vectors $\vec{v}_2$ and $\vec{v}_9$ are perpendicular.

45. Since the plane is normal to the vector $5\vec{i} + \vec{j} - 2\vec{k}$ and passes through the point $(0, 1, -1)$, an equation for the plane is

$$5x + y - 2z = 5 \cdot 0 + 1 \cdot 1 + (-2) \cdot (-1) = 3$$
$$5x + y - 2z = 3.$$

49. (a) Since

$$\vec{PQ} = (3\vec{i} + 5\vec{j} + 7\vec{k}) - (\vec{i} + 2\vec{j} + 3\vec{k}) = 2\vec{i} + 3\vec{j} + 4\vec{k},$$

and

$$\vec{PR} = (2\vec{i} + 5\vec{j} + 3\vec{k}) - (\vec{i} + 2\vec{j} + 3\vec{k}) = \vec{i} + 3\vec{j},$$

$$\vec{PQ} \times \vec{PR} = \begin{vmatrix} \vec{i} & \vec{j} & \vec{k} \\ 2 & 3 & 4 \\ 1 & 3 & 0 \end{vmatrix} = -12\vec{i} + 4\vec{j} + 3\vec{k},$$

which is a vector perpendicular to the plane containing P, Q and R. Since

$$\|\vec{PQ} \times \vec{PR}\| = \sqrt{(-12)^2 + 4^2 + 3^2} = 13,$$

the unit vectors which are perpendicular to a plane containing P, Q, and R are

$$-\frac{12}{13}\vec{i} + \frac{4}{13}\vec{j} + \frac{3}{13}\vec{k},$$

or the unit vector pointing to the opposite direction,

$$\frac{12}{13}\vec{i} - \frac{4}{13}\vec{j} - \frac{3}{13}\vec{k}.$$

(b) The angle between PQ and PR is θ for which

$$\cos\theta = \frac{\overrightarrow{PQ}\cdot\overrightarrow{PR}}{\|\overrightarrow{PQ}\|\cdot\|\overrightarrow{PR}\|} = \frac{2\cdot 1 + 3\cdot 3 + 4\cdot 0}{\sqrt{2^2+3^2+4^2}\cdot\sqrt{1^2+3^2+0^2}} = \frac{11}{\sqrt{290}},$$

so

$$\theta = \cos^{-1}\left(\frac{11}{\sqrt{290}}\right) \approx 49.76°.$$

(c) The area of triangle $PQR = \frac{1}{2}\|\overrightarrow{PQ}\times\overrightarrow{PR}\| = \frac{13}{2}$.

(d) Let d be the distance from R to the line through P and Q (see Figure 13.17), then

$$\frac{1}{2}d\cdot\|\overrightarrow{PQ}\| = \text{the area of } \triangle PQR = \frac{13}{2}.$$

Therefore,

$$d = \frac{13}{\|\overrightarrow{PQ}\|} = \frac{13}{\sqrt{2^2+3^2+4^2}} = \frac{13}{\sqrt{29}}.$$

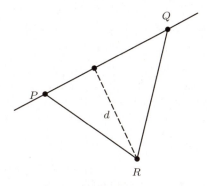

Figure 13.17

53. (a) On the x-axis, $y = z = 0$, so $5x = 21$, giving $x = \frac{21}{5}$. So the only such point is $\left(\frac{21}{5}, 0, 0\right)$.

(b) Other points are $(0, -21, 0)$, and $(0, 0, 3)$. There are many other possible answers.

(c) $\vec{n} = 5\vec{i} - \vec{j} + 7\vec{k}$. It is the normal vector.

(d) The vector between two points in the plane is parallel to the plane. Using the points from part (b), the vector $3\vec{k} - (-21\vec{j}) = 21\vec{j} + 3\vec{k}$ is parallel to the plane.

57. (a) Suppose $\vec{v} = \overrightarrow{OP}$ as in Figure 13.18. The $\vec{i}$ component of $\overrightarrow{OP}$ is the projection of $\overrightarrow{OP}$ on the x-axis:

$$\overrightarrow{OT} = v\cos\alpha\vec{i}.$$

Similarly, the $\vec{j}$ and $\vec{k}$ components of $\overrightarrow{OP}$ are the projections of $\overrightarrow{OP}$ on the y-axis and the z-axis respectively. So:

$$\overrightarrow{OS} = v\cos\beta\vec{j}$$
$$\overrightarrow{OQ} = v\cos\gamma\vec{k}$$

Since $\vec{v} = \overrightarrow{OT} + \overrightarrow{OS} + \overrightarrow{OQ}$, we have

$$\vec{v} = v\cos\alpha\vec{i} + v\cos\beta\vec{j} + v\cos\gamma\vec{k}.$$

(b) Since

$$v^2 = \vec{v}\cdot\vec{v} = (v\cos\alpha\vec{i} + v\cos\beta\vec{j} + v\cos\gamma\vec{k})\cdot$$
$$(v\cos\alpha\vec{i} + v\cos\beta\vec{j} + v\cos\gamma\vec{k})$$
$$= v^2(\cos^2\alpha + \cos^2\beta + \cos^2\gamma)$$

so

$$\cos^2\alpha + \cos^2\beta + \cos^2\gamma = 1.$$

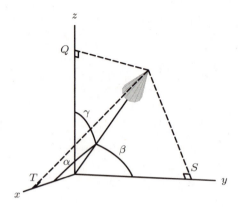

Figure 13.18

CAS Challenge Problems

61. (a) From the geometric definition of the dot product, we have

$$\cos\theta = \frac{|\vec{a}\cdot\vec{b}|}{\|\vec{a}\|\|\vec{b}\|} = \frac{10}{\sqrt{14}\sqrt{9}}.$$

Using $\sin^2\theta = 1 - \cos^2\theta$, we get

$$x + 2y + 3z = 0$$
$$2x + y + 2z = 0$$
$$x^2 + y^2 + z^2 = \|\vec{a}\|^2\|\vec{b}\|^2(1 - \cos^2\theta) = (14)(9)\left(1 - \frac{100}{(14)(9)}\right)$$

Solving these equations we get $x = -1, y = -4, z = 3$ or $x = 1, y = 4$, and $z = -3$. Thus $\vec{c} = -\vec{i} - 4\vec{j} + 3\vec{k}$ or $\vec{c} = \vec{i} + 4\vec{j} - 3\vec{k}$.
   (b) $\vec{a}\times\vec{b} = \vec{i} + 4\vec{j} - 3\vec{k}$. This is the same as one of the answers in part (a). The conditions in part (a) ensured that $\vec{c}$ is perpendicular to $\vec{a}$ and $\vec{b}$ and that it has magnitude $\|\vec{a}\|\|\vec{b}\||\sin\theta|$. The cross product is the solution that, in addition, satisfies the right-hand rule.

CHECK YOUR UNDERSTANDING

1. False. There are exactly two unit vectors: one in the same direction as $\vec{v}$ and the other in the opposite direction. Explicitly, the unit vectors parallel to $\vec{v}$ are $\pm\frac{1}{\|\vec{v}\|}\vec{v}$.

5. False. If $\vec{v}$ and $\vec{w}$ are not parallel, the three vectors $\vec{v}$, $\vec{w}$ and $\vec{v} - \vec{w}$ can be thought of as three sides of a triangle. (If the tails of $\vec{v}$ and $\vec{w}$ are placed together, then $\vec{v} - \vec{w}$ is a vector from the head of $\vec{w}$ to the head of $\vec{v}$.) The length of one side of a triangle is less than the sum of the lengths of the other two sides. Alternatively, a counterexample is $\vec{v} = \vec{i}$ and $\vec{w} = \vec{j}$. Then $\|\vec{i} - \vec{j}\| = \sqrt{2}$ but $\|\vec{i}\| - \|\vec{j}\| = 0$.

9. False. To find the displacement vector *from* $(1, 1, 1)$ *to* $(1, 2, 3)$ we subtract $\vec{i} + \vec{j} + \vec{k}$ from $\vec{i} + 2\vec{j} + 3\vec{k}$ to get $(1 - 1)\vec{i} + (2 - 1)\vec{j} + (3 - 1)\vec{k} = \vec{j} + 2\vec{k}$.

13. True. The cosine of the angle between the vectors is negative when the angle is between $\pi/2$ and π.

17. False. If the vectors are nonzero and perpendicular, the dot product will be zero (e.g. $\vec{i}\cdot\vec{j} = 0$).

21. True. The cross product yields a vector.

25. False. If $\vec{u}$ and $\vec{w}$ are two different vectors both of which are parallel to $\vec{v}$, then $\vec{v}\times\vec{u} = \vec{v}\times\vec{w} = \vec{0}$, but $\vec{u}\neq\vec{w}$. A counterexample is $\vec{v} = \vec{i}$, $\vec{u} = 2\vec{i}$ and $\vec{w} = 3\vec{i}$.

29. True. Any vector $\vec{w}$ that is parallel to $\vec{v}$ will give $\vec{v}\times\vec{w} = \vec{0}$.

CHAPTER FOURTEEN

Solutions for Section 14.1

Exercises

1. If h is small, then

$$f_x(3, 2) \approx \frac{f(3 + h, 2) - f(3, 2)}{h}.$$

With $h = 0.01$, we find

$$f_x(3, 2) \approx \frac{f(3.01, 2) - f(3, 2)}{0.01} = \frac{\frac{3.01^2}{(2+1)} - \frac{3^2}{(2+1)}}{0.01} = 2.00333.$$

With $h = 0.0001$, we get

$$f_x(3, 2) \approx \frac{f(3.0001, 2) - f(3, 2)}{0.0001} = \frac{\frac{3.0001^2}{(2+1)} - \frac{3^2}{(2+1)}}{0.0001} = 2.0000333.$$

Since the difference quotient seems to be approaching 2 as h gets smaller, we conclude

$$f_x(3, 2) \approx 2.$$

To estimate $f_y(3, 2)$, we use

$$f_y(3, 2) \approx \frac{f(3, 2 + h) - f(3, 2)}{h}.$$

With $h = 0.01$, we get

$$f_y(3, 2) \approx \frac{f(3, 2.01) - f(3, 2)}{0.01} = \frac{\frac{3^2}{(2.01+1)} - \frac{3^2}{(2+1)}}{0.01} = -0.99668.$$

With $h = 0.0001$, we get

$$f_y(3, 2) \approx \frac{f(3, 2.0001) - f(3, 2)}{0.0001} = \frac{\frac{3^2}{(2.0001+1)} - \frac{3^2}{(2+1)}}{0.0001} = -0.9999667.$$

Thus, it seems that the difference quotient is approaching -1, so we estimate

$$f_y(3, 2) \approx -1.$$

5. $\partial P/\partial t$: The unit is dollars per month. This is the rate at which payments change as the number of months it takes to pay off the loan changes. The sign is negative because payments decrease as the pay-off time increases.

$\partial P/\partial r$: The unit is dollars per percentage point. This is the rate at which payments change as the interest rate changes. The sign is positive because payments increase as the interest rate increases.

9. Moving right from P in the direction of increasing x increases f, so $f_x(P) > 0$.

Moving up from P in the direction of increasing y increases f, so $f_y(P) > 0$.

13. For $f_w(10, 25)$ we get

$$f_w(10, 25) \approx \frac{f(10 + h, 25) - f(10, 25)}{h}.$$

Choosing $h = 5$ and reading values from Table 12.2 on page 644 of the text, we get

$$f_w(10, 25) \approx \frac{f(15, 25) - f(10, 25)}{5} = \frac{13 - 15}{5} = -0.4°\text{F/mph}.$$

This means that when the wind speed is 10 mph and the true temperature is $25°$F, as the wind speed increases from 10 mph by 1 mph we feel an approximately $0.4°$F drop in temperature. This rate is negative because the temperature you feel drops as the wind speed increases.

Problems

17. The values of z increase as we move in the direction of increasing x-values, so f_x is positive. The values of z decrease as we move in the direction of increasing y-values, so f_y is negative. We see in the contour diagram that $f(2, 1) = 10$. We estimate the partial derivatives:

$$f_x(2, 1) \approx \frac{\Delta z}{\Delta x} = \frac{14 - 10}{4 - 2} = 2,$$

$$f_y(2, 1) \approx \frac{\Delta z}{\Delta y} = \frac{6 - 10}{2 - 1} = -4.$$

21. (a) (i) Near A, the value of z increases as x increases, so $f_x(A) > 0$.

(ii) Near A, the value of z decreases as y increases, so $f_y(A) < 0$.

(b) $f_x(P)$ changes from positive to negative as P moves from A to B along a straight line, because after P crosses the y-axis, z decreases as x increases near P.

$f_y(P)$ does not change sign as P moves from A to B along a straight line; it is negative along AB.

25. (a) Estimating $T(x, t)$ from the figure in the text at $x = 15$, $t = 20$ gives

$$\left.\frac{\partial T}{\partial x}\right|_{(15,20)} \approx \frac{T(23, 20) - T(15, 20)}{23 - 15} = \frac{20 - 23}{8} = -\frac{3}{8} \,^\circ\text{C per m,}$$

$$\left.\frac{\partial T}{\partial t}\right|_{(15,20)} \approx \frac{T(15, 25) - T(15, 20)}{25 - 20} = \frac{25 - 23}{5} = \frac{2}{5} \,^\circ\text{C per min.}$$

At 15 m from heater at time $t = 20$ min, the room temperature decreases by approximately $3/8^\circ$C per meter and increases by approximately $2/5^\circ$C per minute.

(b) We have the estimates,

$$\left.\frac{\partial T}{\partial x}\right|_{(5,12)} \approx \frac{T(7, 12) - T(5, 12)}{7 - 5} = \frac{25 - 27}{2} = -1 \,^\circ\text{C per m,}$$

$$\left.\frac{\partial T}{\partial t}\right|_{(5,12)} \approx \frac{T(5, 40) - T(5, 12)}{40 - 12} = \frac{30 - 27}{28} = \frac{3}{28} \,^\circ\text{C per min.}$$

At $x = 5$, $t = 12$ the temperature decreases by approximately 1°C per meter and increases by approximately $3/28^\circ$C per minute.

29. (a) $\dfrac{\partial p}{\partial c} = f_c(c, s) =$ rate of change in blood pressure as cardiac output increases while systemic vascular resistance remains constant.

(b) Suppose that $p = kcs$. Note that c (cardiac output), a volume, s (SVR), a resistance, and p, a pressure, must all be positive. Thus k must be positive, and our level curves should be confined to the first quadrant. Several level curves are shown in Figure 14.1. Each level curve represents a different blood pressure level. Each point on a given curve is a combination of cardiac output and SVR that results in the blood pressure associated with that curve.

Figure 14.1

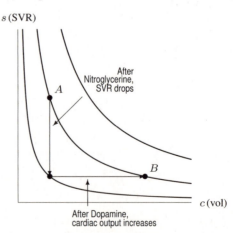

Figure 14.2

(c) Point B in Figure 14.2 shows that if the two doses are correct, the changes in pressure will cancel. The patient's cardiac output will have increased and his SVR will have decreased, but his blood pressure won't have changed.

(d) At point F in Figure 14.3, the patient's blood pressure is normalized, but his/her cardiac output has dropped and his SVR is up.

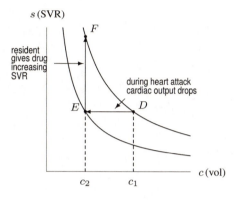

Figure 14.3

Note: c_1 and c_2 are the cardiac outputs before and after the heart attack, respectively.

Solutions for Section 14.2

Exercises

1. (a) Make a difference quotient using the two points $(3, 2)$ and $(3, 2.01)$ that have the same x-coordinate 3 but whose y-coordinates differ by 0.01. We have

$$f_y(3, 2) \approx \frac{f(3, 2.01) - f(3, 2)}{2.01 - 2} = \frac{28.0701 - 28}{0.01} = 7.01.$$

(b) Differentiating gives $f_y(x, y) = x + 2y$, so $f_y(3, 2) = 3 + 2 \cdot 2 = 7$.

5. $\dfrac{\partial z}{\partial x} = \dfrac{\partial}{\partial x}\left[(x^2 + x - y)^7\right] = 7(x^2 + x - y)^6(2x + 1) = (14x + 7)(x^2 + x - y)^6.$

$\dfrac{\partial z}{\partial y} = \dfrac{\partial}{\partial y}\left[(x^2 + x - y)^7\right] = -7(x^2 + x - y)^6.$

9. $\dfrac{\partial}{\partial T}\left(\dfrac{2\pi r}{T}\right) = -\dfrac{2\pi r}{T^2}$

13. $a_v = \dfrac{2v}{r}$

17. $\dfrac{\partial}{\partial r}\left(\dfrac{2\pi r}{v}\right) = \dfrac{2\pi}{v}$

21. $z_x = -\sin x, \quad z_x(2, 3) = -\sin 2 \approx -0.9$

25. $\dfrac{\partial}{\partial M}\left(\dfrac{2\pi r^{3/2}}{\sqrt{GM}}\right) = 2\pi r^{3/2}(-\frac{1}{2})(GM)^{-3/2}(G) = -\pi r^{3/2} \cdot \dfrac{G}{GM\sqrt{GM}} = -\dfrac{\pi r^{3/2}}{M\sqrt{GM}}$

29. $\dfrac{\partial V}{\partial r} = \dfrac{8}{3}\pi rh$ and $\dfrac{\partial V}{\partial h} = \dfrac{4}{3}\pi r^2.$

33. $z_x = 7x^6 + yx^{y-1}$, and $z_y = 2^y \ln 2 + x^y \ln x$

Problems

37. (a) The difference quotient for evaluating $f_w(2,2)$ is

$$f_w(2,2) \approx \frac{f(2+0.01,2) - f(2,2)}{h} = \frac{e^{(2.01)\ln 2} - e^{2\ln 2}}{0.01} = \frac{e^{\ln(2^{2.01})} - e^{\ln(2^2)}}{0.01}$$

$$= \frac{2^{(2.01)} - 2^2}{0.01} \approx 2.78$$

The difference quotient for evaluating $f_z(2,2)$ is

$$f_z(2,2) \approx \frac{f(2, 2+0.01) - f(2,2)}{h}$$

$$= \frac{e^{2\ln(2.01)} - e^{2\ln 2}}{0.01} = \frac{(2.01)^2 - 2^2}{0.01} = 4.01$$

(b) Using the derivative formulas we get

$$f_w = \frac{\partial f}{\partial w} = \ln z \cdot e^{w\ln z} = z^w \cdot \ln z$$

$$f_z = \frac{\partial f}{\partial z} = e^{w\ln z} \cdot \frac{w}{z} = w \cdot z^{w-1}$$

so

$$f_w(2,2) = 2^2 \cdot \ln 2 \approx 2.773$$

$$f_z(2,2) = 2 \cdot 2^{2-1} = 4.$$

41. (a) $\frac{\partial E}{\partial m} = c^2 \left(\frac{1}{\sqrt{1 - v^2/c^2}} - 1 \right)$. We expect this to be positive because energy increases with mass.

(b) $\frac{\partial E}{\partial v} = mc^2 \cdot \left(-\frac{1}{2}\right)(1 - v^2/c^2)^{-3/2}\left(-\frac{2v}{c^2}\right) = \frac{mv}{(1-v^2/c^2)^{3/2}}$. We expect this to be positive because energy increases with velocity.

45. Since $f_x(x,y) = 4x^3y^2 - 3y^4$, we could have

$$f(x,y) = x^4 y^2 - 3xy^4.$$

In that case,

$$f_y(x,y) = \frac{\partial}{\partial y}(x^4 y^2 - 3xy^4) = 2x^4 y - 12xy^3$$

as expected. More generally, we could have $f(x,y) = x^4 y^2 - 3xy^4 + C$, where C is any constant.

Solutions for Section 14.3

Exercises

1. The partial derivatives are

$$z_x = x \quad \text{and} \quad z_y = 4y,$$

so

$$z(2,1) = 4, \quad z_x(2,1) = 2 \quad \text{and} \quad z_y(2,1) = 4.$$

The tangent plane to $z = \frac{1}{2}(x^2 + 4y^2)$ at $(x,y) = (2,1)$ has equation

$$z = z(2,1) + z_x(2,1)(x-2) + z_y(2,1)(y-1)$$

$$= 4 + 2(x-2) + 4(y-1)$$

$$= -4 + 2x + 4y.$$

5. Since

$$z_x = y\cos xy, \text{ we have } z_x\left(2, \frac{3\pi}{4}\right) = \frac{3\pi}{4}\cos\left(2\cdot\frac{3\pi}{4}\right) = 0.$$

$$z_y = x\cos xy, \text{ we have } z_y\left(2, \frac{3\pi}{4}\right) = 2\cos\left(2\cdot\frac{3\pi}{4}\right) = 0.$$

Since

$$z\left(2, \frac{3\pi}{4}\right) = \sin\left(2\cdot\frac{3\pi}{4}\right) = -1,$$

the tangent plane is

$$z = z\left(2, \frac{3\pi}{4}\right) + z_x\left(2, \frac{3\pi}{4}\right)(x-2) + z_y\left(2, \frac{3\pi}{4}\right)\left(y - \frac{3\pi}{4}\right)$$
$$z = -1.$$

9. Since $g_u = 2u + v$ and $g_v = u$, we have

$$dg = (2u + v)\,du + u\,dv$$

13. We have $df = f_x\,dx + f_y\,dy$. Finding the partial derivatives, we have $f_x = e^{-y}$ so $f_x(1,0) = e^{-0} = 1$, and $f_y = -xe^{-y}$ so $f_y(1,0) = -1e^{-0} = -1$. Thus, $df = dx - dy$.

Problems

17. (a) The units are dollars/square foot.
  (b) The price of land 300 feet from the beach and of area near 1000 square feet is greater for larger plots by about $3 per square foot.
  (c) The units are dollars/foot.
  (d) The price of a 1000 square foot plot about 300 feet from the beach is less for plots farther from the beach by about $2 per extra foot from the beach.
  (e) Compared to the 998 ft^2 plot at 295 ft from the beach, the other plot costs about $7 \times 3 = \$21$ more for the extra 7 square feet but about $10 \times 2 = \$20$ less for the extra 10 feet you have to walk to the beach. The net difference is about a dollar, and the smaller plot nearer the beach is cheaper.

21. Since $f_x(x, y) = \frac{x}{\sqrt{x^2+y^3}}$ and $f_y(x,y) = \frac{3y^2}{2\sqrt{x^2+y^3}}$,
$f_x(1,2) = \frac{1}{\sqrt{1^2+2^3}} = \frac{1}{3}$ and $f_y(1,2) = \frac{3\cdot2^2}{2\sqrt{1^2+2^3}} = 2$.
Thus the differential at the point $(1, 2)$ is

$$df = df(1,2) = f_x(1,2)dx + f_y(1,2)dy = \frac{1}{3}dx + 2dy.$$

Using the differential at the point $(1, 2)$, we can estimate $f(1.04, 1.98)$. Since

$$\triangle f \approx f_x(1,2)\triangle x + f_y(1,2)\triangle y$$

where $\triangle f = f(1.04, 1.98) - f(1,2)$ and $\triangle x = 1.04 - 1$ and $\triangle y = 1.98 - 2$, we have

$$f(1.04, 1.98) \approx f(1,2) + f_x(1,2)(1.04 - 1) + f_y(1,2)(1.98 - 2)$$
$$= \sqrt{1^2 + 2^3} + \frac{0.04}{3} - 2(0.02) \approx 2.973.$$

25. (a) The linear approximation gives

$$f(520, 24) \approx 24.20, \quad f(480, 24) \approx 23.18,$$
$$f(500, 22) \approx 25.52, \quad f(500, 26) \approx 21.86.$$

The approximations for $f(520, 24)$ and $f(500, 26)$ agree exactly with the values in the table; the other two do not. The reason for this is that the partial derivatives were estimated using difference quotients with these values.

(b) We could get a more balanced estimate by using a difference quotient that uses the values on both sides. Thus, we could estimate the partial derivatives as follows:

$$f_T(500, 24) \approx \frac{f(520, 24) - f(480, 24)}{40}$$

$$= \frac{(24.20 - 23.19)}{40} = 0.02525,$$

and

$$f_p(500, 24) \approx \frac{f(500, 26) - f(500, 22)}{4}$$

$$= \frac{(21.86 - 25.86)}{4} = -1.$$

This yields the linear approximation

$$V = f(T, p) \approx 23.69 + 0.02525(T - 500) - (p - 24) \text{ ft}^3.$$

This approximation yields values

$$f(520, 24) \approx 24.195, \quad f(480, 24) \approx 23.185,$$
$$f(500, 22) \approx 25.69, \quad f(500, 26) \approx 21.69.$$

Although none of these predictions are accurate, the error in the predictions that were wrong before has been reduced. This new linearization is a better all-round approximation for values near $(500, 24)$.

29. The error in η is approximated by $d\eta$, where

$$d\eta = \frac{\partial \eta}{\partial r} dr + \frac{\partial \eta}{\partial p} dp.$$

We need to find

$$\frac{\partial \eta}{\partial r} = \frac{\pi}{8} \frac{p4r^3}{v}$$

and

$$\frac{\partial \eta}{\partial p} = \frac{\pi}{8} \frac{r^4}{v}.$$

For $r = 0.005$ and $p = 10^5$ we get

$$\frac{\partial \eta}{\partial r}(0.005, 10^5) = 3.14159 \cdot 10^7, \quad \frac{\partial \eta}{\partial p}(0.005, 10^5) = 0.39270,$$

so that

$$d\eta = \frac{\partial \eta}{\partial r} dr + \frac{\partial \eta}{\partial p} dp.$$

is largest when we take all positive values to give

$$d\eta = 3.14159 \cdot 10^7 \cdot 0.00025 + 0.39270 \cdot 1000 = 8246.68.$$

This seems quite large but $\eta(0.005, 10^5) = 39269.9$ so the maximum error represents about 20% of any value computed by the given formula. Notice also the relative error in r is $\pm 5\%$, which means the relative error in r^4 is $\pm 20\%$.

Solutions for Section 14.4

Exercises

1. Since the partial derivatives are

$$\frac{\partial f}{\partial x} = \frac{15}{2} x^4 - 0 = \frac{15}{2} x^4$$

$$\frac{\partial f}{\partial y} = 0 - \frac{24}{7} y^5 = -\frac{24}{7} y^5$$

we have

$$\text{grad } f = \frac{\partial f}{\partial x}\vec{i} + \frac{\partial f}{\partial y}\vec{j} = \left(\frac{15}{2} x^4\right)\vec{i} - \left(\frac{24}{7} y^5\right)\vec{j}.$$

5. Since the partial derivatives are

$$z_x = e^y \quad \text{and} \quad z_y = xe^y + e^y + ye^y,$$

we have

$$\nabla z = e^y \vec{i} + e^y(1 + x + y)\vec{j}.$$

9. Since the partial derivatives are

$$f_r = \sin\theta \quad \text{and} \quad f_\theta = r\cos\theta,$$

we have

$$\nabla f = \sin\theta \vec{i} + r\cos\theta \vec{j}.$$

13. Since the partial derivatives are

$$\frac{\partial f}{\partial \alpha} = \frac{(2\alpha - 3\beta)(2 + 0) - (2 - 0)(2\alpha + 3\beta)}{(2\alpha - 3\beta)^2}$$

$$= \frac{4\alpha - 6\beta - (4\alpha + 6\beta)}{(2\alpha - 3\beta)^2}$$

$$= -\frac{12\beta}{(2\alpha - 3\beta)^2}$$

$$\frac{\partial f}{\partial \beta} = \frac{(2\alpha - 3\beta)(0 + 3) - (0 - 3)(2\alpha + 3\beta)}{(2\alpha - 3\beta)^2}$$

$$= \frac{(6\alpha - 9\beta) + (6\alpha + 9\beta)}{(2\alpha - 3\beta)^2}$$

$$= \frac{12\alpha}{(2\alpha - 3\beta)^2}$$

we have

$$\text{grad } f = \frac{\partial f}{\partial \alpha}\vec{i} + \frac{\partial f}{\partial \beta}\vec{j} = \left(-\frac{12\beta}{(2\alpha - 3\beta)^2}\right)\vec{i} + \left(\frac{12\alpha}{(2\alpha - 3\beta)^2}\right)\vec{j}.$$

17. Since the partial derivatives are

$$f_r = 2\pi(h + r) \quad \text{and} \quad f_h = 2\pi r,$$

we have

$$\nabla f(2, 3) = 10\pi \vec{i} + 4\pi \vec{j}.$$

21. Since the partial derivatives are

$$\frac{\partial f}{\partial x} = \frac{1}{2}(\tan x + y)^{-1/2}\left(\frac{1}{\cos^2 x} + 0\right) = \frac{1}{2\cos^2 x\sqrt{\tan x + y}},$$

and

$$\frac{\partial f}{\partial y} = \frac{1}{2}(\tan x + y)^{-1/2}(0 + 1) = \frac{1}{2\sqrt{\tan x + y}},$$

then

$$\text{grad } f = \frac{\partial f}{\partial x}\vec{i} + \frac{\partial f}{\partial y}\vec{j} = \left(\frac{1}{2\cos^2 x\sqrt{\tan x + y}}\right)\vec{i} + \left(\frac{1}{2\sqrt{\tan x + y}}\right)\vec{j}.$$

Hence we have

$$\text{grad } f\Big|_{(0,1)} = \left(\frac{1}{2(\cos(0))^2\sqrt{\tan(0) + 1}}\right)\vec{i} + \left(\frac{1}{2\sqrt{\tan(0) + 1}}\right)\vec{j}$$

$$= \left(\frac{1}{2(1)^2\sqrt{0 + 1}}\right)\vec{i} + \left(\frac{1}{2\sqrt{0 + 1}}\right)\vec{j}$$

$$= \frac{1}{2}\vec{i} + \frac{1}{2}\vec{j}$$

25. Since $f_x = 2\cos(2x - y)$ and $f_y = -\cos(2x - y)$, at $(1, 2)$ we have grad $f = 2\cos(0)\vec{i} - \cos(0)\vec{j} = 2\vec{i} - \vec{j}$. Thus

$$f_{\vec{u}}(1, 2) = \text{grad } f \cdot \left(\frac{3}{5}\vec{i} - \frac{4}{5}\vec{j}\right) = \frac{2 \cdot 3 - 1(-4)}{5} = \frac{10}{5} = 2.$$

29. Since $df = f_x dx + f_y dy$ and grad $f = f_x\vec{i} + f_y\vec{j}$ we have

$$\text{grad } f = (x + 1)ye^x\vec{i} + xe^x\vec{j}.$$

33. Since the values of z increase as we move in direction $-\vec{i} + \vec{j}$ from the point $(-1, 1)$, the directional derivative is positive.

37. The approximate direction of the gradient vector at point $(0, 2)$ is $\vec{j}$, since the gradient vector is perpendicular to the contour and points in the direction of increasing z-values. Answers may vary since answers are approximate and any positive multiple of the vector given is also correct.

41. The approximate direction of the gradient vector at point $(2, -2)$ is $\vec{i} - \vec{j}$, since the gradient vector is perpendicular to the contour and points in the direction of increasing z-values. Answers may vary since answers are approximate and any positive multiple of the vector given is also correct.

Problems

45. (a) The unit vector $\vec{u}$ in the same direction as $\vec{v}$ is

$$\vec{u} = \frac{1}{\|\vec{v}\|}\vec{v} = \frac{1}{\sqrt{(-1)^2 + 3^2}}\vec{v} = \frac{-1}{\sqrt{10}}\vec{i} + \frac{3}{\sqrt{10}}\vec{j} = -0.316228\vec{i} + 0.948683\vec{j}.$$

The vector $\vec{w}$ of length 0.1 in the direction of $\vec{v}$ is

$$\vec{w} = 0.1\vec{u} = -0.0316228\vec{i} + 0.0948683\vec{j}.$$

The displacement vector from P to Q is $\vec{w}$. Hence

$$Q = (4 - 0.0316228, 5 + 0.0948683) = (3.96838, 5.09487).$$

(b) Since the distance from P to Q is 0.1, the directional derivative of f at P in the direction of Q is approximately

$$f_{\vec{u}} \approx \frac{f(Q) - f(P)}{0.1} = \frac{3.01052 - 3}{0.1} = 0.1052.$$

(c) We have

$$\text{grad } f(x, y) = \frac{1}{2\sqrt{x + y}}\vec{i} + \frac{1}{2\sqrt{x + y}}\vec{j}$$

$$\text{grad } f(4, 5) = \frac{1}{6}\vec{i} + \frac{1}{6}\vec{j}.$$

The directional derivative at $P = (4, 5)$ in the direction of $\vec{u}$ is

$$f_{\vec{u}}(4, 5) = \text{grad } f(4, 5) \cdot \vec{u} = \frac{1}{6}\frac{-1}{\sqrt{10}} + \frac{1}{6}\frac{3}{\sqrt{10}} = \frac{1}{3\sqrt{10}} = 0.1054.$$

49. At points (x, y) where the gradients are defined and are not the zero vector, the level curves of f and g intersect at right angles if and only grad $f \cdot$ grad $g = 0$.

We have grad $f \cdot$ grad $g = (\vec{i} + \vec{j}) \cdot (\vec{i} - \vec{j}) = 0$. The level curves of f and g are straight lines that cross at right angles. See Figure 14.4.

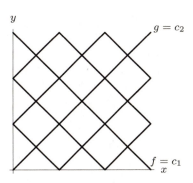

Figure 14.4

53. $f_{\vec{i}}(3, 1)$ means the rate of change of f in the x direction at $(3, 1)$. Thus,

$$f_{\vec{i}}(3, 1) \approx \frac{f(4, 1) - f(3, 1)}{1} = \frac{2 - 1}{1} = 1.$$

57. One way to do this is to estimate the gradient vector and then find grad $f(x, y) \cdot \vec{u}$. This is a useful approach since it is easier to estimate grad f than to estimate $f_{\vec{u}}$ directly. Since grad $f(x, y) = (f_x(x, y), f_y(x, y))$ we can simply estimate the x- and y-derivatives of f at $(3, 1)$ to find grad f at that point. In the x-direction we see that the function is increasing. This implies that the x-derivative is positive. To estimate its value we estimate the slope in the x-direction. Applying the same reasoning to find f_y, we get:

$$f_x(3, 1) \approx \frac{f(3 + 1, 1) - f(3, 1)}{1} = 2 - 1 = 1,$$

$$f_y(3, 1) \approx \frac{f(3, 1) - f(3, 1 - 0.6)}{0.6} \approx \frac{1 - 2}{0.6} \approx -1.67.$$

This gives us our estimated value for grad $f(3, 1) \approx \vec{i} - 1.67\vec{j}$. Now, if $\vec{u} = (-2\vec{i} + \vec{j})/\sqrt{5}$,

$$f_{\vec{u}}(3, 1) = \text{grad } f(3, 1) \cdot (-2\vec{i} + \vec{j})/\sqrt{5}$$
$$\approx (\vec{i} - 1.67\vec{j}) \cdot (-2\vec{i} + \vec{j})/\sqrt{5} \approx -1.64$$

61. First, we check that $3 = 2^2 - 1$. Then let $f(x, y) = y - x^2 + 1$ so that the given curve is the contour $f(x, y) = 0$. Since $f_x = -2x$ and $f_y = 1$, we have grad $f(2, 3) = -4\vec{i} + \vec{j}$. Since gradients are perpendicular to contours, a vector normal to the curve at $(2, 3)$ is $\vec{n} = -4\vec{i} + 1\vec{j}$. Using the normal vector to a line the same way we use the normal vector to a plane, we get that an equation of the tangent line is $-4(x - 2) + (y - 3) = 0$. Notice, if we had instead found the slope of the tangent line using $dy/dx = 2x$, we get $(y - 3) = 4(x - 2)$, which agrees with the equation we got using the gradient.

65. (a) Negative. ∇f is perpendicular to the level curve at the point P, so its x-component which is $\nabla f \cdot \vec{i}$ is negative.
  (b) Positive. The y-component of ∇f is in the same direction as $\vec{j}$ at P and hence the dot product will be positive.
  (c) Positive. The partial derivative with respect to x at Q is positive because the value of f is increasing in the positive x direction at Q. (Note that Q lies between the level curves with values 3 and 4 and that the one with value 4 is further in the positive x direction from Q.)
  (d) Negative. Again, Q lies between the level curves with values 3 and 4 and the one with value 3 is further from Q in the positive y direction, so the partial derivative with respect to y at Q is negative.

69. We see that

$$\text{grad } f = (3y)\vec{i} + (3x + 2y)\vec{j},$$

so at the point $(2, 3)$, we have

$$\text{grad } f = 9\vec{i} + 12\vec{j}.$$

  (a) The directional derivative is $\nabla f \cdot \dfrac{\vec{v}}{\|\vec{v}\|} = \dfrac{(9)(3) + (12)(-1)}{\sqrt{10}} = \dfrac{15}{\sqrt{10}} \approx 4.74.$
  (b) The direction of maximum rate of change is $\nabla f(2, 3) = 9\vec{i} + 12\vec{j}$.
  (c) The maximum rate of change is $\|\nabla f\| = \sqrt{225} = 15.$

73. Directional derivative $= \nabla f \cdot \vec{u}$, where $\vec{u} =$ unit vector. If we move from $(4, 5)$ to $(5, 6)$, we move in the direction $\vec{i} + \vec{j}$ so $\vec{u} = \frac{1}{\sqrt{2}}\vec{i} + \frac{1}{\sqrt{2}}\vec{j}$. So,

$$\nabla f \cdot \vec{u} = f_x \left(\frac{1}{\sqrt{2}}\right) + f_y \left(\frac{1}{\sqrt{2}}\right) = 2.$$

Similarly, if we move from $(4, 5)$ to $(6, 6)$, the direction is $2\vec{i} + \vec{j}$ so $\vec{u} = \frac{2}{\sqrt{5}}\vec{i} + \frac{1}{\sqrt{5}}\vec{j}$. So

$$\nabla f \cdot \vec{u} = f_x \left(\frac{2}{\sqrt{5}}\right) + f_y \left(\frac{1}{\sqrt{5}}\right) = 3.$$

Solving the system of equations for f_x and f_y

$$f_x + f_y = 2\sqrt{2}$$
$$2f_x + f_y = 3\sqrt{5}$$

gives

$$f_x = 3\sqrt{5} - 2\sqrt{2}$$
$$f_y = 4\sqrt{2} - 3\sqrt{5}.$$

Thus at $(4, 5)$,

$$\nabla f = (3\sqrt{5} - 2\sqrt{2})\vec{i} + (4\sqrt{2} - 3\sqrt{5})\vec{j}.$$

77. (a) If $\vec{j}$ points north and $\vec{i}$ points east, then the direction the car is driving is $\vec{j} - \vec{i}$. A unit vector in this direction is

$$\vec{u} = \frac{1}{\sqrt{2}}(\vec{j} - \vec{i}).$$

The gradient of the height function is

$$\text{grad } h = \frac{\partial h}{\partial E}\vec{i} + \frac{\partial h}{\partial N}\vec{j} = 50\vec{i} + 100\vec{j}.$$

So the directional derivative is

$$h_{\vec{u}} = \text{grad } g \cdot \vec{h} = (50\vec{i} + 100\vec{j}) \cdot \frac{1}{\sqrt{2}}(\vec{j} - \vec{i}) = \frac{100 - 50}{\sqrt{2}} = 35.355 \text{ ft/mi.}$$

(b) The car is traveling at v mi/hr, so

$$\text{Rate of change of with respect to time} = v\frac{\text{miles}}{\text{hour}}35.355\frac{\text{ft}}{\text{mile}} = 35.355v \text{ ft/hour.}$$

81. (a) To estimate the change in f, we use the gradient vector to estimate the change in f in moving from P to Q. Because the contours are approximately parallel, moving from P to Q takes you to the same contour as moving from P to R. (See Figure 14.5.) If θ is the angle between $\vec{u}$ and grad $f(a, b)$, then

$$\begin{array}{l} \text{Change in } f \\ \text{between } P \text{ and } Q \end{array} = \text{Change in } f$$

$$= \left(\begin{array}{c}\text{Rate of change}\\\text{in direction } PR\end{array}\right)\left(\begin{array}{c}\text{Distance traveled}\\\text{between } P \text{ and } R\end{array}\right)$$

$$\approx \|\text{ grad } f\|(h\cos\theta).$$

Figure 14.5

(b) Since $\vec{u}$ is a unit vector, we use the definition of $f_{\vec{u}}(a, b)$ to estimate

$$f_{\vec{u}}(a, b) \approx \frac{\text{Change in } f}{h} \approx \frac{\|\operatorname{grad} f(a, b)\| h \cos \theta}{h}$$
$$= \|\operatorname{grad} f(a, b)\| \cos \theta = \|\operatorname{grad} f\| \|\vec{u}\| \cos \theta = \operatorname{grad} f(a, b) \cdot \vec{u}.$$

This approximation gets better as we choose h smaller and smaller, and in the limit we get the formula:

$$f_{\vec{u}}(a, b) = \operatorname{grad} f(a, b) \cdot \vec{u}.$$

Solutions for Section 14.5

Exercises

1. Since $f_x = 2x$, $f_y = 0$ and $f_z = 0$, we have
$$\operatorname{grad} f = 2x\vec{i}.$$

5. Since $f(x, y, z) = \dfrac{1}{xyz}$, we have

$$f_x = -\frac{1}{x^2 yz}, \quad f_y = -\frac{1}{xy^2 z} \quad f_z = -\frac{1}{xyz^2},$$

we have

$$\operatorname{grad} f = -\frac{1}{xyz}\left(\frac{1}{x}\vec{i} + \frac{1}{y}\vec{j} + \frac{1}{z}\vec{k}\right).$$

9. We have $f_x = e^y \sin z$, $f_y = xe^y \sin z$, and $f_z = xe^y \cos z$. Thus
$$\operatorname{grad} f = e^y \sin z\,\vec{i} + xe^y \sin z\,\vec{j} + xe^y \cos z\,\vec{k}.$$

13. We have $f_x = 0$, $f_y = 2yz$ and $f_z = y^2$. Thus $\operatorname{grad} f = 2yz\vec{j} + y^2\vec{k}$ and $\operatorname{grad} f(1, 0, 1) = \vec{0}$.

17. We have
$$\operatorname{grad}(x \ln(yz))\Big|_{(2,1,e)} = \ln(yz)\vec{i} + \frac{x}{y}\vec{j} + \frac{x}{z}\vec{k}\Big|_{(2,1,e)} = \vec{i} + 2\vec{j} + \frac{2}{e}\vec{k}.$$

21. We have grad $f = y\vec{i} + x\vec{j} + 2z\vec{k}$, so grad $f(1,1,0) = \vec{i} + \vec{j}$. A unit vector in the direction we want is $u = (1/\sqrt{2})(-\vec{i} + \vec{k})$. Therefore, the directional derivative is

$$\text{grad } f(1,1,0) \cdot \vec{u} = \frac{1(-1) + 1 \cdot 0 + 0 \cdot 1}{\sqrt{2}} = \frac{-1}{\sqrt{2}}.$$

25. First, we check that $(-1)^2 - (1)^2 + 2^2 = 4$. Then let $f(x,y,z) = x^2 - y^2 + z^2$ so that the given surface is the level surface $f(x,y,z) = 4$. Since $f_x = 2x$, $f_y = -2y$, and $f_z = 2z$, we have grad $f(-1,1,2) = -2\vec{i} - 2\vec{j} + 4\vec{k}$. Since gradients are perpendicular to level surfaces, a vector normal to the surface at $(-1,1,2)$ is $\vec{n} = -2\vec{i} - 2\vec{j} + 4\vec{k}$. Thus an equation for the tangent plane is

$$-2(x+1) - 2(y-1) + 4(z-2) = 0.$$

29. First, we check that $\cos(-1+1) = e^{-2+2}$. Then we let $f(x,y,z) = \cos(x+y) - e^{xz+2}$, so that the given surface is the level surface $f(x,y,z) = 0$. Since $f_x = -\sin(x+y) - ze^{xz+2}$, $f_y = -\sin(x+y)$, and $f_z = -xe^{xz+2}$, we have grad $f(-1,1,2) = -2\vec{i} + \vec{k}$. Since gradients are perpendicular to level surfaces, a vector normal to the surface at $(-1,1,2)$ is $\vec{n} = -2\vec{i} + \vec{k}$. Thus an equation for the tangent plane is

$$-2(x+1) + (z-2) = 0.$$

33. (a) We have $f_x = 2x - yz$, $f_y = 2y - xz$ and $f_z = -xy$ so

$$\text{grad } f = (2x - yz)\vec{i} + (2y - xz)\vec{j} - xy\vec{k}.$$

(b) At the point $(2,3,1)$ we have

$$\text{grad } f(2,3,1) = \vec{i} + 4\vec{j} + 6\vec{k}.$$

Thus an equation of the tangent plane to the level surface at the point $(2,3,1)$ is

$$(x-2) + 4(y-3) + 6(z-1) = 0$$

or

$$x + 4y + 6z = 20.$$

37. Let $f(x,y,z) = x^2 + y^2$ so that the surface is the level surface $f(x,y,z) = 1$. Since

$$\text{grad } f = 2x\vec{i} + 2y\vec{j}$$

we have

$$\text{grad } f(1,0,1) = 2\vec{i}.$$

Thus an equation of the tangent plane at the point $(1,0,1)$ is

$$2(x-1) + 0(y-0) + 0(z-1) = 0$$

or

$$2(x-1) = 0.$$

The tangent plane is given by the equation

$$x = 1.$$

Problems

41. The gradient of (a) is $2x\vec{i} + 2y\vec{j} + 2z\vec{k}$, which points radially outward from the origin, so (a) goes with (III).
The gradient of (c) is parallel to the gradient of (a) but pointing inward, so (c) goes with (IV).
The gradient of (b) is $2x\vec{i} + 2y\vec{j}$, which points radially outward from the z-axis, so (b) goes with (I).
The gradient of (d) is parallel to the gradient of (b) but pointing inward, so (d) goes with (II).

45. The surface is given by $F(x, y, z) = 0$ where $F(x, y, z) = x - y^3 z^7$. The normal direction is

$$\text{grad } F = \frac{\partial F}{\partial x}\vec{i} + \frac{\partial F}{\partial y}\vec{j} + \frac{\partial F}{\partial z}\vec{k} = \vec{i} - 3y^2 z^7\vec{j} - 7y^3 z^6\vec{k}.$$

Thus, at $(1, -1, -1)$ a normal vector is $\vec{i} + 3\vec{j} + 7\vec{k}$. The tangent plane has the equation

$$1(x - 1) + 3(y - (-1)) + 7(z - (-1)) = 0$$
$$x + 3y + 7z = -9.$$

49. (a) A normal to the surface is given by the gradient of the function $f(x, y, z) = x^2 + y^2 + 3z^2$,

$$\text{grad } f = 2x\vec{i} + 2y\vec{j} + 6z\vec{k}.$$

At the point $(0.6, 0.8, 1)$, a normal to the surface and to the tangent plane is

$$\vec{n} = 1.2\vec{i} + 1.6\vec{i} + 6\vec{k}.$$

Since the plane goes through the point $(0.6, 0.8, 1)$, its equation is

$$1.2(x - 0.6) + 1.6(y - 0.8) + 6(z - 1) = 0$$
$$1.2x + 1.6y + 6z = 8.$$

(b) We want to know if there are x, y, z values such that a normal to the surface is parallel to the normal to the plane. That is, is $2x\vec{i} + 2y\vec{j} + 6z\vec{k}$ parallel to $8\vec{i} + 6\vec{j} + 30\vec{k}$? Yes, if $2x = 8t$, $2y = 6t$, $6z = 30t$ for some value of t. That is, if

$$x = 4t, \quad y = 3t, \quad z = 5t.$$

Substituting these equations into the equation for the surface and solving for t, we get

$$(4t)^2 + (3t)^2 + 3(5t)^2 = 4$$
$$100t^2 = 4$$
$$t = \pm\sqrt{\frac{4}{100}} = \pm 0.2.$$

So there are two points on the surface at which the tangent plane is parallel to the plane $8x + 6y + 30z = 1$. They are

$$\pm(0.8, 0.6, 1).$$

53. (a) In the direction of grad F:

$$\text{grad } F \Big|_{(-1,1,1)} = \left((2x + 2xz^2)\vec{i} + (4y^3)\vec{j} + (2x^2 z)\vec{k} \right) \Big|_{(-1,1,1)} = -4\vec{i} + 4\vec{j} + 2\vec{k}.$$

(b) The rate of change in the direction of grad F with respect to distance $= \|\nabla F\| = \sqrt{16 + 16 + 4} = 6$. Now we want rate of change with respect to time. If we move at 4 units/sec:

$$\text{Rate of change of } \frac{\text{Conc}}{\text{Time}} = \text{Rate of change of } \frac{\text{Conc}}{\text{Dist}} \times \text{Rate of change of } \frac{\text{Dist}}{\text{Time}}$$
$$= 6 \times 4 = 24 \text{mg/cm}^3/\text{sec}.$$

57. (a) We have $\nabla G = (2x - 5y)\vec{i} + (-5x + 2yz)\vec{j} + (y^2)\vec{k}$, so $\nabla G(1, 2, 3) = -8\vec{i} + 7\vec{j} + 4\vec{k}$. The rate of change is given by the directional derivative in the direction $\vec{v}$:

$$\text{Rate of change in density} = \nabla G \cdot \frac{\vec{v}}{\|\vec{v}\|} = (-8\vec{i} + 7\vec{j} + 4\vec{k}) \cdot \frac{(2\vec{i} + \vec{j} - 4\vec{k})}{\sqrt{21}}$$
$$= \frac{-16 + 7 - 16}{\sqrt{21}} = \frac{-25}{\sqrt{21}} \approx -5.455.$$

(b) The direction of maximum rate of change is $\nabla G(1, 2, 3) = -8\vec{i} + 7\vec{j} + 4\vec{k}$.

(c) The maximum rate of change is $\|\nabla G(1, 2, 3)\| = \sqrt{(-8)^2 + 7^2 + 4^2} = \sqrt{129} \approx 11.36$.

61. (a) is (V) since $\vec{r} + \vec{a}$ is a vector not a scalar.

(b) is (IV) since $\text{grad}(\vec{r} \cdot \vec{a}) = \text{grad}(a_1 x + a_2 y + a_3 z) = \vec{a}$.

(c) is (V) since $\vec{r} \times \vec{a}$ is a vector not a scalar.

65.

$$\text{grad}(\vec{\mu} \cdot \vec{r}) = \text{grad}(\mu_1 x + \mu_2 y + \mu_3 z)$$
$$= \mu_1 \vec{i} + \mu_2 \vec{j} + \mu_3 \vec{k} = \vec{\mu}.$$

Solutions for Section 14.6

Exercises

1. Using the chain rule we see:

$$\frac{dz}{dt} = \frac{\partial z}{\partial x}\frac{dx}{dt} + \frac{\partial z}{\partial y}\frac{dy}{dt}$$
$$= -y^2 e^{-t} + 2xy \cos t$$
$$= -(\sin t)^2 e^{-t} + 2e^{-t} \sin t \cos t$$
$$= \sin(t)e^{-t}(2\cos t - \sin t)$$

We can also solve the problem using one variable methods:

$$z = e^{-t}(\sin t)^2$$
$$\frac{dz}{dt} = \frac{d}{dt}(e^{-t}(\sin t)^2)$$
$$= \frac{de^{-t}}{dt}(\sin t)^2 + e^{-t}\frac{d(\sin t)^2}{dt}$$
$$= -e^{-t}(\sin t)^2 + 2e^{-t}\sin t \cos t$$
$$= e^{-t}\sin t(2\cos t - \sin t)$$

5. Using the chain rule we see:

$$\frac{dz}{dt} = \frac{\partial z}{\partial x}\frac{dx}{dt} + \frac{\partial z}{\partial y}\frac{dy}{dt}$$
$$= 2t(\sin y + y \cos x) + \frac{1}{t}(x \cos y + \sin x)$$
$$= 2t \sin(\ln t) + 2t \ln(t)\cos(t^2) + t\cos(\ln t) + \frac{\sin t^2}{t}$$

This problem can also be solved using one variable methods. Attempting to solve the problem that way will demonstrate the advantage of using the chain rule.

9. Since z is a function of two variables x and y which are functions of two variables u and v, the two chain rule identities which apply are:

$$\frac{\partial z}{\partial u} = \frac{\partial z}{\partial x}\frac{\partial x}{\partial u} + \frac{\partial z}{\partial y}\frac{\partial y}{\partial u} = e^y(2u) + xe^y(2u)$$
$$= 2ue^y(1+x) = 2ue^{(u^2-v^2)}(1 + u^2 + v^2).$$
$$\frac{\partial z}{\partial v} = \frac{\partial z}{\partial x}\frac{\partial x}{\partial v} + \frac{\partial z}{\partial y}\frac{\partial y}{\partial v} = e^y(2v) + xe^y(-2v)$$
$$= 2ve^y(1-x) = 2ve^{(u^2-v^2)}(1 - u^2 - v^2).$$

13. Since z is a function of two variables x and y which are functions of two variables u and v, the two chain rule identities which apply are:

$$\frac{\partial z}{\partial u} = \frac{\partial z}{\partial x}\frac{\partial x}{\partial u} + \frac{\partial z}{\partial y}\frac{\partial y}{\partial u}$$

$$\frac{\partial z}{\partial v} = \frac{\partial z}{\partial x}\frac{\partial x}{\partial v} + \frac{\partial z}{\partial y}\frac{\partial y}{\partial v}$$

First to find $\partial z/\partial u$

$$\frac{\partial z}{\partial u} = (e^{-y} - ye^{-x})\sin v + (-xe^{-y} + e^{-x})(-v\sin u)$$

$$= (e^{-v\cos u} - v(\cos u)e^{-u\sin v})\sin v - (-u(\sin v)e^{-v\cos u} + e^{-u\sin v})v\sin u$$

Now we find $\partial z/\partial v$ using the same method.

$$\frac{\partial z}{\partial v} = (e^{-y} - ye^{-x})u\cos v + (-xe^{-y} + e^{-x})\cos u$$

$$= (e^{-v\cos u} - v(\cos u)e^{-u\sin v})u\cos v + (-u(\sin v)e^{-v\cos u} + e^{-u\sin v})\cos u$$

Problems

17.

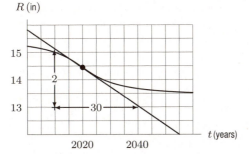

Figure 14.6: Global warming predictions: Rainfall as a function of time

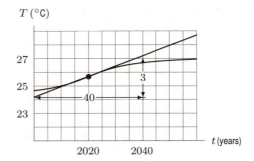

Figure 14.7: Global warming predictions: Temperature as a function of time

We know that, as long as the temperature and rainfall stay close to their current values of $R = 15$ inches and $T = 30°$C, a change, ΔR, in rainfall and a change, ΔT, in temperature produces a change, ΔC, in corn production given by

$$\Delta C \approx 3.3\Delta R - 5\Delta T.$$

Now both R and T are functions of time t (in years), and we want to find the effect of a small change in time, Δt, on R and T. Figure 14.6 shows that the slope of the graph for R versus t is about $-2/30 \approx -0.07$ in/year when $t = 2020$. Similarly, Figure 14.7 shows the slope of the graph of T versus t is about $3/40 \approx 0.08°$C/year when $t = 2020$. Thus, around the year 2020,

$$\Delta R \approx -0.07\Delta t \quad \text{and} \quad \Delta T \approx 0.08\Delta t.$$

Substituting these into the equation for ΔC, we get

$$\Delta C \approx (3.3)(-0.07)\Delta t - (5)(0.08)\Delta t \approx -0.6\Delta t.$$

Since at present $C = 100$, corn production will decline by about 0.6 % between the years 2020 and 2021. Now $\Delta C \approx -0.6\Delta t$ tells us that when $t = 2020$,

$$\frac{\Delta C}{\Delta t} \approx -0.6, \quad \text{and therefore, that} \quad \frac{dC}{dt} \approx -0.6.$$

21.

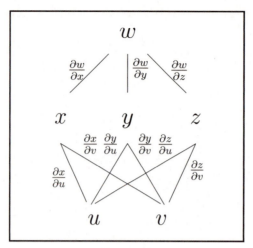

Figure 14.8

The tree diagram in Figure 14.8 tells us that

$$\frac{\partial w}{\partial u} = \frac{\partial w}{\partial x}\frac{\partial x}{\partial u} + \frac{\partial w}{\partial y}\frac{\partial y}{\partial u} + \frac{\partial w}{\partial z}\frac{\partial z}{\partial u},$$

$$\frac{\partial w}{\partial v} = \frac{\partial w}{\partial x}\frac{\partial x}{\partial v} + \frac{\partial w}{\partial y}\frac{\partial y}{\partial v} + \frac{\partial w}{\partial z}\frac{\partial z}{\partial v}.$$

25. All are done using the chain rule.

(a) We have $u = x$, $v = 3$. Thus $du/dx = 1$ and $dv/dx = 0$ so

$$f'(x) = F_u(x, 3)(1) + F_v(x, 3)(0) = F_u(x, 3).$$

(b) We have $u = 3$, $v = x$. Thus $du/dx = 0$ and $dv/dx = 1$ so

$$f'(x) = F_u(3, x)(0) + F_v(3, x)(1) = F_v(3, x).$$

(c) We have $u = x$, $v = x$. Thus $du/dx = dv/dx = 1$ so

$$f'(x) = F_u(x, x)(1) + F_v(x, x)(1) = F_u(x, x) + F_v(x, x).$$

(d) We have $u = 5x$, $v = x^2$. Thus $du/dx = 5$ and $dv/dx = 2x$ so

$$f'(x) = F_u(5x, x^2)(5) + F_v(5x, x^2)(2x).$$

29. We have $z = h(x, y)$ where $h(x, y) = f(x)g(y)$, $x = t$, and $y = t$. The chain rule gives $dz/dt = (\partial h/\partial x)dx/dt + (\partial h/\partial y)dy/dt = \partial h/\partial x + \partial h/\partial y$. Since $\partial h/\partial x = f'(x)g(y)$ and $\partial h/\partial y = f(x)g'(y)$ we have $dz/dt = f'(x)g(y) + f(x)g'(y) = f'(t)g(t) + f(t)g'(t)$.

33. To calculate $\left(\frac{\partial U}{\partial P}\right)_T$, we think of U as a function of P and T, as in $U_1(T, P)$. Thus

$$\left(\frac{\partial U}{\partial P}\right)_T = \frac{\partial U_1}{\partial P}.$$

37. (a) Thinking of V as a function of P and T gives

$$dV = \left(\frac{\partial V}{\partial P}\right)_T dP + \left(\frac{\partial V}{\partial T}\right)_P dT.$$

(b) Substituting for dV in the following expression for dU,

$$dU = \left(\frac{\partial U}{\partial P}\right)_V dP + \left(\frac{\partial U}{\partial V}\right)_P dV,$$

we get

$$dU = \left(\frac{\partial U}{\partial P}\right)_V dP + \left(\frac{\partial U}{\partial V}\right)_P \left(\left(\frac{\partial V}{\partial P}\right)_T dP + \left(\frac{\partial V}{\partial T}\right)_P dT\right).$$

Rearranging terms gives

$$dU = \left(\left(\frac{\partial U}{\partial P}\right)_V + \left(\frac{\partial U}{\partial V}\right)_P \cdot \left(\frac{\partial V}{\partial P}\right)_T\right) dP + \left(\frac{\partial U}{\partial V}\right)_P \cdot \left(\frac{\partial V}{\partial T}\right)_P dT.$$

(c) The formula for dU obtained by thinking of U as a function of P and T is

$$dU = \left(\frac{\partial U}{\partial T}\right)_P dT + \left(\frac{\partial U}{\partial P}\right)_T dP.$$

(d) Comparing coefficients of dP and dT in the two formulas gives

$$\left(\frac{\partial U}{\partial T}\right)_P = \left(\frac{\partial U}{\partial V}\right)_P \cdot \left(\frac{\partial V}{\partial T}\right)_P$$

$$\left(\frac{\partial U}{\partial P}\right)_T = \left(\frac{\partial U}{\partial P}\right)_V + \left(\frac{\partial U}{\partial V}\right)_P \cdot \left(\frac{\partial V}{\partial P}\right)_T.$$

Solutions for Section 14.7

Exercises

1. We have $f_x = 6xy + 5y^3$ and $f_y = 3x^2 + 15xy^2$, so $f_{xx} = 6y$, $f_{xy} = 6x + 15y^2$, $f_{yx} = 6x + 15y^2$, and $f_{yy} = 30xy$.

5. We have $f_x = 2ye^{2xy}$ and $f_y = 2xe^{2xy}$, so $f_{xx} = 4y^2 e^{2xy}$, $f_{xy} = 4xye^{2xy} + 2e^{2xy}$, $f_{yx} = 4xye^{2xy} + 2e^{2xy}$, and $f_{yy} = 4x^2 e^{2xy}$.

9. Since $f(x, y) = \sin(x^2 + y^2)$, we have

$$f_x = (\cos(x^2 + y^2))2x \quad , f_y = (\cos(x^2 + y^2))2y$$
$$f_{xx} = -(\sin(x^2 + y^2))4x^2 + 2\cos(x^2 + y^2)$$
$$f_{xy} = -(\sin(x^2 + y^2))4xy = f_{yx}$$
$$f_{yy} = -(\sin(x^2 + y^2))4y^2 + 2\cos(x^2 + y^2).$$

13. The quadratic Taylor expansion about $(0, 0)$ is given by

$$f(x, y) \approx Q(x, y) = f(0,0) + f_x(0,0)x + f_y(0,0)y + \frac{1}{2}f_{xx}(0,0)x^2 + f_{xy}(0,0)xy + \frac{1}{2}f_{yy}(0,0)y^2.$$

First we find all the relevant derivatives

$$f(x, y) = e^{-2x^2 - y^2}$$
$$f_x(x, y) = -4xe^{-2x^2 - y^2}$$
$$f_y(x, y) = -2ye^{-2x^2 - y^2}$$
$$f_{xx}(x, y) = -4e^{-2x^2 - y^2} + 16x^2 e^{-2x^2 - y^2}$$
$$f_{yy}(x, y) = -2e^{-2x^2 - y^2} + 4y^2 e^{-2x^2 - y^2}$$
$$f_{xy}(x, y) = 8xye^{-2x^2 - y^2}$$

Now we evaluate each of these derivatives at $(0, 0)$ and substitute into the formula to get as our final answer:

$$Q(x, y) = 1 - 2x^2 - y^2$$

17. The quadratic Taylor expansion about $(0, 0)$ is given by

$$f(x, y) \approx Q(x, y) = f(0,0) + f_x(0,0)x + f_y(0,0)y + \frac{1}{2}f_{xx}(0,0)x^2 + f_{xy}(0,0)xy + \frac{1}{2}f_{yy}(0,0)y^2.$$

So first we find all the relevant derivatives:

$$f(x, y) = \ln(1 + x^2 - y)$$
$$f_x(x, y) = \frac{2x}{1 + x^2 - y}$$
$$f_y(x, y) = \frac{-1}{1 + x^2 - y}$$
$$f_{xx}(x, y) = \frac{2(1 + x^2 - y) - 4x^2}{(1 + x^2 - y)^2}$$
$$f_{yy}(x, y) = \frac{-1}{(1 + x^2 - y)^2}$$
$$f_{xy}(x, y) = \frac{2x}{(1 + x^2 - y)^2}$$

Substituting into the formula we get as our answer:

$$Q(x, y) = -y + x^2 - \frac{y^2}{2}$$

21. (a) $f_x(P) < 0$ because f decreases as you go to the right.
(b) $f_y(P) = 0$ because f does not change as you go up.
(c) $f_{xx}(P) < 0$ because f_x decreases as you go to the right (f_x changes from a small negative number to a large negative number).
(d) $f_{yy}(P) = 0$ because f_y does not change as you go up.
(e) $f_{xy}(P) = 0$ because f_x does not change as you go up.

25. (a) $f_x(P) > 0$ because f increases as you go to the right.
(b) $f_y(P) > 0$ because f increases as you go up.
(c) $f_{xx}(P) = 0$ because f_x does not change as you go to the right. (Notice that the level curves are equidistant and parallel, so the partial derivatives of f do not change if you move horizontally or vertically.)
(d) $f_{yy}(P) = 0$ because f_y does not change as you go up.
(e) $f_{xy}(P) = 0$ because f_x does not change as you go up.

29. Since $f(x, y) = \sqrt{1 + 2x - y}$, the first and second derivatives are

$$f_x = \frac{1}{\sqrt{1 + 2x - y}}$$
$$f_y = \frac{-1}{2\sqrt{1 + 2x - y}}$$
$$f_{xx} = \frac{-1}{(1 + 2x - y)^{3/2}}$$
$$f_{xy} = \frac{1}{2(1 + 2x - y)^{3/2}}$$
$$f_{yy} = \frac{-1}{4(1 + x - 2y)^{3/2}},$$

so we find that

$$f(0, 0) = 1$$
$$f_x(0, 0) = 1$$
$$f_y(0, 0) = -1/2$$
$$f_{xx}(0, 0) = -1$$
$$f_{xy}(0, 0) = 1/2$$
$$f_{yy}(0, 0) = -1/4.$$

The best quadratic approximation for $f(x, y)$ for (x, y) near $(0, 0)$ is

$$f(x, y) \approx 1 + x - \frac{1}{2}y - \frac{1}{2}x^2 + \frac{1}{2}xy - \frac{1}{8}y^2.$$

Problems

33. We have $f(1, 0) = 1$ and the relevant derivatives are:

$$f_x = e^{-y} \quad \text{so} \quad f_x(1, 0) = 1$$
$$f_y = -xe^{-y} \quad \text{so} \quad f_y(1, 0) = -1$$
$$f_{xx} = 0 \quad \text{so} \quad f_{xx}(1, 0) = 0$$
$$f_{xy} = -e^{-y} \quad \text{so} \quad f_{xy}(1, 0) = -1$$
$$f_{yy} = xe^{-y} \quad \text{so} \quad f_{yy}(1, 0) = 1.$$

Thus the linear approximation, $L(x, y)$ to $f(x, y)$ at $(1, 0)$, is given by:

$$f(x, y) \approx L(x, y) = f(1, 0) + f_x(1, 0)(x - 1) + f_y(1, 0)(y - 0)$$
$$= 1 + (x - 1) - y.$$

The quadratic approximation, $Q(x, y)$ to $f(x, y)$ near $(1, 0)$, is given by:

$$f(x, y) \approx Q(x, y) = f(1, 0) + f_x(1, 0)(x - 1) + f_y(1, 0)(y - 0) + \frac{1}{2}f_{xx}(1, 0)(x - 1)^2$$
$$+ f_{xy}(1, 0)(x - 1)(y - 0) + \frac{1}{2}f_{yy}(1, 0)(y - 0)^2$$
$$= 1 + (x - 1) - y - (x - 1)y + \frac{1}{2}y^2.$$

The values of the approximations are

$$L(0.9, 0.2) = 1 - 0.1 - 0.2 = 0.7$$
$$Q(0.9, 0.2) = 1 - 0.1 - 0.2 + 0.02 + 0.02 = 0.74$$

and the exact value is

$$f(0.9, 0.2) = (0.9)e^{-0.2} \approx 0.737$$

Observe that the quadratic approximation is closer to the exact value.

37. Differentiating, we get

$$F_x = e^x \sin y + e^y \cos x \quad F_y = e^x \cos y + e^y \sin x$$
$$F_{xx} = e^x \sin y - e^y \sin x \quad F_{yy} = -e^x \sin y + e^y \sin x = -F_{xx}$$

Thus, $F_{xx} + F_{yy} = 0$.

41. Since $z_y = g(x)$, $z_{yy} = 0$, because g is a function of x only.

45. (a) Since P and Q lie on the same level curve, we have $a = k$.
  (b) We have $b = f_x$ and $c = f_y$. Since the gradient of f at P (respectively Q) points toward M or away from M, from the figure, we see $f_x(P)$ and $f_y(P)$ have opposite signs, while $f_x(Q)$ and $f_y(Q)$ have the same signs. Thus Q is the point (x_1, y_1), so P is (x_2, y_2).
  (c) Since $b = f_x(Q) > 0$ and $c = f_y(Q) > 0$, the value of f must increase as we go away from M. Thus, M must be a minimum (the surface is a valley).
  (d) Since M is a minimum, $m = f_x(P) < 0$ and $n = f_y(P) > 0$.

49. If f and g have exactly the same contour diagrams inside a circle about the origin, then any partial derivative of f of any order has the same value at the point $(0, 0)$ as the corresponding partial derivative of g. Since the Taylor polynomials of degree 2 for f and g near $(0, 0)$ are constructed from the values of the partial derivatives at $(0, 0)$, the two Taylor polynomials are identical. It makes no difference what the contour diagrams look like outside the circle. See Figure 14.9 for one possible solution.

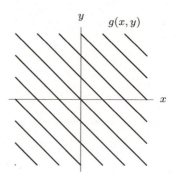

Figure 14.9

Solutions for Section 14.8

Exercises

1. Not differentiable at $(0,0)$.

5. Not differentiable where $x = 0$; that is, not differentiable on y-axis.

9. Not differentiable at $(1,2)$.

Problems

13. (a) The contour diagram for $f(x,y) = xy/\sqrt{x^2 + y^2}$ is shown in Figure 14.10.

Figure 14.10

(b) By the chain rule, f is differentiable at all points (x,y) where $x^2 + y^2 \neq 0$, and so at all points $(x,y) \neq (0,0)$.

(c) The partial derivatives of f are given by

$$f_x(x,y) = \frac{y^3}{(x^2 + y^2)^{3/2}}, \qquad \text{for} \quad (x,y) \neq (0,0),$$

and

$$f_y(x,y) = \frac{x^3}{(x^2 + y^2)^{3/2}}, \qquad \text{for} \quad (x,y) \neq (0,0).$$

Both f_x and f_y are continuous at $(x,y) \neq (0,0)$.

(d) If f were differentiable at $(0,0)$, the chain rule would imply that the function

$$g(t) = \begin{cases} f(t,t), & t \neq 0 \\ 0, & t = 0 \end{cases}$$

would be differentiable at $t = 0$. But

$$g(t) = \frac{t^2}{\sqrt{2t^2}} = \frac{1}{\sqrt{2}} \cdot \frac{t^2}{|t|} = \frac{1}{\sqrt{2}} \cdot |t|,$$

which is not differentiable at $t = 0$. Hence, f is not differentiable at $(0,0)$.

(e) The partial derivatives of f at $(0,0)$ are given by

$$f_x(0,0) = \lim_{x \to 0} \frac{f(x,0) - f(0,0)}{x} = \lim_{x \to 0} \frac{\frac{x \cdot 0}{\sqrt{x^2 + 0^2}} - 0}{x} = \lim_{x \to 0} \frac{0 - 0}{x} = 0,$$

$$f_y(0,0) = \lim_{y \to 0} \frac{f(0,y) - f(0,0)}{y} = \lim_{y \to 0} \frac{\frac{0 \cdot y}{\sqrt{0^2 + y^2}} - 0}{y} = \lim_{y \to 0} \frac{0 - 0}{y} = 0.$$

The limit $\lim_{(x,y) \to (0,0)} f_x(x,y)$ does not exist since if we choose $x = y = t, t \neq 0$, then

$$f_x(x,y) = f_x(t,t) = \frac{t^3}{(2t^2)^{3/2}} = \frac{t^3}{2\sqrt{2} \cdot |t|^3} = \begin{cases} \frac{1}{2\sqrt{2}}, & t > 0, \\ -\frac{1}{2\sqrt{2}}, & t < 0. \end{cases}$$

Thus, f_x is not continuous at $(0,0)$. Similarly, f_y is not continuous at $(0,0)$.

17. (a) The contour diagram of $f(x,y) = \sqrt{|xy|}$ is shown in Figure 14.11.

Figure 14.11

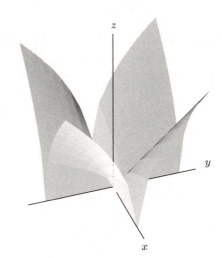

Figure 14.12

(b) The graph of $f(x,y) = \sqrt{|xy|}$ is shown in Figure 14.12.

(c) f is clearly differentiable at (x,y) where $x \neq 0$ and $y \neq 0$. So we need to look at points $(x_0, 0)$, $x_0 \neq 0$ and $(0, y_0)$, $y_0 \neq 0$. At $(x_0, 0)$:

$$f_x(x_0, 0) = \lim_{x \to x_0} \frac{f(x,0) - f(x_0, 0)}{x - x_0} = 0$$

$$f_y(x_0, 0) = \lim_{y \to 0} \frac{f(x_0, y) - f(x_0, 0)}{y} = \lim_{y \to 0} \frac{\sqrt{|x_0 y|}}{y}$$

which does not exist. So f is not differentiable at the points $(x_0, 0)$, $x_0 \neq 0$. Similarly, f is not differentiable at the points $(0, y_0)$, $y_0 \neq 0$.

(d)

$$f_x(0,0) = \lim_{x \to 0} \frac{f(x,0) - f(0,0)}{x} = 0$$

$$f_y(0,0) = \lim_{y \to 0} \frac{f(0,y) - f(0,0)}{y} = 0$$

(e) Let $\vec{u} = (\vec{i} + \vec{j})/\sqrt{2}$:

$$f_{\vec{u}}(0,0) = \lim_{t \to 0^+} \frac{f(\frac{t}{\sqrt{2}}, \frac{t}{\sqrt{2}}) - f(0,0)}{t} = \lim_{t \to 0^+} \frac{\sqrt{\frac{t^2}{2}}}{t} = \frac{1}{\sqrt{2}}.$$

We know that $\nabla f(0,0) = \vec{0}$ because both partial derivatives are 0. But if f were differentiable, $f_{\vec{u}}(0,0) = \nabla f(0,0) \cdot \vec{u} = f_x(0,0) \cdot \frac{1}{\sqrt{2}} + f_y(0,0) \cdot \frac{1}{\sqrt{2}} = 0$. But since, in fact, $f_{\vec{u}}(0,0) = 1/\sqrt{2}$, we conclude that f is not differentiable.

Solutions for Chapter 14 Review

Exercises

1. Vector. Taking the gradient and substituting $x = 1$, $y = 2$ gives

$$\text{grad}(x^3 e^{-y/2})\Big|_{(1,2)} = \left(3x^2 e^{-y/2}\vec{i} - \frac{x^3}{2}e^{-y/2}\vec{j}\right)\Big|_{(1,2)} = 3e^{-1}\vec{i} - \frac{1}{2}e^{-1}\vec{j}.$$

5. Taking the derivative of f with respect to x and treating y as a constant we get $f_x = 2xy + 3x^2 - 7y^6$. To find f_y we take the derivative of f with respect to y and treat x as a constant. Thus, $f_y = x^2 - 42xy^5$.

9. For both partial derivatives we use the quotient rule. Thus,

$$f_x = \frac{2xy(x^2 + y^2) - x^2y2x}{(x^2 + y^2)^2} = \frac{2x^3y + 2xy^3 - 2x^3y}{(x^2 + y^2)^2} = \frac{2xy^3}{(x^2 + y^2)^2},$$

and

$$f_y = \frac{x^2(x^2 + y^2) - x^2y2y}{(x^2 + y^2)^2} = \frac{x^4 + x^2y^2 - 2x^2y^2}{(x^2 + y^2)^2} = \frac{x^4 - x^2y^2}{(x^2 + y^2)^2}.$$

13. Since we take the derivative with respect to N, we use the power rule to obtain $f_N = c\alpha N^{\alpha-1}V^{\beta}$.

17. $z_x = \frac{1}{2ay}(-2)\frac{1}{x^3} + \frac{15x^4 abc}{y} = -\frac{1}{ax^3 y} + \frac{15abcx^4}{y} = \frac{15a^2 bcx^7 - 1}{ax^3 y}$

21. Using the chain rule and the quotient rule,

$$\frac{\partial}{\partial w}\left(\frac{x^2 yw - xy^3 w^7}{w - 1}\right)^{-7/2}$$

$$= -\frac{7}{2}\left(\frac{x^2 yw - xy^3 w^7}{w - 1}\right)^{-9/2}\left(\frac{(w-1)(x^2 y - 7xy^3 w^6) - (x^2 yw - xy^3 w^7)(1)}{(w-1)^2}\right)$$

$$= -\frac{7}{2}\left(\frac{w - 1}{x^2 yw - xy^3 w^7}\right)^{-9/2}\left(\frac{(w-1)(x^2 y - 7xy^3 w^6) - (x^2 yw - xy^3 w^7)}{(w-1)^2}\right)$$

$$= \frac{7}{2}\left(\frac{w - 1}{x^2 yw - xy^3 w^7}\right)^{-9/2} \cdot \frac{x^2 y + 6xy^3 w^7 - 7xy^3 w^6}{(w-1)^2}$$

25. The first order partial derivatives are

$$u_x = e^x \sin y, \quad u_y = e^x \cos y.$$

Thus the second order partials are

$$u_{xx} = e^x \sin y, \quad u_{yy} = -e^x \sin y.$$

29. Since $f_x = 2x$, $f_y = 2y + 3y^2$ and $f_z = 0$, we have

$$\text{grad} f = 2x\vec{i} + (2y + 3y^2)\vec{j}.$$

33. Since $f_x = 2x \sin(x^2 + y^2 + z^2)$, $f_y = 2y \sin(x^2 + y^2 + z^2)$ and $f_z = 2z \sin(x^2 + y^2 + z^2)$, we have

$$\text{grad} f = 2x \sin(x^2 + y^2 + z^2)\vec{i} + 2y \sin(x^2 + y^2 + z^2)\vec{j} + 2z \sin(x^2 + y^2 + z^2)\vec{k}.$$

37. Since the partial derivatives are

$$\frac{\partial f}{\partial x} = \cos(xy) \cdot (y) - \sin(xy) \cdot (y) = y[\cos(xy) - \sin(xy)]$$
$$\frac{\partial f}{\partial y} = \cos(xy) \cdot (x) - \sin(xy) \cdot (x) = x[\cos(xy) - \sin(xy)]$$

we have

$$\text{grad} f = \frac{\partial f}{\partial x}\vec{i} + \frac{\partial f}{\partial y}\vec{j}$$
$$= y[\cos(xy) - \sin(xy)]\vec{i} + x[\cos(xy) - \sin(xy)]\vec{j}.$$

41. We have $\text{grad} f = e^y\vec{i} + xe^y\vec{j}$, so $\text{grad} f(3,0) = \vec{i} + 3\vec{j}$. A unit vector in the direction we want is $\vec{u} = (1/5)(4\vec{i} - 3\vec{j})$. Therefore, the directional derivative is

$$\text{grad} f(3,0) \cdot \vec{u} = \frac{1 \cdot 4 + 3(-3)}{5} = -1.$$

45. We have $\text{grad} f = (e^{x+z} \cos y)\vec{i} - (e^{x+z} \sin y)\vec{j} + (e^{x+z} \cos y)\vec{k}$, so $\text{grad} f(1, 0, -1) = \vec{i} + \vec{k}$. A unit vector in the direction we want is $\vec{u} = (1/\sqrt{3})(\vec{i} + \vec{j} + \vec{k})$. Therefore, the directional derivative is

$$\text{grad} f(1, 0, -1) \cdot \vec{u} = \frac{1 \cdot 1 + 1 \cdot 1}{\sqrt{3}} = \frac{2}{\sqrt{3}}.$$

49. Let $f(x, y, z) = z^2 - 4x^2 - 3y^2$ so that the surface (a hyperboloid) is the level surface $f(x, y, x) = 9$. Since

$$\text{grad} f = -8x\vec{i} - 6y\vec{j} + 2z\vec{k}$$

we have

$$\text{grad} f(1, 1, 4) = -8\vec{i} - 6\vec{j} + 8\vec{k}.$$

Thus an equation of the tangent plane at $(1, 1, 4)$ is

$$-8(x - 1) - 6(y + 6) + 8(z - 4) = 0$$

so

$$-4x - 3y + 4z = 9.$$

53. The quadratic Taylor expansion about $(0, 0)$ is given by

$$f(x, y) \approx Q(x, y) = f(0, 0) + f_x(0, 0)x + f_y(0, 0)y + \frac{1}{2}f_{xx}(0, 0)x^2 + f_{xy}(0, 0)xy + \frac{1}{2}f_{yy}(0, 0)y^2.$$

First we find all the relevant derivatives

$$f(x, y) = (x + 1)^3(y + 2)$$
$$f_x(x, y) = 3(x + 1)^2(y + 2)$$
$$f_y(x, y) = (x + 1)^3$$
$$f_{xx}(x, y) = 6(x + 1)(y + 2)$$
$$f_{yy}(x, y) = 0$$
$$f_{xy}(x, y) = 3(x + 1)^2$$

Now we evaluate each of these derivatives at $(0, 0)$ and substitute into the formula to get as our final answer:

$$Q(x, y) = 2 + 6x + y + 6x^2 + 3xy$$

Notice this is the same as what you get if you expand $(x + 1)^3(y + 2)$ and keep only the terms of degree 2 or less.

Problems

57. (a) Let $f(x, y, z) = 2x^2 - 2xy^2 + az$ so that the given surface is the level surface $f(x, y, z) = a$. Since

$$\text{grad } f = (4x - 2y^2)\vec{i} - 4xy\vec{j} + a\vec{k}$$

we have

$$\text{grad } f(1, 1, 1) = 2\vec{i} - 4\vec{j} + a\vec{k}.$$

Thus an equation of the tangent plane at $(1, 1, 1)$ is

$$2(x - 1) - 4(y - 1) + a(z - 1) = 0$$

or

$$2x - 4y + az = a - 2.$$

(b) Substituting $x = 0$, $y = 0$, $z = 0$ into the equation for the tangent plane in part (a) we have $0 = a - 2$. The tangent plane passes through the origin if $a = 2$.

61. The partial derivative, $\partial Q/\partial b$ is the rate of change of the quantity of beef purchased with respect to the price of beef, when the price of chicken stays constant. If the price of beef increases and the price of chicken stays the same, we expect consumers to buy less beef and more chicken. Thus when b increases, we expect Q to decrease, so $\partial Q/\partial b < 0$.

On the other hand, $\partial Q/\partial c$ is the rate of change of the quantity of beef purchased with respect to the price of chicken, when the price of beef stays constant. An increase in the price of chicken is likely to cause consumers to buy less chicken and more beef. Thus when c increases, we expect Q to increase, so $\partial Q/\partial c > 0$.

65. We need the partial derivatives, $f_x(1, 0)$ and $f_y(1, 0)$. We have

$$f_x(x, y) = 2xe^{xy} + x^2ye^{xy}, \qquad \text{so } f_x(1, 0) = 2$$
$$f_y(x, y) = x^3e^{xy}, \qquad \text{so } f_y(1, 0) = 1.$$

(a) Since $f(1, 0) = 1$, the tangent plane is

$$z = f(1, 0) + 2(x - 1) + 1(y - 0) = 1 + 2(x - 1) + y.$$

(b) The linear approximation can be obtained from the equation of the tangent plane:

$$f(x, y) \approx 1 + 2(x - 1) + y.$$

(c) At $(1, 0)$, the differential is

$$df = f_x dx + f_y dy = 2dx + dy.$$

69. The directional derivative is approximately the change in z (as we move in direction $\vec{v}$) divided by the horizontal change in position. The directional derivative is $f_{\vec{v}} \approx \dfrac{2-1}{4-1} = \dfrac{1}{3} \approx 0.3$.

73. The directional derivative is approximately the change in z (as we move in direction $\vec{v}$) divided by the horizontal change in position. In the direction of $\vec{v}$ we go from point $(3,3)$ to point $(1,4)$. We have the directional derivative
$$\approx \frac{2-3}{\| -2\vec{i} + \vec{j} \|} = \frac{-1}{\sqrt{5}} \approx -0.4.$$

77. (a) Incorrect. $\| \operatorname{grad} H \|$ is not the rate of change of H. In fact, there's no such thing as the rate of change of H, although directional derivatives can give its rate of change in a particular direction. For example, this expression would give the wrong answer if the ant was crawling along a contour of H, since then the rate of change of the temperature it experiences is zero even though $\| \operatorname{grad} H \|$ and $\vec{v}$ might not be zero.

(b) Correct. If $\vec{u} = \vec{v}/\|\vec{v}\|$, then

$$\operatorname{grad} H \cdot \vec{v} = \operatorname{grad} H \cdot \frac{\vec{v}}{\|\vec{v}\|}\|\vec{v}\| = (\operatorname{grad} H \cdot \vec{u})\|\vec{v}\| = H_{\vec{u}}\|\vec{v}\|$$

$$= (\text{Rate of change of } H \text{ in direction } \vec{u} \text{ in deg/cm})(\text{Speed of ant in cm/sec})$$

$$= \text{Rate of change of } H \text{ in deg/sec.}$$

(c) Incorrect, this is the directional derivative, which gives the rate of change with respect to distance, not time.

81. We have

$$\frac{\partial P}{\partial x} = 5e^{-0.1\sqrt{x^2+y^2+z^2}}(-0.1)(2x)\frac{1}{2}(x^2+y^2+z^2)^{-1/2} = -0.5e^{-0.1\sqrt{x^2+y^2+z^2}}\frac{x}{\sqrt{x^2+y^2+z^2}},$$

and similarly

$$\frac{\partial P}{\partial y} = -0.5e^{-0.1\sqrt{x^2+y^2+z^2}}\frac{y}{\sqrt{x^2+y^2+z^2}}, \quad \text{and} \quad \frac{\partial P}{\partial z} = -0.5e^{-0.1\sqrt{x^2+y^2+z^2}}\frac{z}{\sqrt{x^2+y^2+z^2}}.$$

Thus the gradient of P at $(0,0,1)$ is

$$\operatorname{grad} P = \frac{\partial P}{\partial x}\bigg|_{(x,y,z)=(0,0,1)} \vec{i} + \frac{\partial P}{\partial y}\bigg|_{(x,y,z)=(0,0,1)} \vec{j} + \frac{\partial P}{\partial z}\bigg|_{(x,y,z)=(0,0,1)} \vec{k} = -0.5e^{-0.1}\vec{k}.$$

Let $\vec{u}$ be a unit vector in the direction of $\vec{v}$, so

$$\vec{u} = \frac{\vec{v}}{\|\vec{v}\|}.$$

Then

Rate of change of pressure in atm/sec = (Directional derivative in direction $\vec{u}$ in atm/mi)(Speed of spacecraft in mi/sec)

$$= P_{\vec{u}}\|\vec{v}\| = (\operatorname{grad} P \cdot \vec{u})\|\vec{v}\| = \operatorname{grad} P \cdot \frac{\vec{v}}{\|\vec{v}\|}\|\vec{v}\| = \operatorname{grad} P \cdot \vec{v}$$

$$= -0.5e^{-0.1}\vec{k} \cdot (\vec{i} - 2.5\vec{k}) = 0.5 \cdot 2.5e^{-0.1} = 1.131 \text{ atm/sec.}$$

85. (a) The chain rule gives

$$\frac{\partial z}{\partial x} = \frac{\partial z}{\partial u}\frac{\partial u}{\partial x} + \frac{\partial z}{\partial v}\frac{\partial v}{\partial x} = (2u - e^v)1 + (-ue^v)2.$$

At $(x,y) = (1,2)$, we have $u = 1 + 2 \cdot 2 = 5$ and $v = 2 \cdot 1 - 2 = 0$, so

$$\frac{\partial z}{\partial y}\bigg|_{(x,y)=(1,2)} = (2 \cdot 5 - e^0)1 - 5e^0 \cdot 2 = -1.$$

(b) The chain rule gives

$$\frac{\partial z}{\partial y} = \frac{\partial z}{\partial u}\frac{\partial u}{\partial y} + \frac{\partial z}{\partial v}\frac{\partial v}{\partial y} = (2u - e^v)2 + (-ue^v)(-1).$$

At $(x,y) = (1,2)$, we have $(u,v) = (5,0)$, so

$$\frac{\partial z}{\partial y}\bigg|_{(x,y)=(1,2)} = (2 \cdot 5 - e^0)2 + 5e^0 = 23.$$

89. The error, dT, in the period T is given by

$$dT = \frac{\partial T}{\partial l} dl + \frac{\partial T}{\partial g} dg,$$

where

$$\frac{\partial T}{\partial l} = \frac{\sqrt{\frac{l}{g}}\, \pi}{l}$$

and

$$\frac{\partial T}{\partial g} = -\frac{\sqrt{\frac{l}{g}}\, \pi}{g},$$

so that

$$T_l(2, 9.8) = 0.7096, \quad T_g(2, 9.8) = -0.1448.$$

We also have that

$$dl = -0.01, \quad dg = 0.01.$$

The maximum discrepancy in the period is then given by

$$dT = 0.7096(-0.01) - 0.1448(0.01) = -0.008544.$$

93. The directional derivative, or slope, of f at $(2, 1)$ in the direction perpendicular to these contours is the length of the gradient. At the point $(2, 1)$, we have grad $f = -3\vec{i} + 4\vec{j}$, so $\| \text{grad } f \| = 5$. Thus if d is the distance between the contours, we have

$$\frac{7.3 - 7}{d} = \text{Slope in direction perpendicular to contours} = 5.$$

Thus $d = 0.3/5 = 0.06$.

97. By the chain rule,

$$f_r(2, 1) = f_x(2, 1) \cos \theta + f_y(2, 1) \sin \theta$$
$$f_\theta(2, 1) = f_x(2, 1)(-r \sin \theta) + f_y(2, 1)(r \cos \theta).$$

Since $x^2 + y^2 = r^2$, we have $r = \sqrt{2^2 + 1^2} = \sqrt{5}$, and $\cos \theta = x/r = 2/\sqrt{5}$, and $\sin \theta = y/r = 1/\sqrt{5}$. Thus

$$f_r(2, 1) = (-3)\frac{2}{\sqrt{5}} + (4)\frac{1}{\sqrt{5}} = -\frac{2}{\sqrt{5}}$$
$$f_\theta(2, 1) = (-3)(-1) + (4)(2) = 11.$$

For $\vec{u} = (1/\sqrt{5})(2\vec{i} + \vec{j})$, we have $f_{\vec{u}}(2, 1) = -2/\sqrt{5}$. Since $\vec{u}$ is pointing radially out from the origin to the point $(2, 1)$, we expect that $f_{\vec{u}}(2, 1) = f_r(2, 1)$.

101. (a) The first-order Taylor polynomial of a function f about a point (a, b) is equal to

$$f(a, b) + f_x(a, b)(x - a) + f_y(a, b)(y - b).$$

Computing the partial derivatives, we get:

$$f_x = 2(x - 1)e^{(x-1)^2 + (y-3)^2}$$
$$f_y = 2(y - 3)e^{(x-1)^2 + (y-3)^2}$$
$$f_x(0, 0) = 2(-1)e^{(-1)^2 + (-3)^2}$$
$$= -2e^{10}$$
$$f_y(0, 0) = 2(-3)e^{(-1)^2 + (-3)^2}$$
$$= -6e^{10}$$

Thus,

$$f(x, y) \approx e^{10} - 2e^{10}x - 6e^{10}y$$

(b) The second-order Taylor polynomial of a function f about the point $(1, 3)$ is given by

$$f(1,3) + f_x(1,3)(x-1) + f_y(1,3)(y-3)$$
$$+ \frac{1}{2}f_{xx}(1,3)(x-1)^2 + f_{xy}(1,3)(x-1)(y-3) + \frac{1}{2}f_{yy}(1,3)(y-3)^2.$$

Computing the partial derivatives, we get:

$$f_x = 2(x-1)e^{(x-1)^2+(y-3)^2}$$
$$f_y = 2(y-3)e^{(x-1)^2+(y-3)^2}$$
$$f_{xx} = (4(x-1)^2 + 2)e^{(x-1)^2+(y-3)^2}$$
$$f_{xy} = 4(x-1)(y-3)e^{(x-1)^2+(y-3)^2}$$
$$f_{yy} = (4(y-3)^2 + 2)e^{(x-1)^2+(y-3)^2}$$

Substituting in the point $(1, 3)$ to these partial derivatives, we get:

$$f_x(1,3) = 0$$
$$f_y(1,3) = 0$$
$$f_{xy}(1,3) = 0$$
$$f_{xx}(1,3) = (4(0)^2 + 2)e^{0^2+0^2} = 2$$
$$f_{yy}(1,3) = (4(0)^2 + 2)e^{0^2+0^2} = 2$$

Thus,

$$f(x,y) \approx e^0 + 0(x-1) + 0(y-3) + \frac{2}{2}(x-2)^2 + 0(x-1)(y-3) + \frac{2}{2}(y-3)^2$$
$$f(x,y) \approx 1 + (x-1)^2 + (y-3)^2.$$

(c) A vector perpendicular to the level curve is grad f. At the point $(0, 0)$, we have

$$\text{grad } f = f_x(0,0)\vec{i} + f_y(0,0)\vec{j}$$

Computing partial derivatives, we have

$$f_x = 2(x-1)e^{(x-1)^2+(y-3)^2}$$
$$f_y = 2(y-3)e^{(x-1)^2+(y-3)^2}$$
$$f_x(0,0) = 2(-1)e^{(-1)^2+(-3)^2}$$
$$= -2e^{10}$$
$$f_y(0,0) = 2(-3)e^{(-1)^2+(-3)^2}$$
$$= -6e^{10}$$

Therefore, a perpendicular vector is grad $f = -2e^{10}\vec{i} - 6e^{10}\vec{j}$. Any multiple of grad f, say $-2\vec{i} - 6\vec{j}$, will do.

(d) Since the surface can be represented by the level surface

$$F(x,y,z) = f(x,y) - z = 0,$$

a vector perpendicular to the surface at $(0, 0)$ is given by

$$\text{grad } F = f_x(0,0)\vec{i} + f_y(0,0)\vec{j} - \vec{k} = -2e^{10}\vec{i} - 6e^{10}\vec{j} - \vec{k}$$

CAS Challenge Problems

105. (a) We have $f(1,2) = A_0 + A_1 + 2A_2 + A_3 + 2A_4 + 4A_5$, $f_x(1,2) = A_1 + 2A_3 + 2A_4$, and $f_y(1,2) = A_2 + A_4 + 4A_5$, so the linear approximation is

$$L(x,y) = A_0 + A_1 + 2A_2 + A_3 + 2A_4 + 4A_5 + (A_1 + 2A_3 + 2A_4)(x-1) + (A_2 + A_4 + 4A_5)(y-2).$$

Also, $m(t) = 1 + B_1 t$ and $n(t) = 2 + C_1 t$.

(b) Using a CAS to compute the derivatives, we find that they are both the same:

$$\frac{d}{dt}f(x(t), y(t))|_{t=0} = \frac{d}{dt}l(m(t), n(t))|_{t=0} = A_1 B_1 + 2A_3 B_1 + 2A_4 B_1 + A_2 C_1 + A_4 C_1 + 4A_5 C_1.$$

This can be explained using the chain rule and the fact that the derivative of a function at a point is the same as the derivative of its linear approximation there:

$$\frac{d}{dt}f(x(t), y(t))|_{t=0} = \frac{\partial}{\partial x}f(x(0), y(0))x'(0) + \frac{\partial}{\partial y}f(x(0), y(0))y'(0)$$

$$= \frac{\partial}{\partial x}l(x(0), y(0))m'(0) + \frac{\partial}{\partial y}l(x(0), y(0))n'(0) = \frac{d}{dt}l(m(t), n(t))|_{t=0}$$

CHECK YOUR UNDERSTANDING

1. True. This is the instantaneous rate of change of f in the x-direction at the point $(10, 20)$.

5. True. The property $f_x(a, b) > 0$ means that f increases in the positive x-direction near (a, b), so f must decrease in the negative x-direction near (a, b).

9. True. A function with constant $f_x(x, y)$ and $f_y(x, y)$ has constant x-slope and constant y-slope, and therefore has a graph which is a plane.

13. True. Since g is a function of x only, it can be treated like a constant when taking the y partial derivative.

17. False. The equation $z = 2 + 2x(x - 1) + 3y^2(y - 1)$ is not linear. The correct equation is $z = 2 + 2(x - 1) + 3(y - 1)$, which is obtained by evaluating the partial derivatives at the point $(1, 1)$.

21. False. The graph of f is a paraboloid, opening upward. The tangent plane to this surface at any point lies completely under the surface (except at the point of tangency). So the local linearization *underestimates* the value of f at nearby points.

25. True. If f is linear, then $f(x, y) = mx + ny + c$ for some m, n and c. So $f_x = m$ and $f_y = n$ giving $df = m\,dx + n\,dy$, which is linear in the variables dx and dy.

29. False. The gradient is *perpendicular* to the contour of f at (a, b).

33. False. The gradient vector at $(3, 4)$ has no relation to the direction of the vector $3\vec{i} + 4\vec{j}$. For example, if $f(x, y) = x + 2y$, then grad $f = \vec{i} + 2\vec{j}$, which is not perpendicular to $3\vec{i} + 4\vec{j}$. The gradient vector grad $f(3, 4)$ *is* perpendicular to the contour of f passing through the point $(3, 4)$.

37. True. The length of the gradient gives the maximal directional derivative in any direction. The gradient vector is grad $f(0, 0) = \vec{i} + \vec{j}$, which has length $\sqrt{2}$.

41. True. The definition of $f_x(x, y)$ is the limit of the difference quotient

$$f_x(x, y) = \lim_{h \to 0} \frac{f(x + h, y) - f(x, y)}{h}.$$

The symmetry of f gives

$$\frac{f(x + h, y) - f(x, y)}{h} = \frac{f(y, x + h) - f(y, x)}{h}.$$

The definition of $f_y(y, x)$ is the limit of the difference quotient

$$f_y(y, x) = \lim_{h \to 0} \frac{f(y, x + h) - f(y, x)}{h}.$$

Thus $f_x(x, y) = f_y(y, x)$.

45. True. Take the direction perpendicular to grad f at that point. If grad $f = 0$, any direction will do.

CHAPTER FIFTEEN

Solutions for Section 15.1

Exercises

1. The point A is not a critical point and the contour lines look like parallel lines. The point B is a critical point and is a local maximum; the point C is a saddle point.

5. At the origin, the second derivative test gives

$$D = k_{xx}k_{yy} - (k_{xy})^2 = \left((-\sin x \sin y)(-\sin x \sin y) - (\cos x \cos y)^2\right)\Big|_{x=0, y=0}$$
$$= \sin^2 0 \sin^2 0 - \cos^2 0 \cos^2 0$$
$$= -1 < 0.$$

Thus $k(0,0)$ is a saddle point.

9. To find the critical points, we solve $f_x = 0$ and $f_y = 0$ for x and y. Solving

$$f_x = 3x^2 - 6x = 0$$
$$f_y = 2y + 10 = 0$$

shows that $x = 0$ or $x = 2$ and $y = -5$. There are two critical points: $(0, -5)$ and $(2, -5)$.

We have

$$D = (f_{xx})(f_{yy}) - (f_{xy})^2 = (6x - 6)(2) - (0)^2 = 12x - 12.$$

When $x = 0$, we have $D = -12 < 0$, so f has a saddle point at $(0, -5)$. When $x = 2$, we have $D = 12 > 0$ and $f_{xx} = 6 > 0$, so f has a local minimum at $(2, -5)$.

13. Find the critical point(s) by setting

$$f_x = (xy + 1) + (x + y) \cdot y = y^2 + 2xy + 1 = 0,$$
$$f_y = (xy + 1) + (x + y) \cdot x = x^2 + 2xy + 1 = 0,$$

then we get $x^2 = y^2$, and so $x = y$ or $x = -y$.

If $x = y$, then $x^2 + 2x^2 + 1 = 0$, that is, $3x^2 = -1$, and there is no real solution. If $x = -y$, then $x^2 - 2x^2 + 1 = 0$, which gives $x^2 = 1$. Solving it we get $x = 1$ or $x = -1$, then $y = -1$ or $y = 1$, respectively. Hence, $(1, -1)$ and $(-1, 1)$ are critical points.

Since

$$f_{xx}(x, y) = 2y,$$
$$f_{xy}(x, y) = 2y + 2x \quad \text{and}$$
$$f_{yy}(x, y) = 2x,$$

the discriminant is

$$D(x, y) = f_{xx}f_{yy} - f_{xy}^2$$
$$= 2y \cdot 2x - (2y + 2x)^2$$
$$= -4(x^2 + xy + y^2).$$

thus

$$D(1, -1) = -4(1^2 + 1 \cdot (-1) + (-1)^2) = -4 < 0,$$
$$D(-1, 1) = -4((-1)^2 + (-1) \cdot 1 + 1^2) = -4 < 0.$$

Therefore $(1, -1)$ and $(-1, 1)$ are saddle points.

Problems

17. At the critical point $x = 1$, $y = 0$,

$$f_x = 2x + A = 0, \quad \text{so } 2 + A = 0 \text{ or } A = -2$$
$$f_y = 2y = 0.$$

Thus, $f(x, y) = x^2 - 2x + y^2 + B$ has a critical point at $(1, 0)$. Since $f_{xx} = 2$ and $f_{yy} = 2$ and $f_{xy} = 0$ at $(1, 0)$,

$$D = f_{xx}f_{yy} - (f_{xy})^2 = 2 \cdot 2 - 0^2 = 4,$$

so the second derivative test shows the critical point at $(1, 0)$ is a local minimum. The value of the minimum is

$$f(1, 0) = 1^2 - 2 \cdot 1 + 0^2 + B = 20, \quad \text{so } B = 21.$$

21. (a) P is a local maximum.
(b) Q is a saddle point.
(c) R is a local minimum.
(d) S is none of these.

25. At a critical point,

$$f_x = \cos x \sin y = 0 \quad \text{so} \quad \cos x = 0 \text{ or } \sin y = 0;$$

and

$$f_y = \sin x \cos y = 0 \quad \text{so} \quad \sin x = 0 \text{ or } \cos y = 0.$$

Case 1: Assume $\cos x = 0$. This gives

$$x = \cdots - \frac{3\pi}{2}, -\frac{\pi}{2}, \frac{\pi}{2}, \frac{3\pi}{2}, \cdots$$

(This can be written more compactly as: $x = k\pi + \pi/2$, for $k = 0, \pm1, \pm2, \cdots$.)
If $\cos x = 0$, then $\sin x = \pm1 \neq 0$. Thus in order to have $f_y = 0$ we need $\cos y = 0$, giving

$$y = \cdots - \frac{3\pi}{2}, -\frac{\pi}{2}, \frac{\pi}{2}, \frac{3\pi}{2}, \cdots$$

(More compactly, $y = l\pi + \pi/2$, for $l = 0, \pm1, \pm2, \cdots$)
Case 2: Assume $\sin y = 0$. This gives

$$y = \cdots - 2\pi, -\pi, 0, \pi, 2\pi, \cdots$$

(More compactly, $y = l\pi$, for $l = 0, \pm1, \pm2, \cdots$)
If $\sin y = 0$, then $\cos y = \pm1 \neq 0$, so to get $f_y = 0$ we need $\sin x = 0$, giving

$$x = \cdots, -2\pi, -\pi, 0, \pi, 2\pi, \cdots$$

(More compactly, $x = k\pi$ for $k = 0, \pm0, \pm1, \pm2, \cdots$)
Hence we get all the critical points of $f(x, y)$. Those from Case 1 are as follows:

$$\cdots \left(-\frac{\pi}{2}, -\frac{\pi}{2}\right), \left(-\frac{\pi}{2}, \frac{\pi}{2}\right), \left(-\frac{\pi}{2}, \frac{3\pi}{2}\right) \cdots$$

$$\cdots \left(\frac{\pi}{2}, -\frac{\pi}{2}\right), \left(\frac{\pi}{2}, \frac{\pi}{2}\right), \left(\frac{\pi}{2}, \frac{3\pi}{2}\right) \cdots$$

$$\cdots \left(\frac{3\pi}{2}, -\frac{\pi}{2}\right), \left(\frac{3\pi}{2}, \frac{\pi}{2}\right), \left(\frac{3\pi}{2}, \frac{3\pi}{2}\right) \cdots$$

Those from Case 2 are as follows:

$$\cdots (-\pi, -\pi), (-\pi, 0), (-\pi, \pi), (-\pi, 2\pi) \cdots$$

$$\cdots (0, -\pi), (0, 0), (0, \pi), (0, 2\pi) \cdots$$

$$\cdots (\pi, -\pi), (\pi, 0), (\pi, \pi), (\pi, 2\pi) \cdots$$

More compactly these points can be written as,

$$(k\pi, l\pi), \text{ for } k = 0, \pm1, \pm2, \cdots, l = 0, \pm1, \pm2, \cdots$$
$$\text{and } (k\pi + \frac{\pi}{2}, l\pi + \frac{\pi}{2}), \text{ for } k = 0, \pm1, \pm2, \cdots, l = 0, \pm1, \pm2, \cdots$$

To classify the critical points, we find the discriminant. We have

$$f_{xx} = -\sin x \sin y, \qquad f_{yy} = -\sin x \sin y, \qquad \text{and} \qquad f_{xy} = \cos x \cos y.$$

Thus the discriminant is

$$
\begin{aligned}
D(x, y) &= f_{xx}f_{yy} - f_{xy}^2 \\
&= (-\sin x \sin y)(-\sin x \sin y) - (\cos x \cos y)^2 \\
&= \sin^2 x \sin^2 y - \cos^2 x \cos^2 y \\
&= \sin^2 y - \cos^2 x. \quad (\text{Use: } \sin^2 x = 1 - \cos^2 x \text{ and factor.})
\end{aligned}
$$

At points of the form $(k\pi, l\pi)$ where $k = 0, \pm1, \pm2, \cdots; l = 0, \pm1, \pm2, \cdots$, we have
$D(x, y) = -1 < 0$ so $(k\pi, l\pi)$ are saddle points.
At points of the form $(k\pi + \frac{\pi}{2}, l\pi + \frac{\pi}{2})$ where $k = 0, \pm1, \pm2, \cdots; l = 0, \pm1, \pm2, \cdots$
$D(k\pi + \frac{\pi}{2}, l\pi + \frac{\pi}{2}) = 1 > 0$, so we have two cases:
If k and l are both even or k and l are both odd, then
$f_{xx} = -\sin x \sin y = -1 < 0$, so $(k\pi + \frac{\pi}{2}, l\pi + \frac{\pi}{2})$ are local maximum points.
If k is even but l is odd or k is odd but l is even, then
$f_{xx} = 1 > 0$ so $(k\pi + \frac{\pi}{2}, l\pi + \frac{\pi}{2})$ are local minimum points.

29. (a) (a, b) is a critical point. Since the discriminant $D = f_{xx}f_{yy} - f_{xy}^2 = -f_{xy}^2 < 0$, (a, b) is a saddle point.
(b) See Figure 15.1.

Figure 15.1

33. The first order partial derivatives are

$$f_x(x, y) = 2kx - 2y \qquad \text{and} \qquad f_y(x, y) = 2ky - 2x.$$

And the second order partial derivatives are

$$f_{xx}(x, y) = 2k \qquad f_{xy}(x, y) = -2 \qquad f_{yy}(x, y) = 2k$$

Since $f_x(0, 0) = f_y(0, 0) = 0$, the point $(0, 0)$ is a critical point. The discriminant is

$$D = (2k)(2k) - 4 = 4(k^2 - 1).$$

For $k = \pm 2$, the discriminant is positive, $D = 12$. When $k = 2$, $f_{xx}(0,0) = 4$ which is positive so we have a local minimum at the origin. When $k = -2$, $f_{xx}(0,0) = -4$ so we have a local maximum at the origin. In the case $k = 0$, $D = -4$ so the origin is a saddle point.

Lastly, when $k = \pm 1$ the discriminant is zero, so the second derivative test can tell us nothing. Luckily, we can factor $f(x,y)$ when $k = \pm 1$. When $k = 1$,

$$f(x,y) = x^2 - 2xy + y^2 = (x-y)^2.$$

This is always greater than or equal to zero. So $f(0,0) = 0$ is a minimum and the surface is a trough-shaped parabolic cylinder with its base along the line $x = y$.

When $k = -1$,

$$f(x,y) = -x^2 - 2xy - y^2 = -(x+y)^2.$$

This is always less than or equal to zero. So $f(0,0) = 0$ is a maximum. The surface is a parabolic cylinder, with its top ridge along the line $x = -y$.

Figure 15.2

Solutions for Section 15.2

Exercises

1. Mississippi lies entirely within a region designated as 80s so we expect both the maximum and minimum daily high temperatures within the state to be in the 80s. The southwestern-most corner of the state is close to a region designated as 90s, so we would expect the temperature here to be in the high 80s, say 87-88. The northern-most portion of the state is located near the center of the 80s region. We might expect the high temperature there to be between 83-87.

Alabama also lies completely within a region designated as 80s so both the high and low daily high temperatures within the state are in the 80s. The southeastern tip of the state is close to a 90s region so we would expect the temperature here to be about 88-89 degrees. The northern-most part of the state is near the center of the 80s region so the temperature there is 83-87 degrees.

Pennsylvania is also in the 80s region, but it is touched by the boundary line between the 80s and a 70s region. Thus we expect the low daily high temperature to occur there and be about 80 degrees. The state is also touched by a boundary line of a 90s region so the high will occur there and be 89-90 degrees.

New York is split by a boundary between an 80s and a 70s region, so the northern portion of the state is likely to be about 74-76 while the southern portion is likely to be in the low 80s, maybe 81-84 or so.

California contains many different zones. The northern coastal areas will probably have the daily high as low as 65-68, although without another contour on that side, it is difficult to judge how quickly the temperature is dropping off to the west. The tip of Southern California is in a 100s region, so there we expect the daily high to be 100-101.

Arizona will have a low daily high around 85-87 in the northwest corner and a high in the 100s, perhaps 102-107 in its southern regions.

Massachusetts will probably have a high daily high around 81-84 and a low daily high of 70.

5. To maximize $z = x^2 + y^2$, it suffices to maximize x^2 and y^2. We can maximize both of these at the same time by taking the point $(1, 1)$, where $z = 2$. It occurs on the boundary of the square. (Note: We also have maxima at the points $(-1, -1), (-1, 1)$ and $(1, -1)$ which are on the boundary of the square.)

To minimize $z = x^2 + y^2$, we choose the point $(0, 0)$, where $z = 0$. It does not occur on the boundary of the square.

9. The function f has no global maximum or global minimum.

13. Suppose x is fixed. Then for large values of y the sign of f is determined by the highest power of y, namely y^3. Thus,

$$f(x, y) \to \infty \quad \text{as} \quad y \to \infty$$
$$f(x, y) \to -\infty \quad \text{as} \quad y \to -\infty.$$

So f does not have a global maximum or minimum.

Problems

17. Let the sides be x, y, z cm. Then the volume is given by $V = xyz = 32$.

The surface area S is given by

$$S = 2xy + 2xz + 2yz.$$

Substituting $z = 32/(xy)$ gives

$$S = 2xy + \frac{64}{y} + \frac{64}{x}.$$

At a critical point,

$$\frac{\partial S}{\partial x} = 2y - \frac{64}{x^2} = 0$$
$$\frac{\partial S}{\partial y} = 2x - \frac{64}{y^2} = 0,$$

The symmetry of the equations (or by dividing the equations) tells us that $x = y$ and

$$2x - \frac{64}{x^2} = 0$$
$$x^3 = 32$$
$$x = 32^{1/3} = 3.17 \text{ cm.}$$

Thus the only critical point is $x = y = (32)^{1/3}$ cm and $z = 32/\left((32)^{1/3} \cdot (32)^{1/3}\right) = (32)^{1/3}$ cm. At the critical point

$$S_{xx}S_{yy} - (S_{xy})^2 = \frac{128}{x^3} \cdot \frac{128}{y^3} - 2^2 = \frac{(128)^2}{x^3 y^3} - 4.$$

Since $D > 0$ and $S_{xx} > 0$ at this critical point, the critical point $x = y = z = (32)^{1/3}$ is a local minimum. Since $S \to \infty$ as $x, y \to \infty$, the local minimum is a global minimum.

21. The box is shown in Figure 15.3. Cost of four sides $= (2hl + 2wh)(1)¢$. Cost of two bottoms $= (2wl)(2)¢$. Thus the total cost C (in cents) of the box is

$$C = 2(hl + wh) + 4wl.$$

But volume $wlh = 512$, so $l = 512/(wh)$, thus

$$C = \frac{1024}{w} + 2wh + \frac{2048}{h}.$$

To minimize C, find the critical points of C by solving

$$C_h = 2w - \frac{2048}{h^2} = 0,$$

$$C_w = 2h - \frac{1024}{w^2} = 0.$$

We get

$$2wh^2 = 2048$$
$$2hw^2 = 1024.$$

Since $w, h \neq 0$, we can divide the first equation by the second giving

$$\frac{2wh^2}{2hw^2} = \frac{2048}{1024},$$

so

$$\frac{h}{w} = 2,$$

thus

$$h = 2w.$$

Substituting this in $C_h = 0$, we obtain $h^3 = 2048$, so $h = 12.7$ cm. Thus $w = h/2 = 6.35$ cm, and $l = 512/(wh) = 6.35$ cm. Now we check that these dimensions minimize the cost C. We find that

$$D = C_{hh}C_{ww} - C_{hw}^2 = (\frac{4096}{h^3})(\frac{2048}{w^3}) - 2^2,$$

and at $h = 12.7$, $w = 6.35$, $C_{hh} > 0$ and $D = 16 - 4 > 0$, thus C has a local minimum at $h = 12.7$ and $w = 6.35$. Since C increases without bound as $w, h \to 0$ or ∞, this local minimum must be a global minimum.

Therefore, the dimensions of the box that minimize the cost are $w = 6.35$ cm, $l = 6.35$ cm and $h = 12.7$ cm.

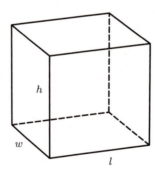

Figure 15.3

25. The total revenue is

$$R = pq = (60 - 0.04q)q = 60q - 0.04q^2,$$

and as $q = q_1 + q_2$, this gives

$$R = 60q_1 + 60q_2 - 0.04q_1^2 - 0.08q_1q_2 - 0.04q_2^2.$$

Therefore, the profit is

$$P(q_1, q_2) = R - C_1 - C_2$$
$$= -13.7 + 60q_1 + 60q_2 - 0.07q_1^2 - 0.08q_2^2 - 0.08q_1q_2.$$

At a local maximum point, we have grad $P = \vec{0}$:

$$\frac{\partial P}{\partial q_1} = 60 - 0.14q_1 - 0.08q_2 = 0,$$

$$\frac{\partial P}{\partial q_2} = 60 - 0.16q_2 - 0.08q_1 = 0.$$

Solving these equations, we find that

$$q_1 = 300 \quad \text{and} \quad q_2 = 225.$$

To see whether or not we have found a local maximum, we compute the second-order partial derivatives:

$$\frac{\partial^2 P}{\partial q_1^2} = -0.14, \quad \frac{\partial^2 P}{\partial q_2^2} = -0.16, \quad \frac{\partial^2 P}{\partial q_1 \partial q_2} = -0.08.$$

Therefore,

$$D = \frac{\partial^2 P}{\partial q_1^2} \frac{\partial^2 P}{\partial q_2^2} - \frac{\partial^2 P}{\partial q_1 \partial q_2} = (-0.14)(-0.16) - (-0.08)^2 = 0.016,$$

and so we have found a local maximum point. The graph of $P(q_1, q_2)$ has the shape of an upside down paraboloid since P is quadratic in q_1 and q_2, hence $(300, 225)$ is a global maximum point.

29. (a) We have $f(2, 1) = 120$.

   (i) If $x > 20$ then $f(x, y) > 10x > 200 > f(2, 1)$.

   (ii) If $y > 20$ then $f(x, y) > 20y > 400 > f(2, 1)$.

   (iii) If $x < 0.01$ and $y \le 20$ then $f(x, y) > 80/(xy) > 80/((0.01)(20)) = 400 > f(2, 1)$.

   (iv) If $y < 0.01$ and $x \le 20$ then $f(x, y) > 80/(xy) > 80/((20)(0.01)) = 400 > f(2, 1)$.

   (b) The continuous function f must achieve a minimum at some point (x_0, y_0) in the closed and bounded region R' : $0.01 \le x \le 20, 0.01 \le y \le 20$. Since $(2, 1)$ is in R', we must have $f(x_0, y_0) \le f(2, 1)$. By part (a), $f(x_0, y_0)$ is less than all values of f in the part of R that is outside R', so $f(x_0, y_0)$ is a minimum for f on all of R. Since (x_0, y_0) is not on the boundary of R, it must be a critical point of f.

   (c) The only critical point of f in R is the point $(2, 1)$, so by part (b) f has a global minimum there.

Solutions for Section 15.3

Exercises

1. Our objective function is $f(x, y) = x + y$ and our equation of constraint is $g(x, y) = x^2 + y^2 = 1$. To optimize $f(x, y)$ with Lagrange multipliers, we solve $\nabla f(x, y) = \lambda \nabla g(x, y)$ subject to $g(x, y) = 1$. The gradients of f and g are

$$\nabla f(x, y) = \vec{i} + \vec{j},$$
$$\nabla g(x, y) = 2x\vec{i} + 2y\vec{j}.$$

So the equation $\nabla f = \lambda \nabla g$ becomes

$$\vec{i} + \vec{j} = \lambda(2x\vec{i} + 2y\vec{j})$$

Solving for λ gives

$$\lambda = \frac{1}{2x} = \frac{1}{2y},$$

which tells us that $x = y$. Going back to our equation of constraint, we use the substitution $x = y$ to solve for y:

$$g(y, y) = y^2 + y^2 = 1$$
$$2y^2 = 1$$
$$y^2 = \frac{1}{2}$$
$$y = \pm\sqrt{\frac{1}{2}} = \pm\frac{\sqrt{2}}{2}.$$

Since $x = y$, our critical points are $(\frac{\sqrt{2}}{2}, \frac{\sqrt{2}}{2})$ and $(-\frac{\sqrt{2}}{2}, -\frac{\sqrt{2}}{2})$. Since the constraint is closed and bounded, maximum and minimum values of f subject to the constraint exist. Evaluating f at the critical points we find that the maximum value is $f(\frac{\sqrt{2}}{2}, \frac{\sqrt{2}}{2}) = \sqrt{2}$ and the minimum value is $f(-\frac{\sqrt{2}}{2}, -\frac{\sqrt{2}}{2}) = -\sqrt{2}$.

5. Our objective function is $f(x, y) = 3x - 2y$ and our equation of constraint is $g(x, y) = x^2 + 2y^2 = 44$. Their gradients are

$$\nabla f(x, y) = 3\vec{i} - 2\vec{j},$$
$$\nabla g(x, y) = 2x\vec{i} + 4y\vec{j}.$$

So the equation $\nabla f = \lambda \nabla g$ becomes $3\vec{i} - 2\vec{j} = \lambda(2x\vec{i} + 4y\vec{j})$. Solving for λ gives us

$$\lambda = \frac{3}{2x} = \frac{-2}{4y},$$

which we can use to find x in terms of y:

$$\frac{3}{2x} = \frac{-2}{4y}$$
$$-4x = 12y$$
$$x = -3y.$$

Using this relation in our equation of constraint, we can solve for y:

$$x^2 + 2y^2 = 44$$
$$(-3y)^2 + 2y^2 = 44$$
$$9y^2 + 2y^2 = 44$$
$$11y^2 = 44$$
$$y^2 = 4$$
$$y = \pm 2.$$

Thus, the critical points are $(-6, 2)$ and $(6, -2)$. Since the constraint is closed and bounded, maximum and minimum values of f subject to the constraint exist. Evaluating f at the critical points, we find that the maximum is $f(6, -2) = 18 + 4 = 22$ and the minimum value is $f(-6, 2) = -18 - 4 = -22$.

9. The objective function is $f(x, y, z) = x + 3y + 5z$ and the equation of constraint is $g(x, y, z) = x^2 + y^2 + z^2 = 1$. Their gradients are

$$\nabla f(x, y, z) = \vec{i} + 3\vec{j} + 5\vec{k},$$
$$\nabla g(x, y, z) = 2x\vec{i} + 2y\vec{j} + 2z\vec{k}.$$

So the equation $\nabla f = \lambda \nabla g$ becomes $\vec{i} + 3\vec{j} + 5\vec{k} = \lambda(2x\vec{i} + 2y\vec{j} + 2z\vec{k})$. Solving for λ we find

$$\lambda = \frac{1}{2x} = \frac{3}{2y} = \frac{5}{2z}.$$

Which provides us with the equations

$$2y = 6x$$
$$10x = 2z.$$

Solving the first equation for y gives us $y = 3x$. Solving the second equation for z gives us $z = 5x$. Substituting these into the equation of constraint, we can find x:

$$x^2 + (3x)^2 + (5x)^2 = 1$$
$$x^2 + 9x^2 + 25x^2 = 1$$
$$35x^2 = 1$$
$$x^2 = \frac{1}{35}$$
$$x = \pm\sqrt{\frac{1}{35}} = \pm\frac{\sqrt{35}}{35}.$$

Since $y = 3x$ and $z = 5x$, the critical points are at $\pm(\frac{\sqrt{35}}{35}, 3\frac{\sqrt{35}}{35}, \frac{\sqrt{35}}{7})$. Since the constraint is closed and bounded, maximum and minimum values of f subject to the constraint exist. Evaluating f at the critical points, we find the maximum is $f(\frac{\sqrt{35}}{35}, 3\frac{\sqrt{35}}{35}, \frac{\sqrt{35}}{7}) = \sqrt{35}\frac{35}{35} = \sqrt{35}$, and the minimum value is $f(-\frac{\sqrt{35}}{35}, -3\frac{\sqrt{35}}{35}, -\frac{\sqrt{35}}{7}) = -\sqrt{35}$.

13. The region $x^2 + y^2 \le 2$ is the shaded disk of radius $\sqrt{2}$ centered at the origin (including the circle $x^2 + y^2 = 2$) as shown in Figure 15.4.

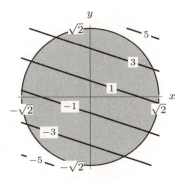

Figure 15.4

We first find the local maxima and minima of f in the interior of our disk. So we need to find the extrema of

$$f(x, y) = x + 3y, \quad \text{in the region} \quad x^2 + y^2 < 2.$$

As

$$f_x = 1$$
$$f_y = 3$$

f does not have critical points. Now let's find the local extrema of f on the boundary of the disk. We want to find the extrema of $f(x, y) = x + 3y$ subject to the constraint $g(x, y) = x^2 + y^2 - 2 = 0$. We use Lagrange multipliers

$$\operatorname{grad} f = \lambda \operatorname{grad} g \quad \text{and} \quad x^2 + y^2 = 2,$$

which give

$$1 = 2\lambda x$$
$$3 = 2\lambda y$$
$$x^2 + y^2 = 2.$$

As λ cannot be zero, we solve for x and y in the first two equations and get $x = \frac{1}{2\lambda}$ and $y = \frac{3}{2\lambda}$. Plugging into the third equation gives

$$8\lambda^2 = 10$$

so $\lambda = \pm\frac{\sqrt{5}}{2}$ and we get the solutions $(\frac{1}{\sqrt{5}}, \frac{3}{\sqrt{5}})$ and $(-\frac{1}{\sqrt{5}}, -\frac{3}{\sqrt{5}})$. Evaluating f at these points gives

$$f(\frac{1}{\sqrt{5}}, \frac{3}{\sqrt{5}}) = 2\sqrt{5} \quad \text{and}$$

$$f(-\frac{1}{\sqrt{5}}, -\frac{3}{\sqrt{5}}) = -2\sqrt{5}$$

The region $x^2 + y^2 \le 2$ is closed and bounded, so maximum and minimum values of f in the region exist. Therefore $(\frac{1}{\sqrt{5}}, \frac{3}{\sqrt{5}})$ is a global maximum of f and $(-\frac{1}{\sqrt{5}}, -\frac{3}{\sqrt{5}})$ is a global minimum of f on the whole region $x^2 + y^2 \le 2$.

17. We first find the critical points of f:

$$f_x = 2xy = 0, \quad f_y = x^2 + 6y - 1 = 0.$$

From the first equation, we get either $x = 0$ or $y = 0$. If $x = 0$, from the second equation we get $6y - 1 = 0$ so $y = 1/6$. If instead $y = 0$, then from the second equation $x = \pm 1$. We conclude that the critical points are $(0, 1/6)$, $(1, 0)$, and $(-1, 0)$. All three critical points satisfy the constraint $x^2 + y^2 \le 10$.

The Lagrange conditions, grad $f = \lambda$ grad g, are:

$$2xy = \lambda 2x, \qquad x^2 + 6y - 1 = \lambda 2y$$

From the first equation, when $x \neq 0$, we divide by x to get $\lambda = y$. Substituting into the second equation, we get

$$x^2 + 6y - 1 = 2y^2.$$

Then using $x^2 = 10 - y^2$ from the constraint, we have

$$10 - y^2 + 6y - 1 = 2y^2,$$

so $3y^2 - 6y - 9 = 0$. Factoring, we get $3(y-3)(y+1) = 0$. From the constraint, we get $x = \pm 1$ when $y = 3$ and $x = \pm 3$ for $y = -1$. If instead $x = 0$, so that we cannot divide by x in the first Lagrange equation, then from the constraint, $y = \pm\sqrt{10}$. Summarizing, the following points are either critical points or satisfy the Lagrange conditions:

$$(1,0), (-1,0), (0,1/6), (\pm 1, 3), (\pm 3, -1), (0, \pm\sqrt{10}).$$

These are the candidates for global maximum or minimum points. The corresponding values for $f(x,y) = x^2 y + 3y^2 - y$ are:

$$0, 0, -1/12, 27, -5, 30 \mp \sqrt{10}.$$

The largest value is $30 + \sqrt{10}$ at the point $(0, -\sqrt{10})$ and the smallest value is -5 at $(\pm 3, -1)$.

Problems

21. (a) C. The vectors $\vec{v}$ and $\vec{w}$ may, or may not, be parallel.

(b) F. Since $\vec{v} \cdot \vec{w} = 2 = ||\vec{v}|| \cdot ||\vec{w}|| \cos\theta$, we must have $\cos\theta > 0$, so $0 \leq \theta < \pi/2$. Thus, the statement $\pi/2 < \theta < \pi$ is false.

(c) T. Since $||\vec{v}|| \cdot ||\vec{w}|| \cos\theta = 2$ and $\cos\theta \leq 1$, it is true that either $||\vec{v}||$ or $||\vec{w}||$ or both must be greater than 1.

(d) F. Since P is the hottest point on R, we know grad $g = \vec{v}$ is perpendicular to R at P.

(e) F. Since $\vec{v}$ is perpendicular to R and $\vec{w}$ is not perpendicular to $\vec{v}$ (because $\vec{v} \cdot \vec{w} \neq 0$), we know that $\vec{w}$ is not tangent to R at P.

25. The maximum and minimum values change by approximately $\lambda\Delta c$. The Lagrange conditions give:

$$3 = \lambda 2x, \quad -2 = \lambda 4y.$$

Solving for λ and setting the expressions equal, we get $x = -3y$. Substituting into the constraint, we get $y = \pm 2$, so the points satisfying the Lagrange conditions are $(-6, 2)$ and $(6, -2)$. The corresponding values of $f(x,y) = 3x - 2y$ are -22 and 22. From the first equation, we have $\lambda = 3/(2x)$. Thus the minimum value changes by $3/(-12)\Delta c = -\Delta c/4$ and the maximum changes by $3/(12)\Delta c = \Delta c/4$.

29. (a) We want to minimize C subject to $g = x + y = 39$. Solving $\nabla C = \lambda\nabla g$ gives

$$10x + 2y = \lambda$$
$$2x + 6y = \lambda$$

so $y = 2x$. Solving with $x + y = 39$ gives $x = 13, y = 26, \lambda = 182$. Therefore $C = \$4349$.

(b) Since $\lambda = 182$, increasing production by 1 will cause costs to increase by approximately $\$182$. (because $\lambda = \dfrac{||\nabla C||}{||\nabla g||}$ = rate of change of C with g). Similarly, decreasing production by 1 will save approximately $\$182$.

33. (a) Let $g(x,y) = x + y$. We are minimizing $f(x,y) = x^2 + 2y^2$ subject to the constraint $g(x,y) = c$.

The method of Lagrange multipliers is to solve the equations

$$f_x = \lambda g_x \qquad f_y = \lambda g_y \qquad g = c,$$

which are

$$2x = \lambda \qquad 4y = \lambda \qquad x + y = c.$$

We have

$$x = \frac{2c}{3} \qquad y = \frac{c}{3} \qquad \lambda = \frac{4c}{3},$$

so there is a critical point at $(2c/3, c/3)$. Since moving away from the origin increases values of f in Figure 15.5, we see that f has a minimum on the constraint. The minimum value is

$$m(c) = f\left(\frac{2c}{3}, \frac{c}{3}\right) = \frac{2c^2}{3}.$$

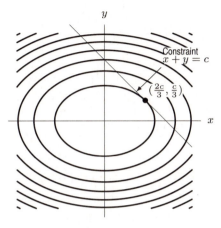

Figure 15.5

(b) Calculations in part (a) showed that $\lambda = 4c/3$.

(c) The multiplier λ is the rate of change of $m(c)$ as c increases and the constraint moves: that is, $\lambda = m'(c)$.

37. (a) The objective function is the complementary energy, $\dfrac{f_1^2}{2k_1} + \dfrac{f_2^2}{2k_2}$, and the constraint is $f_1 + f_2 = mg$. The Lagrangian function is

$$\mathcal{L}(f_1, f_2, \lambda) = \frac{f_1^2}{2k_1} + \frac{f_2^2}{2k_2} - \lambda(f_1 + f_2 - mg).$$

We look for solutions to the system of equations we get from grad $\mathcal{L} = \vec{0}$:

$$\frac{\partial \mathcal{L}}{\partial f_1} = \frac{f_1}{k_1} - \lambda = 0$$

$$\frac{\partial \mathcal{L}}{\partial f_2} = \frac{f_2}{k_2} - \lambda = 0$$

$$\frac{\partial \mathcal{L}}{\partial \lambda} = -(f_1 + f_2 - mg) = 0.$$

Combining $\dfrac{\partial \mathcal{L}}{\partial f_1} - \dfrac{\partial \mathcal{L}}{\partial f_2} = \dfrac{f_1}{k_1} - \dfrac{f_2}{k_2} = 0$ with $\dfrac{\partial \mathcal{L}}{\partial \lambda} = 0$ gives the two equation system

$$\frac{f_1}{k_1} - \frac{f_2}{k_2} = 0$$
$$f_1 + f_2 = mg.$$

Substituting $f_2 = mg - f_1$ into the first equation leads to

$$f_1 = \frac{k_1}{k_1 + k_2} mg$$
$$f_2 = \frac{k_2}{k_1 + k_2} mg.$$

(b) Hooke's Law states that for a spring

Force of spring = Spring constant · Distance stretched or compressed from equilibrium.

Since $f_1 = k_1 \cdot \lambda$ and $f_2 = k_2 \cdot \lambda$, the Lagrange multiplier λ equals the distance the mass stretches the top spring and compresses the lower spring.

41. (a) The objective function $f(x, y) = px + qy$ gives the cost to buy x units of input 1 at unit price p and y units of input 2 at unit price q.

   The constraint $g(x, y) = u$ tells us that we are only considering the cost of inputs x and y that can be used to produce quantity u of the product.

   Thus the number $C(p, q, u)$ gives the minimum cost to the company of producing quantity u if the inputs it needs have unit prices p and q.

(b) The Lagrangian function is

$$\mathcal{L}(x, y, \lambda) = px + qy - \lambda(xy - u).$$

We look for solutions to the system of equations we get from grad $\mathcal{L} = \vec{0}$:

$$\frac{\partial \mathcal{L}}{\partial x} = p - \lambda y = 0$$

$$\frac{\partial \mathcal{L}}{\partial y} = q - \lambda x = 0$$

$$\frac{\partial \mathcal{L}}{\partial \lambda} = -(xy - u) = 0.$$

We see that $\lambda = p/y = q/x$ so $y = px/q$. Substituting for y in the constraint $xy = u$ leads to $x = \sqrt{qu/p}$, $y = \sqrt{pu/q}$ and $\lambda = \sqrt{pq/u}$. The minimum cost is thus

$$C(p, q, u) = p\sqrt{\frac{qu}{p}} + q\sqrt{\frac{pu}{q}} = 2\sqrt{pqu}.$$

45. (a) If the prices are p_1 and p_2 and the budget is b, the quantities consumed are constrained by

$$p_1 x_1 + p_2 x_2 = b.$$

We want to maximize

$$u(x_1, x_2) = a \ln x_1 + (1 - a) \ln x_2$$

subject to the constraint

$$p_1 x_1 + p_2 x_2 = b.$$

Using Lagrange multipliers, we solve

$$\frac{\partial u}{\partial x_1} = \frac{a}{x_1} = \lambda p_1$$

$$\frac{\partial u}{\partial x_2} = \frac{1 - a}{x_2} = \lambda p_2,$$

giving $x_1 = a/(\lambda p_1)$ and $x_2 = (1 - a)/(\lambda p_2)$. Substituting into the constraint, we get

$$\frac{a}{\lambda} + \frac{(1 - a)}{\lambda} = b$$

so

$$\lambda = \frac{1}{b}.$$

Thus

$$x_1 = \frac{ab}{p_1} \qquad x_2 = \frac{(1 - a)b}{p_2}$$

so the maximum satisfaction is given by

$$S = u(x_1, x_2) = u\left(\frac{ab}{p_1}, \frac{(1 - a)b}{p_2}\right) = a \ln \left(\frac{ab}{p_1}\right) + (1 - a) \ln \left(\frac{(1 - a)b}{p_2}\right)$$

$$= a \ln a + a \ln b - a \ln p_1 + (1 - a) \ln(1 - a) + (1 - a) \ln b - (1 - a) \ln p_2$$

$$= a \ln a + (1 - a) \ln(1 - a) + \ln b - a \ln p_1 - (1 - a) \ln p_2.$$

(b) We want to calculate the value of b needed to achieve $u(x_1, x_2) = c$. Thus, we solve for b in the equation

$$c = a \ln a + (1 - a) \ln(1 - a) + \ln b - a \ln p_1 - (1 - a) \ln p_2.$$

Since
$$\ln b = c - a \ln a - (1 - a) \ln(1 - a) + a \ln p_1 + (1 - a) \ln p_2,$$

we have
$$b = \frac{e^c \cdot e^{a \ln p_1} \cdot e^{(1-a) \ln p_2}}{e^{a \ln a} \cdot e^{(1-a) \ln(1-a)}} = \frac{e^c p_1^a p_2^{1-a}}{a^a (1 - a)^{(1-a)}}.$$

Solutions for Chapter 15 Review

Exercises

1. The critical points of f are obtained by solving $f_x = f_y = 0$, that is
$$f_x(x, y) = 2y^2 - 2x = 0 \quad \text{and} \quad f_y(x, y) = 4xy - 4y = 0,$$

so
$$2(y^2 - x) = 0 \quad \text{and} \quad 4y(x - 1) = 0$$

The second equation gives either $y = 0$ or $x = 1$. If $y = 0$ then $x = 0$ by the first equation, so $(0, 0)$ is a critical point. If $x = 1$ then $y^2 = 1$ from which $y = 1$ or $y = -1$, so two further critical points are $(1, -1)$, and $(1, 1)$.

Since
$$D = f_{xx} f_{yy} - (f_{xy})^2 = (-2)(4x - 4) - (4y)^2 = 8 - 8x - 16y^2,$$

we have
$$D(0, 0) = 8 > 0, \quad D(1, 1) = D(1, -1) = -16 < 0,$$

and $f_{xx} = -2 < 0$. Thus, $(0, 0)$ is a local maximum; $(1, 1)$ and $(1, -1)$ are saddle points.

5. The partial derivatives are
$$f_x = \cos x + \cos(x + y).$$
$$f_y = \cos y + \cos(x + y).$$

Setting $f_x = 0$ and $f_y = 0$ gives
$$\cos x = \cos y$$

For $0 < x < \pi$ and $0 < y < \pi$, $\cos x = \cos y$ only if $x = y$. Then, setting $f_x = f_y = 0$:
$$\cos x + \cos 2x = 0,$$
$$\cos x + 2 \cos^2 x - 1 = 0,$$
$$(2 \cos x - 1)(\cos x + 1) = 0.$$

So $\cos x = 1/2$ or $\cos x = -1$, that is $x = \pi/3$ or $x = \pi$. For the given domain $0 < x < \pi$, $0 < y < \pi$, we only consider the solution when $x = \pi/3$ then $y = x = \pi/3$. Therefore, the critical point is $(\frac{\pi}{3}, \frac{\pi}{3})$.

Since
$$f_{xx}(x, y) = -\sin x - \sin(x + y) \quad f_{xx}(\tfrac{\pi}{3}, \tfrac{\pi}{3}) = -\sin \tfrac{\pi}{3} - \sin \tfrac{2\pi}{3} = -\sqrt{3}$$
$$f_{xy}(x, y) = -\sin(x + y) \quad\quad\quad f_{xy}(\tfrac{\pi}{3}, \tfrac{\pi}{3}) = -\sin \tfrac{2\pi}{3} \quad\quad = -\tfrac{\sqrt{3}}{2}$$
$$f_{yy}(x, y) = -\sin y - \sin(x + y) \quad f_{yy}(\tfrac{\pi}{3}, \tfrac{\pi}{3}) = -\sin \tfrac{\pi}{3} - \sin \tfrac{2\pi}{3} = -\sqrt{3}$$

the discriminant is
$$D(x, y) = f_{xx} f_{yy} - f_{xy}^2$$
$$= (-\sqrt{3})(-\sqrt{3}) - (-\tfrac{\sqrt{3}}{2})^2 = \tfrac{9}{4} > 0.$$

Since $f_{xx}(\tfrac{\pi}{3}, \tfrac{\pi}{3}) = -\sqrt{3} < 0$, $(\tfrac{\pi}{3}, \tfrac{\pi}{3})$ is a local maximum.

9. The partial derivatives are
$$f_x = y + \frac{1}{x}, f_y = x + 2y.$$

For critical points, solve $f_x = 0$ and $f_y = 0$ simultaneously. From $f_y = x + 2y = 0$ we get that $x = -2y$. Substituting into $f_x = 0$, we have
$$y + \frac{1}{x} = y - \frac{1}{2y} = \frac{1}{y}(y^2 - \frac{1}{2}) = 0$$

Since $\frac{1}{y} \neq 0$, $y^2 - \frac{1}{2} = 0$, therefore
$$y = \pm\frac{1}{\sqrt{2}} = \pm\frac{\sqrt{2}}{2},$$

and $x = \mp\sqrt{2}$. So the critical points are $\left(-\sqrt{2}, \frac{\sqrt{2}}{2}\right)$ and $\left(\sqrt{2}, -\frac{\sqrt{2}}{2}\right)$. But x must be greater than 0, so $\left(-\sqrt{2}, \frac{\sqrt{2}}{2}\right)$ is not in the domain.

The contour diagram for f in Figure 15.6 (drawn by computer), shows that $\left(\sqrt{2}, -\frac{\sqrt{2}}{2}\right)$ is a saddle point of $f(x, y)$.

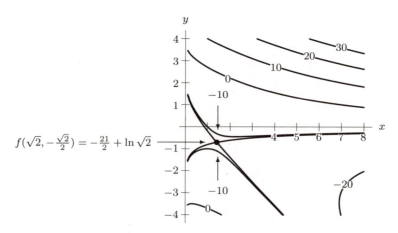

Figure 15.6: Contour map of $f(x, y) = xy + \ln x + y^2 - 10$

We can also see that $\left(\sqrt{2}, -\frac{\sqrt{2}}{2}\right)$ is a saddle point analytically.

Since $f_{xx} = -\frac{1}{x^2}, f_{yy} = 2, f_{xy} = 1$, the discriminant is:

$$D(x, y) = f_{xx}f_{yy} - f_{xy}^2$$
$$= -\frac{2}{x^2} - 1.$$

$D\left(\sqrt{2}, -\frac{\sqrt{2}}{2}\right) = -2 < 0$, so $\left(\sqrt{2}, -\frac{\sqrt{2}}{2}\right)$ is a saddle point.

13. The objective function is $f(x, y) = x^2 - xy + y^2$ and the equation of constraint is $g(x, y) = x^2 - y^2 = 1$. The gradients of f and g are

$$\nabla f(x, y) = (2x - y)\vec{i} + (-x + 2y)\vec{j},$$
$$\nabla g(x, y) = 2x\vec{i} - 2y\vec{j}.$$

Therefore the equation $\nabla f(x, y) = \lambda \nabla g(x, y)$ gives

$$2x - y = 2\lambda x$$
$$-x + 2y = -2\lambda y$$
$$x^2 - y^2 = 1.$$

Let us suppose that $\lambda = 0$. Then $2x = y$ and $2y = x$ give $x = y = 0$. But $(0, 0)$ is not a solution of the third equation, so we conclude that $\lambda \neq 0$. Now let's multiply the first two equations

$$-2\lambda y(2x - y) = 2\lambda x(-x + 2y).$$

As $\lambda \neq 0$, we can cancel it in the equation above and after doing the algebra we get

$$x^2 - 4xy + y^2 = 0$$

which gives $x = (2 + \sqrt{3})y$ or $x = (2 - \sqrt{3})y$.

If $x = (2 + \sqrt{3})y$, the third equation gives

$$(2 + \sqrt{3})^2 y^2 - y^2 = 1$$

so $y \approx \pm 0.278$ and $x \approx \pm 1.038$. These give the critical points $(1.038, 0.278)$, $(-1.038, -0.278)$.

If $x = (2 - \sqrt{3})y$, from the third equation we get

$$(2 - \sqrt{3})^2 y^2 - y^2 = 1.$$

But $(2 - \sqrt{3})^2 - 1 \approx -0.928 < 0$ so the equation has no solution. Evaluating f gives

$$f(1.038, 0.278) = f(-1.038, -0.278) \approx 0.866$$

Since $y \to \infty$ on the constraint, rewriting f as

$$f(x, y) = \left(x - \frac{y}{2}\right)^2 + \frac{3}{4}y^2$$

shows that f has no maximum on the constraint. The minimum value of f is 0.866. See Figure 15.7.

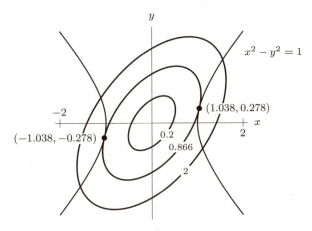

Figure 15.7

17. Our objective function is $f(x, y, z) = x^2 - 2y + 2z^2$ and our equation of constraint is $g(x, y, z) = x^2 + y^2 + z^2 - 1 = 0$. To optimize $f(x, y, z)$ with Lagrange multipliers, we solve $\nabla f(x, y, z) = \lambda \nabla g(x, y, z)$ subject to $g(x, y, z) = 0$. The gradients of f and g are

$$\nabla f(x, y, z) = 2x\vec{i} - 2\vec{j} + 4z\vec{k},$$
$$\nabla g(x, y) = 2x\vec{i} + 2y\vec{j} + 2z\vec{k}.$$

We get,

$$x = \lambda x$$
$$-1 = \lambda y$$
$$2z = \lambda z$$
$$x^2 + y^2 + z^2 = 1.$$

From the first equation we get $x = 0$ or $\lambda = 1$.

If $x = 0$ we have

$$-1 = \lambda y$$
$$2z = \lambda z$$
$$y^2 + z^2 = 1.$$

From the second equation $z = 0$ or $\lambda = 2$. So if $z = 0$, we have $y = \pm 1$ and we get the solutions $(0, 1, 0), (0, -1, 0)$. If $z \neq 0$ then $\lambda = 2$ and $y = -\frac{1}{2}$. So $z^2 = \frac{3}{4}$ which gives the solutions $(0, -\frac{1}{2}, \frac{\sqrt{3}}{2}), (0, -\frac{1}{2}, -\frac{\sqrt{3}}{2})$.

If $x \neq 0$, then $\lambda = 1$, so $y = -1$, which implies, from the equation $x^2 + y^2 + z^2 = 1$, that $x = 0$, which contradicts the assumption.

Since the constraint is closed and bounded, maximum and minimum values of f subject to the constraint exist. Therefore, evaluating f at the critical points, we get $f(0, 1, 0) = -2$, $f(0, -1, 0) = 2$ and $f(0, -\frac{1}{2}, \frac{\sqrt{3}}{2}) = f(0, -\frac{1}{2}, -\frac{\sqrt{3}}{2}) = 4$. So the maximum value of f is 4 and the minimum is -2.

21. The region $x + y \geq 1$ is the shaded half plane (including the line $x + y = 1$) shown in Figure 15.8.

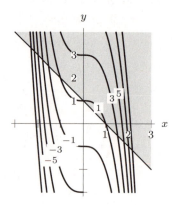

Figure 15.8

Let's look for the critical points of f in the interior of the region. As

$$f_x = 3x^2$$
$$f_y = 1$$

there are no critical points inside the shaded region. Now let's find the extrema of f on the boundary of our region. We want the extrema of $f(x, y) = x^3 + y$ subject to the constraint $g(x, y) = x + y - 1 = 0$. We use Lagrange multipliers

$$\text{grad } f = \lambda \text{ grad } g \quad \text{and} \quad x + y = 1,$$

which give

$$3x^2 = \lambda$$
$$1 = \lambda$$
$$x + y = 1.$$

From the first two equations we get $3x^2 = 1$, so the solutions are

$$\left(\frac{1}{\sqrt{3}}, 1 - \frac{1}{\sqrt{3}}\right) \quad \text{and} \quad \left(-\frac{1}{\sqrt{3}}, 1 + \frac{1}{\sqrt{3}}\right).$$

Evaluating f at these points we get

$$f\left(\frac{1}{\sqrt{3}}, 1 - \frac{1}{\sqrt{3}}\right) = 1 - \frac{2}{3\sqrt{3}}$$
$$f\left(-\frac{1}{\sqrt{3}}, 1 + \frac{1}{\sqrt{3}}\right) = 1 + \frac{2}{3\sqrt{3}}.$$

From the contour diagram in Figure 15.8, we see that $\left(\frac{1}{\sqrt{3}}, 1 - \frac{1}{\sqrt{3}}\right)$ is a local minimum and $\left(-\frac{1}{\sqrt{3}}, 1 + \frac{1}{\sqrt{3}}\right)$ is a local maximum of f on $x + y = 1$. Are they global extrema as well?

If we take x very big and $y = 1 - x$ then $f(x, y) = x^3 + y = x^3 - x + 1$ which can be made as big as we want (if we choose x big enough). So there will be no global maximum.

Similarly, taking x negative with big absolute value and $y = 1 - x$, $f(x, y) = x^3 + y = x^3 - x + 1$ can be made as small as we want (if we choose x small enough). So there is no global minimum. This can also be seen from Figure 15.8.

25. If $xy = 10$, then $f(x, y) = x^2 - y^2 = x^2 - 100/x^2$ and x can take any nonzero value. Since

$$\lim_{x \to \infty} \left(x^2 - \frac{100}{x^2}\right) = \infty,$$

we see f has no maximum on the constraint. Since

$$\lim_{x \to 0} \left(x^2 - \frac{100}{x^2}\right) = -\infty,$$

we see f has no minimum on the constraint.

Problems

29. (a) **(i)** Suppose $N = kA^p$. Then the rule of thumb tells us that if A is multiplied by 10, the value of N doubles. Thus

$$2N = k(10A)^p = k10^p A^p.$$

Thus, dividing by $N = kA^p$, we have

$$2 = 10^p$$

so taking logs to base 10 we have

$$p = \log 2 = 0.3010.$$

(where $\log 2$ means $\log_{10} 2$). Thus,

$$N = kA^{0.3010}.$$

(ii) Taking natural logs gives

$$\ln N = \ln(kA^p)$$
$$\ln N = \ln k + p \ln A$$
$$\ln N \approx \ln k + 0.301 \ln A$$

Thus, $\ln N$ is a linear function of $\ln A$.

(b) Table 15.1 contains the natural logarithms of the data:

Table 15.1 $\ln N$ *and* $\ln A$

Island	$\ln A$	$\ln N$
Redonda	1.1	1.6
Saba	3.0	2.2
Montserrat	2.3	2.7
Puerto Rico	9.1	4.3
Jamaica	9.3	4.2
Hispaniola	11.2	4.8
Cuba	11.6	4.8

Using a least squares fit we find the line:

$$\ln N = 1.20 + 0.32 \ln A$$

This yields the power function:

$$N = e^{1.20} A^{0.32} = 3.32 A^{0.32}$$

Since 0.32 is pretty close to $\log 2 \approx 0.301$, the answer does agree with the biological rule.

33. (a) The problem is to maximize

$$V = 1000 D^{0.6} N^{0.3}$$

subject to the budget constraint in dollars

$$40000D + 10000N \leq 600000$$

or (in thousand dollars)

$$40D + 10N \leq 600$$

(b) Let $B = 40D + 10N = 600$ (thousand dollars) be the budget constraint. At the optimum

$$\nabla V = \lambda \nabla B,$$

so

$$\frac{\partial V}{\partial D} = \lambda \frac{\partial B}{\partial D} = 40\lambda$$

$$\frac{\partial V}{\partial N} = \lambda \frac{\partial B}{\partial N} = 10\lambda.$$

Thus

$$\frac{\partial V / \partial D}{\partial V / \partial N} = 4.$$

Therefore, at the optimum point, the rate of increase in the number of visits with respect to an increase in the number of doctors is four times the corresponding rate for nurses. This factor of four is the same as the ratio of the salaries.

(c) Differentiating and setting $\nabla V = \lambda \nabla B$ yields

$$600D^{-0.4}N^{0.3} = 40\lambda$$
$$300D^{0.6}N^{-0.7} = 10\lambda$$

Thus, we get

$$\frac{600D^{-0.4}N^{0.3}}{40} = \lambda = \frac{300D^{0.6}N^{-0.7}}{10}$$

So

$$N = 2D.$$

To solve for D and N, substitute in the budget constraint:

$$600 - 40D - 10N = 0$$

$$600 - 40D - 10 \cdot (2D) = 0$$

So $D = 10$ and $N = 20$.

$$\lambda = \frac{600(10^{-0.4})(20^{0.3})}{40} \approx 14.67$$

Thus the clinic should hire 10 doctors and 20 nurses. With that staff, the clinic can provide

$$V = 1000(10^{0.6})(20^{0.3}) \approx 9{,}779 \text{ visits per year.}$$

(d) From part c), the Lagrange multiplier is $\lambda = 14.67$. At the optimum, the Lagrange multiplier tells us that about 14.67 extra visits can be generated through an increase of $\$1{,}000$ in the budget. (If we had written out the constraint in dollars instead of thousands of dollars, the Lagrange multiplier would tell us the number of extra visits per dollar.)

(e) The marginal cost, MC, is the cost of an additional visit. Thus, at the optimum point, we need the reciprocal of the Lagrange multiplier:

$$\text{MC} = \frac{1}{\lambda} \approx \frac{1}{14.67} \approx 0.068 \text{ (thousand dollars)},$$

that is, at the optimum point, an extra visit costs the clinic 0.068 thousand dollars, or $\$68$.

This production function exhibits declining returns to scale (e.g. doubling both inputs less than doubles output, because the two exponents add up to less than one). This means that for large V, increasing V will require increasing D and N by more than when V is small. Thus the cost of an additional visit is greater for large V than for small. In other words, the marginal cost will rise with the number of visits.

37. The objective function is

$$f(x, y, z) = \sqrt{(x - a)^2 + (y - b)^2 + (z - c)^2},$$

and the constraint is

$$g(x, y, z) = Ax + By + Cz + D = 0.$$

Partial derivatives of f and g are

$$f_x = \frac{\frac{1}{2} \cdot 2 \cdot (x - a)}{f(x, y, z)} = \frac{x - a}{f(x, y, z)},$$

$$f_y = \frac{\frac{1}{2} \cdot 2 \cdot (y - b)}{f(x, y, z)} = \frac{y - b}{f(x, y, z)},$$

$$f_z = \frac{\frac{1}{2} \cdot 2 \cdot (z - c)}{f(x, y, z)} = \frac{z - c}{f(x, y, z)},$$

$$g_x = A, \quad g_y = B, \quad \text{and} \quad g_z = C.$$

Using Lagrange multipliers, we need to solve the equations

$$\operatorname{grad} f = \lambda \operatorname{grad} g$$

where grad $f = f_x \vec{i} + f_y \vec{j} + f_z \vec{k}$ and grad $g = g_x \vec{i} + g_y \vec{j} + g_z \vec{k}$. This gives a system of equations:

$$\frac{x-a}{f(x,y,z)} = \lambda A$$

$$\frac{y-b}{f(x,y,z)} = \lambda B$$

$$\frac{z-c}{f(x,y,z)} = \lambda C$$

$$Ax + By + Cz + D = 0.$$

Now $\frac{x-a}{A} = \frac{y-b}{B} = \frac{z-c}{C} = \lambda f(x,y,z)$ gives

$$x = \frac{A}{B}(y-b) + a,$$

$$z = \frac{C}{B}(y-b) + c,$$

Substitute into the constraint,

$$A\left(\frac{A}{B}(y-b) + a\right) + By + C\left(\frac{C}{B}(y-b) + c\right) + D = 0,$$

$$\left(\frac{A^2}{B} + B + \frac{C^2}{B}\right)y = \frac{A^2}{B}b - Aa + \frac{C^2}{B}b - Cc - D.$$

Hence

$$y = \frac{(A^2 + C^2)b - B(Aa + Cc + D)}{A^2 + B^2 + C^2},$$

$$y - b = \frac{-B(Aa + Bb + Cc + D)}{A^2 + B^2 + C^2}$$

$$x - a = \frac{A}{B}(y - b)$$

$$= \frac{-A(Aa + Bb + Cc + D)}{A^2 + B^2 + C^2}$$

$$z - c = \frac{C}{B}(y - b)$$

$$= \frac{-C(Aa + Bb + Cc + D)}{A^2 + B^2 + C^2}$$

Thus the minimum $f(x,y,z)$ is

$$f(x,y,z) = \sqrt{(x-a)^2 + (y-b)^2 + (z-c)^2}$$

$$= \left[\left(\frac{-A(Aa + Bb + Cc + D)}{A^2 + B^2 + C^2}\right)^2 + \left(\frac{-B(Aa + Bb + Cc + D)}{A^2 + B^2 + C^2}\right)^2 \right.$$

$$\left. + \left(\frac{-C(Aa + Bb + Cc + D)}{A^2 + B^2 + C^2}\right)^2\right]^{1/2}$$

$$= \frac{|Aa + Bb + Cc + D|}{\sqrt{A^2 + B^2 + C^2}}.$$

The geometric meaning is finding the shortest distance from a point (a, b, c) to the plane $Ax + By + Cz + D = 0$.

41. Cost of production, C, is given by $C = p_1 W + p_2 K = b$. At the optimal point, $\nabla q = \lambda \nabla C$.
Since $\nabla q = \left(c(1-a)W^{-a}K^a\right)\vec{i} + \left(caW^{1-a}K^{a-1}\right)\vec{j}$ and $\nabla C = p_1\vec{i} + p_2\vec{j}$, we get

$$c(1-a)W^{-a}K^a = \lambda p_1 \text{ and } caW^{1-a}K^{a-1} = \lambda p_2.$$

Now, marginal productivity of labor is given by $\frac{\partial q}{\partial W} = c(1-a)W^{-a}K^a$ and marginal productivity of capital is given by $\frac{\partial q}{\partial K} = caW^{1-a}K^{a-1}$, so their ratio is given by

$$\frac{\frac{\partial q}{\partial W}}{\frac{\partial q}{\partial K}} = \frac{c(1-a)W^{-a}K^a}{caW^{1-a}K^{a-1}} = \frac{\lambda p_1}{\lambda p_2} = \frac{p_1}{p_2}$$

which is the ratio of the cost of one unit of labor to the cost of one unit of capital.

45. You should try to anticipate your opponent's choice. After you choose a value λ, your opponent will use calculus to find the point (x, y) that maximizes the function $f(x, y) = 10 - x^2 - y^2 - 2x - \lambda(2x + 2y)$. At that point, we have $f_x = -2x - 2 - 2\lambda = 0$ and $f_y = -2y - 2\lambda = 0$, so your opponent will choose $x = -1 - \lambda$ and $y = -\lambda$. This gives a value $\mathcal{L}(-1 - \lambda, -\lambda, \lambda) = 10 - (-1 - \lambda)^2 - (-\lambda)^2 - 2(-1 - \lambda) - \lambda(2(-1 - \lambda) + 2(-\lambda)) = 11 + 2\lambda + 2\lambda^2$ which you want to make as small as possible. You should choose λ to minimize the function $h(\lambda) = 11 + 2\lambda + 2\lambda^2$. You choose λ so that $h'(\lambda) = 2 + 4\lambda = 0$, or $\lambda = -1/2$. Your opponent then chooses $(x, y) = (-1 - \lambda, -\lambda) = (-1/2, 1/2)$, giving a final score of $\mathcal{L}(-1/2, 1/2, -1/2) = 10.5$. No choice of λ that you can make can force the value of $\mathcal{L}$ below 10.5. But your choice of $\lambda = -1/2$ makes it impossible for your opponent to force the value of $\mathcal{L}$ above 10.5.

CAS Challenge Problems

49. (a) We have grad $f = 3\vec{i} + 2\vec{j}$ and grad $g = (4x - 4y)\vec{i} + (-4x + 10y)\vec{j}$, so the Lagrange multiplier equations are

$$3 = \lambda(4x - 4y)$$
$$2 = \lambda(-4x + 10y)$$
$$2x^2 - 4xy + 5y^2 = 20$$

Solving these with a CAS we get $\lambda = -0.4005, x = -3.9532, y = -2.0806$ and $\lambda = 0.4005, x = 3.9532, y = 2.0806$. We have $f(-3.9532, -2.0806) = -11.0208$, and $f(3, 9532, 2.0806) = 21.0208$. The constraint equation is $2x^2 - 4xy + 5y^2 = 20$, or, completing the square, $2(x - y)^2 + 3y^2 = 20$. This has the shape of a skewed ellipse, so the constraint curve is bounded, and therefore the local maximum is a global maximum. Thus the maximum value is 21.0208.

(b) The maximum value on $g = 20.5$ is $\approx 21.0208 + 0.5(0.4005) = 21.2211$. The maximum value on $g = 20.2$ is $\approx 21.0208 + 0.2(0.4005) = 21.1008$.

(c) We use the same commands in the CAS from part (a), with 20 replaced by 20.5 and 20.2, and get the maximum values 21.2198 for $g = 20.5$ and 21.1007 for $g = 20.2$. These agree with the approximations we found in part (b) to 2 decimal places.

CHECK YOUR UNDERSTANDING

1. True. By definition, a critical point is either where the gradient of f is zero or does not exist.

5. True. The graph of this function is a cone that opens upward with its vertex at the origin.

9. False. For example, the linear function $f(x, y) = x + y$ has no local extrema at all.

13. False. For example, the linear function $f(x, y) = x + y$ has neither a global minimum or global maximum on all of 2-space.

17. False. On the given region the function f is always less than one. By picking points closer and closer to the circle $x^2 + y^2 = 1$ we can make f larger and larger (although never larger than one). There is no point in the open disk that gives f its largest value.

21. True. The point (a, b) must lie on the constraint $g(x, y) = c$, so $g(a, b) = c$.

25. False. Since grad f and grad g point in opposite directions, they are parallel. Therefore (a, b) could be a local maximum or local minimum of f constrained to $g = c$. However the information given is not enough to determine that it is a minimum. If the contours of g near (a, b) increase in the opposite direction as the contours of f, then at a point with grad $f(a, b) = \lambda$grad $g(a, b)$ we have $\lambda \leq 0$, but this can be a local maximum or minimum.

For example, $f(x, y) = 4 - x^2 - y^2$ has a local maximum at $(1, 1)$ on the constraint $g(x, y) = x + y = 2$. Yet at this point, grad $f = -2\vec{i} - 2\vec{j}$ and grad $g = \vec{i} + \vec{j}$, so grad f and grad g point in opposite directions.

29. True. Since $f(a, b) = M$, we must satisfy the Lagrange conditions that $f_x(a, b) = \lambda g(a, b)$ and $f_y(a, b) = \lambda g_y(a, b)$, for some λ. Thus $f_x(a, b)/f_y(a, b) = g_x(a, b)/g_y(a, b)$.

33. False. The value of λ at a minimum point gives the proportional change in m for a change in c. If $\lambda > 0$ and the change in c is positive, the change in m will also be positive.

CHAPTER SIXTEEN

Solutions for Section 16.1

Exercises

1. Mark the values of the function on the plane, as shown in Figure 16.1, so that you can guess respectively at the smallest and largest values the function takes on each small rectangle.

$$\text{Lower sum} = \sum f(x_i, y_i)\Delta x \Delta y$$
$$= 4\Delta x \Delta y + 6\Delta x \Delta y + 3\Delta x \Delta y + 4\Delta x \Delta y$$
$$= 17\Delta x \Delta y$$
$$= 17(0.1)(0.2) = 0.34.$$

$$\text{Upper sum} = \sum f(x_i, y_i)\Delta x \Delta y$$
$$= 7\Delta x \Delta y + 10\Delta x \Delta y + 6\Delta x \Delta y + 8\Delta x \Delta y$$
$$= 31\Delta x \Delta y$$
$$= 31(0.1)(0.2) = 0.62.$$

Figure 16.1

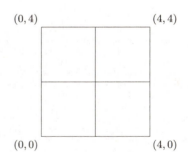

Figure 16.2

5. (a) If we take the partition of R consisting of just R itself, we get

$$\text{Lower bound for integral } = \min_R f \cdot A_R = 0 \cdot (4-0)(4-0) = 0.$$

Similarly, we get

$$\text{Upper bound for integral } = \max_R f \cdot A_R = 4 \cdot (4-0)(4-0) = 64.$$

(b) The estimates asked for are just the upper and lower sums. We partition R into subrectangles $R_{(a,b)}$ of width 2 and height 2, where (a, b) is the lower-left corner of $R_{(a,b)}$. The subrectangles are then $R_{(0,0)}, R_{(2,0)}, R_{(0,2)}$, and $R_{(2,2)}$, as in Figure 16.2. Then we find the lower sum

$$\text{Lower sum} = \sum_{(a,b)} A_{R_{(a,b)}} \cdot \min_{R_{(a,b)}} f = \sum_{(a,b)} 4 \cdot (\text{Min of } f \text{ on } R_{(a,b)})$$
$$= 4 \sum_{(a,b)} (\text{Min of } f \text{ on } R_{(a,b)})$$
$$= 4(f(0,0) + f(2,0) + f(0,2) + f(2,2))$$
$$= 4(\sqrt{0 \cdot 0} + \sqrt{2 \cdot 0} + \sqrt{0 \cdot 2} + \sqrt{2 \cdot 2})$$
$$= 8.$$

Similarly, the upper sum is

$$\text{Upper sum} = 4\sum_{(a,b)}(\text{Max of } f \text{ on } R_{(a,b)})$$

$$= 4(f(2,2) + f(4,2) + f(2,4) + f(4,4))$$

$$= 4(\sqrt{2 \cdot 2} + \sqrt{4 \cdot 2} + \sqrt{2 \cdot 4} + \sqrt{4 \cdot 4})$$

$$= 24 + 16\sqrt{2} \approx 46.63.$$

The upper sum is an overestimate and the lower sum is an underestimate, so we can get a better estimate by averaging them to get $16 + 8\sqrt{2} \approx 27.3$.

Problems

9. The function being integrated is $f(x,y) = 1$, which is positive everywhere. Thus, its integral over any region is positive.

13. The region D is symmetric both with respect to x and y axes. The function being integrated is $f(x,y) = 5x$, which is an odd function in x. Since D is symmetric with respect to x, the contributions to the integral cancel out. Thus, the integral of the function over the region D is zero.

17. The function being integrated, $f(x,y) = y - y^3$ is always negative in the region B since in that region $-1 < y < 0$ and $|y^3| < |y|$. Thus, the integral is negative.

21. The function $f(x,y) = e^x$ is positive for any value of x. Thus, its integral is always positive for any region, such as D, with nonzero area.

25. The question is asking which graph has more volume under it, and from inspection, it appears that it would be the graph for the mosquitos.

29. Let R be the region $0 \leq x \leq 60, \quad 0 \leq y \leq 8$. Then

$$\text{Volume} = \int_R w(x,y)\,dA$$

Lower estimate: $10 \cdot 2(1+4+8+10+10+8+0+3+4+6+6+4+0+1+2+3+3+2+0+0+1+1+1+1) = 1580$.
Upper estimate:
$10 \cdot 2(8+13+16+17+17+16+4+8+10+11+11+10+3+4+6+7+7+6+1+2+3+4+4+3) = 3820$.
The average of the two estimates is 2700 cubic feet.

Solutions for Section 16.2

Exercises

1. See Figure 16.3.

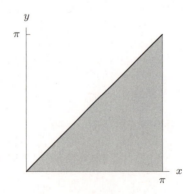

Figure 16.3

5. We evaluate the inside integral first:

$$\int_0^4 (4x + 3y)\, dx = (2x^2 + 3yx)\Big|_0^4 = 32 + 12y.$$

Therefore, we have

$$\int_0^3 \int_0^4 (4x + 3y)\, dxdy = \int_0^3 (32 + 12y)\, dy = (32y + 6y^2)\Big|_0^3 = 150.$$

9. Calculating the inner integral first, we have

$$\int_0^1 \int_0^1 ye^{xy}\, dx\, dy = \int_0^1 \left(e^{xy}\Big|_0^1 \right) dy = \int_0^1 \left(e^y - e^0 \right) dy = \int_0^1 (e^y - 1)\, dy = (e^y - y)\Big|_0^1 = e^1 - 1 - (e^0 - 0) = e - 2.$$

13. This region lies between $x = 0$ and $x = 4$ and between the lines $y = 3x$ and $y = 12$, and so the iterated integral is

$$\int_0^4 \int_{3x}^{12} f(x, y)\, dydx.$$

Alternatively, we could have set up the integral as follows:

$$\int_0^{12} \int_0^{y/3} f(x, y)\, dxdy.$$

17. The line connecting $(1, 0)$ and $(4, 1)$ is

$$y = \frac{1}{3}(x - 1)$$

So the integral is

$$\int_1^4 \int_{(x-1)/3}^2 f\, dy\, dx$$

21.

$$\int_1^4 \int_{\sqrt{y}}^{y} x^2 y^3\, dxdy = \int_1^4 y^3 \frac{x^3}{3}\Big|_{\sqrt{y}}^{y}\, dy$$

$$= \frac{1}{3} \int_1^4 (y^6 - y^{\frac{9}{2}})\, dy$$

$$= \frac{1}{3} \left(\frac{y^7}{7} - \frac{y^{11/2}}{11/2} \right)\Big|_1^4$$

$$= \frac{1}{3} \left[\left(\frac{4^7}{7} - \frac{4^{11/2} \times 2}{11} \right) - \left(\frac{1}{7} - \frac{2}{11} \right) \right] \approx 656.082$$

See Figure 16.4.

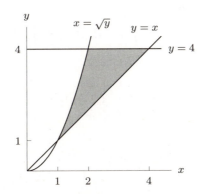

Figure 16.4

25. In the other order, the integral is

$$\int_0^1 \int_0^2 \sqrt{x+y}\, dy\, dx.$$

First we keep x fixed and calculate the inside integral with respect to y:

$$\int_0^2 \sqrt{x+y}\, dy = \frac{2}{3}(x+y)^{3/2}\Big|_{y=0}^{y=2}$$

$$= \frac{2}{3}\left[(x+2)^{3/2} - x^{3/2}\right].$$

Then the outside integral becomes

$$\int_0^1 \frac{2}{3}\left[(x+2)^{3/2} - x^{3/2}\right]\, dx = \frac{2}{3}\left[\frac{2}{5}(x+2)^{5/2} - \frac{2}{5}x^{5/2}\right]\Big|_0^1$$

$$= \frac{2}{3}\cdot\frac{2}{5}\left[3^{5/2} - 1 - 2^{5/2}\right] = 2.38176$$

Note that the answer is the same as the one we got in Exercise 24.

Problems

29. The region is bounded by $x = 1$, $x = 4$, $y = 2$, and $y = 2x$. Thus

$$\text{Volume} = \int_1^4 \int_2^{2x} (6x^2 y)\, dy dx.$$

To evaluate this integral, we evaluate the inside integral first:

$$\int_2^{2x} (6x^2 y)\, dy = (3x^2 y^2)\Big|_2^{2x} = 3x^2 (2x)^2 - 3x^2 (2^2) = 12x^4 - 12x^2.$$

Therefore, we have

$$\int_1^4 \int_2^{2x} (6x^2 y)\, dy dx = \int_1^4 (12x^4 - 12x^2)\, dx = \left(\frac{12}{5}x^5 - 4x^3\right)\Big|_1^4 = 2203.2.$$

The volume of this object is 2203.2.

33. (a) The line $x = y/2$ is the line $y = 2x$, and $y = x$ and $y = 2x$ intersect at $x = 0$. Thus, R is the shaded region in Figure 16.5. One expression for the integral is

$$\int_R f\, dA = \int_0^3 \int_x^{2x} x^2 e^{x^2}\, dy\, dx.$$

Another expression is obtained by reversing the order of integration. When we do this, it is necessary to split R into two regions on a line parallel to the x-axis along the point of intersection of $y = x$ and $x = 3$; this line is $y = 3$. Then we obtain

$$\int_R f\, dA = \int_0^3 \int_{y/2}^y x^2 e^{x^2}\, dx\, dy + \int_3^6 \int_{y/2}^3 x^2 e^{x^2}\, dx\, dy.$$

(b) We evaluate the first integral. Integrating with respect to y first:

$$\int_0^3 \int_x^{2x} x^2 e^{x^2}\, dy\, dx = \int_0^3 (x^2 e^{x^2} y)\Big|_x^{2x}\, dx = \int_0^3 x^3 e^{x^2}\, dx.$$

We use integration by parts with $u = x^2$, $v' = xe^{x^2}$. Then $u' = 2x$ and $v = \frac{1}{2}e^{x^2}$, so

$$\int_0^3 x^3 e^{x^2}\, dx = \frac{1}{2}x^2 e^{x^2}\Big|_0^3 - \int_0^3 xe^{x^2}\, dx = \frac{1}{2}x^2 e^{x^2}\Big|_0^3 - \frac{1}{2}e^{x^2}\Big|_0^3$$

$$= \left(\frac{1}{2}(9)e^9 - 0\right) - \left(\frac{1}{2}e^9 - \frac{1}{2}\right) = \frac{1}{2} + 4e^9.$$

Figure 16.5

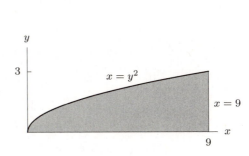

Figure 16.6

37. As given, the region of integration is as shown in Figure 16.6.
Reversing the limits gives

$$\int_0^9 \int_0^{\sqrt{x}} y \sin\left(x^2\right) dy dx = \int_0^9 \left(\left. \frac{y^2 \sin\left(x^2\right)}{2} \right|_0^{\sqrt{x}} \right) dx$$

$$= \frac{1}{2} \int_0^9 x \sin\left(x^2\right) dx$$

$$= \left. -\frac{\cos\left(x^2\right)}{4} \right|_0^9$$

$$= \frac{1}{4} - \frac{\cos\left(81\right)}{4} = 0.056.$$

41. (a) The contour $f(x, y) = 1$ lies in the xy-plane and has equation

$$2e^{-(x-1)^2 - y^2} = 1,$$

so

$$-(x - 1)^2 - y^2 = \ln(1/2)$$
$$(x - 1)^2 + y^2 = \ln 2 = 0.69.$$

This is the equation of a circle centered at $(1, 0)$ in the xy-plane.
Other contours are of the form

$$2e^{-(x-1)^2 - y^2} = c$$
$$-(x - 1)^2 - y^2 = \ln(c/2).$$

Thus, all the contours are circles centered at the point $(1, 0)$.

(b) The cross-section has equation $z = f(1, y) = e^{-y^2}$. If $x = 1$, the base region in the xy-plane extends from $y = -\sqrt{3}$ to $y = \sqrt{3}$. See Figure 16.7, which shows the circular region below W in the xy-plane. So

$$\text{Area} = \int_{-\sqrt{3}}^{\sqrt{3}} e^{-y^2} dy.$$

(c) Slicing parallel to the y-axis, we get

$$\text{Volume} = \int_{-2}^{2} \int_{-\sqrt{4-x^2}}^{\sqrt{4-x^2}} e^{-(x-1)^2 - y^2} \, dy \, dx.$$

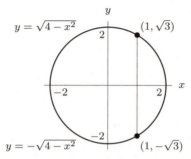

Figure 16.7: Region beneath W in the xy-plane

45.

$$\text{Volume} = \int_0^2 \int_0^2 xy \, dy \, dx = \int_0^2 \frac{1}{2}xy^2 \Big|_0^2 \, dx$$

$$= \int_0^2 2x \, dx$$

$$= x^2 \Big|_0^2$$

$$= 4$$

49. Let R be the triangle with vertices $(1,0)$, $(2,2)$ and $(0,1)$. Note that $(3x + 2y + 1) - (x + y) = 2x + y + 1 > 0$ for $x, y > 0$, so $z = 3x + 2y + 1$ is above $z = x + y$ on R. We want to find

$$\text{Volume} = \int_R ((3x + 2y + 1) - (x + y)) \, dA = \int_R (2x + y + 1) \, dA.$$

We need to express this in terms of double integrals.

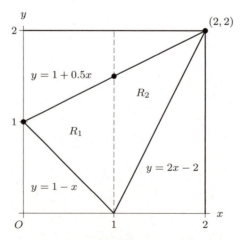

Figure 16.8

To do this, divide R into two regions with the line $x = 1$ to make regions R_1 for $x \leq 1$ and R_2 for $x \geq 1$. See Figure 16.8. We want to find

$$\int_R (2x + y + 1) \, dA = \int_{R_1} (2x + y + 1) \, dA + \int_{R_2} (2x + y + 1) \, dA.$$

Note that the line connecting $(0, 1)$ and $(1, 0)$ is $y = 1 - x$, and the line connecting $(0, 1)$ and $(2, 2)$ is $y = 1 + 0.5x$. So

$$\int_{R_1} (2x + y + 1) \, dA = \int_0^1 \int_{1-x}^{1+0.5x} (2x + y + 1) \, dy \, dx.$$

The line between $(1, 0)$ and $(2, 2)$ is $y = 2x - 2$, so

$$\int_{R_2} (2x + y + 1) \, dA = \int_1^2 \int_{2x-2}^{1+0.5x} (2x + y + 1) \, dy \, dx.$$

We can now compute the double integral for R_1:

$$\int_0^1 \int_{1-x}^{1+0.5x} (2x + y + 1) \, dy \, dx = \int_0^1 \left(2xy + \frac{y^2}{2} + y \right) \Big|_{1-x}^{1+0.5x} dx$$

$$= \int_0^1 \left(\frac{21}{8} x^2 + 3x \right) \, dx$$

$$= \left(\frac{7}{8} x^3 + \frac{3}{2} x^2 \right) \Big|_0^1 dx$$

$$= \frac{19}{8},$$

and the double integral for R_2:

$$\int_1^2 \int_{2x-2}^{1+0.5x} (2x + y + 1) \, dy \, dx = \int_1^2 (2xy + y^2/2 + y) \Big|_{2x-2}^{1+0.5x} dx$$

$$= \int_0^1 \left(-\frac{39}{8} x^2 + 9x + \frac{3}{2} \right) \, dx$$

$$= \left(-\frac{13}{8} x^3 + \frac{9}{2} x^2 + \frac{3}{2} x \right) \Big|_1^2$$

$$= \frac{29}{8}.$$

So, Volume $= \dfrac{19}{8} + \dfrac{29}{8} = \dfrac{48}{8} = 6.$

53. Assume the length of the two legs of the right triangle are a and b, respectively. See Figure 16.9. The line through $(a, 0)$ and $(0, b)$ is given by $\frac{y}{b} + \frac{x}{a} = 1$. So the area of this triangle is

$$A = \frac{1}{2} ab.$$

Thus the average distance from the points in the triangle to the y-axis (one of the legs) is

$$\text{Average distance} = \frac{1}{A} \int_0^a \int_0^{-\frac{b}{a}x+b} x \, dy \, dx$$

$$= \frac{2}{ab} \int_0^a \left(-\frac{b}{a} x^2 + bx \right) \, dx$$

$$= \frac{2}{ab} \left(-\frac{b}{3a} x^3 + \frac{b}{2} x^2 \right) \Big|_0^a$$

$$= \frac{2}{ab} \left(\frac{a^2 b}{6} \right) = \frac{a}{3}.$$

Similarly, the average distance from the points in the triangle to the x-axis (the other leg) is

$$\text{Average distance} = \frac{1}{A} \int_0^b \int_0^{-\frac{a}{b}y+a} y \, dx \, dy$$

$$= \frac{2}{ab} \int_0^b \left(-\frac{a}{b}y^2 + ay\right) dy$$

$$= \frac{2}{ab} \left(\frac{ab^2}{6}\right) = \frac{b}{3}.$$

Figure 16.9

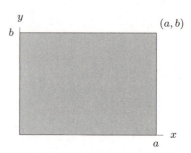

Figure 16.10

57. The force, ΔF, acting on ΔA, a small piece of area, is given by

$$\Delta F \approx p\Delta A,$$

where p is the pressure at that point. Thus, if R is the rectangle, the total force is given by

$$F = \int_R p \, dA.$$

We choose coordinates with the origin at one corner of the plate. See Figure 16.10. Suppose p is proportional to the square of the distance from the corner represented by the origin. Then we have

$$p = k(x^2 + y^2), \quad \text{for some positive constant } k.$$

Thus, we want to compute $\int_R k(x^2 + y^2) dA$. Rewriting as an iterated integral, we have

$$F = \int_R k(x^2 + y^2) \, dA = \int_0^b \int_0^a k(x^2 + y^2) \, dx dy = k \int_0^b \left(\frac{x^3}{3} + xy^2 \Big|_0^a\right) dy$$

$$= k \int_0^b \left(\frac{a^3}{3} + ay^2\right) dy = k \left(\frac{a^3 y}{3} + a\frac{y^3}{3} \Big|_0^b\right)$$

$$= \frac{k}{3}(a^3 b + ab^3).$$

Solutions for Section 16.3

Exercises

1.

$$\int_W f\, dV = \int_0^2 \int_{-1}^1 \int_2^3 (x^2 + 5y^2 - z)\, dz\, dy\, dx$$

$$= \int_0^2 \int_{-1}^1 \left(x^2 z + 5y^2 z - \frac{1}{2}z^2\right)\Big|_2^3 dy\, dx$$

$$= \int_0^2 \int_{-1}^1 \left(x^2 + 5y^2 - \frac{5}{2}\right) dy\, dx$$

$$= \int_0^2 \left(x^2 y + \frac{5}{3}y^3 - \frac{5}{2}y\right)\Big|_{-1}^1 dx$$

$$= \int_0^2 \left(2x^2 + \frac{10}{3} - 5\right) dx$$

$$= \left(\frac{2}{3}x^3 - \frac{5}{3}x\right)\Big|_0^2$$

$$= \frac{16}{3} - \frac{10}{3} = 2$$

5. The region is the half cylinder in Figure 16.11.

9. The region is the cylinder in Figure 16.12.

13. The region is the quarter sphere in Figure 16.13.

Figure 16.11

Figure 16.12

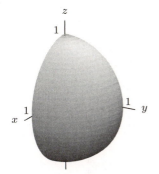

Figure 16.13

Problems

17. The sphere $x^2 + y^2 + z^2 = 9$ intersects the plane $z = 2$ in the circle

$$x^2 + y^2 + 2^2 = 9$$
$$x^2 + y^2 = 5.$$

The upper half of the sphere is given by $z = \sqrt{9 - x^2 - y^2}$. Thus, using the limits from Figure 16.14 gives

$$V = \int_{-\sqrt{5}}^{\sqrt{5}} \int_{-\sqrt{5-x^2}}^{\sqrt{5-x^2}} \int_2^{\sqrt{9-x^2-y^2}} 1\, dz\, dy\, dx.$$

The order of integration of x and y can be reversed.

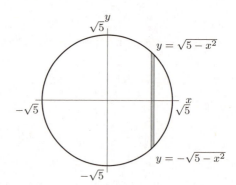

Figure 16.14

21. A slice through W for a fixed value of x is a semi-circle the boundary of which is $z^2 = r^2 - x^2 - y^2$, for $z \geq 0$, so the inner integral is

$$\int_0^{\sqrt{r^2 - x^2 - y^2}} f(x, y, z)\, dz.$$

Lining up these stacks parallel to y-axis gives a slice from $y = -\sqrt{r^2 - x^2}$ to $y = \sqrt{r^2 - x^2}$ giving

$$\int_{-\sqrt{r^2 - x^2}}^{\sqrt{r^2 - x^2}} \int_0^{\sqrt{r^2 - x^2 - y^2}} f(x, y, z)\, dz\, dy.$$

Finally, there is a slice for each x between 0 and r, so the integral we want is

$$\int_0^r \int_{-\sqrt{r^2 - x^2}}^{\sqrt{r^2 - x^2}} \int_0^{\sqrt{r^2 - x^2 - y^2}} f(x, y, z)\, dz\, dy\, dx.$$

25. The pyramid is shown in Figure 16.15. The planes $y = 0$, and $y - x = 4$, and $2x + y + z = 4$ intersect the plane $z = -6$ in the lines $y = 0$, $y - x = 4$, $2x + y = 10$ on the $z = -6$ plane as shown in Figure 16.16.

These three lines intersect at the points $(-4, 0, -6)$, $(5, 0, -6)$, and $(2, 6, -6)$. Let R be the triangle in the planes $z = -6$ with the above three points as vertices. Then, the volume of the solid is

$$
\begin{aligned}
V &= \int_0^6 \int_{y-4}^{(10-y)/2} \int_{-6}^{4-2x-y} dz\, dx\, dy \\
&= \int_0^6 \int_{y-4}^{(10-y)/2} (10 - 2x - y)\, dx\, dy = 162 \\
&= \int_0^6 (10x - x^2 - xy)\Big|_{y-4}^{(10-y)/2}\, dy \\
&= \int_0^6 \left(\frac{9y^2}{4} - 27y + 81\right) dy \\
&= 162
\end{aligned}
$$

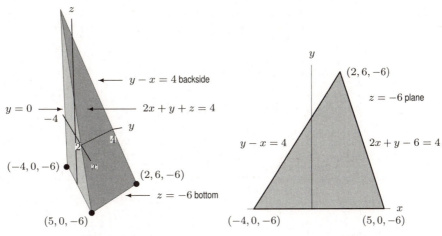

Figure 16.15   **Figure 16.16**

29. The required volume, V, is given by

$$V = \int_0^5 \int_0^3 \int_0^{x^2} dz\,dy\,dx$$

$$= \int_0^5 \int_0^3 x^2\,dy\,dx$$

$$= \int_0^5 x^2 y \Big|_{y=0}^{y=3} dx$$

$$= \int_0^5 3x^2\,dx$$

$$= 125.$$

33. (a) The vectors $\vec{u} = \vec{i} - \vec{j}$ and $\vec{v} = \vec{i} - \vec{k}$ lie in the required plane so $\vec{p} = \vec{u} \times \vec{v} = \vec{i} + \vec{j} + \vec{k}$ is perpendicular to this plane. Let (x, y, z) be a point in the plane, then $(x - 1)\vec{i} + y\vec{j} + z\vec{k}$ is perpendicular to $\vec{p}$, so $((x - 1)\vec{i} + y\vec{j} + z\vec{k}) \cdot (\vec{i} + \vec{j} + \vec{j}) = 0$ and so

$$(x - 1) + y + z = 0.$$

Therefore, the equation of the required plane is $x + y + z = 1$.

(b) The required volume, V, is given by

$$V = \int_0^1 \int_0^{1-x} \int_0^{1-x-y} dz\,dy\,dx$$

$$= \int_0^1 \int_0^{1-x} (1 - x - y)\,dy\,dx$$

$$= \int_0^1 y - xy - \frac{1}{2}y^2 \Big|_0^{1-x} dx$$

$$= \int_0^1 \left(1 - x - x(1 - x) - \frac{1}{2}(1 - x)^2\right) dx$$

$$= \int_0^1 \frac{1}{2}(1 - x)^2 dx$$

$$= \frac{1}{6}.$$

37. From the problem, we know that (x, y, z) is in the cube which is bounded by the three coordinate planes, $x = 0$, $y = 0$, $z = 0$ and the planes $x = 2$, $y = 2$, $z = 2$. We can regard the value $x^2 + y^2 + z^2$ as the density of the cube. The average value of $x^2 + y^2 + z^2$ is given by

$$\text{average value} = \frac{\int_V (x^2 + y^2 + z^2)\, dV}{\text{volume}(V)}$$

$$= \frac{\int_0^2 \int_0^2 \int_0^2 (x^2 + y^2 + z^2)\, dx dy dz}{8}$$

$$= \frac{\int_0^2 \int_0^2 \left(\frac{x^3}{3} + (y^2 + z^2)x\right) \Big|_0^2 dy dz}{8}$$

$$= \frac{\int_0^2 \int_0^2 \left(\frac{8}{3} + 2y^2 + 2z^2\right) dy dz}{8}$$

$$= \frac{\int_0^2 \left(\frac{8}{3}y + \frac{2}{3}y^3 + 2z^2 y\right) \Big|_0^2 dz}{8}$$

$$= \frac{\int_0^2 \left(\frac{16}{3} + \frac{16}{3} + 4z^2\right) dz}{8}$$

$$= \frac{\left(\frac{32}{3}z + \frac{4}{3}z^3\right) \Big|_0^2}{8}$$

$$= \frac{\left(\frac{64}{3} + \frac{32}{3}\right)}{8} = 4.$$

41. Zero. The value of y is positive on the half of the cone above the second and third quadrants and negative (of equal absolute value) on the half of the cone above the third and fourth quadrants. The integral of y over the entire solid cone is zero because the integrals over the four quadrants cancel.

45. Negative. If (x, y, z) is any point inside the cone then $z < 2$. Hence the function $z - 2$ is negative on W and so is its integral.

49. Positive. The value of x is positive on the half-cone, so its integral is positive.

53. Zero. Write the triple integral as an iterated integral, say integrating first with respect to y. For fixed x and z, the y-integral is over an interval symmetric about 0. The integral of y over such an interval is zero. If any of the inner integrals in an iterated integral is zero, then the triple integral is zero.

57. The intersection of two cylinders $x^2 + z^2 = 1$ and $y^2 + z^2 = 1$ is shown in Figure 16.17. This region is bounded by four surfaces:

$$z = -\sqrt{1 - x^2}, \quad z = \sqrt{1 - x^2}, \quad y = -\sqrt{1 - z^2}, \quad \text{and} \quad y = \sqrt{1 - z^2}$$

So the volume of the given solid is

$$V = \int_{-1}^{1} \int_{-\sqrt{1-x^2}}^{\sqrt{1-x^2}} \int_{-\sqrt{1-z^2}}^{\sqrt{1-z^2}} dy\, dz\, dx$$

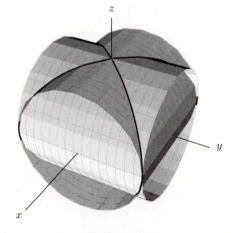

Figure 16.17

61. The volume V of the solid is $1 \cdot 2 \cdot 3 = 6$. We need to compute

$$
\frac{m}{6} \int_W x^2 + y^2 \, dV = \frac{m}{6} \int_0^1 \int_0^2 \int_0^3 x^2 + y^2 \, dz \, dy \, dx
$$

$$
= \frac{m}{6} \int_0^1 \int_0^2 3(x^2 + y^2) \, dy \, dx
$$

$$
= \frac{m}{2} \int_0^1 (x^2 y + y^3/3) \Big|_0^2 \, dx
$$

$$
= \frac{m}{2} \int_0^1 (2x^2 + 8/3) \, dx = 5m/3
$$

Solutions for Section 16.4

Exercises

1. $\displaystyle \int_0^{2\pi} \int_0^{\sqrt{2}} f \, r \, dr \, d\theta$

5. A circle is best described in polar coordinates. The radius is 5, so r goes from 0 to 5. To include the entire circle, we need θ to go from 0 to 2π. The integral is

$$
\int_0^{2\pi} \int_0^5 f(r \cos \theta, r \sin \theta) \, r \, dr \, d\theta.
$$

9. See Figure 16.18.

13. See Figure 16.19.

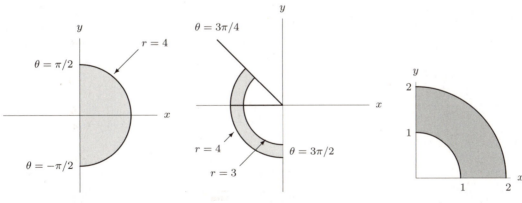

Figure 16.18	Figure 16.19	Figure 16.20

17. The region is pictured in Figure 16.20. Using polar coordinates, we get

$$
\int_R (x^2 - y^2) dA = \int_0^{\pi/2} \int_1^2 r^2 (\cos^2 \theta - \sin^2 \theta) r \, dr \, d\theta = \int_0^{\pi/2} (\cos^2 \theta - \sin^2 \theta) \cdot \frac{1}{4} r^4 \Big|_1^2 \, d\theta
$$

$$
= \frac{15}{4} \int_0^{\pi/2} (\cos^2 \theta - \sin^2 \theta) \, d\theta
$$

$$
= \frac{15}{4} \int_0^{\pi/2} \cos 2\theta \, d\theta
$$

$$
= \frac{15}{4} \cdot \frac{1}{2} \sin 2\theta \Big|_0^{\pi/2} = 0.
$$

Problems

21. From the given limits, the region of integration is in Figure 16.21.

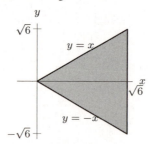

Figure 16.21

In polar coordinates, $-\pi/4 \le \theta \le \pi/4$. Also, $\sqrt{6} = x = r \cos \theta$. Hence, $0 \le r \le \sqrt{6}/\cos \theta$. The integral becomes

$$\int_0^{\sqrt{6}} \int_{-x}^{x} dy\, dx = \int_{-\pi/4}^{\pi/4} \int_0^{\sqrt{6}/\cos\theta} r\, dr\, d\theta$$

$$= \int_{-\pi/4}^{\pi/4} \left(\frac{r^2}{2} \bigg|_0^{\sqrt{6}/\cos\theta} \right) d\theta = \int_{-\pi/4}^{\pi/4} \frac{6}{2\cos^2\theta}\, d\theta$$

$$= 3\tan\theta \bigg|_{-\pi/4}^{\pi/4} = 3 \cdot (1 - (-1)) = 6.$$

Notice that we can check this answer because the integral gives the area of the shaded triangular region which is $\frac{1}{2} \cdot \sqrt{6} \cdot (2\sqrt{6}) = 6$.

25. The average value of the function r on the disc R of radius a is

$$\text{Average of } r = \frac{1}{\text{Area of } R} \int_R r\, dA = \frac{1}{\pi a^2} \int_0^{2\pi} \int_0^{a} r\cdot r\, dr\, d\theta = \frac{1}{\pi a^2} \int_0^{2\pi} \frac{a^3}{3}\, d\theta = \frac{1}{\pi a^2} 2\pi \frac{a^3}{3} = \frac{2a}{3}.$$

29. A rough graph of the base of the spring is in Figure 16.22, where the coil is roughly of width 0.01 inches. The volume is equal to the product of the base area and the height. To calculate the area we use polar coordinates, taking the following integral:

$$\text{Area} = \int_0^{4\pi} \int_{0.25+0.04\theta}^{0.26+0.04\theta} r\, dr\, d\theta$$

$$= \frac{1}{2} \int_0^{4\pi} (0.26 - 0.04\theta)^2 - (0.25 - 0.04\theta)^2\, d\theta$$

$$= \frac{1}{2} \int_0^{4\pi} 0.01 \cdot (0.51 + 0.08\theta)\, d\theta$$

$$= 0.0051 \cdot 2\pi + \frac{1}{4}(0.0008\theta^2) \bigg|_0^{4\pi} = 0.0636$$

Therefore, the volume$= 0.0636 \cdot 0.2 = 0.0127 \text{ in}^3$.

Figure 16.22

33. (a) The curve $r = 1/(2\cos\theta)$ or $r\cos\theta = 1/2$ is the line $x = 1/2$. The curve $r = 1$ is the circle of radius 1 centered at the origin. See Figure 16.23.

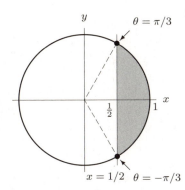

Figure 16.23

(b) The line intersects the circle where $2\cos\theta = 1$, so $\theta = \pm\pi/3$. From Figure 16.23 we see that

$$\text{Area} = \int_{-\pi/3}^{\pi/3}\int_{1/(2\cos\theta)}^{1} r\, dr\, d\theta.$$

Evaluating gives

$$\text{Area} = \int_{-\pi/3}^{\pi/3}\left(\frac{r^2}{2}\bigg|_{1/(2\cos\theta)}^{1}\right) d\theta = \frac{1}{2}\int_{-\pi/3}^{\pi/3}\left(1 - \frac{1}{4\cos^2\theta}\right) d\theta$$

$$= \frac{1}{2}\left(\theta - \frac{\tan\theta}{4}\right)\bigg|_{-\pi/3}^{\pi/3} = \frac{1}{2}\left(2\frac{\pi}{3} - 2\frac{\sqrt{3}}{4}\right) = \frac{4\pi - 3\sqrt{3}}{12}.$$

Solutions for Section 16.5

Exercises

1. (a) is (IV); (b) is (II); (c) is (VII); (d) is (VI); (e) is (III); (f) is (V).

5. The plane has equation $\theta = \pi/4$.

9.

$$\int_W f\, dV = \int_{-1}^{3}\int_0^{2\pi}\int_0^1 (\sin(r^2))\, r\, dr\, d\theta\, dz$$

$$= \int_{-1}^{3}\int_0^{2\pi}\left(-\frac{1}{2}\cos r^2\right)\bigg|_0^1 d\theta\, dz$$

$$= -\frac{1}{2}\int_{-1}^{3}\int_0^{2\pi}(\cos 1 - \cos 0)\, d\theta\, dz$$

$$= -\pi\int_{-1}^{3}(\cos 1 - 1)\, dz = -4\pi(\cos 1 - 1) = 4\pi(1 - \cos 1)$$

13. Using cylindrical coordinates, we get:

$$\int_0^1 \int_0^{2\pi} \int_0^4 f \cdot r dr \, d\theta \, dz$$

17. We use Cartesian coordinates, oriented as shown in Figure 16.24. The slanted top has equation $z = mx$, where m is the slope in the x-direction, so $m = 1/5$. Then if f is an arbitrary function, the triple integral is

$$\int_0^5 \int_0^2 \int_0^{x/5} f \, dz dy dx.$$

Other answers are possible.

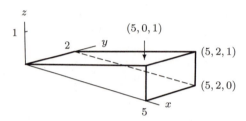

Figure 16.24

Problems

21. We want the volume of the region above the cone $\phi = \pi/3$ and below the sphere $\rho = 3$:

$$V = \int_0^{2\pi} \int_0^{\pi/3} \int_0^3 \rho^2 \sin \phi \, d\rho \, d\phi \, d\theta.$$

The order of integration can be altered and other coordinates can be used.

25. We use cylindrical coordinates since the sphere $x^2 + y^2 + z^2 = 10$, or $r^2 + z^2 = 10$, and the plane $z = 1$ can both be simply expressed. The plane cuts the sphere in the circle $r^2 + 1^2 = 10$, or $r = 3$. Thus

$$V = \int_0^{2\pi} \int_0^3 \int_1^{\sqrt{10-r^2}} r \, dz \, dr \, d\theta,$$

or

$$V = \int_0^{2\pi} \int_1^{\sqrt{10}} \int_0^{\sqrt{10-z^2}} r \, dr \, dz \, d\theta.$$

Order of integration can be altered and other coordinates can be used.

29. In rectangular coordinates, a cone has equation $z = k\sqrt{x^2 + y^2}$ for some constant k. Since $z = 4$ when $\sqrt{x^2 + y^2} = \sqrt{2^2} = 2$, we have $k = 2$. Thus, the integral is

$$\int_{-2}^2 \int_{-\sqrt{4-x^2}}^{\sqrt{4-x^2}} \int_{2\sqrt{x^2+y^2}}^4 h(x, y, z) \, dz \, dy \, dx.$$

33. The region is a solid cylinder of height 1, radius 1 with base on the xy-plane and axis on the z-axis. We have:

$$\int_0^1 \int_{-1}^1 \int_{-\sqrt{1-x^2}}^{\sqrt{1-x^2}} \frac{1}{(x^2 + y^2)^{1/2}} \, dy \, dx \, dz = \int_0^1 \int_0^{2\pi} \int_0^1 \frac{1}{r} r \, dr \, d\theta \, dz$$

$$= \int_0^1 \int_0^{2\pi} r \Big|_0^1 \, d\theta \, dz$$

$$= \int_0^1 \int_0^{2\pi} d\theta \, dz = 2\pi.$$

Note that the integral is improper, but it can be shown that the result is correct.

37. The cone is centered along the positive x-axis and intersects the sphere in the circle

$$(y^2 + z^2) + y^2 + z^2 = 4$$
$$y^2 + z^2 = 2.$$

We use spherical coordinates with ϕ measured from the x-axis and θ measured in the yz-plane. (Alternatively, the volume we want is equal to the volume between the cone $z = \sqrt{x^2 + y^2}$ and the sphere $x^2 + y^2 + z^2 = 4$.) The cone is given by $\phi = \pi/4$. The sphere has equation $\rho = 2$. Thus

$$\text{Volume} = \int_0^{2\pi} \int_0^{\pi/4} \int_0^2 \rho^2 \sin\phi \, d\rho \, d\phi \, d\theta$$

$$= \int_0^{2\pi} \int_0^{\pi/4} \frac{\rho^3}{3} \sin\phi \Big|_0^2 \, d\phi \, d\theta$$

$$= \int_0^{2\pi} \int_0^{\pi/4} \frac{8}{3} \sin\phi \, d\phi \, d\theta$$

$$= \int_0^{2\pi} -\frac{8}{3} \cos\phi \Big|_0^{\pi/4} \, d\theta = \int_0^{2\pi} \frac{8}{3} \left(1 - \frac{1}{\sqrt{2}} \right) \, d\theta$$

$$= \frac{16\pi}{3} \left(1 - \frac{1}{\sqrt{2}} \right).$$

41. The region whose volume we want is shown in Figure 16.25:

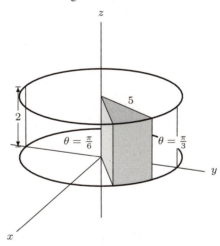

Figure 16.25

Using cylindrical coordinates, the volume is given by the integral:

$$V = \int_0^2 \int_{\pi/6}^{\pi/3} \int_0^5 r \, dr \, d\theta \, dz$$

$$= \int_0^2 \int_{\pi/6}^{\pi/3} \frac{r^2}{2} \Big|_0^5 \, d\theta \, dz$$

$$= \frac{25}{2} \int_0^2 \int_{\pi/6}^{\pi/3} \, d\theta \, dz$$

$$= \frac{25}{2} \int_0^2 \left(\frac{\pi}{3} - \frac{\pi}{6} \right) \, dz$$

$$= \frac{25}{2} \cdot \frac{\pi}{6} \cdot 2 = \frac{25\pi}{6}.$$

45. The cylinder has radius 2. Using cylindrical coordinates to find the mass and integrating with respect to r first, we have

$$\text{Mass} = \int_0^{2\pi} \int_0^3 \int_0^2 (1+r)r \, dr \, dz \, d\theta = \int_0^{2\pi} \int_0^3 \left(\frac{r^2}{2} + \frac{r^3}{3} \right) \Big|_0^2 dz \, d\theta = 2\pi \cdot 3 \cdot \left(\frac{4}{2} + \frac{8}{3} \right) = 28\pi \text{ gm}.$$

49. (a) We use spherical coordinates. Since $\delta = 9$ where $\rho = 6$ and $\delta = 11$ where $\rho = 7$, the density increases at a rate of 2 gm/cm^3 for each cm increase in radius. Thus, since density is a linear function of radius, the slope of the linear function is 2. Its equation is

$$\delta - 11 = 2(\rho - 7) \quad \text{so} \quad \delta = 2\rho - 3.$$

(b) Thus,

$$\text{Mass} = \int_0^{2\pi} \int_0^\pi \int_6^7 (2\rho - 3)\rho^2 \sin \phi \, d\rho \, d\phi \, d\theta.$$

(c) Evaluating the integral, we have

$$\text{Mass} = 2\pi \left(-\cos\phi \Big|_0^\pi \right) \left(\frac{2\rho^4}{4} - \frac{3\rho^3}{3} \Big|_6^7 \right) = 2\pi \cdot 2(425.5) = 1702\pi \text{ gm} = 5346.991 \text{ gm}.$$

53. We must first decide on coordinates. We pick spherical coordinates with the common center of the two spheres as the origin. We imagine the half-melon with the flat side horizontal and the positive z-axis going through the curved surface. See Figure 16.26. The volume is given by the integral

$$\text{Volume} = \int_0^{2\pi} \int_0^{\pi/2} \int_a^b \rho^2 \sin \phi \, d\rho d\phi d\theta.$$

Evaluating gives

$$\text{Volume} = \int_0^{2\pi} \int_0^{\pi/2} \sin\phi \frac{\rho^3}{3} \Big|_{p=a}^{p=b} d\phi d\theta = 2\pi(-\cos\phi) \Big|_0^{\pi/2} \left(\frac{b^3}{3} - \frac{a^3}{3} \right) = \frac{2\pi}{3}(b^3 - a^3).$$

To check our answer, notice that the volume is the difference between the volumes of two half spheres of radius a and b. These half spheres have volumes $2\pi b^3/3$ and $2\pi a^3/3$, respectively.

Figure 16.26

57. We first need to find the mass of the solid, using cylindrical coordinates:

$$m = \int_0^{2\pi} \int_0^1 \int_0^{\sqrt{z/a}} r \, dr \, dz \, d\theta$$

$$= \int_0^{2\pi} \int_0^1 \frac{z}{2a} \, dz \, d\theta$$

$$= \int_0^{2\pi} \frac{1}{4a} \, d\theta = \frac{\pi}{2a}$$

It makes sense that the mass would vary inversely with a, since increasing a makes the paraboloid skinnier. Now for the z-coordinate of the center of mass, again using cylindrical coordinates:

$$\bar{z} = \frac{2a}{\pi} \int_0^{2\pi} \int_0^1 \int_0^{\sqrt{z/a}} zr \, dr \, dz \, d\theta$$

$$= \frac{2a}{\pi} \int_0^{2\pi} \int_0^1 \frac{z^2}{2a} \, dz \, d\theta$$

$$= \frac{2a}{\pi} \int_0^{2\pi} \frac{1}{6a} \, d\theta = \frac{2}{3}$$

61. Assume the base of the cylinder sits on the xy-plane with center at the origin. Because the cylinder is symmetric about the z-axis, the force in the horizontal x or y direction is 0. Thus we need only compute the vertical z component of the force. We are going to use cylindrical coordinates; since the force is $G \cdot \text{mass}/(\text{distance})^2$, a piece of the cylinder of volume dV located at (r, θ, z) exerts on the unit mass a force with magnitude $G(\delta \, dV)/(r^2 + z^2)$. See Figure 16.27.

$$\begin{array}{c} \text{Vertical component} \\ \text{of force} \end{array} = \frac{G(\delta \, dV)}{r^2 + z^2} \cdot \cos\phi = \frac{G\delta \, dV}{r^2 + z^2} \cdot \frac{z}{\sqrt{r^2 + z^2}} = \frac{G\delta z \, dV}{(r^2 + z^2)^{3/2}}.$$

Adding up all the contributions of all the dV's, we obtain

$$\text{Vertical force} = \int_0^H \int_0^{2\pi} \int_0^R \frac{G\delta zr}{(r^2 + z^2)^{3/2}} \, dr \, d\theta \, dz$$

$$= \int_0^H \int_0^{2\pi} (G\delta z) \left(-\frac{1}{\sqrt{r^2 + z^2}} \right) \Bigg|_0^R \, d\theta \, dz$$

$$= \int_0^H \int_0^{2\pi} (G\delta z) \cdot \left(-\frac{1}{\sqrt{R^2 + z^2}} + \frac{1}{z} \right) \, d\theta \, dz$$

$$= \int_0^H 2\pi G\delta \left(1 - \frac{z}{\sqrt{R^2 + z^2}} \right) \, dz$$

$$= 2\pi G\delta(z - \sqrt{R^2 + z^2}) \Bigg|_0^H$$

$$= 2\pi G\delta(H - \sqrt{R^2 + H^2} + R) = 2\pi G\delta(H + R - \sqrt{R^2 + H^2})$$

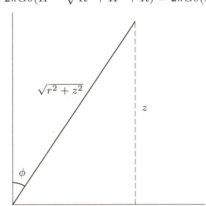

Figure 16.27

65. Use cylindrical coordinates, with the z-axis being the axis of the cable. Consider a piece of cable of length 1. Then

$$\text{Stored energy} = \frac{1}{2} \int_a^b \int_0^1 \int_0^{2\pi} \epsilon E^2 r \, d\theta \, dz \, dr = \frac{q^2}{8\pi^2\epsilon} \int_a^b \int_0^1 \int_0^{2\pi} \frac{1}{r} \, d\theta \, dz \, dr$$

$$= \frac{q^2}{4\pi\epsilon} \int_a^b \frac{1}{r} \, dr = \frac{q^2}{4\pi\epsilon}(\ln b - \ln a) = \frac{q^2}{4\pi\epsilon} \ln \frac{b}{a}.$$

So the stored energy is proportional to $\ln(b/a)$ with constant of proportionality $q^2/4\pi\epsilon$.

Solutions for Section 16.6

Exercises

1. We have $p(x, y) = 0$ for all points (x, y) satisfying $x \geq 3$, since all such points lie outside the region R. Therefore the fraction of the population satisfying $x \geq 3$ is 0.

5. Since $p(x, y) = 0$ for all (x, y) outside the rectangle R, the population is given by the volume under the graph of p over the region inside the rectangle R and to the right of the line $x = y$. Therefore the fraction of the population is given by the double integral:

$$\int_0^1 \int_y^2 xy \, dx \, dy = \int_0^1 \left. \frac{x^2 y}{2} \right|_y^2 dy = \int_0^1 \left(2y - \frac{y^3}{2} \right) dy = \left. \left(y^2 - \frac{y^4}{8} \right) \right|_0^1 = \frac{7}{8}.$$

9. No, p is not a joint density function. Since $p(x, y) = 0$ outside the region R, the volume under the graph of p is the same as the volume under the graph of p over the region R, which is 2 not 1.

13. Yes, p is a joint density function. In the region R we have $1 \geq x^2 + y^2$, so $p(x, y) = (2/\pi)(1 - x^2 - y^2) \geq 0$ for all x and y in R, and $p(x, y) = 0$ for all other (x, y). To check that p is a joint density function, we check that the total volume under the graph of p over the region R is 1. Using polar coordinates, we get:

$$\int_R p(x, y) dA = \frac{2}{\pi} \int_0^{2\pi} \int_0^1 (1 - r^2) r \, dr \, d\theta = \frac{2}{\pi} \int_0^{2\pi} \left. \left(\frac{r^2}{2} - \frac{r^4}{4} \right) \right|_0^1 d\theta = \frac{2}{\pi} \int_0^{2\pi} \frac{1}{4} d\theta = 1.$$

Problems

17. (a) For a density function,

$$1 = \int_{-\infty}^{\infty} \int_{-\infty}^{\infty} f(x, y) \, dy \, dx = \int_0^2 \int_0^1 kx^2 \, dy \, dx$$
$$= \int_0^2 kx^2 \, dx$$
$$= \left. \frac{kx^3}{3} \right|_0^2 = \frac{8k}{3}.$$

So $k = 3/8$.

(b)
$$\int_0^1 \int_0^{2-y} \frac{3}{8} x^2 \, dx \, dy = \int_0^1 \frac{1}{8} (2 - y)^3 \, dy = \left. \frac{-1}{32} (2 - y)^4 \right|_0^1 = \frac{15}{32}$$

(c)
$$\int_0^{1/2} \int_0^1 \frac{3}{8} x^2 \, dx \, dy = \int_0^{1/2} \left. \frac{1}{8} x^3 \right|_0^1 dy = \int_0^{1/2} \frac{1}{8} dy = \frac{1}{16}.$$

21. (a) If $t \leq 0$, then $F(t) = 0$ because the average of two positive numbers can not be negative. If $1 < t$ then $F(t) = 1$ because the average of two numbers each at most 1 is certain to be less than or equal to 1. For any t, we have $F(t) = \int_R p(x, y) dA$ where R is the region of the plane defined by $(x + y)/2 \leq t$. Since $p(x, y) = 0$ outside the unit square, we need integrate only over the part of R that lies inside the square, and since $p(x, y) = 1$ inside the square, the integral equals the area of that part of the square. Thus, we can calculate the area using area formulas. For $0 \leq t \leq 1$, we draw the line $(x + y)/2 = t$, which has x- and y-intercepts of $2t$. Figure 16.28 shows that for $0 < t \leq 1/2$,

$$F(t) = \text{Area of triangle} = \frac{1}{2} \cdot 2t \cdot 2t = 2t^2.$$

In Figure 16.29, when $x = 1$, we have $y = 2t - 1$. Thus, the vertical side of the unshaded triangle is $1 - (2t - 1) = 2 - 2t$. The horizontal side is the same length, so for $1/2 < t \leq 1$,

$$F(t) = \text{Area of Square} - \text{Area of triangle} = 1^2 - \frac{1}{2}(2 - 2t)^2 - 1 - 2(1 - t)^2.$$

The final result is:

$$F(t) = \begin{cases} 0 & \text{if } t \leq 0 \\ 2t^2 & \text{if } 0 < t \leq 1/2 \\ 1 - 2(1-t)^2 & \text{if } 1/2 < t \leq 1 \\ 1 & \text{if } 1 < t \end{cases}.$$

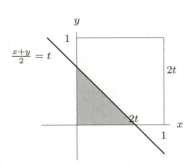

Figure 16.28: For $0 < t \leq \frac{1}{2}$

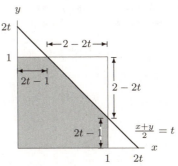

Figure 16.29: For $\dfrac{1}{2} < t \leq 1$

(b) The probability density function $p(t)$ of z is the derivative of its cumulative distribution function. We have

$$p(t) = \begin{cases} 0 & \text{if } t \leq 0 \\ 4t & \text{if } 0 < t \leq 1/2 \\ 4 - 4t & \text{if } 1/2 < t \leq 1 \\ 0 & \text{if } 1 < t \end{cases}.$$

See Figure 16.30.

(c) The values of x and y and equally likely to be near $0, 1/2$, and 1. Notice from the graph of the density function in Figure 16.30 that even though x and y separately are equally likely to be anywhere between 0 and 1, their average $z = (x+y)/2$ is more likely to be near $1/2$ than to be near 0 or 1.

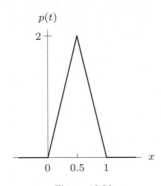

Figure 16.30

Solutions for Section 16.7

Exercises

1. We have

$$\frac{\partial(x,y)}{\partial(s,t)} = \begin{vmatrix} x_s & x_t \\ y_s & y_t \end{vmatrix} = \begin{vmatrix} 5 & 2 \\ 3 & 1 \end{vmatrix} = -1.$$

Therefore,

$$\left| \frac{\partial(x,y)}{\partial(s,t)} \right| = 1.$$

5. We have

$$\frac{\partial(x, y, z)}{\partial(s, t, u)} = \begin{vmatrix} x_s & x_t & x_u \\ y_s & y_t & y_u \\ z_s & z_t & z_u \end{vmatrix} = \begin{vmatrix} 3 & 1 & 2 \\ 1 & 5 & -1 \\ 2 & -1 & 1 \end{vmatrix}.$$

This 3×3 determinant is computed the same way as for the cross product, with the entries $3, 1, 2$ in the first row playing the same role as $\vec{i}, \vec{j}, \vec{k}$. We get

$$\frac{\partial(x, y, z)}{\partial(s, t, u)} = ((5)(1) - (-1)(-1))3 + ((-1)(2) - (1)(1))1 + ((1)(-1) - (2)(5))2 = -13.$$

9. The square T is defined by the inequalities

$$0 \le s = ax \le 1 \qquad 0 \le t = by \le 1$$

that correspond to the inequalities

$$0 \le x \le 1/a = 50 \qquad 0 \le y \le 1/b = 10$$

that define R. Thus $a = 1/50$ and $b = 1/10$.

13. Inverting the change of variables gives $x = s - at, y = t$.

The four edges of R are

$$y = 0, y = 5, y = -\frac{1}{3}x, y = -\frac{1}{3}(x - 10).$$

The change of variables transforms the edges to

$$t = 0, t = 5, t = -\frac{1}{3}s + \frac{1}{3}at, t = -\frac{1}{3}s + \frac{1}{3}at + \frac{10}{3}.$$

These are equations for the edges of a rectangle in the st-plane if the last two are of the form: $s = $ (Constant). This happens when the t terms drop out, or $a = 3$. With $a = 3$ the change of variables gives

$$\int \int_T \left| \frac{\partial(x, y)}{\partial(s, t)} \right| \, ds \, dt$$

over the rectangle

$$T : 0 \le t \le 5, 0 \le s \le 10.$$

Problems

17. Given

$$\begin{cases} x = \rho \sin \phi \cos \theta \\ y = \rho \sin \phi \sin \theta \\ z = \rho \cos \phi, \end{cases}$$

$$\frac{\partial(x, y, z)}{\partial(\rho, \phi, \theta)} = \begin{vmatrix} \frac{\partial x}{\partial \rho} & \frac{\partial x}{\partial \phi} & \frac{\partial x}{\partial \theta} \\ \frac{\partial y}{\partial \rho} & \frac{\partial y}{\partial \phi} & \frac{\partial y}{\partial \theta} \\ \frac{\partial z}{\partial \rho} & \frac{\partial z}{\partial \phi} & \frac{\partial z}{\partial \theta} \end{vmatrix} = \begin{vmatrix} \sin \phi \cos \theta & \rho \cos \phi \cos \theta & -\rho \sin \phi \sin \theta \\ \sin \phi \sin \theta & \rho \cos \phi \sin \theta & \rho \sin \phi \cos \theta \\ \cos \phi & -\rho \sin \phi & 0 \end{vmatrix}$$

$$= \cos \phi \begin{vmatrix} \rho \cos \phi \cos \theta & -\rho \sin \phi \sin \theta \\ \rho \cos \phi \sin \theta & \rho \sin \phi \cos \theta \end{vmatrix} + \rho \sin \phi \begin{vmatrix} \sin \phi \cos \theta & -\rho \sin \phi \sin \theta \\ \sin \phi \sin \theta & \rho \sin \phi \cos \theta \end{vmatrix}$$

$$= \cos \phi (\rho^2 \cos^2 \theta \cos \phi \sin \phi + \rho^2 \sin^2 \theta \cos \phi \sin \phi)$$

$$\quad + \rho \sin \phi (\rho \sin^2 \phi \cos^2 \theta + \rho \sin^2 \phi \sin^2 \theta)$$

$$= \rho^2 \cos^2 \phi \sin \phi + \rho^2 \sin^3 \phi$$

$$= \rho^2 \sin \phi.$$

21. The area of the ellipse is $\int\int_R dx\,dy$ where R is the region $x^2 + 2xy + 2y^2 \leq 1$. We must change variables in both the area element $dA = dx\,dy$ and the region R.

Inverting the variable change gives $x = s - t$, $y = t$. Thus

$$\frac{\partial(x,y)}{\partial(s,t)} = \begin{vmatrix} \frac{\partial x}{\partial s} & \frac{\partial x}{\partial t} \\ \frac{\partial y}{\partial s} & \frac{\partial y}{\partial t} \end{vmatrix} = \begin{vmatrix} 1 & -1 \\ 0 & 1 \end{vmatrix} = 1.$$

Therefore

$$dx\,dy = \left| \frac{\partial(x,y)}{\partial(s,t)} \right| ds\,dt = ds\,dt.$$

The region of integration is

$$x^2 + 2xy + 2y^2 = (s-t)^2 + 2(s-t)t + 2t^2 = s^2 + t^2 \leq 1.$$

Let T be the unit disc $s^2 + t^2 \leq 1$. We have

$$\int\int_R dx\,dy = \int\int_T ds\,dt = \text{Area of } T = \pi.$$

25. Given

$$\begin{cases} s = xy \\ t = xy^2, \end{cases}$$

we have

$$\frac{\partial(s,t)}{\partial(x,y)} = \begin{vmatrix} \frac{\partial s}{\partial x} & \frac{\partial s}{\partial y} \\ \frac{\partial t}{\partial x} & \frac{\partial t}{\partial y} \end{vmatrix} = \begin{vmatrix} y & x \\ y^2 & 2xy \end{vmatrix} = xy^2 = t.$$

Since

$$\frac{\partial(s,t)}{\partial(x,y)} \cdot \frac{\partial(x,y)}{\partial(s,t)} = 1,$$

$$\frac{\partial(x,y)}{\partial(s,t)} = t \qquad \text{so} \qquad \left| \frac{\partial(x,y)}{\partial(s,t)} \right| = \frac{1}{t}$$

So

$$\int_R xy^2\,dA = \int_T t\left| \frac{\partial(x,y)}{\partial(s,t)} \right| ds\,dt = \int_T t(\frac{1}{t})\,ds\,dt = \int_T ds\,dt,$$

where T is the region bounded by $s = 1$, $s = 4$, $t = 1$, $t = 4$.

Then

$$\int_R xy^2\,dA = \int_1^4 ds \int_1^4 dt = 9.$$

Solutions for Chapter 16 Review

Exercises

1. We use Cartesian coordinates, oriented so that the cube is in the first quadrant. See Figure 16.31. Then, if f is an arbitrary function, the integral is

$$\int_0^2 \int_0^3 \int_0^5 f\,dx\,dy\,dz.$$

Other answers are possible. In particular, the order of integration can be changed.

Figure 16.31

5. See Figure 16.32.

9. The region is the half cylinder in Figure 16.33.

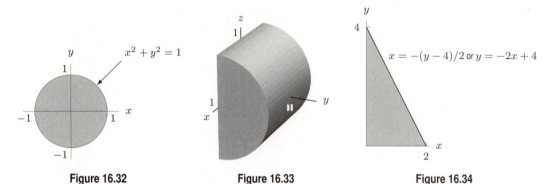

| **Figure 16.32** | **Figure 16.33** | **Figure 16.34** |

13. (a) See Figure 16.34.

(b) $\int_0^2 \int_0^{-2x+4} g(x, y) \, dy \, dx$.

17. First use integration by parts, with y as the variable, $u = x^2 y$, $u' = x^2$, $v = \frac{\sin(xy)}{x}$, $v' = \cos(xy)$. Then,

$$\int_3^4 \int_0^1 x^2 y \cos(xy) \, dy \, dx = \int_3^4 \left([xy \sin(xy)]_0^1 - \int_0^1 x \sin(xy) \, dy \right) \, dx$$

$$= \int_3^4 \left(x \sin x + [\cos(xy)]_0^1 \right) \, dx$$

$$= \int_3^4 (x \sin x + \cos x - 1) \, dx.$$

Now use integration by parts again, with $u = x$, $u' = 1$, $v = -\cos x$, $v' = \sin x$. Then,

$$\int_3^4 (x \sin x + \cos x - 1) \, dx = [-x \cos x]_3^4 + \int_3^4 \cos x \, dx + \int_3^4 (\cos x - 1) \, dx$$

$$= (-x \cos x + 2 \sin x - x)|_3^4$$

$$= -4 \cos 4 + 2 \sin 4 + 3 \cos 3 - 2 \sin 3 - 1.$$

Thus,

$$\int_3^4 \int_0^1 x^2 y \cos(xy) \, dy \, dx = -4 \cos 4 + 2 \sin 4 + 3 \cos 3 - 2 \sin 3 - 1.$$

21. (a) A vertical plane perpendicular to the x-axis: $x = 2$.

(b) A cylinder: $r = 3$.

(c) A sphere: $\rho = \sqrt{3}$.

(d) A cone: $\phi = \pi/4$.

(e) A horizontal plane: $z = -5$.

(f) A vertical half-plane: $\theta = \pi/4$.

25. From Figure 16.35, we have the following iterated integrals:

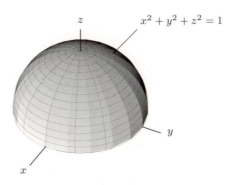

$x^2 + y^2 + z^2 = 1$

Figure 16.35

(a) $\displaystyle\int_R f \, dV = \int_{-1}^{1} \int_{-\sqrt{1-x^2}}^{\sqrt{1-x^2}} \int_{0}^{\sqrt{1-x^2-y^2}} f(x,y,z) \, dzdydx$

(b) $\displaystyle\int_R f \, dV = \int_{-1}^{1} \int_{-\sqrt{1-y^2}}^{\sqrt{1-y^2}} \int_{0}^{\sqrt{1-x^2-y^2}} f(x,y,z) \, dzdxdy$

(c) $\displaystyle\int_R f \, dV = \int_{-1}^{1} \int_{0}^{\sqrt{1-y^2}} \int_{-\sqrt{1-y^2-z^2}}^{\sqrt{1-y^2-z^2}} f(x,y,z) \, dxdzdy$

(d) $\displaystyle\int_R f \, dV = \int_{-1}^{1} \int_{0}^{\sqrt{1-x^2}} \int_{-\sqrt{1-x^2-z^2}}^{\sqrt{1-x^2-z^2}} f(x,y,z) \, dydzdx$

(e) $\displaystyle\int_R f \, dV = \int_{0}^{1} \int_{-\sqrt{1-z^2}}^{\sqrt{1-z^2}} \int_{-\sqrt{1-x^2-z^2}}^{\sqrt{1-x^2-z^2}} f(x,y,z) \, dydxdz$

(f) $\displaystyle\int_R f \, dV = \int_{0}^{1} \int_{-\sqrt{1-z^2}}^{\sqrt{1-z^2}} \int_{-\sqrt{1-y^2-z^2}}^{\sqrt{1-y^2-z^2}} f(x,y,z) \, dxdydz$

Problems

29. The region is a hollow half-sphere, with inner radius $\sqrt{3}$ and outer radius $\sqrt{4} = 2$. See Figure 16.36. In spherical coordinates, the integral is

$$\int_0^\pi \int_0^\pi \int_{\sqrt{3}}^2 \rho^2 \sin \phi \, d\rho \, d\phi \, d\theta.$$

The order of integration can be altered.

Figure 16.36

33. W is a cylindrical shell, so cylindrical coordinates should be used. See Figure 16.37.

Figure 16.37

$$
\int_W \frac{z}{(x^2 + y^2)^{3/2}}\, dV = \int_0^4 \int_0^{2\pi} \int_1^2 \frac{z}{r^3}\, r\,dr\,d\theta\,dz
$$

$$
= \int_0^4 \int_0^{2\pi} \int_1^2 \frac{z}{r^2}\, dr\,d\theta\,dz
$$

$$
= \int_0^4 \int_0^{2\pi} \left(-\frac{z}{r} \right)\Bigg|_1^2 \, d\theta\,dz
$$

$$
= \int_0^4 \int_0^{2\pi} \frac{z}{2}\, d\theta\,dz
$$

$$
= \int_0^4 \frac{z}{2} \cdot 2\pi\, dz = \frac{1}{2}\pi \cdot z^2 \Bigg|_0^4 = 8\pi
$$

37. Can't tell, since y^3 is both positive and negative for $x < 0$.

41. Zero. You can see this in several ways. One way is to observe that xy is positive on the part of the sphere above and below the first and third quadrants (where x and y are of the same sign) and negative (of equal absolute value) on the part of the sphere above and below the second and fourth quadrants (where x and y have opposite signs). These add up to zero in the integral of xy over all of W.

Another way to see that the integral is zero is to write the triple integral as an iterated integral, say integrating first with respect to x. For fixed y and z, the x-integral is over an interval symmetric about 0. The integral of x over such an interval is zero. If any of the inner integrals in an iterated integral is zero, then the triple integral is zero.

45. Negative. Since $z^2 - 1 \le 0$ in the sphere, its integral is negative.

49. The depth of the lake is given in meters and the diameter in kilometers. We should work with a single unit of length. In this solution we work with kilometers, but meters would work just as well.

The shape of the lake suggests integration in polar coordinates, with r km measured from the center of the island. Thus $t = r - 1$ is the distance in kilometers from the island when r varies between 1 and 5. The depth of the lake r km from the center of the island is

$$
\text{Depth} = \frac{100(r - 1)(4 - (r - 1))}{1000} = -\frac{1}{2} + \frac{3r}{5} - \frac{r^2}{10} \text{ km.}
$$

$$
\text{Volume of the lake} = \int_0^{2\pi} \int_1^5 \left(-\frac{1}{2} + \frac{3r}{5} - \frac{r^2}{10} \right) r\,dr\,d\theta = \frac{32\pi}{5} = 20.1 \text{ km}^3.
$$

53. (a) $\displaystyle\int_0^{2\pi} \int_0^{\pi} \int_1^2 \rho^2 \sin\phi\, d\rho\,d\phi\,d\theta.$

(b) $\displaystyle\int_0^{2\pi} \int_0^2 \int_{-\sqrt{4-r^2}}^{\sqrt{4-r^2}} r\,dz\,dr\,d\theta - \int_0^{2\pi} \int_0^1 \int_{-\sqrt{1-r^2}}^{\sqrt{1-r^2}} r\,dz\,dr\,d\theta.$

57. The region of integration is shown in Figure 16.38, and the mass of the given solid is given by

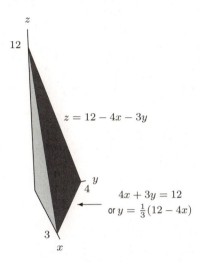

$z = 12 - 4x - 3y$

y
4

$4x + 3y = 12$
or $y = \frac{1}{3}(12 - 4x)$

3

Figure 16.38

$$\text{Mass} = \int_R \delta \, dV$$

$$= \int_0^3 \int_0^{\frac{1}{3}(12-4x)} \int_0^{12-4x-3y} x^2 \, dz\,dy\,dx$$

$$= \int_0^3 \int_0^{\frac{1}{3}(12-4x)} x^2 z \Big|_{z=0}^{z=12-4x-3y} dy\,dx$$

$$= \int_0^3 \int_0^{\frac{1}{3}(12-4x)} x^2(12 - 4x - 3y) \, dy\,dx$$

$$= \int_0^3 x^2\left(12y - 4xy - \frac{3}{2}y^2\right)\Big|_0^{y=\frac{1}{3}(12-4x)} dx$$

$$= \left(8x^3 - 4x^4 + \frac{8}{15}x^5\right)\Big|_0^3$$

$$= \frac{108}{5}.$$

61. The plane $(x/p) + (y/q) + (z/r) = 1$ cuts the axes at the points $(p,0,0)$; $(0,q,0)$; $(0,0,r)$. Since p, q, r are positive, the region between this plane and the coordinate planes is a pyramid in the first octant. Solving for z gives

$$z = r\left(1 - \frac{x}{p} - \frac{y}{q}\right) = r - \frac{rx}{p} - \frac{ry}{q}.$$

The volume, V, is given by the double integral

$$V = \int_R \left(r - \frac{rx}{p} - \frac{ry}{q}\right) dA,$$

where R is the region shown in Figure 16.39. Thus

$$V = \int_0^p \int_0^{q-qx/p} \left(r - \frac{rx}{p} - \frac{ry}{q}\right) dy\,dx$$

$$= \int_0^p \left(\left(ry - \frac{rxy}{p} - \frac{ry^2}{2q} \right) \bigg|_{y=0}^{y=q-qx/p} \right) dx$$

$$= \int_0^p \left(r\left(q - \frac{qx}{p} \right) - \frac{r}{p}x\left(q - \frac{qx}{p} \right) - \frac{r}{2q}\left(q - \frac{qx}{p} \right)^2 \right) dx$$

$$= \int_0^p rq - \frac{2rqx}{p} + \frac{rqx^2}{p^2} - \frac{rq}{2q}\left(1 - \frac{2x}{p} + \frac{x^2}{p^2} \right) dx$$

$$= \left(rqx - \frac{rqx^2}{p} + \frac{rqx^3}{p^23} - \frac{rqx}{2} + \frac{rqx^2}{p2} - \frac{rqx^3}{2p^23} \right) \bigg|_0^p$$

$$= pqr - pqr + \frac{pqr}{3} - \frac{pqr}{2} + \frac{pqr}{2} - \frac{pqr}{6} = \frac{pqr}{6}.$$

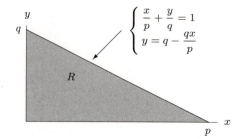

Figure 16.39

65. We must first decide on coordinates. We pick Cartesian coordinates with the smaller sphere centered at the origin, the larger one centered at $(0, 0, -1)$. A vertical cross-section of the region in the xz-plane is shown in Figure 16.40. The smaller sphere has equation $x^2 + y^2 + z^2 = 1$. The larger sphere has equation $x^2 + y^2 + (z+1)^2 = 2$.

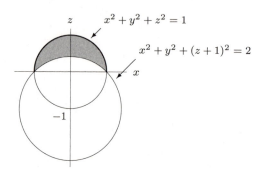

Figure 16.40

Let R represent the region in the xy-plane which lies directly underneath (or above) the region whose volume we want. The curve bounding this region is a circle, and we find its equation by solving the system:

$$x^2 + y^2 + z^2 = 1$$
$$x^2 + y^2 + (z+1)^2 = 2$$

Subtracting the equations gives

$$(z+1)^2 - z^2 = 1$$
$$2z + 1 = 1$$
$$z = 0.$$

Since $z = 0$, the two surfaces intersect in the xy-plane in the circle $x^2 + y^2 = 1$. Thus R is $x^2 + y^2 \le 1$.

The top half of the small sphere is represented by $z = \sqrt{1 - x^2 - y^2}$; the top half of the large sphere is represented by $z = -1 + \sqrt{2 - x^2 - y^2}$. Thus the volume is given by

$$\text{Volume} = \int_{-1}^{1} \int_{-\sqrt{1-x^2}}^{\sqrt{1-x^2}} \int_{-1+\sqrt{2-x^2-y^2}}^{\sqrt{1-x^2-y^2}} dz\,dy\,dx.$$

Starting to evaluate the integral, we get

$$\text{Volume} = \int_{-1}^{1} \int_{-\sqrt{1-x^2}}^{\sqrt{1-x^2}} (\sqrt{1 - x^2 - y^2} + 1 - \sqrt{2 - x^2 - y^2})\,dy\,dx.$$

We simplify the integral by converting to polar coordinates

$$\text{Volume} = \int_{0}^{2\pi} \int_{0}^{1} \left(\sqrt{1 - r^2} + 1 - \sqrt{2 - r^2} \right) r\,dr\,d\theta$$

$$= \int_{0}^{2\pi} \left(-\frac{(1 - r^2)^{3/2}}{3} + \frac{r^2}{2} + \frac{(2 - r^2)^{3/2}}{3} \right) \Big|_{0}^{1} d\theta$$

$$= 2\pi \left(\frac{1}{2} + \frac{1}{3} - \left(-\frac{1}{3} + \frac{2^{3/2}}{3} \right) \right) = 2\pi \left(\frac{7}{6} - \frac{2\sqrt{2}}{3} \right) = 1.41.$$

69. The outer circle is a semicircle of radius 4. This is shown in Figure 16.41, with center at D. Thus, $CE = 2$ and $DC = 2$, while $AD = 4$. Notice that angle ADO is a right angle.

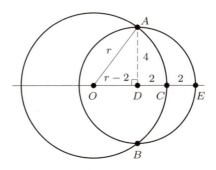

Figure 16.41

Suppose the large circle has center O and radius r. Then $OA = r$ and $OD = OC - DC = r - 2$. Applying Pythagoras' Theorem to triangle OAD gives

$$r^2 = 4^2 + (r - 2)^2$$
$$r^2 = 16 + r^2 - 4r + 4$$
$$r = 5.$$

If we put the origin at O, the equation of the large circle is $x^2 + y^2 = 25$. In the same coordinates, the equation of the small circle, which has center at $D = (3, 0)$, is $(x - 3)^2 + y^2 = 16$. The right hand side of the two circles are given by

$$x = \sqrt{25 - y^2} \quad \text{and} \quad x = 3 + \sqrt{16 - y^2}.$$

Since the y-coordinate of A is 4 and the y-coordinate of B is -4, we have

$$\text{Area} = \int_{-4}^{4} \int_{\sqrt{25-y^2}}^{3+\sqrt{16-y^2}} 1\,dx\,dy$$

$$= \int_{-4}^{4} (3 + \sqrt{16 - y^2} - \sqrt{25 - y^2})\,dy$$

$$= 13.95.$$

CAS Challenge Problems

73. In Cartesian coordinates the integral is

$$\int_D \sqrt[3]{x^2 + y^2}\, dA = \int_{-1}^{1} \int_{-\sqrt{1-x^2}}^{\sqrt{1-x^2}} \sqrt[3]{x^2 + y^2}\, dydx.$$

In polar coordinates it is

$$\int_D \sqrt[3]{x^2 + y^2}\, dA = \int_0^{2\pi} \int_0^1 \sqrt[3]{r^2}\, r\, dr d\theta = \int_0^{2\pi} \int_0^1 r^{5/3}\, dr d\theta$$

$$= \int_0^{2\pi} \frac{3}{8}\, d\theta = \frac{6\pi}{7}$$

The Cartesian coordinate version requires the use of a computer algebra system. Some CASs may be able to handle it and may give the answer in terms of functions called hypergeometric functions. To compare the answers are the same you may need to ask the CAS to give a numerical value for the answer. It's possible your CAS will not be able to handle the integral at all.

CHECK YOUR UNDERSTANDING

1. False. For example, if $f(x, y) < 0$ for all (x, y) in the region R, then $\int_R f\, dA$ is negative.

5. True. The double integral is the limit of the sum $\sum_{\Delta A \to 0} \rho(x, y)\Delta A$. Each of the terms $\rho(x, y)\Delta A$ is an approximation of the total population inside a small rectangle of area ΔA. Thus the limit of the sum of all of these numbers as $\Delta A \to \infty$ gives the total population of the region R.

9. False. There is no reason to expect this to be true, since the behavior of f on one half of R can be completely unrelated to the behavior of f on the other half. As a counterexample, suppose that f is defined so that $f(x, y) = 0$ for points (x, y) lying in S, and $f(x, y) = 1$ for points (x, y) lying in the part of R that is not in S. Then $\int_S f\, dA = 0$, since $f = 0$ on all of S. To evaluate $\int_R f\, dA$, note that $f = 1$ on the square S_1 which is $0 \le x \le 1, 1 \le y \le 2$. Then $\int_R f\, dA = \int_{S_1} f\, dA = \text{Area}(S_1) = 1$, since $f = 0$ on S.

13. True. For any point in the region of integration we have $1 \le x \le 2$, and so y is between the positive numbers 1 and 8.

17. False. The given limits describe only the upper half disk where $y \ge 0$. The correct limits are $\int_{-a}^{a} \int_{-\sqrt{a^2-x^2}}^{\sqrt{a^2-x^2}} f\, dydx$.

21. False. The integral gives the total mass of the material contained in W.

25. True. Both sets of limits describe the solid region lying above the rectangle $-1 \le x \le 1, 0 \le y \le 1, z = 0$ and below the parabolic cylinder $z = 1 - x^2$.

29. False. As a counterexample, let W_1 be the solid cube $0 \le x \le 1, 0 \le y \le 1, 0 \le z \le 1$, and let W_2 be the solid cube $-\frac{1}{2} \le x \le 0, -\frac{1}{2} \le y \le 0, -\frac{1}{2} \le z \le 0$. Then volume$(W_1) = 1$ and volume$(W_2) = \frac{1}{8}$. Now if $f(x, y, z) = -1$, then $\int_{W_1} f\, dV = 1 \cdot -1$ which is less than $\int_{W_2} f\, dV = \frac{1}{8} \cdot -1$.

CHAPTER SEVENTEEN

Solutions for Section 17.1

Exercises

1. We want the bottom half of a semicircle of radius 1 centered at $(0, 1)$. The equations $x = \cos t$, $y = 1 + \sin t$ describe clockwise motion in this circle, passing $(-1, 1)$ when $t = \pi$ and $(1, 1)$ when $t = 2\pi$. So a possible parameterization is

$$x = \cos t, \quad y = 1 + \sin t, \quad \pi \le t \le 2\pi.$$

5. Since we are moving on the y-axis, $x = 0$, and y goes from -2 to 1. Thus a possible parameterization is

$$x = 0, \quad y = t, \quad -2 \le t \le 1.$$

9. One possible parameterization is

$$x = -3 + 2t, \quad y = 4 + 2t, \quad z = -2 - 3t.$$

13. The displacement vector from the first point to the second is $\vec{v} = 4\vec{i} - 5\vec{j} - 3\vec{k}$. The line through point $(1, 5, 2)$ and with direction vector $\vec{v} = 4\vec{i} - 5\vec{j} - 3\vec{k}$ is given by parametric equations

$$x = 1 + 4t,$$
$$y = 5 - 5t,$$
$$z = 2 - 3t.$$

Other parameterizations of the same line are also possible.

17. The line passes through $(3, 0, 0)$ and $(0, 0, -5)$. The displacement vector from the first of these points to the second is $\vec{v} = -3\vec{i} - 5\vec{k}$. The line through point $(3, 0, 0)$ and with direction vector $\vec{v} = -3\vec{i} - 5\vec{k}$ is given by parametric equations

$$x = 3 - 3t,$$
$$y = 0,$$
$$z = -5t.$$

Other parameterizations of the same line are also possible.

21. The circle lies in the plane $z = 2$, so one possible answer is

$$x = 3\cos t, \quad y = 3\sin t, \quad z = 2.$$

25. The xy-plane is $z = 0$, so a possible answer is

$$x = t^2, \quad y = t, \quad z = 0.$$

29. Since its diameters are parallel to the y and z-axes and its center is in the yz-plane, the ellipse must lie in the yz-plane, $x = 0$. The ellipse with the same diameters centered at the origin would have its y-coordinate range between $-5/2$ and $5/2$ and its z-coordinate range between -1 and 1. Thus this ellipse has equation

$$x = 0, \quad y = \frac{5}{2}\cos t, \quad z = \sin t.$$

To move the center to $(0, 1, -2)$, we add 1 to the equation for y and -2 to the equation for z, so one possible answer for our ellipse is

$$x = 0, \quad y = 1 + \frac{5}{2}\cos t, \quad z = -2 + \sin t.$$

33. The displacement vector between the points is $\vec{u} = 3\vec{i} + 5\vec{k}$, so a possible parameterization of the line is

$$x = -1 + 3t, \quad y = 2, \quad z = -3 + 5t.$$

37. The line segment PQ has length 10, so it must be a diameter of the circle. The center of the circle is therefore the midpoint of PQ, which is the point $(5, 0)$. The upper arc of the circle between P and Q can be parameterized as follows:

$$\vec{r}(t) = 5\vec{i} + 5(-\cos t\vec{i} + \sin t\vec{j}), \quad 0 \le t \le \pi.$$

The lower arc can be parameterized as follows:

$$\vec{r}(t) = 5\vec{i} + 5(\cos t\vec{i} + \sin t\vec{j}), \quad \pi \le t \le 2\pi.$$

Problems

41. We find the parameterization in terms of the displacement vector $\overrightarrow{OP} = 2\vec{i} + 5\vec{j}$ from the origin to the point P and the displacement vector $\overrightarrow{PQ} = 10\vec{i} + 4\vec{j}$ from P to Q.
$$\vec{r}(t) = \overrightarrow{OP} + (t - 10)\overrightarrow{PQ} \text{ or } \vec{r}(t) = (2 + (t - 10)10)\vec{i} + (5 + (t - 10)4)\vec{j}$$

45. These equations parameterize a line. Since $(3 + t) + (2t) + 3(1 - t) = 6$, we have $x + y + 3z = 6$. Similarly, $x - y - z = (3 + t) - 2t - (1 - t) = 2$. That is, the curve lies entirely in the plane $x + y + 3z = 6$ and in the plane $x - y - z = 2$. Since the normals to the two planes, $\vec{n_1} = \vec{i} + \vec{j} + 3\vec{k}$ and $\vec{n_2} = \vec{i} - \vec{j} - \vec{k}$ are not parallel, the line is the intersection of two nonparallel planes, which is a straight line in 3-dimensional space.

49. The coefficients of t in the parameterizations show that line $\vec{r}_1$ is parallel to the vector $-3\vec{i} + 2\vec{j} + \vec{k}$ and line $\vec{r}_2$ is parallel to $-6\vec{i} + 4\vec{j} + 3\vec{k}$. Since these vectors are not parallel, the lines are not parallel, so the lines are different.

53. Add the two equations to get $2x + 3z = 5$, or $x = -\frac{3}{2}z + \frac{5}{2}$. Subtract the two equations to get $2y - z = 1$, or $y = \frac{1}{2}z + \frac{1}{2}$. So a possible parameterization is

$$x = -\frac{3}{2}t + \frac{5}{2}, \quad y = \frac{1}{2}t + \frac{1}{2}, \quad z = t.$$

57. (a) Both paths are straight lines, the first passes through the point $(-1, 4, -1)$ in the direction of the vector $\vec{i} - \vec{j} + 2\vec{k}$ and the second passes through $(-7, -6, -1)$ in the direction of the vector $2\vec{i} + 2\vec{j} + \vec{k}$. The two paths are not parallel.

(b) Is there a time t when the two particles are at the same place at the same time? If so, then their coordinates will be the same, so equating coordinates we get

$$-1 + t = -7 + 2t$$
$$4 - t = -6 + 2t$$
$$-1 + 2t = -1 + t.$$

Since the first equation is solved by $t = 6$, the second by $t = 10/3$, and the third by $t = 0$, no value of t solves all three equations. The two particles never arrive at the same place at the same time, and so they do not collide.

(c) Are there any times t_1 and t_2 such that the position of the first particle at time t_1 is the same as the position of the second particle at time t_2? If so then

$$-1 + t_1 = -7 + 2t_2$$
$$4 - t_1 = -6 + 2t_2$$
$$-1 + 2t_1 = -1 + t_2.$$

We solve the first two equations and get $t_1 = 2$ and $t_2 = 4$. This is a solution for the third equation as well, so the three equations are satisfied by $t_1 = 2$ and $t_2 = 4$. At time $t = 2$ the first particle is at the point $(1, 2, 3)$, and at time $t = 4$ the second is at the same point. The paths cross at the point $(1, 2, 3)$, and the first particle gets there first.

61. The helices wind around a cylinder of radius α, which explains the significance of α. As t increases from 0 to 2π, the helix winds once around the cylinder, climbing upward a distance of $2\pi\beta$. Thus β controls how stretched out the helix is in the vertical direction. See Figure 17.1 and Figure 17.2.

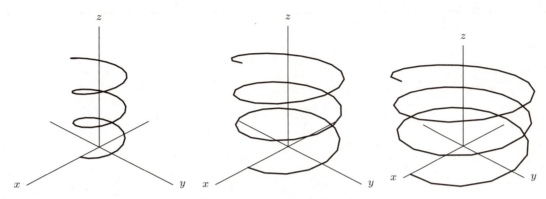

Figure 17.1: Three values of α with the same β

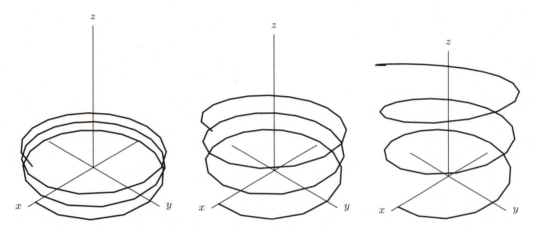

Figure 17.2: Three values of β with the same α

65. The line $\vec{r} = \vec{a} + t\vec{b}$ is parallel to the vector $\vec{b}$ and through the point with position vector $\vec{a}$.

(a) is (vii). The equation $\vec{b} \cdot \vec{r} = 0$ is a plane perpendicular to $\vec{b}$ and satisfied by $(0, 0, 0)$.

(b) is (ii). For any constant k, the equation $\vec{b} \cdot \vec{r} = k$ is a plane perpendicular to $\vec{b}$. If $k = ||\vec{a}|| \neq 0$, the plane does not contain the origin.

(c) is (iv). The equation $(\vec{a} \times \vec{b}) \cdot (\vec{r} - \vec{a}) = 0$ is the equation of a plane which is satisfied by $\vec{r} = \vec{a}$, so the point with position vector $\vec{a}$ lies on the plane. Since $\vec{a} \times \vec{b}$ is perpendicular to $\vec{b}$, the plane is parallel to the line, and therefore it contains the line.

69. (a) If $\vec{n} \cdot \vec{v} = 0$, then $\vec{n}$ and $\vec{v}$ are perpendicular. Since P_1 is perpendicular to $\vec{n}$ and L is parallel to $\vec{v}$, we see that P_1 and L are parallel. In fact, L may lie in the plane.

(b) Since $\vec{n} \times \vec{v}$ is perpendicular to $\vec{n}$ and to $\vec{v}$, the vector $\vec{n} \times \vec{v}$ is parallel to P_1 and perpendicular to L. Thus, P_2, which is perpendicular to $\vec{n} \times \vec{v}$, is

  (i) Perpendicular to P_1.

  (ii) Parallel to L.

Solutions for Section 17.2

Exercises

1. The velocity vector $\vec{v}$ is given by:

$$\vec{v} = \frac{d(t)}{dt}\vec{i} + \left(\frac{d}{dt}(t^3 - t)\right)\vec{j} = \vec{i} + (3t^2 - 1)\vec{j}.$$

The acceleration vector $\vec{a}$ is given by:

$$\vec{a} = \frac{d\vec{v}}{dt} = \frac{d(1)}{dt}\vec{i} + \left(\frac{d}{dt}(3t^2 - 1)\right)\vec{j} = 6t\vec{j}.$$

5. The velocity vector $\vec{v}$ is given by:

$$\vec{v} = \frac{d}{dt}(3\cos t)\vec{i} + \frac{d}{dt}(4\sin t)\vec{j} = -3\sin t\vec{i} + 4\cos t\vec{j}.$$

The acceleration vector $\vec{a}$ is given by:

$$\vec{a} = \frac{d\vec{v}}{dt} = \frac{d}{dt}(-3\sin t)\vec{i} + \frac{d}{dt}(4\cos t)\vec{j} = -3\cos t\vec{i} - 4\sin t\vec{j}.$$

9. The velocity vector $\vec{v}$ is given by:

$$\vec{v} = \frac{d}{dt}(t)\vec{i} + \frac{d}{dt}(t^2)\vec{j} + \frac{d}{dt}(t^3)\vec{k}$$
$$= \vec{i} + 2t\vec{j} + 3t^2\vec{k}.$$

The speed is given by:

$$\|\vec{v}\| = \sqrt{1 + 4t^2 + 9t^4}.$$

Now $\|\vec{v}\|$ is never zero since $1 + 4t^2 + 9t^4 \geq 1$ for all t. Thus, the particle never stops.

13. We have

$$\text{Length} = \int_1^2 \sqrt{(x'(t))^2 + (y'(t))^2 + (z'(t))^2}\, dt = \int_1^2 \sqrt{5^2 + 4^2 + (-1)^2}\, dt = \sqrt{42}.$$

This is the length of a straight line from the point $(8, 5, 2)$ to $(13, 9, 1)$.

17. The velocity vector $\vec{v}$ is

$$\vec{v} = \frac{dx}{dt}\vec{i} + \frac{dy}{dt}\vec{j} + \frac{dz}{dt}\vec{k} = 0\vec{i} + 2(3)\cos(3t)\vec{j} + 2(3)(-\sin(3t))\vec{k}$$
$$= 6\cos(3t)\vec{j} - 6\sin(3t)\vec{k}.$$

The acceleration vector $\vec{a}$ is

$$\vec{a} = \frac{d^2x}{dt^2}\vec{i} + \frac{d^2y}{dt^2}\vec{j} + \frac{d^2z}{dt^2}\vec{k} = 6(3)(-\sin(3t))\vec{j} - 6(3)\cos(3t)\vec{k}$$
$$= -18\sin(3t)\vec{j} - 18\cos(3t)\vec{k}.$$

To check that $\vec{v}$ and $\vec{a}$ are perpendicular, we check that the dot product is zero:

$$\vec{v} \cdot \vec{a} = (6\cos(3t)\vec{j} - 6\sin(3t)\vec{k}) \cdot (-18\sin(3t)\vec{j} - 18\cos(3t)\vec{k})$$
$$= -108\cos(3t)\sin(3t) + 108\sin(3t)\cos(3t) = 0.$$

The speed is

$$\|\vec{v}\| = \|6\cos(3t)\vec{j} - 6\sin(3t)\vec{k}\| = 6\sqrt{\sin^2(3t) + \cos^2(3t)} = 6,$$

and so is constant. The magnitude of the acceleration is

$$\|\vec{a}\| = \|-18\sin(3t)\vec{j} - 18\cos(3t)\vec{k}\| = 18\sqrt{\sin^2(3t) + \cos^2(3t)} = 18,$$

which is also constant.

Problems

21. Plotting the positions on the xy plane and noting their times gives the graph shown in Figure 17.3.

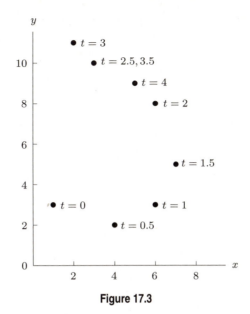

Figure 17.3

(a) We approximate dx/dt by $\Delta x/\Delta t$ calculated between $t = 1.5$ and $t = 2.5$:

$$\frac{dx}{dt} \approx \frac{\Delta x}{\Delta t} = \frac{3 - 7}{2.5 - 1.5} = \frac{-4}{1} = -4.$$

Similarly,

$$\frac{dy}{dt} \approx \frac{\Delta y}{\Delta t} = \frac{10 - 5}{2.5 - 1.5} = \frac{5}{1} = 5.$$

So,

$$\vec{v}(2) \approx -4\vec{i} + 5\vec{j} \quad \text{and} \quad \text{Speed} = \|\vec{v}\| = \sqrt{41}.$$

(b) The particle is moving vertically at about time $t = 1.5$. Note that the particle is momentarily stopped at about $t = 3$; however it is not moving parallel to the y-axis at this instant.

(c) The particle stops at about time $t = 3$ and reverses course.

25. (a) The vector $\vec{PQ}$ between the points is given by

$$\vec{PQ} = 2\vec{i} + 5\vec{j} + 3\vec{k}.$$

Since $\|\vec{PQ}\| = \sqrt{2^2 + 5^2 + 3^2} = \sqrt{38}$, the velocity vector of the motion is

$$\vec{v} = \frac{5}{\sqrt{38}}(2\vec{i} + 5\vec{j} + 3\vec{k}).$$

(b) The motion is along a line starting at the point $(3, 2, -5)$ and with the velocity vector from part (a). The equation of the line is

$$\vec{r} = 3\vec{i} + 2\vec{j} - 5\vec{k} + t\vec{v} = 3\vec{i} + 2\vec{j} - 5\vec{k} + \frac{5}{\sqrt{38}}(2\vec{i} + 5\vec{j} + 3\vec{k})t,$$

so

$$x = 3 + \frac{10}{\sqrt{38}}t, \qquad y = 2 + \frac{25}{\sqrt{38}}t, \qquad z = -5 + \frac{15}{\sqrt{38}}t.$$

29. (a) Since $z = 90$ feet when $t = 0$, the tower is 90 feet high.
   (b) The child reaches the bottom when $z = 0$, so $t = 90/5 = 18$ minutes.
   (c) Her velocity is given by

$$\vec{v} = \frac{d\vec{r}}{dt} = -(10\sin t)\vec{i} + (10\cos t)\vec{j} - 5\vec{k},$$

so

$$\text{Speed} = ||\vec{v}|| = \sqrt{(-10\sin t)^2 + (10\cos t)^2 + (-5)^2} = \sqrt{10^2 + 5^2} = \sqrt{125} \text{ ft/min}.$$

   (d) Her acceleration is given by

$$\vec{a} = \frac{d\vec{v}}{dt} = -(10\cos t)\vec{i} - (10\sin t)\vec{j} \text{ ft/min}^2.$$

33. (a) For any positive constant k, the parameterization

$$x = -5\sin(kt) \quad y = 5\cos(kt)$$

moves counterclockwise on a circle of radius 5 starting at the point $(0, 5)$. We choose k to make the period 8 seconds. If $k \cdot 8 = 2\pi$, then $k = \pi/4$ and the parameterization is

$$x = -5\sin\left(\frac{\pi t}{4}\right) \quad y = 5\cos\left(\frac{\pi t}{4}\right).$$

   (b) Since it takes 8 seconds for the particle to go around the circle

$$\text{Speed} = \frac{\text{Circumference of circle}}{8} = \frac{2\pi(5)}{8} = \frac{5\pi}{4} \text{ cm/sec}.$$

37. (a) The center of the wheel moves horizontally, so its y-coordinate will never change; it will equal 1 at all times. In one second, the wheel rotates 1 radian, which corresponds to 1 meter on the rim of a wheel of radius 1 meter, and so the rolling wheel advances at a rate of 1 meter/sec. Thus the x-coordinate of the center, which equals 0 at $t = 0$, will equal t at time t. At time t the center will be at the point $(x, y) = (t, 1)$.
   (b) By time t the spot on the rim will have rotated t radians clockwise, putting it at angle $-t$ as in Figure 17.4. The coordinates of the spot with respect to the center of the wheel are $(\cos(-t), \sin(-t))$. Adding these to the coordinates $(t, 1)$ of the center gives the location of the spot as $(x, y) = (t + \cos t, 1 - \sin t)$. See Figure 17.5.

Figure 17.4

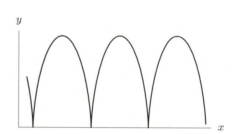

Figure 17.5

41. (a) Let x represent horizontal displacement (in cm) from some starting point and y the distance (in cm) above the ground. Since

$$25 \text{ km/hr} = \frac{25 \cdot 10^5}{60^2} = 694.444 \text{ cm/sec},$$

if t is in seconds, the motion of the center of the pedal is given by

$$x\vec{i} + y\vec{j} = 694.444t\vec{i} + 30\vec{j}.$$

The circular motion of your foot relative to the center is described by

$$h\vec{i} + k\vec{j} = 20\cos(2\pi t)\vec{i} + 20\sin(2\pi t)\vec{j},$$

so the motion of the light on your foot relative to the ground is described by

$$x\vec{i} + y\vec{j} = (694.444t + 20\cos(2\pi t))\vec{i} + (30 + 20\sin(2\pi t))\vec{j}.$$

(b) See Figure 17.6.

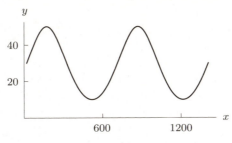

Figure 17.6

(c) Suppose your pedal is rotating with angular velocity ω radians/sec, so that the motion is described by

$$x\vec{i} + y\vec{j} = (694.444t + 20\cos\omega t)\vec{i} + (30 + 20\sin\omega t)\vec{j}\,.$$

The light moves backward if dx/dt is negative. Since

$$\frac{dx}{dt} = 694.444 - 20\omega\sin\omega t,$$

the minimum value of dx/dt occurs when $\omega t = \pi/2$, and then

$$\frac{dx}{dt} = 694.444 - 20\omega < 0$$

giving

$$\omega \geq 34.722 \text{ radians/sec.}$$

Since there are 2π radians in a complete revolution, an angular velocity of 34.722 radians/sec means $34.722/2\pi \approx$ 5.526 revolutions/sec.

45. (a) Using the product rule for differentiation we get

$$\frac{d}{dt}(\vec{r}\cdot\vec{r}) = \vec{r}\cdot\frac{d\vec{r}}{dt} + \frac{d\vec{r}}{dt}\cdot\vec{r} = 2\vec{r}\cdot\frac{d\vec{r}}{dt}.$$

(b) Since $\vec{a}$ is a constant, $d\vec{a}/dt = 0$ so the product rule gives

$$\frac{d}{dt}(\vec{a}\times\vec{r}) = \vec{a}\times\frac{d\vec{r}}{dt}.$$

(c) The product rule gives

$$\frac{d}{dt}(r^3\vec{r}) = r^3\frac{d\vec{r}}{dt} + \frac{d}{dt}(r^3)\vec{r} = r^3\frac{d\vec{r}}{dt} + 3r^2\vec{r}\,.$$

Solutions for Section 17.3

Exercises

1. (a) Parallel to y-axis.
   (b) Length increasing in x-direction.
   (c) Length not dependent on y.
5. $\vec{V} = x\vec{i}$
9. $\vec{V} = -y\vec{i} + x\vec{j}$

13. See Figure 17.7.

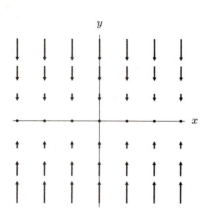

Figure 17.7: $\vec{F}(x,y) = -y\vec{j}$

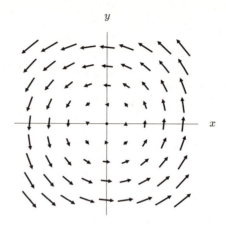

Figure 17.8: $\vec{F}(x,y) = -y\vec{i} + x\vec{j}$

17. See Figure 17.8.

Problems

21. $\vec{F}(x,y) = a\vec{i} + b\vec{j}$ for any real numbers a and b is a constant vector field. For example, $\vec{F}(x,y) = 3\vec{i} - 4\vec{j}$.

25. If $\vec{F}(x,y) = f(x,y)((1+y^2)\vec{i} - (x+y)\vec{j})$ where $f(x,y)$ is any function, then $\vec{F} \cdot \vec{G} = 0$, which shows that $\vec{F}$ is perpendicular to $\vec{G}$.
   For example $\vec{F}(x,y) = (y + \cos x)((1+y^2)\vec{i} - (x+y)\vec{j})$.

29. One possible solution is $\vec{F}(x,y) = x\vec{i}$. See Figure 17.9.

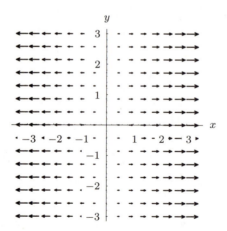

Figure 17.9

33. (a) The line l is parallel to the vector $\vec{v} = \vec{i} - 2\vec{j} - 3\vec{k}$. The vector field $\vec{F}$ is parallel to the line when $\vec{F}$ is a multiple of $\vec{v}$. Taking the multiple to be 1 and solving for x, y, z we find a point at which this occurs:

$$x = 1$$
$$x + y = -2$$
$$x - y + z = -3$$

gives $x = 1$, $y = -3$, $z = -7$, so a point is $(1, -3, -7)$. Other answers are possible.

(b) The line and vector field are perpendicular if $\vec{F} \cdot \vec{v} = 0$, that is

$$(x\vec{i} + (x+y)\vec{j} + (x-y+z)\vec{k}) \cdot (\vec{i} - 2\vec{j} - 3\vec{k}) = 0$$
$$x - 2x - 2y - 3x + 3y - 3z = 0$$
$$-4x + y - 3z = 0.$$

One point which satisfies this equation is $(0, 0, 0)$. There are many others.

(c) The equation for this set of points is $-4x + y - 3z = 0$. This is a plane through the origin.

37. (a) Dividing a vector $\vec{F}$ by its magnitude always produces the unit vector in the same direction as $\vec{F}$.

(b) Since

$$\|\vec{N}\| = \|(1/F)(-v\vec{i} + u\vec{j})\| = (1/F)\sqrt{v^2 + u^2} = (1/F)F = 1,$$

then $\vec{N}$ is a unit vector. We check that $\vec{N}$ is perpendicular to $\vec{F}$ using the dot product of $\vec{N}$ and $\vec{F}$:

$$\vec{N} \cdot \vec{F} = (1/F)(-v\vec{i} + u\vec{j}) \cdot (u\vec{i} + v\vec{j}) = 0.$$

Which side of $\vec{F}$ does $\vec{N}$ point? The vector $\vec{k}$ is pointing out of the diagram. Since the cross product $\vec{k} \times \vec{F}$ is perpendicular to both $\vec{k}$ and $\vec{F}$, then $\vec{N}$ lies in the xy-plane and points at a right angle to the direction of $\vec{F}$. By the right-hand rule, $\vec{N}$ points to the left as shown in the figure.

Solutions for Section 17.4

Exercises

1. Since $x'(t) = 3$ and $y'(t) = 0$, we have $x = 3t + x_0$ and $y = y_0$. Thus, the solution curves are $y = $ constant.

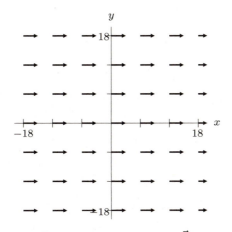

Figure 17.10: The field $\vec{v} = 3\vec{i}$

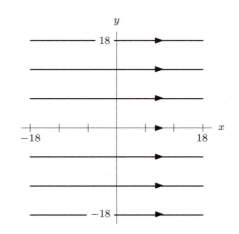

Figure 17.11: The flow $y =$ constant

5. As

$$\vec{v}(t) = \frac{dx}{dt}\vec{i} + \frac{dy}{dt}\vec{j},$$

the system of differential equations is

$$\begin{cases} \frac{dx}{dt} = x \\ \frac{dy}{dt} = 0. \end{cases}$$

Since

$$\frac{d}{dt}(x(t)) = \frac{d}{dt}(ae^t) = x$$

and

$$\frac{d}{dt}(y(t)) = \frac{d}{dt}(b) = 0,$$

the given flow satisfies the system. The solution curves are the horizontal lines $y = b$. See Figures 17.12 and 17.13.

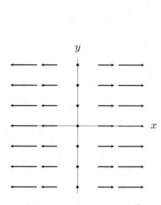

Figure 17.12: $\vec{v}(t) = x\vec{i}$

Figure 17.13: The flow $x(t) = ae^t$, $y(t) = b$

9. As

$$\vec{v}(t) = \frac{dx}{dt}\vec{i} + \frac{dy}{dt}\vec{j},$$

the system of differential equations is

$$\begin{cases} \frac{dx}{dt} = y \\ \frac{dy}{dt} = x. \end{cases}$$

Since

$$\frac{dx(t)}{dt} = \frac{d}{dt}[a(e^t + e^{-t})] = a(e^t - e^{-t}) = y(t)$$

and

$$\frac{dy(t)}{dt} = \frac{d}{dt}[a(e^t - e^{-t})] = a(e^t + e^{-t}) = x(t),$$

the given flow satisfies the system. By eliminating the parameter t in $x(t)$ and $y(t)$, the solution curves obtained are $x^2 - y^2 = 4a^2$. See Figures 17.14 and 17.15.

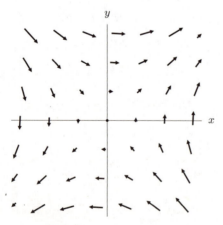

Figure 17.14: $\vec{v}(t) = y\vec{i} + x\vec{j}$

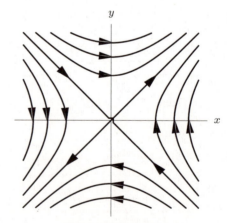

Figure 17.15: The flow
$x(t) = a(e^t + e^{-t})$, $y(t) = a(e^t - e^{-t})$

Problems

13. This corresponds to area C in Figure 17.16.

Figure 17.16

17. (a) Perpendicularity is indicated by zero dot product. We have $\vec{v} \cdot \text{grad } H = (-H_y\vec{i} + H_x\vec{j}) \cdot (H_x\vec{i} + H_y\vec{j}) = 0$.
  (b) If $\vec{r}(t) = x(t)\vec{i} + y(t)\vec{j}$ is a flow line we have, using the chain rule,

$$\frac{d}{dt} H(x(t), y(t)) = H_x\frac{dx}{dt} + H_y\frac{dy}{dt} = H_x(-H_y) + H_y(H_x) = 0.$$

Thus $H(x(t), y(t))$ is constant which shows that a flow line stays on a single level curve of H.

For a different solution, use geometric reasoning. The vector field $\vec{v}$ is tangent to the level curves of H because, by part (a), $\vec{v}$ and the level curves are both perpendicular to the same vector field grad H. Thus the level curves of H and the flowlines of $\vec{v}$ run in the same direction.

21. Let $\vec{r}(t) = x(t)\vec{i} + y(t)\vec{j}$ be a flow line of $\vec{v}$. If $f(x, y)$ has the same value at all points $(x(t), y(t))$ then the flow line lies on a level curve of f. We can check whether

$$g(t) = f(x(t), y(t)) = x(t)^2 - y(t)^2$$

is constant by computing the derivative $g'(t)$. Since $\vec{v} = y\vec{i} + x\vec{j}$, we have $dx/dt = y$ and $dy/dt = x$. Thus,

$$g'(t) = 2x\frac{dx}{dt} - 2y\frac{dy}{dt} = 2xy - 2yx = 0$$

and g is constant. This means that the flow line lies on a level curve of f. The flow lines are parameterized hyperbolas with equation $x^2 - y^2 = c$.

Solutions for Section 17.5

Exercises

1. A horizontal disk of radius 5 in the plane $z = 7$.

5. Since $z = r = \sqrt{x^2 + y^2}$, we have a cone around the z-axis. Since $0 \le r \le 5$, we have $0 \le z \le 5$, so the cone has height and maximum radius of 5.

9. The top half of the sphere ($z \ge 0$).

Problems

13. Two vectors in the plane containing $P = (0, 0, 0)$, $Q = (1, 2, 3)$, and $R = (2, 1, 0)$ are the displacement vectors

$$\vec{v}_1 = \vec{PQ} = \vec{i} + 2\vec{j} + 3\vec{k}$$
$$\vec{v}_2 = \vec{PR} = 2\vec{i} + \vec{j}.$$

Letting $\vec{r}_0 = 0\vec{i} + 0\vec{j} + 0\vec{k} = \vec{0}$ we have the parameterization

$$\vec{r}(s, t) = \vec{r}_0 + s\vec{v}_1 + t\vec{v}_2$$
$$= (s + 2t)\vec{i} + (2s + t)\vec{j} + 3s\vec{k}.$$

17. To parameterize the plane we need two nonparallel vectors $\vec{v}_1$ and $\vec{v}_2$ that are parallel to the plane. Such vectors are perpendicular to the normal vector to the plane, $\vec{n} = \vec{i} + 2\vec{j} + 3\vec{k}$. We can choose any vectors $\vec{v}_1$ and $\vec{v}_2$ such that $\vec{v}_1 \cdot \vec{n} = \vec{v}_2 \cdot \vec{n} = 0$.

One choice is

$$\vec{v}_1 = 2\vec{i} - j \qquad \vec{v}_2 = 3\vec{i} - \vec{k}.$$

Letting $\vec{r}_0 = 5\vec{i} + \vec{j} + 4\vec{k}$ we have the parameterization

$$\vec{r}(s, t) = \vec{r}_0 + s\vec{v}_1 + t\vec{v}_2$$
$$= (5 + 2s + 3t)\vec{i} + (1 - s)\vec{j} + (4 - t)\vec{k}.$$

21. Since you walk 5 blocks east and 1 block west, you walk 5 blocks in the direction of $\vec{v}_1$, and 1 block in the opposite direction. Thus,

$$s = 5 - 1 = 4,$$

Similarly,

$$t = 4 - 2 = 2.$$

Hence

$$x\vec{i} + y\vec{j} + z\vec{k} = (x_0\vec{i} + y_0\vec{j} + z_0\vec{k}) + 4\vec{v_1} + 2\vec{v_2}$$
$$= (x_0\vec{i} + y_0\vec{j} + z_0\vec{k}) + 4(2\vec{i} - 3\vec{j} + 2\vec{k}) + 2(\vec{i} + 4\vec{j} + 5\vec{k})$$
$$= (x_0 + 10)\vec{i} + (y_0 - 4)\vec{j} + (z_0 + 18)\vec{k}.$$

Thus the coordinates are:

$$x = x_0 + 10, \quad y = y_0 - 4, \quad z = z_0 + 18.$$

25. Set up the coordinates as in Figure 17.17. The surface is the revolution surface obtained by revolving the curve shown in Figure 17.18 about the z axis. From the measurements given, we obtain the equation of the curve in Figure 17.18:

$$x = \cos\left(\frac{\pi}{3}z\right) + 3, \qquad 0 \le z \le 48$$

(a) Rotating this around the z-axis, and taking $z = t$ as the parameter, we get the parametric equations

$$x = \left(\cos\left(\frac{\pi}{3}t\right) + 3\right)\cos\theta$$
$$y = \left(\cos\left(\frac{\pi}{3}t\right) + 3\right)\sin\theta$$
$$z = t \qquad 0 \le \theta \le 2\pi, 0 \le t \le 48$$

(b) We know that the points in the curve consists of cross-sections of circles parallel to the xy plane and of radius $\cos((\pi/3)z + 3)$. Thus,

$$\text{Area of cross-section} = \pi\left(\cos\left(\frac{\pi}{3}z + 3\right)\right)^2$$

Integrating over z, we get

$$\text{Volume} = \pi \int_0^{48} \left(\cos\frac{\pi}{3}z + 3\right)^2 dz$$
$$= \pi \int_0^{48} \left(\cos^2\frac{\pi}{3}z + 6\cos\frac{\pi}{3}z + 9\right) dz$$
$$= 456\pi \text{ in}^3.$$

Figure 17.17   **Figure 17.18**

29. Let $(\theta, \pi/2)$ be the original coordinates. If $\theta < \pi$, then the new coordinates will be $(\theta + \pi, \pi/4)$. If $\theta \geq \pi$, then the new coordinates will be $(\theta - \pi, \pi/4)$.

33. The vase obtained by rotating the curve $z = 10\sqrt{x - 1}$, $1 \leq x \leq 2$, around the z-axis is shown in Figure 17.19. At height z, the cross-section is a horizontal circle of radius a. Thus, a point on this horizontal circle is given by

$$\vec{r} = a\cos\theta\vec{i} + a\sin\theta\vec{j} + z\vec{k}.$$

However, the radius a varies, so we need to express it in terms of the other parameters θ and z. If you look at the xz-plane, the radius of this circle is given by x, so solving for x in $z = 10\sqrt{x - 1}$ gives

$$a = x = \left(\frac{z}{10}\right)^2 + 1.$$

Thus, a parameterization is

$$\vec{r} = \left(\left(\frac{z}{10}\right)^2 + 1\right)\cos\theta\vec{i} + \left(\left(\frac{z}{10}\right)^2 + 1\right)\sin\theta\vec{j} + z\vec{k}$$

so

$$x = \left(\left(\frac{z}{10}\right)^2 + 1\right)\cos\theta, \quad y = \left(\left(\frac{z}{10}\right)^2 + 1\right)\sin\theta, \quad z = z,$$

where $0 \leq \theta \leq 2\pi$, $0 \leq z \leq 10$.

$z = 10\sqrt{x - 1}$

Figure 17.19

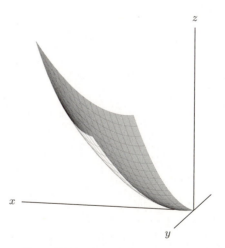

Figure 17.20: The surface $x = s + t$, $y = s - t$, $z = s^2 + t^2$ for $0 \leq s \leq 1$, $0 \leq t \leq 1$

37. (a) From the first two equations we get:

$$s = \frac{x+y}{2}, \qquad t = \frac{x-y}{2}.$$

Hence the equation of our surface is:

$$z = \left(\frac{x+y}{2}\right)^2 + \left(\frac{x-y}{2}\right)^2 = \frac{x^2}{2} + \frac{y^2}{2},$$

which is the equation of a paraboloid.

The conditions: $0 \le s \le 1, 0 \le t \le 1$ are equivalent to: $0 \le x + y \le 2, 0 \le x - y \le 2$. So our surface is defined by:

$$z = \frac{x^2}{2} + \frac{y^2}{2}, \qquad 0 \le x + y \le 2 \quad 0 \le x - y \le 2$$

(b) The surface is shown in Figure 17.20.

Solutions for Chapter 17 Review

Exercises

1. The line has equation

$$\vec{r} = 2\vec{i} - \vec{j} + 3\vec{k} + t(5\vec{i} + 4\vec{j} - \vec{k}),$$

or, equivalently

$$x = 2 + 5t$$
$$y = -1 + 4t$$
$$z = 3 - t.$$

5. The parameterization $x\vec{i} + y\vec{j} = (4 + 4\cos t)\vec{i} + (4 + 4\sin t)\vec{j}$ gives the correct circle, but starts at $(8, 4)$. To start on the x-axis we need

$$x\vec{i} + y\vec{j} = (4 + 4\cos(t - \frac{\pi}{2}))\vec{i} + (4 + 4\sin(t - \frac{\pi}{2}))\vec{j} = (4 + 4\sin t)\vec{i} + (4 - 4\cos t)\vec{j}.$$

9. Since the vector $\vec{n} = \text{grad}(2x - 3y + 5z) = 2\vec{i} - 3\vec{j} + 5\vec{k}$ is perpendicular to the plane, this vector is parallel to the line. Thus the equation of the line is

$$x = 1 + 2t, \quad y = 1 - 3t, \quad z = 1 + 5t.$$

13. See Figure 17.21. The parameterization is

$$\vec{r} = 10\cos\left(\frac{2\pi t}{30}\right)\vec{i} - 10\sin\left(\frac{2\pi t}{30}\right)\vec{j} + 7\vec{k}.$$

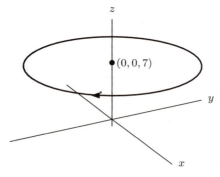

Figure 17.21

17. The velocity vector $\vec{v}$ is given by:

$$\vec{v} = \frac{d}{dt}(2 + 3t^2)\vec{i} + \frac{d}{dt}(4 + t^2)\vec{j} + \frac{d}{dt}(1 - t^2)\vec{k} = 6t\vec{i} + 2t\vec{j} - 2t\vec{k}.$$

21. Vector. Differentiating using the chain rule gives

$$\text{Velocity} = \left(\frac{-\cos t}{2\sqrt{3 + \sin t}}\right)\vec{i} + \left(\frac{-\sin t}{2\sqrt{3 + \cos t}}\right)\vec{j}.$$

25. The direction vectors of the lines, $-\vec{i} + 4\vec{j} - 2\vec{k}$ and $2\vec{i} - 8\vec{j} + 4\vec{k}$, are multiplies of each other (the second is -2 times the first). Thus the lines are parallel. To see if they are the same line, we take the point corresponding to $t = 0$ on the first line, which has position vector $3\vec{i} + 3\vec{j} - \vec{k}$, and see if it is on the second line. So we solve

$$(1 + 2t)\vec{i} + (11 - 8t)\vec{j} + (4t - 5)\vec{k} = 3\vec{i} + 3\vec{j} - \vec{k}.$$

This has solution $t = 1$, so the two lines have a point in common and must be the same line, parameterized in two different ways.

29. The vector field points in a clockwise direction around the origin. Since

$$\left\| \left(\frac{y}{\sqrt{x^2 + y^2}}\right)\vec{i} - \left(\frac{x}{\sqrt{x^2 + y^2}}\right)\vec{j} \right\| = \frac{\sqrt{x^2 + y^2}}{\sqrt{x^2 + y^2}} = 1,$$

the length of the vectors is constant everywhere.

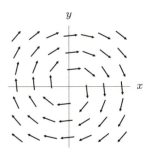

Figure 17.22

Problems

33. (a) A vector field associates a vector to every point in a region of the space. In other words, a vector field is a vector-valued function of position given by $\vec{v} = \vec{f}(\vec{r}) = \vec{f}(x, y, z)$

    (b)   (i) Yes, $\vec{r} + \vec{a} = (x + a_1)\vec{i} + (y + a_2)\vec{j} + (z + a_3)\vec{k}$ is a vector-valued function of position.

        (ii) No, $\vec{r} \cdot \vec{a}$ is a scalar.

        (iii) Yes.

        (iv) $x^2 + y^2 + z^2$ is a scalar.

37. The displacement vector from $(1, 1, 1)$ to $(2, -1, 3)$ is $\vec{d} = (2\vec{i} - \vec{j} + 3\vec{k}) - (\vec{i} + \vec{j} + \vec{k}) = \vec{i} - 2\vec{j} + 2\vec{k}$ meters. The velocity vector has the same direction as $\vec{d}$ and is given by

$$\vec{v} = \frac{\vec{d}}{5} = 0.2\vec{i} - 0.4\vec{j} + 0.4\vec{k} \text{ meters/sec.}$$

Since $\vec{v}$ is constant, the acceleration $\vec{a} = \vec{0}$.

41. (a) In order for the particle to stop, its velocity $\vec{v} = (dx/dt)\vec{i} + (dy/dt)\vec{j}$ must be zero, so we solve for t such that $dx/dt = 0$ and $dy/dt = 0$, that is

$$\frac{dx}{dt} = 3t^2 - 3 = 3(t-1)(t+1) = 0,$$

$$\frac{dy}{dt} = 2t - 2 = 2(t-1) = 0.$$

The value $t = 1$ is the only solution. Therefore, the particle stops when $t = 1$ at the point $(t^3 - 3t,\ t^2 - 2t)|_{t=1} = (-2, -1)$.

(b) In order for the particle to be traveling straight up or down, the x-component of the velocity vector must be 0. Thus, we solve $dx/dt = 3t^2 - 3 = 0$ and obtain $t = \pm 1$. However, at $t = 1$ the particle has no vertical motion, as we saw in part (a). Thus, the particle is moving straight up or down only when $t = -1$. Since the velocity at time $t = -1$ is

$$\vec{v}(-1) = \left.\frac{dx}{dt}\right|_{t=-1}\vec{i} + \left.\frac{dy}{dt}\right|_{t=-1}\vec{j} = -4\vec{j},$$

the motion is straight down. The position at that time is $(t^3 - 3t,\ t^2 - 2t)|_{t=-1} = (2, 3)$.

(c) For horizontal motion we need $dy/dt = 0$. That happens when $dy/dt = 2t - 2 = 0$, and so $t = 1$. But from part (a) we also have $dx/dt = 0$ also at $t = 1$, so the particle is not moving at all when $t = 1$. Thus, there is no time when the motion is horizontal.

45. (a) $f_x = \dfrac{\left[2x(x^2 + y^2) - 2x(x^2 - y^2)\right]}{(x^2 + y^2)^2} = \dfrac{4xy^2}{(x^2 + y^2)^2}.$

$f_y = \dfrac{\left[-2y(x^2 + y^2) - 2y(x^2 - y^2)\right]}{(x^2 + y^2)^2} = \dfrac{-4yx^2}{(x^2 + y^2)^2}.$

$\nabla f(1, 1) = \vec{i} - \vec{j}$, i.e., south-east.

(b) We need a vector $\vec{u}$ such that $\nabla f(1, 1) \cdot \vec{u} = 0$, i.e., such that $(\vec{i} - \vec{j}) \cdot \vec{u} = 0$. The vector $\vec{u} = \vec{i} + \vec{j}$ clearly works; so does $\vec{u} = -\vec{i} - \vec{j}$. Dividing by the length to get a unit vector, we have $\vec{u} = \frac{1}{\sqrt{2}}\vec{i} + \frac{1}{\sqrt{2}}\vec{j}$ or $\vec{u} = -\frac{1}{\sqrt{2}}\vec{i} - \frac{1}{\sqrt{2}}\vec{j}$.

(c) f is a function of x and y, which in turn are functions of t. Thus, the chain rule can be used to show how f changed with t.

$$\frac{df}{dt} = \frac{\partial f}{\partial x} \cdot \frac{dx}{dt} + \frac{\partial f}{\partial y} \cdot \frac{dy}{dt} = \frac{4xy^2}{(x^2 + y^2)^2} \cdot 2e^{2t} - \frac{4x^2 y}{(x^2 + y^2)^2} \cdot (6t^2 + 6).$$

At $t = 0$, $x = 1, y = 1$; so, $\dfrac{df}{dt} = \dfrac{4}{4} \cdot 2 - \dfrac{4}{4} \cdot 6 = -4.$

49. Note that uniform circular motion is possible with the given conditions, since $\vec{v} \cdot \vec{a} = 0$ shows that the velocity and acceleration vectors are perpendicular.

For uniform circular motion in an orbit of radius R, we have $\|\vec{a}\| = \|\vec{v}\|^2 / R$. Thus $R = \|\vec{v}\|^2 / \|\vec{a}\| = 52/\sqrt{13}$ for both parts (a) and (b).

The center of the orbit is at distance R in the direction of the acceleration vector from the point P on the orbit. The vector

$$R\frac{\vec{a}}{\|\vec{a}\|} = \frac{52}{\sqrt{13}} \frac{2\vec{i} + 3\vec{j}}{\sqrt{13}} = 8\vec{i} + 12\vec{j}$$

thus extends from the point P to the center of the orbit.

(a) The center of the orbit is at the point $(0 + 8, 0 + 12) = (8, 12)$

(b) The center of the orbit is at the point $(10 + 8, 50 + 12) = (18, 62)$

53. At time t the particle has polar coordinates $r = \|\vec{r}(t)\| = at$ and $\theta = \omega t$. At time t, the ray from the origin to the particle is at angle ωt radians from the positive x-axis. The ray is therefore rotating at a rate of ω radians per unit time. The parameter ω is the rate of change of the polar angle θ of the particle measured in radians per unit time. The larger ω is, the quicker the particle completes a complete revolution (a $360°$ trip) around the origin. At time t, the particle is at distance at from the origin. Thus a equals the rate of change of the particle's distance from the origin. The larger a is, the faster the particle moves away from the origin.

57. (a) The cone of height h, maximum radius a, vertex at the origin and opening upward is shown in Figure 17.23.

Figure 17.23

By similar triangles, we have

$$\frac{r}{z} = \frac{a}{h},$$

so

$$z = \frac{hr}{a}.$$

Therefore, one parameterization is

$$
\begin{aligned}
x &= r\cos\theta, && 0 \le r \le a, \\
y &= r\sin\theta, && 0 \le \theta < 2\pi, \\
z &= \frac{hr}{a}.
\end{aligned}
$$

(b) Since $r = az/h$, we can write the parameterization in part (a) as

$$
\begin{aligned}
x &= \frac{az}{h}\cos\theta, && 0 \le z \le h, \\
y &= \frac{az}{h}\sin\theta, && 0 \le \theta < 2\pi, \\
z &= z.
\end{aligned}
$$

61. The parameterization for a sphere of radius a using spherical coordinates is

$$x = a\sin\phi\cos\theta, \quad y = a\sin\phi\sin\theta, \quad z = a\cos\phi.$$

Think of the ellipsoid as a sphere whose radius is different along each axis and you get the parameterization:

$$
\begin{cases}
x = a\sin\phi\cos\theta, & 0 \le \phi \le \pi, \\
y = b\sin\phi\sin\theta, & 0 \le \theta \le 2\pi, \\
z = c\cos\phi.
\end{cases}
$$

To check this parameterization, substitute into the equation for the ellipsoid:

$$\frac{x^2}{a^2} + \frac{y^2}{b^2} + \frac{z^2}{c^2} = \frac{a^2\sin^2\phi\cos^2\theta}{a^2} + \frac{b^2\sin^2\phi\sin^2\theta}{b^2} + \frac{c^2\cos^2\phi}{c^2}$$

$$= \sin^2\phi(\cos^2\theta + \sin^2\theta) + \cos^2\phi = 1.$$

CAS Challenge Problems

65. (a)

$$
\begin{aligned}
\vec{r}\cdot\vec{F} &= (x\vec{i} + y\vec{j})\cdot(-y(1-y^2)\vec{i} + x(1-y^2)\vec{j}) \\
&= -xy(1-y^2) + yx(1-y^2) = 0
\end{aligned}
$$

This means that the tangent line to the flow line at a point is always perpendicular to the vector from the origin to that point. Hence the flow lines are circles centered at the origin.

(b) The circle $\vec{r}(t) = \cos t\,\vec{i} + \sin t\,\vec{j}$ has velocity vector $\vec{v}(t) = -\sin t\,\vec{i} + \cos t\,\vec{j} = -y\vec{i} + x\vec{j} = (1-y^2)\vec{F}$. Thus the velocity vector is a scalar multiple of $\vec{F}$, and hence parallel to $\vec{F}$. However, since $\vec{v}(t)$ is not equal to $\vec{F}(\vec{r}(t))$, it is not a flow line.

(c) Using a CAS, we find

$$\vec{v}\,(t) = -\frac{t}{(1+t^2)^{3/2}}\vec{i} + \left(-\frac{t^2}{(1+t^2)^{3/2}} + \frac{1}{\sqrt{1+t^2}}\right)\vec{j} = -\frac{t}{(1+t^2)^{3/2}}\vec{i} + \frac{1}{(1+t^2)^{3/2}}\vec{j}$$

and

$$\vec{F}\,(\vec{r}\,(t)) = -\left(\frac{t\left(1 - \frac{t^2}{1+t^2}\right)}{\sqrt{1+t^2}}\right)\vec{i} + \frac{1 - \frac{t^2}{1+t^2}}{\sqrt{1+t^2}}\vec{j} = -\frac{t}{(1+t^2)^{3/2}}\vec{i} + \frac{1}{(1+t^2)^{3/2}}\vec{j} = \vec{v}\,(t).$$

Although the circle parameterized in part (b) has velocity vectors parallel to $\vec{F}$ at each point of the circle, its speed is not equal to the magnitude of the vector field. The circle in part (c) is parameterized at the correct speed to be the flow line.

CHECK YOUR UNDERSTANDING

1. False. The y coordinate is zero when $t = 0$, but when $t = 0$ we have $x = 2$ so the curve never passes through $(0, 0)$.

5. False. When $t = 0$, we have $(x, y) = (0, -1)$. When $t = \pi/2$, we have $(x, y) = (-1, 0)$. Thus the circle is being traced out clockwise.

9. True. To find an intersection point, we look for values of s and t that make the coordinates in the first line the same as the coordinates in the second. Setting $x = t$ and $x = 2s$ equal, we see that $t = 2s$. Setting $y = 2 + t$ equal to $y = 1 - s$, we see that $t = -1 - s$. Solving both $t = 2s$ and $t = -1 - s$ yields $t = -\frac{2}{3}$, $s = -\frac{1}{3}$. These values of s and t will give equal x and y coordinates on both lines. We need to check if the z coordinates are equal also. In the first line, setting $t = -\frac{2}{3}$ gives $z = \frac{7}{3}$. In the second line, setting $s = -\frac{1}{3}$ gives $z = -\frac{1}{3}$. As these are not the same, the lines do not intersect.

13. False. The velocity vector is $\vec{v}\,(t) = \vec{r}\,'(t) = 2t\vec{i} - \vec{j}$. Then $\vec{v}\,(-1) = -2\vec{i} - \vec{j}$ and $\vec{v}\,(1) = 2\vec{i} - \vec{j}$, which are not equal.

17. False. As a counterexample, consider the curve $\vec{r}\,(t) = t^2\vec{i} + t^2\vec{j}$ for $0 \le t \le 1$. In this case, when t is replaced by $-t$, the parameterization is the same, and is not reversed.

21. True, since the vectors $x\vec{j}$ are parallel to the y-axis.

25. True. Any flow line which stays in the first quadrant has $x, y \to \infty$.

29. True. If (x, y) were a point where the y-coordinate along a flow line reached a relative maximum, then the tangent vector to the flow line, namely $\vec{F}\,(x, y)$, there would have to be horizontal (or $\vec{0}$), that is its $\vec{j}$ component would have to be 0. But the $\vec{j}$ component of $\vec{F}$ is always 2.

33. False. There is only one parameter, s. The equations parameterize a line.

37. True. If the surface is parameterized by $\vec{r}\,(s, t)$ and the point has parameters (s_0, t_0) then the parameter curves $\vec{r}\,(s_0, t)$ and $\vec{r}\,(s, t_0)$.

41. False. Suppose $\vec{r}\,(t) = t\vec{i} + t\vec{j}$. Then $\vec{r}\,'(t) = \vec{i} + \vec{j}$ and

$$\vec{r}\,'(t) \cdot \vec{r}\,(t) = (t\vec{i} + t\vec{j}) \cdot (\vec{i} + \vec{j}) = 2t.$$

So $\vec{r}\,'(t) \cdot \vec{r}\,(t) \ne 0$ for $t \ne 0$.

CHAPTER EIGHTEEN

Solutions for Section 18.1

Exercises

1. Positive, because the vectors are longer on the portion of the path that goes in the same direction as the vector field.

5. Negative, because the vector field points in the opposite direction to the path.

9. Since $\vec{F}$ is a constant vector field and the curve is a line, $\int_C \vec{F} \cdot d\vec{r} = \vec{F} \cdot \Delta\vec{r}$, where $\Delta\vec{r} = 7\vec{j}$. Therefore,

$$\int_C \vec{F} \cdot d\vec{r} = (3\vec{i} + 4\vec{j}) \cdot 7\vec{j} = 28$$

13. At every point, the vector field is parallel to segments $\Delta\vec{r} = \Delta x\vec{i}$ of the curve. Thus,

$$\int_C \vec{F} \cdot d\vec{r} = \int_2^6 x\vec{i} \cdot dx\vec{i} = \int_2^6 x\,dx = \left.\frac{x^2}{2}\right|_2^6 = 16.$$

17. Since the curve is along the y-axis, only the $\vec{j}$ component of the vector field contributes to the integral:

$$\int_C (2\vec{j} + 3\vec{k}) \cdot d\vec{r} = \int_C 2\vec{j} \cdot d\vec{r} = 2 \cdot \text{ Length of } C = 2 \cdot 10 = 20.$$

Problems

21. Since it appears that C_1 is everywhere perpendicular to the vector field, all of the dot products in the line integral are zero, hence $\int_{C_1} \vec{F} \cdot d\vec{r} \approx 0$. Along the path C_2 the dot products of $\vec{F}$ with $\Delta\vec{r_i}$ are all positive, so their sum is positive and we have $\int_{C_1} \vec{F} \cdot d\vec{r} < \int_{C_2} \vec{F} \cdot d\vec{r}$. For C_3 the vectors $\Delta\vec{r_i}$ are in the opposite direction to the vectors of $\vec{F}$, so the dot products $\vec{F} \cdot \Delta\vec{r_i}$ are all negative; so, $\int_{C_3} \vec{F} \cdot d\vec{r} < 0$. Thus, we have

$$\int_{C_3} \vec{F} \cdot d\vec{r} < \int_{C_1} \vec{F} \cdot d\vec{r} < \int_{C_2} \vec{F} \cdot d\vec{r}$$

25. The line integral along C_1 is negative, the line integral along C_2 is negative, and the line integral along C_3 appears to be zero.

29. The vector field $\vec{F}$ is in the same direction as C if $b > 0$, so we want $b < 0$. No restriction is needed on c.

33. This vector field is illustrated in Figure 18.1. It is perpendicular to C_2 and C_4 at every point, since $\vec{F}(x, y) \cdot \vec{r}(x, y) = 0$ and C_2 and C_4 are radial line segments, then

$$\int_{C_2} \vec{F} \cdot d\vec{r} = \int_{C_4} \vec{F} \cdot d\vec{r} = 0.$$

Since C_3 is longer than C_1, and the vector field is larger in magnitude along C_3, the line integral along C_3 has greater absolute value than that along C_1. The line integral along C_3 is positive and the line integral along C_1 is negative, so

$$\int_C \vec{F} \cdot d\vec{r} = \int_{C_3} \vec{F} \cdot d\vec{r} + \int_{C_1} \vec{F} \cdot d\vec{r} > 0.$$

See Figure 18.1.

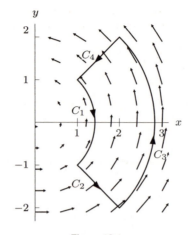

Figure 18.1

37. (a) See Figure 18.2.

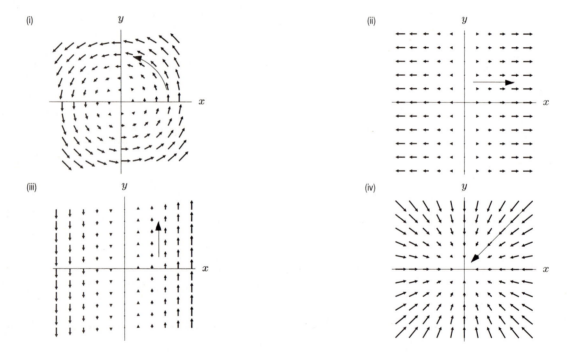

Figure 18.2

(b) For (i) and (iii) a closed curve can be drawn; not for the others.

41. See Figure 18.3. The example chosen is the vector field $\vec{F}(x, y) = y\vec{j}$ and the path C is the line from $(0, -1)$ to $(0, 1)$. Since the vectors are symmetric about the x-axis, the dot products $\vec{F} \cdot \Delta\vec{r}$ cancel out along C to give 0 for the line integral. Many other answers are possible.

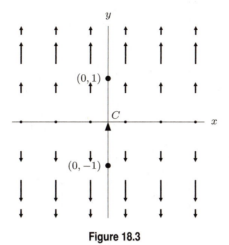

Figure 18.3

45. Let $r = \|\vec{r}\|$. Since $\Delta\vec{r}$ points outward, in the opposite direction to $\vec{F}$, we expect the answer to be negative.

$$\int_C \vec{F} \cdot d\vec{r} = \int_C -\frac{GMm\vec{r}}{r^3} \cdot d\vec{r} = \int_{8000}^{10000} -\frac{GMm}{r^2}\,dr$$

$$= \frac{GMm}{r}\bigg|_{8000}^{10000} = GMm\left(\frac{1}{10000} - \frac{1}{8000}\right)$$

$$= -2.5 \cdot 10^{-5} GMm.$$

49. In Problem 48 we saw that the surface where the potential is zero is a sphere of radius a. Let S be any sphere centered at the origin, and let P_1 be a point on S, and C_1 a path from P_0 to P_1. If P is any point on S, then P can be reached from P_0 by a path, C, consisting of C_1 followed by C_2, where C_2 is a path from P_1 to P lying entirely on the sphere, S. Then $\int_{C_2} \vec{E} \cdot d\vec{r} = 0$, since $\vec{E}$ is perpendicular to the sphere. So

$$\phi(P) = -\int_C \vec{E} \cdot d\vec{r} = -\int_{C_1} \vec{E} \cdot d\vec{r} - \int_{C_2} \vec{E} \cdot d\vec{r} = -\int_{C_1} \vec{E} \cdot d\vec{r} = \phi(P_1).$$

Thus, ϕ is constant on S. The equipotential surfaces are spheres centered at the origin.

Solutions for Section 18.2

Exercises

1. Only the $\vec{i}$-component contributes to the line integral, so $d\vec{r} = \vec{i}\,dx$ and

$$\int_C (2x\vec{i} + 3y\vec{j}) \cdot d\vec{r} = \int_{(1,0,0)}^{(5,0,0)} (2x\vec{i} + 3y\vec{j}) \cdot \vec{i}\,dx = \int_1^5 2x\,dx = x^2\bigg|_1^5 = 24.$$

5. Since $\vec{F} = (x^2 + y)\vec{i} + y^3\vec{j}$, the line integral along the third segment, which is parallel to the z-axis, is zero. On the first segment, which is parallel to the y-axis, only the $\vec{j}$-component contributes. On the second segment, which is parallel to the x-axis, only the $\vec{i}$-component contributes. On the first segment $x = 4$ and y varies from 0 to 3; on the second segment $y = 3$ and x varies from 4 to 0. Thus, we have

$$\int_C \vec{F} \cdot d\vec{r} = \int_0^3 ((4^2 + y)\vec{i} + y^3\vec{j}) \cdot \vec{j}\,dy + \int_4^0 ((x^2 + 3)\vec{i} + 3^3\vec{j}) \cdot \vec{i}\,dx$$

$$= \int_0^3 y^3\,dy + \int_4^0 (x^2 + 3)\,dx = \frac{y^4}{4}\bigg|_0^3 - \left(\frac{x^3}{3} + 3x\right)\bigg|_0^4 = \frac{81}{4} - \frac{64}{3} - 12 = -\frac{157}{12}.$$

9. The line can be parameterized by $(1 + 2t, 2 + 2t)$, for $0 \le t \le 1$, so the integral looks like

$$
\begin{aligned}
\int_C \vec{F} \cdot d\vec{r} &= \int_0^1 \vec{F}\,(1 + 2t, 2 + 2t) \cdot (2\vec{i} + 2\vec{j}\,)\,dt \\
&= \int_0^1 [(1 + 2t)^2\vec{i} + (2 + 2t)^2\vec{j}\,] \cdot (2\vec{i} + 2\vec{j}\,)\,dt \\
&= \int_0^1 2(1 + 4t + 4t^2) + 2(4 + 8t + 4t^2)\,dt \\
&= \int_0^1 (10 + 24t + 16t^2)\,dt \\
&= (10t + 12t^2 + 16t^3/3)\Big|_0^1 \\
&= 10 + 12 + 16/3 - (0 + 0 + 0) = 82/3
\end{aligned}
$$

13. The curve C is parameterized by

$$
\vec{r} = \cos t\vec{i} + \sin t\vec{j}\,, \qquad \text{for } 0 \le t \le 2\pi,
$$

so,

$$
\vec{r}\,'(t) = -\sin t\vec{i} + \cos t\vec{j}\,.
$$

Thus,

$$
\begin{aligned}
\int_C \vec{F} \cdot d\vec{r} &= \int_0^{2\pi} (2\sin t\vec{i} - \sin(\sin t)\vec{j}\,) \cdot (-\sin t\vec{i} + \cos t\vec{j}\,)\,dt \\
&= \int_0^{2\pi} (-2\sin^2 t - \sin(\sin t)\cos t)\,dt \\
&= \sin t \cos t - t + \cos(\sin t)\Big|_0^{2\pi} \\
&= -2\pi.
\end{aligned}
$$

17. Since $\vec{r} = x(t)\vec{i} + y(t)\vec{j} + z(t)\vec{k} = t\vec{i} + t^2\vec{j} + t^3\vec{k}$, for $1 \le t \le 2$,
we have $\vec{r}\,'(t) = x'(t)\vec{i} + y'(t)\vec{j} + z'(t)\vec{k} = \vec{i} + 2t\vec{j} + 3t^2\vec{k}$. Then

$$
\begin{aligned}
\int_C \vec{F} \cdot d\vec{r} &= \int_1^2 (t\vec{i} + 2t^3t^2\vec{j} + t\vec{k}) \cdot (\vec{i} + 2t\vec{j} + 3t^2\vec{k}\,)\,dt \\
&= \int_1^2 (t + 4t^6 + 3t^3)\,dt \\
&= \frac{t^2}{2} + \frac{4t^7}{7} + \frac{3t^4}{4}\Big|_1^2 = \frac{2389}{28} \approx 85.32
\end{aligned}
$$

21. $\int_C 3x\,dx - y\sin x\,dy$

25. From $x = t^2$ and $y = t^3$ we get $dx = 2t\,dt$ and $dy = 3t^2\,dt$. Hence

$$
\int_C y\,dx + x\,dy = \int_1^5 t^3(2t)\,dt + t^2(3t^2)\,dt = \int_1^5 5t^4\,dt = 5^5 - 1 = 3124.
$$

Problems

29. (a) Figure 18.4 shows the curves.

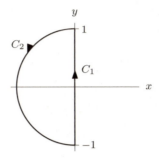

Figure 18.4

(b) On C_1, only the $\vec{j}$-component of $\vec{F}$ contributes to the integral. There $d\vec{r} = \vec{j}\, dy$, so

$$\int_{C_1} \vec{F} \cdot d\vec{r} = \int_{-1}^{1} y\vec{j} \cdot \vec{j}\, dy = \int_{-1}^{1} y\, dy = \left. \frac{y^2}{2} \right|_{-1}^{1} = 0.$$

On C_2, we have $\vec{r}\,'(t) = -\sin t\vec{i} + \cos t\vec{j}$, so

$$\int_{C_2} \vec{F} \cdot d\vec{r} = \int_{\pi/2}^{3\pi/2} ((\cos t + 3\sin t)\vec{i} + \sin t\vec{j}) \cdot (-\sin t\vec{i} + \cos t\vec{j})\, dt$$

$$= \int_{\pi/2}^{3\pi/2} -\cos t \sin t - 3\sin^2 t + \cos t \sin t\, dt = \int_{\pi/2}^{3\pi/2} -3\sin^2 t\, dt$$

$$= -3 \left(\frac{t}{2} - \frac{\sin t \cos t}{2} \right) \Bigg|_{\pi/2}^{3\pi/2} = -\frac{3\pi}{2}.$$

33. (a) The line integral $\int_C (xy\vec{i} + x\vec{j}) \cdot d\vec{r}$ is positive. This follows from the fact that all of the vectors of $xy\vec{i} + x\vec{j}$ at points along C point approximately in the same direction as C (meaning the angles between the vectors and the direction of C are less than $\pi/2$).

(b) Using the parameterization $x(t) = t$, $y(t) = 3t$, with $x'(t) = 1$, $y'(t) = 3$, we have

$$\int_C \vec{F} \cdot d\vec{r} = \int_0^4 \vec{F}(t, 3t) \cdot (\vec{i} + 3\vec{j})\, dt$$

$$= \int_0^4 (3t^2\vec{i} + t\vec{j}) \cdot (\vec{i} + 3\vec{j})\, dt$$

$$= \int_0^4 (3t^2 + 3t)\, dt$$

$$= \left(t^3 + \frac{3}{2}t^2 \right) \Bigg|_0^4$$

$$= 88.$$

(c) Figure 18.5 shows the oriented path C', with the "turn around" points P and Q. The particle first travels from the origin to the point P (call this path C_1), then backs up from P to Q (call this path C_2), then goes from Q to the point $(4, 12)$ in the original direction (call this path C_3). See Figure 18.6. Thus, $C' = C_1 + C_2 + C_3$. Along the parts of C_1 and C_2 that overlap, the line integrals cancel, so we are left with the line integral over the part of C_1 that does not overlap with C_2, followed by the line integral over C_3. Thus, the line integral over C' is the same as the line integral over the direct route from the point $(0, 0)$ to the point $(4, 12)$.

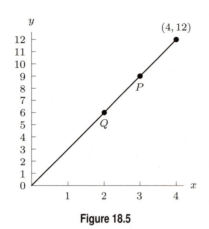

Figure 18.5   **Figure 18.6**

(d) The parameterization

$$(x(t), y(t)) = \left(\frac{1}{3}(t^3 - 6t^2 + 11t), (t^3 - 6t^2 + 11t)\right)$$

has $(x(0), y(0)) = (0,0)$ and $(x(4), y(4)) = (4, 12)$. The form of the parameterization we were given shows that the second coordinate is always three times the first. Thus all points on the parameterized curve lie on the line $y = 3x$.

We have to do a bit more work to guarantee that all points on the curve lie on the line *between* the point $(0, 0)$ and the point $(4, 12)$; it is possible that they might shoot off to, say, $(100, 300)$ before returning to $(4, 12)$. Let's investigate the maximum and minimum values of $f(t) = t^3 - 6t^2 + 11t$ on the interval $0 \le t \le 4$. We can do this on a graphing calculator or computer, or use single-variable calculus. We already know the values of f at the endpoints, namely 0 and 12. We'll look for local extrema:

$$0 = f'(t) = 3t^2 - 12t + 11$$

which has roots at $t = 2 \pm \frac{1}{\sqrt{3}}$. These are the values of t where the particle changes direction: $t = 2 - \frac{1}{\sqrt{3}}$ corresponds to point P and $t = 2 + \frac{1}{\sqrt{3}}$ corresponds to point Q of C'. At these values of t we have $f(2 - \frac{1}{\sqrt{3}}) \approx 6.4$, and $f(2 + \frac{1}{\sqrt{3}}) \approx 5.6$. The fact that these values are between 0 and 12 shows that f takes on its maximum and minimum values at the endpoints of the interval and not in between.

(e) Using the parameterization given in part (d), we have

$$\vec{r}'(t) = x'(t)\vec{i} + y'(t)\vec{j} = \frac{1}{3}(3t^2 - 12t + 11)\vec{i} + (3t^2 - 12t + 11)\vec{j}.$$

Thus,

$$\int_{C'} \vec{F} \cdot d\vec{r}$$

$$= \int_0^4 \vec{F}\left(\frac{1}{3}(t^3 - 6t^2 + 11t), t^3 - 6t^2 + 11t\right) \cdot \left(\frac{1}{3}(3t^2 - 12t + 11)\vec{i} + (3t^2 - 12t + 11)\vec{j}\right) dt$$

$$= \int_0^4 \left(\frac{1}{3}(t^3 - 6t^2 + 11t)^2\vec{i} + \frac{1}{3}(t^3 - 6t^2 + 11t)\vec{j}\right) \cdot \left(\frac{1}{3}(3t^2 - 12t + 11)\vec{i} + (3t^2 - 12t + 11)\vec{j}\right) dt$$

$$= \int_0^4 \frac{1}{3}(t^3 - 6t^2 + 11t)(3t^2 - 12t + 11) \left\{((t^3 - 6t^2 + 11t)\vec{i} + \vec{j}) \cdot (\frac{1}{3}\vec{i} + \vec{j})\right\} dt$$

$$= \int_0^4 \frac{1}{3}(t^3 - 6t^2 + 11t)(3t^2 - 12t + 11) \left\{\frac{1}{3}(t^3 - 6t^2 + 11t) + 1\right\} dt$$

$$= \frac{1}{9}\int_0^4 (t^3 - 6t^2 + 11t)(3t^2 - 12t + 11)(t^3 - 6t^2 + 11t + 3) dt$$

Numerical integration yields an answer of 88, which agrees with the answer found in part b).

37. The integral corresponding to $A(t) = (t, t)$ is

$$\int_0^1 3t\, dt.$$

The integral corresponding to $D(t) = (e^t - 1, e^t - 1)$ is

$$3\int_0^{\ln 2} (e^{2t} - e^t)\, dt.$$

The substitution $s = e^t - 1$ has $ds = e^t\, dt$. Also $s = 0$ when $t = 0$ and $s = 1$ when $t = \ln 2$. Thus, substituting into the integral corresponding to $D(t)$ and using the fact that $e^{2t} = e^t \cdot e^t$ gives

$$3\int_0^{\ln 2} (e^{2t} - e^t)\, dt = 3\int_0^{\ln 2} (e^t - 1)e^t\, dt = \int_0^1 3s\, ds.$$

The integral on the right-hand side is the same as the integral corresponding to $A(t)$. Therefore we have

$$3\int_0^{\ln 2} (e^{2t} - e^t)\, dt = \int_0^1 3s\, ds = \int_0^1 3t\, dt.$$

Alternatively, the substitution $t = e^w - 1$ converts the integral corresponding to $A(t)$ into the integral corresponding to $B(t)$.

Solutions for Section 18.3

Exercises

1. Since $\vec{F}$ is a gradient field, with $\vec{F} = \operatorname{grad} f$ where $f(x, y) = x^2 + y^4$, we use the Fundamental Theorem of Line Integrals. The starting point of the path C is $(2, 0)$ and the end is $(0, 2)$. Thus,

$$\int_C \vec{F} \cdot d\vec{r} = f(0, 2) - f(2, 0) = 16 - 4 = 12.$$

5. Path-independent, because the vector field appears constant.

9. Since $\vec{F} = 3x^2\vec{i} + 4y^3\vec{j} = \operatorname{grad}(x^3 + y^4)$, we take $f(x, y) = x^3 + y^4$. Then by the Fundamental Theorem of Line Integrals,

$$\int_C \vec{F} \cdot d\vec{r} = f(-1, 0) - f(1, 0) = (-1)^3 - 1^3 = -2.$$

13. Since $\vec{F} = y\sin(xy)\vec{i} + x\sin(xy)\vec{j} = \operatorname{grad}(-\cos(xy))$, the Fundamental Theorem of Line Integrals gives

$$\int_C \vec{F} \cdot d\vec{r} = -\cos(xy)\Big|_{(1,2)}^{(3,18)} = -\cos(54) + \cos(2) = \cos(2) - \cos(54).$$

17. Since $\vec{F} = 2xy^2ze^{x^2y^2z}\vec{i} + 2x^2yze^{x^2y^2z}\vec{j} + x^2y^2e^{x^2y^2z}\vec{k} = \operatorname{grad}(e^{x^2y^2z})$ and the curve C is closed, the Fundamental Theorem of Line Integrals tells us that $\int_C \vec{F} \cdot d\vec{r} = 0$, since

$$\int_C \vec{F} \cdot d\vec{r} = e^{x^2y^2z}\Big|_{(1,0,1)}^{(1,0,1)} = e^0 - e^0 = 0.$$

Problems

21. Yes. If $f(x, y) = \frac{1}{2}x^2$, then grad $f = x\vec{i}$.

25. (a) To find the change in f by computing a line integral, we first choose a path C between the points; the simplest is a line. We parameterize the line by $(x(t), y(t)) = (t, \pi t/2)$, with $0 \le t \le 1$. Then $(x'(t), y'(t)) = (1, \pi/2)$, so the Fundamental Theorem of Line Integrals tells us that

$$
\begin{aligned}
f\left(1, \frac{\pi}{2}\right) - f(0, 0) &= \int_C \text{grad } f \cdot d\vec{r} \\
&= \int_0^1 \text{grad } f\left(t, \frac{\pi t}{2}\right) \cdot \left(\vec{i} + \frac{\pi}{2}\vec{j}\right) dt \\
&= \int_0^1 \left(2te^{t^2} \sin\left(\frac{\pi t}{2}\right)\vec{i} + e^{t^2} \cos\left(\frac{\pi t}{2}\right)\vec{j}\right) \cdot \left(\vec{i} + \frac{\pi}{2}\vec{j}\right) dt \\
&= \int_0^1 \left(2te^{t^2} \sin\left(\frac{\pi t}{2}\right) + \frac{\pi e^{t^2}}{2} \cos\left(\frac{\pi t}{2}\right)\right) dt \\
&= \int_0^1 \frac{d}{dt}\left(e^{t^2} \sin\left(\frac{\pi t}{2}\right)\right) dt \\
&= e^{t^2} \sin\left(\frac{\pi t}{2}\right) \Big|_0^1 = e = 2.718.
\end{aligned}
$$

This integral can also be approximated numerically.

(b) The other way to find the change in f between these two points is to first find f. To do this, observe that

$$
2xe^{x^2} \sin y\vec{i} + e^{x^2} \cos y\vec{j} = \frac{\partial}{\partial x}\left(e^{x^2} \sin y\right)\vec{i} + \frac{\partial}{\partial y}\left(e^{x^2} \sin y\right)\vec{j} = \text{grad}\left(e^{x^2} \sin y\right).
$$

So one possibility for f is $f(x, y) = e^{x^2} \sin y$. Thus,

$$
\text{Change in } f\Big|_{(0,0)}^{(1,\pi/2)} = e^{x^2} \sin y\Big|_{(0,0)}^{(1,\pi/2)} = e^1 \sin\left(\frac{\pi}{2}\right) - e^0 \sin 0 = e.
$$

The exact answer confirms our calculations in part (a) which show that the answer is e.

29. This vector field is not a gradient field, so we evaluate the line integral directly. Let C_1 be the path along the x-axis from $(0, 0)$ to $(3, 0)$ and let C_2 be the path from $(3, 0)$ to $(3/\sqrt{2}, 3/\sqrt{2})$ along $x^2 + y^2 = 9$. Then

$$
\int_C \vec{H} \cdot d\vec{r} = \int_{C_1} \vec{H} \cdot d\vec{r} + \int_{C_2} \vec{H} \cdot d\vec{r}.
$$

On C_1, the vector field has only a $\vec{j}$ component (since $y = 0$), and $\vec{H}$ is therefore perpendicular to the path. Thus,

$$
\int_{C_1} \vec{H} \cdot d\vec{r} = 0.
$$

On C_2, the vector field is tangent to the path. The path is one eighth of a circle of radius 3 and so has length $2\pi(3/8) = 3\pi/4$.

$$
\int_{C_2} \vec{H} \cdot d\vec{r} = \|\vec{H}\| \cdot \text{Length of path} = 3 \cdot \left(\frac{3\pi}{4}\right) = \frac{9\pi}{4}.
$$

Thus,

$$
\int_C \vec{H} \cdot d\vec{r} = \frac{9\pi}{4}.
$$

33. Although this curve is complicated, the vector field is a gradient field since

$$
\vec{F} = \sin\left(\frac{x}{2}\right) \sin\left(\frac{y}{2}\right)\vec{i} - \cos\left(\frac{x}{2}\right) \cos\left(\frac{y}{2}\right)\vec{j} = \text{grad}\left(-2\cos\left(\frac{x}{2}\right) \sin\left(\frac{y}{2}\right)\right).
$$

Thus, only the endpoints of the curve, P and Q, are needed. Since $P = (-3\pi/2, 3\pi/2)$ and $Q = (-3\pi/2, -3\pi/2)$ and $\vec{F} = \text{grad}(-2\cos(x/2)\sin(y/2))$, we have

$$\int_C \vec{F} \cdot d\vec{r} = -2\cos\left(\frac{x}{2}\right)\sin\left(\frac{y}{2}\right)\Bigg|_{P=(-3\pi/2,3\pi/2)}^{Q=(-3\pi/2,-3\pi/2)}$$

$$= -2\cos\left(-\frac{3\pi}{4}\right)\sin\left(-\frac{3\pi}{4}\right) + 2\cos\left(-\frac{3\pi}{4}\right)\sin\left(\frac{3\pi}{4}\right)$$

$$= 2\cos\left(\frac{3\pi}{4}\right)\sin\left(\frac{3\pi}{4}\right) + 2\cos\left(\frac{3\pi}{4}\right)\sin\left(\frac{3\pi}{4}\right)$$

$$= -2\cdot\frac{1}{\sqrt{2}}\cdot\frac{1}{\sqrt{2}} - 2\cdot\frac{1}{\sqrt{2}}\cdot\frac{1}{\sqrt{2}} = -2.$$

37. (a) By the Fundamental Theorem of Line Integrals

$$\int_{(0,2)}^{(3,4)} \text{grad } f \cdot d\vec{r} = f(3,4) - f(0,2) = 66 - 57 = 9.$$

(b) By the Fundamental Theorem of Line Integrals, since C is a closed path, $\int_C \text{grad } f \cdot d\vec{r} = 0$.

41. (a) Work done by the force is the line integral, so

$$\text{Work done against force} = -\int_C \vec{F} \cdot d\vec{r} = -\int_C (-mg\vec{k}) \cdot d\vec{r}.$$

Since $\vec{r} = (\cos t)\vec{i} + (\sin t)\vec{j} + t\vec{k}$, we have $\vec{r}\,' = -(\sin t)\vec{i} + (\cos t)\vec{j} + \vec{k}$,

$$\text{Work done against force} = \int_0^{2\pi} mg\vec{k} \cdot (-\sin t\vec{i} + \cos t\vec{j} + \vec{k})dt$$

$$= \int_0^{2\pi} mg\, dt = 2\pi\, mg.$$

(b) We know from physical principles that the force is conservative. (Because the work done depends only on the vertical distance moved, not on the path taken.) Alternatively, we see that

$$\vec{F} = -mg\vec{k} = \text{grad}(-mgz),$$

so $\vec{F}$ is a gradient field and therefore path independent, or conservative.

45. (a) We have

$$\text{grad } h = \vec{i}$$
$$\text{grad } \phi = 2y\vec{i} + 2x\vec{j}$$
$$\vec{F} - \text{grad } \phi = -y\vec{i} = -y\,\text{grad } h.$$

Thus, $\vec{F} - \text{grad } \phi$ is a multiple of $\text{grad } h$.

(b) By part (a) the vector fields $\vec{F}$ and $\text{grad } \phi$ have the same components perpendicular to $\text{grad } h$, which is to say the same components in the direction of the level curve C of h. Thus, the line integrals of $\vec{F}$ and $\text{grad } \phi$ along C are equal. Using the Fundamental Theorem of Calculus for Line Integrals, we have

$$\int_C \vec{F} \cdot d\vec{r} = \int_C \text{grad } \phi \cdot d\vec{r} = \phi(Q) - \phi(P) = 60 - 30 = 30.$$

49. (a) By the chain rule

$$\frac{dh}{dt} = \frac{\partial f}{\partial x}\frac{dx}{dt} + \frac{\partial f}{\partial y}\frac{dy}{dt} = f_x x'(t) + f_y y'(t),$$

which is the result we want.

(b) Using the parameterization of C that we were given,

$$\int_C \text{grad } f \cdot d\vec{r} = \int_a^b (f_x(x(t), y(t))\vec{i} + f_y(x(t), y(t))\vec{j}) \cdot (x'(t)\vec{i} + y'(t)\vec{j})dt$$

$$= \int_a^b (f_x(x(t), y(t))x'(t) + f_y(x(t), y(t))y'(t))dt.$$

Using the result of part (a), this gives us

$$\int_C \text{grad } f \cdot d\vec{r} = \int_a^b h'(t)dt$$

$$= h(b) - h(a) = f(Q) - f(P).$$

Solutions for Section 18.4

Exercises

1. We know that

$$\frac{\partial f}{\partial x} = 2xy \quad \text{and} \quad \frac{\partial f}{\partial y} = x^2,$$

so, integrating with respect to x, thinking of y as a constant gives

$$f(x, y) = x^2 y + C(y).$$

Differentiating with respect to y gives

$$\frac{\partial f}{\partial y} = x^2 + C'(y),$$

so we take $C(y) = k$ for some constant K^2. Thus

$$f(x, y) = x^2 y + K.$$

5. Yes, since $\vec{F} = 2xy\vec{i} + x^2\vec{j} = \text{grad}(x^2 y)$.

9. The domain of the vector field $\vec{F} = (2xy^3 + y)\vec{i} + (3x^2 y^2 + x)\vec{j}$ is the whole xy-plane. We apply the curl test:

$$\frac{\partial F_1}{\partial y} = 6xy^2 + 1 = \frac{\partial F_2}{\partial x}$$

so $\vec{F}$ is the gradient of a function f. In order to compute f we first integrate

$$\frac{\partial f}{\partial x} = 2xy^3 + y$$

with respect to x thinking of y as a constant. We get

$$f(x, y) = x^2 y^3 + xy + C(y)$$

Differentiating with respect to y and using the fact that $\partial f/\partial y = 3x^2 y^2 + x$ gives

$$\frac{\partial f}{\partial y} = 3x^2 y^2 + x + C'(y) = 3x^2 y^2 + x$$

Thus $C'(y) = 0$ so C is constant and

$$f(x, y) = x^2 y^3 + xy + C.$$

13. We have

$$\frac{\partial F_1}{\partial y} = \frac{(x^2 + y^2)1 - y(2y)}{(x^2 + y^2)^2} = \frac{x^2 - y^2}{(x^2 + y^2)^2}$$

$$\frac{\partial F_2}{\partial x} = -\frac{(x^2 + y^2)1 - x(2x)}{(x^2 + y^2)^2} = -\frac{y^2 - x^2}{(x^2 + y^2)^2} = \frac{x^2 - y^2}{(x^2 + y^2)^2}.$$

Thus $\frac{\partial F_1}{\partial y} = \frac{\partial F_2}{\partial x}$. However, the domain of the vector field contains a "hole" at the origin, so the curl test does not apply. This is not a gradient field. See Example 7 on page 957 of the text.

17. By Green's Theorem, with R representing the interior of the circle,

$$\int_C \vec{F} \cdot d\vec{r} = \int_R \left(\frac{\partial}{\partial x}(xy) - \frac{\partial}{\partial y}(3y) \right) dA = \int_R (y - 3)\, dA.$$

The integral of y over the interior of the circle is 0, by symmetry, because positive contributions of y from the top half of the circle cancel those from the bottom half. Thus

$$\int_R y\, dA = 0.$$

So

$$\int_C \vec{F} \cdot d\vec{r} = \int_R (y - 3)\, dA = \int_R -3\, dA = -3 \cdot \text{Area of circle} = -3 \cdot \pi(1)^2 = -3\pi.$$

Problems

21. The curve is closed, so we can use Green's Theorem. If R represents the interior of the region

$$\int_C \vec{F} \cdot d\vec{r} = \int_R \left(\frac{\partial F_2}{\partial x} - \frac{\partial F_1}{\partial y} \right) dA = \int_R \left(\frac{\partial(x)}{\partial x} - \frac{\partial(x - y)}{\partial y} \right) dA$$

$$= \int_R (1 - (-1))\, dA = \int_R 2\, dA = 2 \cdot \text{Area of sector}.$$

Since R is $1/8$ of a circle, R has area $\pi(3^2)/8$. Thus,

$$\int_C \vec{F} \cdot d\vec{r} = 2 \cdot \frac{9\pi}{8} = \frac{9\pi}{4}.$$

25. Since $\vec{F} = x\vec{j}$, we have $\partial F_2/\partial x = 1$ and $\partial F_1/\partial y = 0$. Thus, using Green's Theorem if R is the region enclosed by the closed curve C, we have

$$\int_C \vec{F} \cdot d\vec{r} = \int_R \left(\frac{\partial F_2}{\partial x} - \frac{\partial F_1}{\partial y} \right) dx\, dy = \int_R 1\, dx\, dy = \text{Area of } R$$

29. (a) The curve, C, is closed and oriented in the correct direction for Green's Theorem. See Figure 18.7. Writing R for the interior of the circle, we have

$$\int_C \left((x^2 - y)\vec{i} + (y^2 + x)\vec{j} \right) \cdot d\vec{r} = \int_R \left(\frac{\partial(y^2 + x)}{\partial x} - \frac{\partial(x^2 - y)}{\partial y} \right) dx\, dy$$

$$= \int_R (1 - (-1))\, dx\, dy = 2 \int_R dx\, dy$$

$$= 2 \cdot \text{Area of circle} = 2(\pi \cdot 3^2) = 18\pi.$$

(b) The circle given has radius R and center (a, b). The argument in part (a) works for any circle of radius R, oriented counterclockwise. So the line integral has the value $2\pi R^2$.

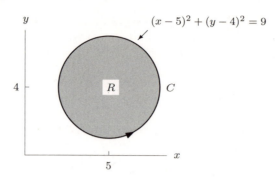

Figure 18.7

33. (a) We see that $\vec{F}, \vec{G}, \vec{H}$ are all gradient vector fields, since

$$\mathrm{grad}(xy) = \vec{F} \quad \text{for all } x, y$$
$$\mathrm{grad}(\arctan(x/y)) = \vec{G} \quad \text{except where } y = 0$$
$$\mathrm{grad}\left((x^2 + y^2)^{1/2}\right) = \vec{H} \quad \text{except at } (0, 0).$$

Other answer are possible. For example $\mathrm{grad}(-\arctan(y/x)) = \vec{G}$ for $x \neq 0$.

(b) Parameterizing the unit circle, C, by $x = \cos t, y = \sin t, 0 \le t \le \pi$, we have $\vec{r}\,'(t) = -\sin t\vec{i} + \cos t\vec{j}$, so

$$\int_C \vec{F} \cdot d\vec{r} = \int_0^{2\pi} ((\sin t)\vec{i} + (\cos t)\vec{j}) \cdot ((-\sin t)\vec{i} + (\cos t)\vec{j})\, dt = \int_0^{2\pi} \cos(2t)\, dt = 0.$$

The vector field $\vec{G}$ is tangent to the circle, pointing in the opposite direction to the parameterization, and of length 1 everywhere. Thus

$$\int_C \vec{G} \cdot d\vec{r} = -1 \cdot \text{ Length of circle } = -2\pi.$$

The vector field $\vec{H}$ points radially outward, so it is perpendicular to the circle everywhere. Thus

$$\int_C \vec{H} \cdot d\vec{r} = 0.$$

(c) Green's Theorem does not apply to the computation of the line integrals for $\vec{G}$ and $\vec{H}$ because their domains do not include the origin, which is in the interior, R, of the circle. Green's Theorem does apply to $\vec{F} = y\vec{i} + x\vec{j}$.

$$\int_C \vec{F} \cdot d\vec{r} = \int_R \left(\frac{\partial F_2}{\partial x} - \frac{\partial F_1}{\partial y}\right) dx\, dy = \int 0\, dx\, dy = 0.$$

37. (a) We use Green's Theorem. Let R be the region enclosed by the circle C. Then

$$\int_C \vec{F} \cdot d\vec{r} = \int_R \left(\frac{\partial F_2}{\partial x} - \frac{\partial F_1}{\partial y}\right) dA = \int_R \left(\frac{\partial}{\partial x}(e^{y^2} + 12x) - \frac{\partial}{\partial y}(3x^2 y + y^3 + e^x)\right) dA$$
$$= \int_R (12 - (3x^2 + 3y^2))\, dA = \int_R (12 - 3(x^2 + y^2))\, dA.$$

Converting to polar coordinates, we have

$$\int_C \vec{F} \cdot d\vec{r} = \int_0^{2\pi} \int_0^1 (12 - 3r^2) r\, dr\, d\theta = 2\pi \left(6r^2 - \frac{3}{4}r^4\right)\Big|_0^1 = 2\pi \left(6 - \frac{3}{4}\right) = \frac{21\pi}{2}.$$

(b) The integrand of the integral over the disk R is $12 - 3(x^2 + y^2)$. Since the integrand is positive for $x^2 + y^2 < 4$ and negative for $x^2 + y^2 > 4$, the integrand is positive inside the circle of radius 2 and negative outside that circle. Thus, the integral over R increases with a until $a = 2$ and then decreases. The maximum value of the line integral occurs when $a = 2$.

Solutions for Chapter 18 Review

Exercises

1. The angle between the vector field and the curve is more than $90°$ at all points on C, so the line integral is negative.

5. Since $\vec{F} = 6\vec{i} - 7\vec{j}$, consider the function f

$$f(x, y) = 6x - 7y.$$

Then we see that $\operatorname{grad} f = 6\vec{i} - 7\vec{j}$, so we use the Fundamental Theorem of Calculus for Line Integrals:

$$\int_C \vec{F} \cdot d\vec{r} = \int_C \operatorname{grad} f \cdot d\vec{r}$$
$$= f(4, 4) - f(2, -6) = (-4) - (54) = -58.$$

9. Only the $\vec{j}$ component contributes to the integral. On the y-axis, $x = 0$, so

$$\int_C \vec{F} \cdot d\vec{r} = \int_3^5 y^2 \vec{j} \cdot \vec{j} \, dy = \frac{y^3}{3} \bigg|_3^5 = \frac{98}{3}.$$

13. The triangle C consists of the three paths shown in Figure 18.8.

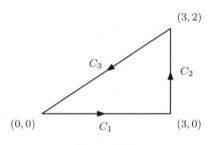

Figure 18.8

Write $C = C_1 + C_2 + C_3$ where C_1, C_2, and C_3 are parameterized by

$$C_1 : (t, 0) \text{ for } 0 \le t \le 3; \quad C_2 : (3, t) \text{ for } 0 \le t \le 2; \quad C_3 : (3 - 3t, 2 - 2t) \text{ for } 0 \le t \le 1.$$

Then

$$\int_C \vec{F} \cdot d\vec{r} = \int_{C_1} \vec{F} \cdot d\vec{r} + \int_{C_2} \vec{F} \cdot d\vec{r} + \int_{C_3} \vec{F} \cdot d\vec{r}$$

where

$$\int_{C_1} \vec{F} \cdot d\vec{r} = \int_0^3 \vec{F}(t, 0) \cdot \vec{i} \, dt = \int_0^3 (2t + 4) dt = (t^2 + 4t) \big|_0^3 = 21$$

$$\int_{C_2} \vec{F} \cdot d\vec{r} = \int_0^2 \vec{F}(3, t) \cdot \vec{j} \, dt = \int_0^2 (5t + 3) dt = (5t^2/2 + 3t) \big|_0^2 = 16$$

$$\int_{C_3} \vec{F} \cdot d\vec{r} = \int_0^1 \vec{F}(3 - 3t, 2 - 2t) \cdot (-3\vec{i} - 2\vec{j}) dt$$

$$= \int_0^1 ((-4t + 8)\vec{i} + (-19t + 13)\vec{j}) \cdot (-3\vec{i} - 2\vec{j}) dt$$

$$= 50 \int_0^1 (t - 1) dt = -25.$$

So
$$\int_C \vec{F} \, d\vec{r} = 21 + 16 - 25 = 12.$$

17. The domain is all 3-space. Since $F_1 = y$,
$$\text{curl } y\vec{i} = \left(\frac{\partial F_3}{\partial y} - \frac{\partial F_2}{\partial z}\right)\vec{i} + \left(\frac{\partial F_1}{\partial z} - \frac{\partial F_3}{\partial x}\right)\vec{j} + \left(\frac{\partial F_2}{\partial x} - \frac{\partial F_1}{\partial y}\right)\vec{k} = -\vec{k} \neq \vec{0},$$
so $\vec{F}$ is not path-independent.

21. The domain is all 3-space. Since $F_1 = y$, $F_2 = x$,
$$\text{curl } y\vec{i} + x\vec{j} = \left(\frac{\partial F_3}{\partial y} - \frac{\partial F_2}{\partial z}\right)\vec{i} + \left(\frac{\partial F_1}{\partial z} - \frac{\partial F_3}{\partial x}\right)\vec{j} + \left(\frac{\partial F_2}{\partial x} - \frac{\partial F_1}{\partial y}\right)\vec{k} = \vec{0},$$
so $\vec{F}$ is path-independent

25. Since the line is parallel to the x-axis, only the $\vec{i}$-component contributes to the line integral. On C, we have $d\vec{r} = \vec{i} \, dx$, so
$$\int_C \vec{F} \cdot d\vec{r} = \int_2^{12} 5x\vec{i} \cdot \vec{i} \, dx = \frac{5}{2}x^2 \Big|_2^{12} = 350.$$

29. We can calculate this line integral either by calculating a separate line integral for each side, or by adding a line segment, C_1, from $(1, 4)$ to $(1, 1)$ to form the closed curve $C + C_1$. Since we now have a closed curve, we can use Green's Theorem:
$$\int_{C+C_1} \vec{F} \cdot d\vec{r} = \int_{C+C_1} (5x\vec{i} + 3x\vec{j}) \cdot d\vec{r} = \int_R \left(\frac{\partial}{\partial x}(3x) - \frac{\partial}{\partial y}(5y)\right) dx \, dy$$
$$= \int_R 3 \, dx \, dy = 3 \cdot \text{ Area of region} = 3\left(2 \cdot 3 + \frac{1}{2}3 \cdot 4\right) = 36.$$

Since $d\vec{r} = -\vec{j} \, dy$ on C_1, we have
$$\int_{C_1} \vec{F} \cdot d\vec{r} = \int_4^1 3 \cdot 1\vec{j} \cdot (-\vec{j} \, dy) = -3 \cdot 3 = -9.$$

Since
$$\int_{C+C_1} \vec{F} \cdot d\vec{r} = \int_C \vec{F} \cdot d\vec{r} + \int_{C_1} \vec{F} \cdot d\vec{r} = \int_C \vec{F} \cdot d\vec{r} - 9 = 36$$
we have
$$\int_C \vec{F} \cdot d\vec{r} = 45.$$

Problems

33. (a) The vector field is everywhere perpendicular to the radial line from the origin to $(2, 3)$, so the line integral is 0.

(b) Since the path is parallel to the x-axis, only the $\vec{i}$ component of the vector field contributes to the line integral. The $\vec{i}$ component is $-3\vec{i}$ on this line, and the displacement along this line is $-2\vec{i}$, so
$$\text{Line integral} = (-3\vec{i}) \cdot (-2\vec{i}) = 6.$$

(c) The circle of radius 5 has equation $x^2 + y^2 = 25$. On this curve, $\|\vec{F}\| = \sqrt{(-y^2) + x^2} = \sqrt{25} = 5$. In addition, $\vec{F}$ is everywhere tangent to the circle, and the path is $3/4$ of the circle. Thus
$$\text{Line integral} = \|\vec{F}\| \cdot \text{ Length of curve} = 5 \cdot \frac{3}{4} \cdot 2\pi(5) = \frac{75}{2}\pi.$$

(d) Use Green's Theorem. Writing C for the curve around the boundary of the triangle, we have
$$\frac{\partial F_2}{\partial x} - \frac{\partial F_1}{\partial y} = 1 - (-1) = 2,$$
so
$$\int_C \vec{F} \cdot d\vec{r} = \int_{\text{Triangle}} 2 \, dA = 2 \cdot \text{ Area of triangle} = 2 \cdot 7 = 14.$$

37. (a) Since $\vec{F} = (6x + y^2)\vec{i} + 2xy\vec{j} = \text{grad}(3x^2 + xy^2)$, the vector field $\vec{F}$ is path independent, so

$$\int_{C_1} \vec{F} \cdot d\vec{r} = 0.$$

(b) Since C_1 is closed, we use Green's Theorem, so

$$\int_{C_1} \vec{G} \cdot d\vec{r} = \int_{\text{Interior of } C_1} \left(\frac{\partial}{\partial x}(x + y) - \frac{\partial}{\partial y}(x - y) \right) dA$$

$$= 2 \int_{C_1} dA = 2 \cdot \text{Area inside } C_1 = 2 \cdot \frac{1}{2} \cdot 2 \cdot 2 = 4.$$

(c) Since $\vec{F} = \text{grad}(3x^2 + xy^2)$, using the Fundamental Theorem of Line Integrals gives

$$\int_{C_2} \vec{F} \cdot d\vec{r} = (3x^2 + xy^2)\Big|_{(2,0)}^{(0,-2)} = 0 - 3 \cdot 2^2 = -12.$$

(d) Parameterizing the circle by

$$x = 2\cos t \qquad y = 2\sin t \qquad 0 \le t \le \frac{3\pi}{2},$$

gives

$$x' = -2\sin t \qquad y' = 2\cos t,$$

so the integral is

$$\int_{C_2} \vec{G} \cdot d\vec{r} = \int_0^{3\pi/2} \left((2\cos t - 2\sin t)\vec{i} + (2\cos t + 2\sin t)\vec{j} \right) \cdot (-2\sin t\vec{i} + 2\cos t\vec{j}) \, dt$$

$$= \int_0^{3\pi/2} 4(-\cos t\sin t + \sin^2 t + \cos^2 t + \sin t\cos t) \, dt$$

$$= 4\int_0^{3\pi/2} dt = 4 \cdot \frac{3\pi}{2} = 6\pi.$$

41. (a) Since $\frac{\partial}{\partial x}(y^5 + x) - \frac{\partial}{\partial y}(x^3 - y) = 1 + 1 = 2$, any closed curve oriented counterclockwise will do. See Figure 18.9.

Figure 18.9   Figure 18.10

(b) Since $\frac{\partial}{\partial x}(y^5 - xy) - \frac{\partial}{\partial y}(x^3) = -y$, any closed curve in the lower half-plane oriented counterclockwise or any closed curve in the upper half-plane oriented clockwise will do. See Figure 18.10. Other answers are possible.

45. Since $||\vec{F}|| \le 7$, the line integral cannot be larger than 7 times the length of the curve. Thus

$$\int_C \vec{F} \cdot d\vec{r} \le 7 \cdot \text{Circumference of circle} = 7 \cdot 2\pi = 14\pi.$$

The line integral is equal to 14π if $\vec{F}$ is everywhere of magnitude 7, tangent to the curve, and pointing in the direction in which the curve is traversed.

The smallest possible value occurs if the vector field is everywhere of magnitude 7, tangent to the curve and pointing opposite to the direction in which the curve is transversed. Thus

$$\int_C \vec{F} \cdot d\vec{r} \geq -14\pi.$$

49. The free vortex appears to starts at about $r = 200$ meters (that's where the graph changes its behavior) and the tangential velocity at this point is about 200 km/hr $= 2 \cdot 10^5$ meters/hr.

Since $\vec{v} = \omega(-y\vec{i} + x\vec{j})$ for $\sqrt{x^2 + y^2} \leq 200$, at $r = 200$ we have

$$\|\vec{v}\| = \omega\sqrt{(-y)^2 + x^2} = \omega(200) = 2 \cdot 10^5 \text{ meters/hr},$$

so

$$\omega = 10^3 \text{ rad/hr}.$$

Since $\vec{v} = K(x^2 + y^2)^{-1}(-y\vec{i} + x\vec{j})$ for $\sqrt{x^2 + y^2} \geq 200$, at $r = 200$ we have

$$\|\vec{v}\| = K(200^2)^{-1}(200) = \frac{K}{200} = 2 \cdot 10^5 \text{ meters/hr}$$

so

$$K = 4 \cdot 10^7 \text{ m}^2\cdot\text{rad/hr}.$$

CAS Challenge Problems

53. We have

$$\int_{C_1} \vec{F} \cdot d\vec{r} = \int_0^3 \vec{F}\,(\vec{r}\,(t)) \cdot \vec{r}\,'(t)dt$$

$$= \int_0^3 \left(2\left(2at + bt^2\right) + 2t\left(2ct + dt^2\right)\right)dt = 18a + 18b + 36c + (81d/2)$$

and

$$\int_{C_2} \vec{F} \cdot d\vec{r} = \int_0^3 \vec{F}\,(\vec{r}\,(t)) \cdot \vec{r}\,'(t)dt$$

$$= \int_0^3 \left(-2\left(2a\,(3-t) + b(3-t)^2\right) - 2\left(2c\,(3-t) + d(3-t)^2\right)(3-t)\right)dt$$

$$= -18a - 18b - 36c - (81d/2)$$

The second integral is the negative of the first. This is because C_2 is the same curve as C_1 but traveling in the opposite direction.

CHECK YOUR UNDERSTANDING

1. A path-independent vector field must have zero circulation around all closed paths. Consider a vector field like $\vec{F}(x, y) = |x|\vec{j}$, shown in Figure 18.11.

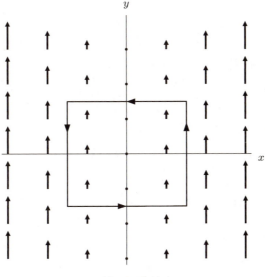

Figure 18.11

A rectangular path that is symmetric about the y-axis will have zero circulation: on the horizontal sides, the field is perpendicular, so the line integral is zero. The line integrals on the vertical sides are equal in magnitude and opposite in sign, so they cancel out, giving a line integral of zero. However, this field is not path-independent, because it is possible to find two paths with the same endpoints but different values of the line integral of $\vec{F}$. For example, consider the two points $(0, 0)$ and $(0, 1)$. The path C_1 in Figure 18.12 along the y axis gives zero for the line integral, because the field is 0 along the y axis, whereas a path like C_2 will have a nonzero line integral. Thus the line integral depends on the path between the points, so $\vec{F}$ is not path-independent.

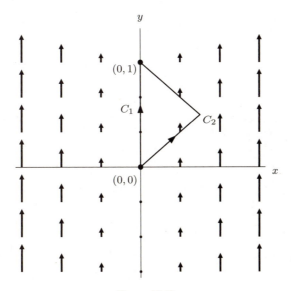

Figure 18.12

5. False. Because $\vec{F} \cdot \Delta \vec{r}$ is a scalar quantity, $\int_C \vec{F} \cdot d\vec{r}$ is also a scalar quantity.

9. False. We can calculate a line integral of any vector field.

13. False. The relative sizes of the line integrals along C_1 and C_2 depend on the behavior of the vector field $\vec{F}$ along the curves. As a counterexample, take the vector field $\vec{F} = \vec{i}$, and C_1 to be the line from the origin to $(0, 2)$, while C_2 is the line from the origin to $(1, 0)$. Then the length of C_1 is 2, which is greater than the length of C_2, which is 1. However $\int_{C_1} \vec{F} \cdot d\vec{r} = 0$ (since $\vec{F}$ is perpendicular to C_1) while $\int_{C_2} \vec{F} \cdot d\vec{r} > 0$ (since $\vec{F}$ points along C_2).

17. False. The relation between these two line integrals depends on the behavior of the vector field along each of the curves, so there is no reason to expect one to be the negative of the other. As an example, if $\vec{F}(x, y) = y\vec{i}$, then, by symmetry, both line integrals are equal to the same negative number.

21. True. The dot product of the integrand $4\vec{i}$ with $\vec{r}'(t) = \vec{i} + 2t\vec{j}$ is 4, so the integral has value $\int_0^2 4\,dt = 8$.

25. False. As a counterexample, consider the vector field $\vec{F} = x\vec{i}$. Then if we parameterize C_1 by $\vec{r}(t) = t\vec{i}$, with $0 \le t \le 1$, we get

$$\int_{C_1} x\vec{i} \cdot d\vec{r} = \int_0^1 t\vec{i} \cdot \vec{i}\,dt = \int_0^1 t\,dt = \left.\frac{t^2}{2}\right|_0^1 = \frac{1}{2}.$$

A similar computation for C_2 gives a line integral with value 2.

29. True. By the Fundamental Theorem for Line Integrals, if C is a path from P to Q, then $\int_C \operatorname{grad} f \cdot d\vec{r} = f(Q) - f(P)$, so the value of the line integral $\int_C \operatorname{grad} f \cdot d\vec{r}$ depends only on the endpoints and not the path.

33. True. Since a gradient field is path-independent, and C_1 and C_2 have the same initial and final points, the two line integrals are equal.

37. True. The value of $\dfrac{\partial F_2}{\partial x} - \dfrac{\partial F_1}{\partial y}$ is $0 - 0 = 0$, so the field is path-independent.

41. True. This vector field has components $F_1 = x$, $F_2 = y$, and $F_3 = z$. Using the 3-space curl test gives zero for all of the components of $\operatorname{curl}\vec{F}$, so the field is path-independent.

CHAPTER NINETEEN

Solutions for Section 19.1

Exercises

1. Scalar. Only the $\vec{j}$-component of the vector field contributes to the flux and $d\vec{A} = -\vec{j}\, dA$, so

$$\int_S (3\vec{i} + 4\vec{j}) \cdot d\vec{A} = -4 \cdot \text{Area of disk} = -4 \cdot \pi 5^2 = -100\pi.$$

5. On the surface, $d\vec{A} = \vec{k}\, dA$, so only the $\vec{k}$ component of $\vec{v}$ contributes to the flux:

$$\text{Flux} = \int_S \vec{v} \cdot d\vec{A} = \int_S (\vec{i} - \vec{j} + 3\vec{k}) \cdot \vec{k}\, dA = 3 \cdot \text{Area of disk} = 3 \cdot \pi 2^2 = 12\pi.$$

9. $\vec{v} \cdot \vec{A} = (2\vec{i} + 3\vec{j} + 5\vec{k}) \cdot \vec{k} = 5.$

13. The disk has area 25π, so its area vector is $25\pi\vec{j}$. Thus

$$\text{Flux} = (2\vec{i} + 3\vec{j}) \cdot 25\pi\vec{j} = 75\pi.$$

17. Since the square, S, is in the plane $y = 0$ and oriented in the negative y-direction, $d\vec{A} = -\vec{j}\, dxdz$ and

$$\int_S \vec{F} \cdot d\vec{A} = \int_S (0 + 3)\vec{j} \cdot (-\vec{j}\, dxdz) = -3 \int_S dxdz = -3 \cdot \text{Area of square} = -3(2^2) = -12.$$

21. The only contribution to the flux is from the $\vec{j}$-component, and since $d\vec{A} = \vec{j}\, dx\, dz$ on the square, S, we have

$$\text{Flux} = \int_S (6\vec{i} + x^2\vec{j} - \vec{k}) \cdot d\vec{A} = \int_{-2}^{2} \int_{-2}^{2} x^2\vec{j} \cdot \vec{j}\, dx\, dz = \int_{-2}^{2} \left.\frac{x^3}{3}\right|_{-2}^{2} dz = \frac{16}{3} \cdot 4 = \frac{64}{3}.$$

25. We have $d\vec{A} = \vec{i}\, dA$, so

$$\int_S \vec{F} \cdot d\vec{A} = \int_S (2z\vec{i} + x\vec{j} + x\vec{k}) \cdot \vec{i}\, dA = \int_S 2z\, dA$$

$$= \int_0^2 \int_0^3 2z\, dzdy = 18.$$

29. Since the disk is oriented in the positive x-direction, $d\vec{A} = \vec{i}\, dydz$, so we have

$$\text{Flux} = \int_{\text{Disk}} \vec{F} \cdot d\vec{A} = \int_{\text{Disk}} e^{y^2 + z^2}\vec{i} \cdot \vec{i}\, dydz = \int_{\text{Disk}} e^{y^2 + z^2}\, dydz.$$

To calculate this integral, we use polar coordinates with $y = r\cos\theta$ and $z = r\sin\theta$. Then $r^2 = y^2 + z^2$ and

$$\int_{\text{Disk}} \vec{F} \cdot d\vec{A} = \int_0^{2\pi} \int_0^2 e^{r^2} \cdot r\, drd\theta = 2\pi \cdot \left.\frac{e^{r^2}}{2}\right|_0^2 = \pi(e^4 - 1).$$

33. Since $\vec{r}$ is perpendicular to S and $\|\vec{r}\| = 3$ on S, we have

$$\int_S \vec{r} \cdot d\vec{A} = 3 \cdot \text{Area of surface} = 3 \cdot 4\pi 3^2 = 108\pi.$$

Problems

37. Since this vector field points radially out from the origin, it is everywhere parallel to the vector representing the surface area, $d\vec{A}$. Thus since $\|\vec{F}(\vec{r})\| = 1/R^2$ on the surface, S,

$$\vec{F}(\vec{r}) \cdot d\vec{A} = \frac{1}{R^2} \, dA,$$

so

$$\int_S \vec{F}(\vec{r}) \cdot d\vec{A} = \frac{1}{R^2} \cdot \text{Surface area of sphere} = \frac{1}{R^2}(4\pi R^2) = 4\pi.$$

41. By the symmetry of the sphere, the $\vec{i}$ and $\vec{j}$ components of $\vec{F}$ do not contribute to the flux; only the $\vec{k}$ component contributes. The vector normal to S has a negative $\vec{k}$ component, so we need $c > 0$. There are no conditions on a and b.

45. The square of side 2 in the plane $x = 5$, oriented in the positive x-direction, has area vector $\vec{A} = 4\vec{i}$. Since the vector field is constant

$$\text{Flux} = (a\vec{i} + b\vec{j} + c\vec{k}) \cdot 4\vec{i} = 4a = 24.$$

Thus, $a = 6$ and we cannot say anything about the values of b and c.

49. (a) (i) The integral $\int_W \rho \, dV$ represents the total charge in the volume W.

(ii) The integral $\int_S \vec{J} \cdot d\vec{A}$ represents the total current flowing out of the surface S.

(b) The total current flowing out of the surface S is the rate at which the total charge inside the surface S (i.e., in the volume W) is decreasing. In other words,

$$\text{Rate current flowing out of } S = -\frac{\partial}{\partial t}(\text{charge in } W),$$

so

$$\int_S \vec{J} \cdot d\vec{A} = -\frac{\partial}{\partial t}\left(\int_W \rho \, dV\right).$$

Solutions for Section 19.2

Exercises

1. Only the z-component of the vector field contributes to the flux. Since $d\vec{A} = \vec{k} \, dx \, dy$ on the surface, we have

$$\int_S (3\vec{i} + 4\vec{j} + xy\vec{k}) \cdot d\vec{A} = \int_0^7 \int_0^5 xy \, dx \, dy = \int_0^7 \frac{x^2 y}{2}\Big|_0^5 \, dy = \int_0^7 \frac{25}{2}y \, dy = \frac{25y^2}{4}\Big|_0^7 = \frac{1225}{4}.$$

5. Writing the surface S as $z = f(x,y) = -y + 1$, we have

$$d\vec{A} = (-f_x\vec{i} - f_y\vec{j} + \vec{k})dxdy.$$

Thus,

$$\int_S \vec{F} \cdot d\vec{A} = \int_R \vec{F}(x, y, f(x,y)) \cdot (-f_x\vec{i} - f_y\vec{j} + \vec{k}) \, dxdy$$

$$= \int_0^1 \int_0^1 (2x\vec{j} + y\vec{k}) \cdot (\vec{j} + \vec{k}) \, dxdy$$

$$= \int_0^1 \int_0^1 (2x + y) \, dxdy = \int_0^1 (x^2 + xy)\Big|_0^1 \, dy$$

$$= \int_0^1 (1 + y) \, dy = (y + \frac{y^2}{2})\Big|_0^1 = \frac{3}{2}.$$

9. On the curved side of the cylinder, only the components $x\vec{i} + z\vec{k}$ contribute to the flux. Since $x\vec{i} + z\vec{k}$ is perpendicular to the curved surface and $||x\vec{i} + z\vec{k}|| = 2$ there (because the cylinder has radius 2), we have

$$\text{Flux through sides} = 2 \cdot \text{Area of curved surface} = 2 \cdot 2\pi \cdot 2 \cdot 6 = 48\pi.$$

On the flat ends, only $y\vec{j}$ contributes to the flux. On one end, $y = 3$ and $d\vec{A} = \vec{j}\, dA$; on the other end, $y = -3$ and $d\vec{A} = -\vec{j}\, dA$. Thus

$$\text{Flux through ends} = \text{Flux through top} + \text{Flux through bottom}$$
$$= 3\vec{j} \cdot \vec{j}\,\pi(2^2) + (-3\vec{j}) \cdot (-\vec{j}\,\pi(2^2)) = 24\pi.$$

So,

$$\text{Total flux} = 48\pi + 24\pi = 72\pi.$$

13. Using $z = 1 - x - y$, the upward pointing area element is $d\vec{A} = (\vec{i} + \vec{j} + \vec{k})\, dx\, dy$, so the downward one is $d\vec{A} = (-\vec{i} - \vec{j} - \vec{k})\, dx\, dy$. Since S is oriented downward, we have

$$\int_S \vec{F} \cdot d\vec{A} = \int_S (x\vec{i} + y\vec{j} + z\vec{k}) \cdot d\vec{A}$$
$$= \int_0^3 \int_0^2 (x\vec{i} + y\vec{j} + (1 - x - y)\vec{k}) \cdot (-\vec{i} - \vec{j} - \vec{k})\, dx\, dy$$
$$= \int_0^3 \int_0^2 (-x - y - 1 + x + y)\, dx\, dy = -6.$$

17. Since $y = f(x, z) = x^2 + z^2$, we have

$$d\vec{A} = (-f_x\vec{i} + \vec{j} - f_z\vec{k})\, dx\, dz = (-2x\vec{i} + \vec{j} - 2z\vec{k})\, dx\, dz.$$

Thus, substituting $y = x^2 + z^2$ into $\vec{F}$, we have

$$\int_S \vec{F} \cdot d\vec{A} = \int_{x^2+z^2 \le 1} ((x^2 + z^2)\vec{i} + \vec{j} - xz\vec{k}) \cdot (-2x\vec{i} + \vec{j} - 2z\vec{k})\, dx\, dz$$
$$= \int_{x^2+z^2 \le 1} (-2x^3 - 2xz^2 + 1 + 2xz^2)\, dx\, dz$$
$$= \int_{-1}^1 \int_{-\sqrt{1-z^2}}^{\sqrt{1-z^2}} (1 - 2x^3)\, dx\, dz$$
$$= \int_{-1}^1 \int_{-\sqrt{1-z^2}}^{\sqrt{1-z^2}} dx\, dz - \int_{-1}^1 \int_{-\sqrt{1-z^2}}^{\sqrt{1-z^2}} 2x^3\, dx\, dz$$
$$= \text{Area of disk} - \int_{-1}^1 \left(\frac{x^4}{2} \Big|_{-\sqrt{1-z^2}}^{\sqrt{1-z^2}} \right) dz = \pi - 0 = \pi$$

21. The flux of $\vec{F}$ through S is given by

$$\int_S \vec{F} \cdot d\vec{A} = \int_0^{2\pi} \int_0^{\pi/2} (2\cos\phi\vec{k}) \cdot (\sin\phi\cos\theta\vec{i} + \sin\phi\sin\theta\vec{j} + \cos\phi\vec{k})2^2 \sin\phi\, d\phi\, d\theta$$
$$= \int_{\theta=0}^{2\pi} \int_{\phi=0}^{\pi/2} 8\sin\phi\cos^2\phi\, d\phi\, d\theta = 16\pi \left(\frac{-\cos^3\phi}{3} \right) \Big|_{\phi=0}^{\pi/2} = \frac{16\pi}{3}.$$

Problems

25. The $\vec{k}$-component of $\vec{F}$ does not contribute to the flux as it is perpendicular to the surface. The vector field $x\vec{i} + y\vec{j}$ is everywhere perpendicular to S and has constant magnitude $||x^2 + y^2|| = 1$ on the surface S. Thus

$$\int_S \vec{F} \cdot d\vec{A} = \int_S (x\vec{i} + y\vec{j}) \cdot d\vec{A} = 1 \cdot \text{Area of } S = 1\frac{\pi}{2} = \frac{\pi}{2}.$$

Alternatively, the flux can be computed by integrating with respect to x and z, treating y as a function of x and z. A parameterization of S is given by $y = \sqrt{1 - x^2}, 0 \le x \le 1, 0 \le y \le 1, 0 \le z \le 1$. Thus,

$$\int_S \vec{F} \cdot d\vec{A} = \int_0^1 \int_0^1 (x\vec{i} + \sqrt{1 - x^2}\vec{j} + z\vec{k}) \cdot (-y_x\vec{i} + \vec{j} - y_z\vec{k}) \, dx \, dz$$

$$= \int_0^1 \int_0^1 (x\vec{i} + \sqrt{1 - x^2}\vec{j} + z\vec{k}) \cdot \left(\frac{x}{\sqrt{1 - x^2}}\vec{i} + \vec{j} + 0\vec{k} \right) \, dx \, dz$$

$$= \int_0^1 \int_0^1 \left(\frac{x^2}{\sqrt{1 - x^2}} + \sqrt{1 - x^2} \right) \, dx \, dz$$

$$= \int_0^1 \int_0^1 \frac{1}{\sqrt{1 - x^2}} \, dx \, dz$$

$$= 1 \cdot \arcsin x \Big|_0^1 = \frac{\pi}{2}.$$

29. The plane is $x - z = 0$ over region $0 \le x \le \sqrt{2}, 0 \le y \le 2$. See Figure 19.1.

Figure 19.1

$$\text{Flux} = \int_0^2 \int_0^{\sqrt{2}} \left((e^{xy} + 3z + 5)\vec{i} + (e^{xy} + 5z + 3)\vec{j} + (3z + e^{xy})\vec{k} \right) \cdot (\vec{i} - \vec{k}) \, dx \, dy$$

$$= \int_0^2 \int_0^{\sqrt{2}} (e^{xy} + 3z + 5 - 3z - e^{xy}) \, dx \, dy = 5(2)(\sqrt{2}) = 10\sqrt{2}$$

Alternatively, since a unit normal to the surface is $\vec{n}/\sqrt{2} = (\vec{i} - \vec{j})/\sqrt{2}$, writing $dA = ||d\vec{A}||$, we have

$$\text{Flux} = \int_S \vec{H} \cdot d\vec{A} = \int \vec{H} \cdot \frac{\vec{i} - \vec{k}}{\sqrt{2}} \, dA = \int \frac{5}{\sqrt{2}} \, dA$$

$$= \frac{5}{\sqrt{2}}(\text{Area of slanted square}) = \frac{5}{\sqrt{2}}4 = 10\sqrt{2}.$$

Solutions for Section 19.3

Exercises

1. Since S is given by

$$\vec{r}(s,t) = (s+t)\vec{i} + (s-t)\vec{j} + (s^2+t^2)\vec{k},$$

we have

$$\frac{\partial \vec{r}}{\partial s} = \vec{i} + \vec{j} + 2s\vec{k} \quad \text{and} \quad \frac{\partial \vec{r}}{\partial t} = \vec{i} - \vec{j} + 2t\vec{k},$$

and

$$\frac{\partial \vec{r}}{\partial s} \times \frac{\partial \vec{r}}{\partial t} = \begin{vmatrix} \vec{i} & \vec{j} & \vec{k} \\ 1 & 1 & 2s \\ 1 & -1 & 2t \end{vmatrix} = (2s+2t)\vec{i} + (2s-2t)\vec{j} - 2\vec{k}.$$

Since the $\vec{i}$ component of this vector is positive for $0 < s < 1, 0 < t < 1$, it points away from the z-axis, and so has the opposite orientation to the one specified. Thus, we use

$$d\vec{A} = -\frac{\partial \vec{r}}{\partial s} \times \frac{\partial \vec{r}}{\partial t}\, ds\, dt,$$

and so we have

$$\int_S \vec{F} \cdot d\vec{A} = -\int_0^1 \int_0^1 (s^2+t^2)\vec{k} \cdot \left((2s+2t)\vec{i} + (2s-2t)\vec{j} - 2\vec{k} \right) ds\, dt$$

$$= 2\int_0^1 \int_0^1 (s^2+t^2)\, ds\, dt = 2\int_0^1 \left(\frac{s^3}{3} + st^2 \right) \Big|_{s=0}^{s=1} dt$$

$$= 2\int_0^1 \left(\frac{1}{3} + t^2 \right) dt = 2\left(\frac{1}{3}t + \frac{t^3}{3}\right)\Big|_0^1 = 2\left(\frac{1}{3} + \frac{1}{3}\right) = \frac{4}{3}.$$

5. The cross product $\partial \vec{r}/\partial s \times \partial \vec{r}/\partial t$ is given by

$$\frac{\partial \vec{r}}{\partial s} \times \frac{\partial \vec{r}}{\partial t} = \begin{vmatrix} \vec{i} & \vec{j} & \vec{k} \\ 2s & 2 & 0 \\ 0 & 2t & 5 \end{vmatrix} = 10\vec{i} - 10s\vec{j} + 4st\vec{k}.$$

Since the z-component, $4st$, of the vector $\partial \vec{r}/\partial s \times \partial \vec{r}/\partial t$ is positive for $0 < s \leq 1, 1 \leq t \leq 3$, we see that $\partial \vec{r}/\partial s \times \partial \vec{r}/\partial t$ points upward, in the direction of the orientation of S we were given. Thus, we use

$$d\vec{A} = \left(\frac{\partial \vec{r}}{\partial s} \times \frac{\partial \vec{r}}{\partial t} \right) ds\, dt,$$

and so we have

$$\int_S \vec{F} \cdot d\vec{A} = \int_0^1 \int_1^3 (5t\vec{i} + s^2\vec{j}) \cdot (10\vec{i} - 10s\vec{j} + 4st\vec{k})\, dt\, ds$$

$$= \int_0^1 \int_1^3 (50t - 10s^3)\, dt\, ds = \int_0^1 (25t^2 - 10s^3 t)\Big|_{t=1}^{t=3} ds$$

$$= \int_0^1 (200 - 20s^3)\, ds = (200s - 5s^4)\Big|_0^1$$

$$= 200 - 5 = 195.$$

9. A parameterization of S is

$$x = s, \quad y = t, \quad z = 3s + 2t, \quad \text{for } 0 \le s \le 10, \quad 0 \le t \le 20.$$

We compute

$$\frac{\partial \vec{r}}{\partial s} \times \frac{\partial \vec{r}}{\partial t} = (\vec{i} + 3\vec{k}) \times (\vec{j} + 2\vec{k}) = -3\vec{i} - 2\vec{j} + \vec{k}$$

$$\left\| \frac{\partial \vec{r}}{\partial \theta} \times \frac{\partial \vec{r}}{\partial t} \right\| = \sqrt{14}$$

$$\text{Surface area} = \int_S dA = \int_R \left\| \frac{\partial \vec{r}}{\partial \theta} \times \frac{\partial \vec{r}}{\partial t} \right\| dA = \int_{t=0}^{20} \int_{s=0}^{10} \sqrt{14}\, ds\, dt = 200\sqrt{14}.$$

Problems

13. The elliptic cylindrical surface is parameterized by

$$\vec{r} = x\vec{i} + y\vec{j} + z\vec{k} = a \cos \theta \vec{i} + b \sin \theta \vec{j} + z\vec{k} \qquad \text{where } 0 \le \theta \le 2\pi, -c \le z \le c.$$

We have

$$\frac{\partial \vec{r}}{\partial \theta} \times \frac{\partial \vec{r}}{\partial z} = \begin{vmatrix} \vec{i} & \vec{j} & \vec{k} \\ -a \sin \theta & b \cos \theta & 0 \\ 0 & 0 & 1 \end{vmatrix} = b \cos \theta \vec{i} + a \sin \theta \vec{j}.$$

This vector points away from the z-axis, so we use $d\vec{A} = (b \cos \theta \vec{i} + a \sin \theta \vec{j})\, d\theta dz$, giving

$$\int_S \vec{F} \cdot d\vec{A} = \int_{-c}^{c} \int_0^{2\pi} \left(\frac{b}{a}(a \cos \theta)\vec{i} + \frac{a}{b}(b \sin \theta \vec{j}) \right) \cdot (b \cos \theta \vec{i} + a \sin \theta \vec{j})\, d\theta\, dz$$

$$= \int_{-c}^{c} \int_0^{2\pi} (b^2 \cos^2 \theta + a^2 \sin^2 \theta)\, d\theta dz$$

$$= 2\pi c(a^2 + b^2).$$

17. The surface of S is parameterized by

$$\vec{r}(\theta, \phi) = x\vec{i} + y\vec{j} + z\vec{k},$$

where

$$\begin{cases} x = a + d \sin \phi \cos \theta, \\ y = b + d \sin \phi \sin \theta, \quad \text{for} \quad 0 \le \phi \le \pi, 0 \le \theta \le 2\pi. \\ z = c + d \cos \phi, \end{cases}$$

The vector $\partial \vec{r}/\partial \phi \times \partial \vec{r}/\partial \theta$ points outward by the right-hand rule, so

$$d\vec{A} = \left(\frac{\partial \vec{r}}{\partial \phi} \times \frac{\partial \vec{r}}{\partial \theta} \right) d\phi d\theta.$$

Thus,

$$\vec{F} \cdot d\vec{A} = \vec{F} \cdot \left(\frac{\partial \vec{r}}{\partial \phi} \times \frac{\partial \vec{r}}{\partial \theta} \right) d\phi d\theta$$

$$= \begin{vmatrix} x^2 & y^2 & z^2 \\ \frac{\partial x}{\partial \phi} & \frac{\partial y}{\partial \phi} & \frac{\partial z}{\partial \phi} \\ \frac{\partial x}{\partial \theta} & \frac{\partial y}{\partial \theta} & \frac{\partial z}{\partial \theta} \end{vmatrix} d\phi d\theta$$

$$= \begin{vmatrix} (a + d \sin \phi \cos \theta)^2 & (b + d \sin \phi \sin \theta)^2 & (c + d \cos \phi)^2 \\ d \cos \phi \cos \theta & d \cos \phi \sin \theta & -d \sin \phi \\ -d \sin \phi \sin \theta & d \sin \phi \cos \theta & 0 \end{vmatrix} d\phi d\theta.$$

Hence,

$$
\int_S \vec{F} \cdot d\vec{A} = d^2 \int_0^{2\pi} \int_0^{\pi} (\ a^2 \sin^2 \phi \cos \theta + 2ad \sin^3 \phi \cos^2 \theta + d^2 \sin^4 \phi \cos^3 \theta
$$
$$
+ \ b^2 \sin^2 \phi \sin \theta + 2bd \sin^3 \phi \sin^2 \theta + d^2 \sin^4 \phi \sin^3 \theta
$$
$$
+ \ c^2 \sin \phi \cos \phi + 2cd \sin \phi \cos^2 \phi + d^2 \sin \phi \cos^3 \phi)\ d\phi d\theta.
$$

Since

$$
\int_0^{2\pi} \cos \theta \, d\theta = \int_0^{2\pi} \sin \theta \, d\theta = \int_0^{2\pi} \cos^3 \theta \, d\theta = \int_0^{2\pi} \sin^3 \theta \, d\theta = 0,
$$

and

$$
\int_0^{\pi} \sin \phi \cos \phi \, d\phi = \int_0^{\pi} \sin \phi \cos^3 \phi \, d\phi = 0,
$$

we have

$$
\int_S \vec{F} \cdot \vec{A} \quad d^2 \int_0^{2\pi} \int_0^{\pi} (2ad \sin^3 \phi \cos^2 \theta + 2bd \sin^3 \phi \sin^2 \theta + 2cd \sin \phi \cos^2 \phi)\ d\phi d\theta
$$

$$
= 2\pi d^3 \int_0^{\pi} (a \sin^3 \phi + b \sin^3 \phi + 2c \sin \phi \cos^2 \phi)\ d\phi
$$

$$
= 4\pi d^3 \int_0^{\pi/2} (a \sin^3 \phi + b \sin^3 \phi + 2c \sin \phi \cos^2 \phi)\ d\phi
$$

$$
= \frac{8}{3}\pi d^3 (a + b + c).
$$

21. In terms of the st-parameterization,

$$
d\vec{A} = \frac{\partial \vec{r}}{\partial s} \times \frac{\partial \vec{r}}{\partial t} \, ds \, dt.
$$

By the chain rule, we have

$$
\frac{\partial \vec{r}}{\partial s} = \frac{\partial \vec{r}}{\partial u} \frac{\partial u}{\partial s} + \frac{\partial \vec{r}}{\partial v} \frac{\partial v}{\partial s}
$$
$$
\frac{\partial \vec{r}}{\partial t} = \frac{\partial \vec{r}}{\partial u} \frac{\partial u}{\partial t} + \frac{\partial \vec{r}}{\partial v} \frac{\partial v}{\partial t}.
$$

So taking the cross product, we get

$$
\frac{\partial \vec{r}}{\partial s} \times \frac{\partial \vec{r}}{\partial t} = \left(\frac{\partial \vec{r}}{\partial u} \frac{\partial u}{\partial s} + \frac{\partial \vec{r}}{\partial v} \frac{\partial v}{\partial s} \right) \times \left(\frac{\partial \vec{r}}{\partial u} \frac{\partial u}{\partial t} + \frac{\partial \vec{r}}{\partial v} \frac{\partial v}{\partial t} \right)
$$
$$
= \left(\frac{\partial u}{\partial s} \frac{\partial v}{\partial t} - \frac{\partial u}{\partial t} \frac{\partial v}{\partial s} \right) \frac{\partial \vec{r}}{\partial u} \times \frac{\partial \vec{r}}{\partial v}.
$$

Now suppose we are going to change variables in a double integral from uv-coordinates to st-coordinates. The Jacobian is

$$
\frac{\partial(u, v)}{\partial(s, t)} = \begin{vmatrix} \frac{\partial u}{\partial s} & \frac{\partial v}{\partial s} \\ \frac{\partial u}{\partial t} & \frac{\partial v}{\partial t} \end{vmatrix} = \frac{\partial u}{\partial s} \frac{\partial v}{\partial t} - \frac{\partial u}{\partial t} \frac{\partial v}{\partial s}.
$$

Since the Jacobian is assumed to be positive, converting from a uv-integral to an st-integral gives:

$$
\int_T \vec{F} \cdot \frac{\partial \vec{r}}{\partial u} \times \frac{\partial \vec{r}}{\partial v} \, du dv = \int_R \vec{F} \cdot \frac{\partial \vec{r}}{\partial u} \times \frac{\partial \vec{r}}{\partial v} \frac{\partial(u, v)}{\partial(s, t)} \, ds dt
$$
$$
= \int_R \vec{F} \cdot \frac{\partial \vec{r}}{\partial u} \times \frac{\partial \vec{r}}{\partial v} \left(\frac{\partial u}{\partial s} \frac{\partial v}{\partial t} - \frac{\partial u}{\partial t} \frac{\partial v}{\partial s} \right) \, ds dt.
$$

However, we know that this gives us

$$
\int_T \vec{F} \cdot \frac{\partial \vec{r}}{\partial u} \times \frac{\partial \vec{r}}{\partial v} \, du \, dv = \int_R \vec{F} \cdot \frac{\partial \vec{r}}{\partial u} \times \frac{\partial \vec{r}}{\partial v} \left(\frac{\partial u}{\partial s} \frac{\partial v}{\partial t} - \frac{\partial u}{\partial t} \frac{\partial v}{\partial s} \right) \, ds dt = \int_R \vec{F} \cdot \frac{\partial \vec{r}}{\partial s} \times \frac{\partial \vec{r}}{\partial t} \, ds dt.
$$

Thus, the flux integral in uv-coordinates equals the flux integral in st-coordinates.

Solutions for Chapter 19 Review

Exercises

1. Scalar. Since the surface is closed and the vector field is constant, the flux in one side equals the flux out on the other side, so the net flux through the surface is 0.

5. The only contribution to the flux is from the face $z = 3$, since the vector field is zero or parallel to the other faces. On this face, $\vec{H} = 3x\vec{k}$. The vector field is everywhere perpendicular to the face $z = 3$ but varies in magnitude from point to point. On this surface, $d\vec{A} = \vec{k}\ dx\,dy$. Thus

$$\text{Flux} = \int_0^2 \int_0^1 3x\vec{k}\cdot\vec{k}\ dx\,dy = \int_0^2 \int_0^1 3x\,dx\,dy = \left.\frac{3x^2}{2}\right|_0^1 \cdot \left.y\right|_0^2 = 3.$$

9. Only the $\vec{i}$-component contributes to the flux, so

$$\text{Flux} = 2\pi \cdot \text{Area of surface} = 2\pi(\pi 3^2) = 18\pi^2.$$

13. Since the vector field is everywhere perpendicular to the surface of the sphere, and $\|\vec{F}\| = \pi$ on the surface, we have

$$\int_S \vec{F}\cdot d\vec{A} = \|\vec{F}\| \cdot \text{Area of sphere} = \pi \cdot 4\pi(\pi)^2 = 4\pi^4.$$

17. The area vector of the surface is $-49\vec{j}$, so

$$\int_S \vec{G}\cdot d\vec{A} = (2\vec{j} + 3\vec{k})\cdot(-49\vec{j}) = -98.$$

21. Only the $\vec{k}$ component contributes to the flux. In the plane $z = 4$, we have $\vec{F} = 2\vec{i} + 3\vec{j} + 4\vec{k}$. On the square $d\vec{A} = \vec{k}\ dA$, so we have

$$\text{Flux} = \int \vec{F}\cdot d\vec{A} = 4\vec{k}\cdot(\vec{k}\ \text{Area of square}) = 4(5^2) = 100.$$

25. See Figure 19.2. The vector field is a vortex going around the z-axis, and the square is centered on the x-axis, so the flux going across one half of the square is balanced by the flux coming back across the other half. Thus, the net flux is zero, so

$$\int_S \vec{F}\cdot d\vec{A} = 0.$$

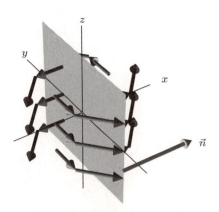

Figure 19.2

29. On the sphere of radius 2, the vector field has $||\vec{F}|| = 10$ and points inward everywhere (opposite to the orientation of the surface). So

$$\text{Flux} = \int_S \vec{F} \cdot d\vec{A} = -||\vec{F}|| \cdot \text{Area of sphere} = -10 \cdot 4\pi 2^2 = -160\,\pi.$$

33. On the surface S, y is constant, $y = -1$, and $d\vec{A} = -\vec{j}\,dA$, so,

$$\int_S \vec{F} \cdot d\vec{A} = \int_S (x^2\vec{i} + (x + e^{-1})\vec{j} - \vec{k}) \cdot (-\vec{j})\,dA = -\int_S (x + e^{-1})\,dA$$

$$= -\int_0^4 \int_0^2 (x + e^{-1})\,dx\,dz = -4(2 + 2e^{-1}) = -8(1 + e^{-1}).$$

37. First we have

$$z_x = \frac{x}{\sqrt{x^2 + y^2}} \qquad z_y = \frac{y}{\sqrt{x^2 + y^2}}.$$

Although z is not a smooth function of x and y at $(0,0)$, the improper integral that we get converges:

$$\int_S \vec{F} \cdot d\vec{A} = \int_S (x^2\vec{i} + y^2\vec{j} + \sqrt{x^2 + y^2}\,\vec{k}) \cdot (-\frac{x}{\sqrt{x^2 + y^2}}\vec{i} - \frac{y}{\sqrt{x^2 + y^2}}\vec{j} + \vec{k})\,dA$$

$$= \int_S \left(-\frac{x^3 + y^3}{\sqrt{x^2 + y^2}} + \sqrt{x^2 + y^2}\right)\,dA$$

Changing to polar coordinates we have

$$\int_S \vec{F} \cdot d\vec{A} = \int_0^{\pi/2} \int_0^1 (-r^2\cos^3\theta - r^2\sin^3\theta + r)r\,dr\,d\theta$$

$$= \int_0^{\pi/2} \left(-\frac{r^4}{4}(\cos^3\theta + \sin^3\theta) + \frac{1}{3}r^3\Big|_{r=0}^{r=1}\right)\,d\theta$$

$$= \int_0^{\pi/2} \left(-\frac{1}{4}(\cos^3\theta + \sin^3\theta) + \frac{1}{3}\right)\,d\theta$$

$$= \int_0^{\pi/2} \left(-\frac{1}{4}(\cos\theta - \cos\theta\sin^2\theta + \sin\theta - \sin\theta\cos^2\theta) + \frac{1}{3}\right)\,d\theta$$

$$= -\frac{1}{4}(\sin\theta - \frac{1}{3}\sin^3\theta - \cos\theta + \frac{1}{3}\cos^3\theta) + \frac{\theta}{3}\Big|_0^{\pi/2}$$

$$= \frac{\pi}{6} - \frac{1}{3}.$$

Problems

41. (a) We have grad $f = (y + yze^{xyz})\vec{i} + (x + xze^{xyz})\vec{j} + xye^{xyz}\vec{k}$.

(b) By the Fundamental Theorem of Line Integrals, we have

$$\int_C \text{grad } f \cdot d\vec{r} = (xy + e^{xyz})\Big|_{(1,1,1)}^{(2,3,4)} = (2 \cdot 3 + e^{2 \cdot 3 \cdot 4}) - (1 \cdot 1 + e^{1 \cdot 1 \cdot 1}) = 5 + e^{24} - e^1.$$

(c) Only the k-component of grad f contributes to the flux integral. On the xy-plane, $d\vec{A} = \vec{k}\,dx\,dy$ and $z = 0$, so

$$\int_S \text{grad } f \cdot d\vec{A} = \int_0^2 \int_0^{\sqrt{4-x^2}} xye^0\,dy\,dx = \int_0^2 \frac{xy^2}{2}\Big|_0^{\sqrt{4-x^2}} = \int_0^2 \frac{x}{2}(4 - x^2)\,dx = \left(x^2 - \frac{x^4}{8}\right)\Big|_0^2 = 2.$$

45. The vector field $\vec{D}$ has constant magnitude on S, equal to $Q/4\pi R^2$, and points radially outward, so

$$\int_S \vec{D} \cdot d\vec{A} = \frac{Q}{4\pi R^2} \cdot 4\pi R^2 = Q.$$

49. (a) Consider two opposite faces of the cube, S_1 and S_2. The corresponding area vectors are $\vec{A}_1 = 4\vec{i}$ and $\vec{A}_2 = -4\vec{i}$ (since the side of the cube has length 2). Since $\vec{E}$ is constant, we find the flux by taking the dot product, giving

$$\text{Flux through } S_1 = \vec{E} \cdot \vec{A}_1 = (a\vec{i} + b\vec{j} + c\vec{k}) \cdot 4\vec{i} = 4a.$$

$$\text{Flux through } S_2 = \vec{E} \cdot \vec{A}_2 = (a\vec{i} + b\vec{j} + c\vec{k}) \cdot (-4\vec{i}) = -4a.$$

Thus the fluxes through S_1 and S_2 cancel. Arguing similarly, we conclude that, for any pair of opposite faces, the sum of the fluxes of $\vec{E}$ through these faces is zero. Hence, by addition, $\int_S \vec{E} \cdot d\vec{A} = 0$.

(b) The basic idea is the same as in part (a), except that we now need to use Riemann sums. First divide S into two hemispheres H_1 and H_2 by the equator C located in a plane perpendicular to $\vec{E}$. For a tiny patch S_1 in the hemisphere H_1, consider the patch S_2 in the opposite hemisphere which is symmetric to S_1 with respect to the center O of the sphere. The area vectors $\Delta\vec{A}_1$ and $\Delta\vec{A}_2$ satisfy $\Delta\vec{A}_2 = -\Delta\vec{A}_1$, so if we consider S_1 and S_2 to be approximately flat, then $\vec{E} \cdot \Delta\vec{A}_1 = -\vec{E} \cdot \Delta\vec{A}_2$. By decomposing H_1 and H_2 into small patches as above and using Riemann sums, we get

$$\int_{H_1} \vec{E} \cdot d\vec{A} = -\int_{H_2} \vec{E} \cdot d\vec{A}, \quad \text{so} \quad \int_S \vec{E} \cdot d\vec{A} = 0.$$

(c) The reasoning in part (b) can be used to prove that the flux of $\vec{E}$ through any surface with a center of symmetry is zero. For instance, in the case of the cylinder, cut it in half with a plane $z = 1$ and denote the two halves by H_1 and H_2. Just as before, take patches in H_1 and H_2 with $\Delta A_1 = -\Delta A_2$, so that $\vec{E} \cdot \Delta A_1 = -\vec{E} \cdot \Delta\vec{A}_2$. Thus, we get

$$\int_{H_1} \vec{E} \cdot d\vec{A} = -\int_{H_2} \vec{E} \cdot d\vec{A},$$

which shows that

$$\int_S \vec{E} \cdot d\vec{A} = 0.$$

CAS Challenge Problems

53. (a) When $x > 0$, the vector $x\vec{i}$ points in the positive x-direction, and when $x < 0$ it points in the negative x-direction. Thus it always points from the inside of the ellipsoid to the outside, so we expect the flux integral to be positive. The upper half of the ellipsoid is the graph of $z = f(x,y) = \frac{1}{\sqrt{2}}(1 - x^2 - y^2)$, so the flux integral is

$$\int_S \vec{F} \cdot d\vec{A} = \int_{-1/2}^{1/2} \int_{-1/2}^{1/2} x\vec{i} \cdot (-f_x\vec{i} - f_y\vec{j} + \vec{k}) \, dx \, dy$$

$$= \int_{-1/2}^{1/2} \int_{-1/2}^{1/2} (-xf_x) \, dx \, dy = \int_{-1/2}^{1/2} \int_{-1/2}^{1/2} \frac{x^2}{\sqrt{1 - x^2 - y^2}} \, dx \, dy$$

$$= \frac{-\sqrt{2} + 11 \arcsin(\frac{1}{\sqrt{3}}) + 10 \arctan(\frac{1}{\sqrt{2}}) - 8 \arctan(\frac{5}{\sqrt{2}})}{12} = 0.0958.$$

Different CASs may give the answer in different forms. Note that we could have predicted the integral was positive without evaluating it, since the integrand is positive everywhere in the region of integration.

(b) For $x > -1$, the quantity $x + 1$ is positive, so the vector field $(x + 1)\vec{i}$ always points in the direction of the positive x-axis. It is pointing into the ellipsoid when $x < 0$ and out of it when $x > 0$. However, its magnitude is smaller when $-1/2 < x < 0$ than it is when $0 < x < 1/2$, so the net flux out of the ellipsoid should be positive. The flux integral is

$$\int_S \vec{F} \cdot d\vec{A} = \int_{-1/2}^{1/2} \int_{-1/2}^{1/2} (x + 1)\vec{i} \cdot (-f_x\vec{i} - f_y\vec{j} + \vec{k}) \, dx \, dy$$

$$= \int_{-1/2}^{1/2} \int_{-1/2}^{1/2} -(x + 1)f_x \, dx \, dy = \int_{-1/2}^{1/2} \int_{-1/2}^{1/2} \frac{x(1 + x)}{\sqrt{1 - x^2 - y^2}} \, dx \, dy$$

$$= \frac{\sqrt{2} - 11 \ \arcsin(\frac{1}{\sqrt{3}}) - 10 \ \arctan(\frac{1}{\sqrt{2}}) + 8 \ \arctan(\frac{5}{\sqrt{2}})}{12} = 0.0958$$

The answer is the same as in part (a). This makes sense because the difference between the integrals in parts (a) and (b) is the integral of $\int_{-1/2}^{1/2} \int_{-1/2}^{1/2} (x/\sqrt{1 - x^2 - y^2}) \, dxdy$, which is zero because the integrand is odd with respect to x.

(c) This integral should be positive for the same reason as in part (a). The vector field $y\vec{j}$ points in the positive y-direction when $y > 0$ and in the negative y-direction when $y < 0$, thus it always points out of the ellipsoid. Evaluating the integral we get

$$\int_S \vec{F} \cdot d\vec{A} = \int_{-1/2}^{1/2} \int_{-1/2}^{1/2} y\vec{j} \cdot (-f_x\vec{i} - f_y\vec{j} + \vec{k}) \, dxdy$$

$$= \int_{-1/2}^{1/2} \int_{-1/2}^{1/2} (-yf_y) \, dxdy = \int_{-1/2}^{1/2} \int_{-1/2}^{1/2} \frac{y^2}{\sqrt{1 - x^2 - y^2}} \, dxdy$$

$$= \frac{\sqrt{2} - 2 \ \arcsin(\frac{1}{\sqrt{3}}) - 19 \ \arctan(\frac{1}{\sqrt{2}}) + 8 \ \arctan(\frac{5}{\sqrt{2}})}{12} = 0.0958.$$

The symbolic answer appears different but has the same numerical value as in parts (a) and (b). In fact the answer is the same because the integral here is the same as in part (a) except that the roles of x and y have been exchanged. Different CASs may give different symbolic forms.

CHECK YOUR UNDERSTANDING

1. True. By definition, the flux integral is the limit of a sum of dot products, hence is a scalar.

5. True. The flow of this field is in the same direction as the orientation of the surface everywhere on the surface, so the flux is positive.

9. True. In the sum defining the flux integral for $\vec{F}$, we have terms like $\vec{F} \cdot \Delta\vec{A} = (2\vec{G}) \cdot \Delta\vec{A} = 2(\vec{G} \cdot \Delta\vec{A})$. So each term in the sum approximating the flux of $\vec{F}$ is twice the corresponding term in the sum approximating the flux of $\vec{G}$, making the sum for $\vec{F}$ twice that of the sum for $\vec{G}$. Thus the flux of $\vec{F}$ is twice the flux of $\vec{G}$.

13. False. Both surfaces are oriented upward, so $\vec{A}(x, y)$ and $\vec{B}(x, y)$ both point upward. But they could point in different directions, since the graph of $z = -f(x, y)$ is the graph of $z = f(x, y)$ turned upside down.

CHAPTER TWENTY

Solutions for Section 20.1

Exercises

1. Scalar. Since

$$\text{div}\left(\frac{y\vec{i} - x\vec{j}}{x^2 + y^2}\right) = \frac{-y \cdot 2x}{(x^2 + y^2)^2} + \frac{x \cdot 2y}{(x^2 + y^2)^2} = 0.$$

5. $\text{div}\,\vec{F} = \dfrac{\partial}{\partial x}(-x + y) + \dfrac{\partial}{\partial y}(y + z) + \dfrac{\partial}{\partial z}(-z + x) = -1 + 1 - 1 = -1$

9. Using the formula for $\vec{a} \times \vec{r}$ in Cartesian coordinates, we get

$$\text{div}\,\vec{F} = \frac{\partial}{\partial x}(a_2 z - a_3 y) + \frac{\partial}{\partial y}(a_3 x - a_1 z) + \frac{\partial}{\partial z}(a_1 y - a_2 x) = 0$$

13. Two vector fields that have positive divergence everywhere are in Figures 20.1 and 20.2.

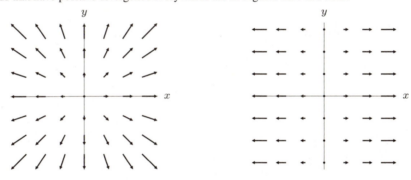

Figure 20.1   Figure 20.2

Problems

17. Since $\text{div}\,F(1, 2, 3)$ is the flux density out of a small region surrounding the point $(1, 2, 3)$, we have

$$\text{div}\,\vec{F}(1, 2, 3) \approx \frac{\text{Flux out of small region around } (1, 2, 3)}{\text{Volume of region.}}$$

So

$$\text{Flux out of region} \approx (\text{div}\,\vec{F}(1, 2, 3)) \cdot \text{Volume of region}$$

$$= 5 \cdot \frac{4}{3}\pi(0.01)^3$$

$$= \frac{0.00002\pi}{3}.$$

21. Using flux: On S_1, $x = a$ and normal is in negative x-direction, so

$$\vec{F} \cdot \Delta\vec{A} = ((3a + 2)\vec{i} + 4a\vec{j} + (5a + 1)\vec{k}) \cdot (-\Delta A\,\vec{i}) = -(3a + 2)\Delta A$$

Thus

$$\int_{S_1} \vec{F} \cdot d\vec{A} = \int_{S_1} -(3a + 2)dA = -(3a + 2)(\text{Area of } S_1) = -(3a + 2)w^2.$$

On S_2, $x = a + w$ and normal is in the positive x-direction, so

$$\vec{F} \cdot \Delta\vec{A} = [(3(a + w) + 2)\,\vec{i} + 4(a + w)\vec{j} + (5(a + w) + 1)\,\vec{k}] \cdot (\Delta A\,\vec{i}) = (3a + 3w + 2)\Delta A.$$

Thus

$$\int_{S_2} \vec{F} \cdot d\vec{A} = \int_{S_2} (3a + 3w + 2)dA = (3a + 3w + 2)(\text{Area of } S_2) = (3a + 3w + 2)w^2.$$

Next, we have $\int_{S_3} \vec{F} \cdot d\vec{A} = \int_{S_3} -4x dA$ and $\int_{S_4} \vec{F} \cdot d\vec{A} = \int_{S_4} 4x dA$. Since these two are integrated over the same region in the xz-plane, the two integrals cancel. Similarly, $\int_{S_5} \vec{F} \cdot d\vec{A} = \int_{S_5} -(5x + 1)d\vec{A}$ cancels out $\int_{S_6} \vec{F} \cdot d\vec{A} = \int_{S_6} (5x + 1)d\vec{A}$. Therefore,

Total flux

$$= \int_{S_1} \vec{F} \cdot d\vec{A} + \int_{S_2} \vec{F} \cdot d\vec{A} + \int_{S_3} \vec{F} \cdot d\vec{A} + \int_{S_4} \vec{F} \cdot d\vec{A} + \int_{S_5} \vec{F} \cdot d\vec{A} + \int_{S_6} \vec{F} \cdot d\vec{A}$$

$$= -(3a + 2)w^2 + (3a + 3w + 2)w^2 + \int_{S_3} -4x dA + \int_{S_4} 4x dA$$

$$+ \int_{S_5} -(5x + 1)dA + \int_{S_6} (5x + 1)dA = 3w^3.$$

To find div $\vec{F}$ at the point (a, b, c), let the box shrink to the point by letting $w \to 0$. Then

$$\text{div} \, \vec{F} = \lim_{w \to 0} \left(\frac{\text{Flux through box}}{\text{Volume of box}} \right)$$

$$= \lim_{w \to 0} \left(\frac{3w^3}{w^3} \right) = 3.$$

Using partial derivatives:

$$\text{div} \, \vec{F} = \frac{\partial}{\partial x}(3x + 2) + \frac{\partial}{\partial y}(4x) + \frac{\partial}{\partial z}(5x + 1) = 3$$

25. (a) div $\vec{B} = \frac{\partial}{\partial x}(-y) + \frac{\partial}{\partial y}(x) + \frac{\partial}{\partial z}(x + y) = 0$, so this could be a magnetic field.

(b) div $\vec{B} = \frac{\partial}{\partial x}(-z) + \frac{\partial}{\partial y}(y) + \frac{\partial}{\partial z}(x) = 0 + 1 + 0 = 1$, so this could not be a magnetic field.

(c) div $\vec{B} = \frac{\partial}{\partial x}(x^2 - y^2 - x) + \frac{\partial}{\partial y}(y - 2xy) + \frac{\partial}{\partial z}(0) = 2x - 1 + 1 - 2x + 0 = 0$, so this could be a magnetic field.

29. Let $\vec{F} = F_1 \vec{i} + F_2 \vec{j} + F_3 \vec{k}$. Then

$$\text{div}(g\vec{F}) = \text{div}(gF_1 \vec{i} + gF_2 \vec{j} + gF_3 \vec{k})$$

$$= \frac{\partial}{\partial x}(gF_1) + \frac{\partial}{\partial y}(gF_2) + \frac{\partial}{\partial z}(gF_3)$$

$$= \frac{\partial g}{\partial x}F_1 + g\frac{\partial F_1}{\partial x} + \frac{\partial g}{\partial y}F_2 + g\frac{\partial F_2}{\partial y} + \frac{\partial g}{\partial z}F_3 + g\frac{\partial F_3}{\partial z}$$

$$= \frac{\partial g}{\partial x}F_1 + \frac{\partial g}{\partial y}F_2 + \frac{\partial g}{\partial z}F_3 + g\left(\frac{\partial F_1}{\partial x} + \frac{\partial F_2}{\partial y} + \frac{\partial F_3}{\partial z} \right)$$

$$= (\text{grad} \, g) \cdot \vec{F} + g \, \text{div} \, \vec{F}.$$

33. Using $\text{div}(g\vec{F}) = (\text{grad} \, g) \cdot \vec{F} + g \, \text{div} \, \vec{F}$, we have

$$\text{div} \, \vec{G} = \text{grad}(\vec{b} \cdot \vec{r}) \cdot (\vec{a} \times \vec{r}) + \vec{b} \cdot \vec{r} \, \text{div}(\vec{a} \times \vec{r}) = \vec{b} \cdot (\vec{a} \times \vec{r}) + \vec{v} \cdot \vec{r} 0 = \vec{b} \cdot (\vec{a} \times \vec{r}).$$

37. (a) The velocity vector for the traffic flow would look like:

(b) When $0 \le x < 2000$, the velocity is decreasing linearly from 55 to 15, so its formula is $(55 - x/50)\vec{i}$ mph. Then, when $2000 \le x < 7000$, the speed is constant, so $\vec{v}(x) = 15\vec{i}$ mph. Next, when $7000 \le x < 8000$, the velocity is increasing linearly from 15 to 55, so $\vec{v}(x) = (15 + (x - 7000)/25)\vec{i}$ mph. Finally, when $x \ge 8000$, the speed is constant, so $\vec{v}(x) = 55\vec{i}$ mph.

(c) div $\vec{v} = dv(x)/dx$.

   At $x = 1000$, $v(x) = 55 - x/50$, so div $\vec{v} = -1/50$.

   At $x = 5000$, $v(x) = 15$, so div $\vec{v} = 0$.

   At $x = 7500$, $v(x) = 15 + (x - 7000)/25$, so div $\vec{v} = 1/25$.

   At $x = 10,000$, $v(x) = 55$, so div $\vec{v} = 0$.

   In each case the units of div $\vec{v}$ are $\dfrac{\text{miles/hour}}{\text{feet}}$.

41. (a) At any point $\vec{r} = x\vec{i} + y\vec{j}$, the direction of the vector field $\vec{v}$ is pointing toward the origin, which means it is of the form $\vec{v} = f\vec{r}$ for some negative function f whose value can vary depending on $\vec{r}$. The magnitude of $\vec{v}$ depends only on the distance r, thus f must be a function depending only on r, which is equivalent to depending only on r^2 since $r \geq 0$. So $\vec{v} = f(r^2)\vec{r} = \left(f(x^2 + y^2)\right)(x\vec{i} + y\vec{j})$.

(b) At $(x, y) \neq (0, 0)$ the divergence of $\vec{v}$ is

$$\text{div } \vec{v} = \frac{\partial(K(x^2 + y^2)^{-1}x)}{\partial x} + \frac{\partial(K(x^2 + y^2)^{-1}y)}{\partial y} = \frac{Ky^2 - Kx^2}{(x^2 + y^2)^2} + \frac{Kx^2 - Ky^2}{(x^2 + y^2)^2} = 0.$$

Therefore, $\vec{v}$ is a point sink at the origin.

(c) The magnitude of $\vec{v}$ is

$$\|\vec{v}\| = |K|(x^2 + y^2)^{-1}|x\vec{i} + y\vec{j}| = |K|(x^2 + y^2)^{-1}(x^2 + y^2)^{1/2} = |K|(x^2 + y^2)^{-1/2} = \frac{|K|}{r}.$$

(remember, $K < 0$)

(d) The vector field looks like the following:

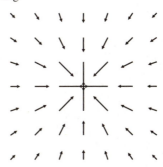

Figure 20.3

(e) We need to show that grad $\phi = \vec{v}$.

$$\text{grad } \phi = \frac{\partial}{\partial x}(\frac{K}{2} \log(x^2 + y^2))\vec{i} + \frac{\partial}{\partial y}(\frac{K}{2} \log(x^2 + y^2))\vec{j}$$

$$= \frac{Kx}{x^2 + y^2}\vec{i} + \frac{Ky}{x^2 + y^2}\vec{j}$$

$$= K(x^2 + y^2)^{-1}(x\vec{i} + y\vec{j})$$

$$= \vec{v}$$

Solutions for Section 20.2

Exercises

1. First directly: On the faces $x = 0$, $y = 0$, $z = 0$, the flux is zero. On the face $x = 2$, a unit normal is $\vec{i}$ and $d\vec{A} = dA\vec{i}$. So

$$\int_{S_{x=2}} \vec{r} \cdot d\vec{A} = \int_{S_{x=2}} (2\vec{i} + y\vec{j} + z\vec{k}) \cdot (dA\vec{i})$$

(since on that face, $x = 2$)

$$= \int_{S_{x=2}} 2dA = 2 \cdot (\text{Area of face}) = 2 \cdot 4 = 8.$$

In exactly the same way, you get

$$\int_{S_{y=2}} \vec{r} \cdot d\vec{A} = \int_{S_{z=2}} \vec{r} \cdot d\vec{A} = 8,$$

so

$$\int_S \vec{r} \cdot d\vec{A} = 3 \cdot 8 = 24.$$

Now using divergence:

$$\operatorname{div} \vec{F} = \frac{\partial x}{\partial x} + \frac{\partial y}{\partial y} + \frac{\partial z}{\partial z} = 3,$$

so

$$\text{Flux} = \int_0^2 \int_0^2 \int_0^2 3 \, dx \, dy \, dz = 3 \cdot (\text{Volume of Cube}) = 3 \cdot 8 = 24$$

5. The location of the pyramid has not been completely specified. For instance, where is it centered on the xy plane? How is base oriented with respect to the axes? Thus, we cannot compute the flux by direct integration with the information we have. However, we can calculate it using the divergence theorem. First we calculate the divergence of $\vec{F}$.

$$\operatorname{div} \vec{F} = \frac{\partial(-z)}{\partial x} + \frac{\partial 0}{\partial y} + \frac{\partial x}{\partial z} = 0 + 0 + 0 = 0$$

Thus for any closed surface the flux will be zero, so the flux through our pyramid, regardless of its location or orientation, is zero.

9. Since $\operatorname{div} \vec{G} = 1$, if W is the interior of the box, the Divergence Theorem gives

$$\text{Flux} = \int_W 1 \, dV = 1 \cdot \text{ Volume of box } = 1 \cdot 2 \cdot 3 \cdot 4 = 24.$$

13. We have

$$\operatorname{div}((3x + 4y)\vec{i} + (4y + 5z)\vec{j} + (5z + 3x)\vec{k}) = 3 + 4 + 5 = 12.$$

Let W be the interior of the cube. Then by the divergence theorem,

$$\int_S ((3x + 4y)\vec{i} + (4y + 5z)\vec{j} + (5z + 3x)\vec{k}) \cdot d\vec{A} = \int_W 12 \, dV = 12 \cdot \text{Volume of cube} = 12 \cdot (2 \cdot 3 \cdot 4) = 288.$$

Problems

17. Since $\operatorname{div} \vec{F} = 3x^2 + 3y^2 + 3z^2$, the Divergence Theorem gives

$$\text{Flux} = \int_S \vec{F} \cdot d\vec{A} = \int_W (3x^2 + 3y^2 + 3z^2) \, dV.$$

In spherical coordinates, the region W lies between the spheres $\rho = 2$ and $\rho = 3$ and inside the cone $\phi = \pi/4$. Since $3x^2 + 3y^2 + 3z^2 = 3\rho^2$, we have

$$\text{Flux} = \int_S \vec{F} \cdot d\vec{A} = \int_0^{2\pi} \int_0^{\pi/4} \int_2^3 3\rho^2 \cdot \rho^2 \sin\phi \, d\rho \, d\phi \, d\theta$$

$$= 2\pi \cdot \frac{3}{5}\rho^5 \Big|_2^3 (-\cos\phi) \Big|_0^{\pi/4} = \frac{633(2 - \sqrt{2})}{5} \pi = 232.98.$$

21. By the Divergence Theorem, $\int_S \vec{F} \cdot d\vec{A} = \int_W \operatorname{div}\vec{F} \, dV = \int_W 0 \, dV = 0$ for a closed surface S, where W is the region enclosed by S.

25. We consider a sphere of radius R centered at the origin and compute the flux of $\vec{r}$ through its surface. Since $\vec{r}$ and $d\vec{A}$ both point radially outward, $\vec{r} \cdot d\vec{A} = \|\vec{r}\| \|d\vec{A}\| = RdA$ on the surface of the sphere, so

$$\int_S \vec{F} \cdot d\vec{A} = \int_S RdA = R \int_S dA = R(\text{Surface area of a sphere}) = R(4\pi R^2) = 4\pi R^3.$$

Therefore, volume of sphere $= \frac{1}{3} \int_S \vec{F} \cdot d\vec{A} = \frac{4}{3}\pi R^3$.

29. (a) Since $\vec{F}$ is radial, it is everywhere parallel to the area vector, $\Delta \vec{A}$. Also, $\|\vec{F}\| = 1$ on the surface of the sphere $x^2 + y^2 + z^2 = 1$, so

$$\text{Flux through the sphere} = \int_S \vec{F} \cdot d\vec{A} = \lim_{\|\Delta \vec{A}\| \to 0} \sum \vec{F} \cdot \Delta \vec{A}$$

$$= \lim_{\|\Delta \vec{A}\| \to 0} \sum \|\vec{F}\| \|\Delta \vec{A}\| = \lim_{\|\Delta \vec{A}\| \to 0} \sum \|\Delta \vec{A}\|$$

$$= \text{Surface area of sphere} = 4\pi \cdot 1^2 = 4\pi.$$

(b) In Cartesian coordinates,

$$\vec{F}(x, y, z) = \frac{x}{(x^2 + y^2 + z^2)^{3/2}} \vec{i} + \frac{y}{(x^2 + y^2 + z^2)^{3/2}} \vec{j} + \frac{z}{(x^2 + y^2 + z^2)^{3/2}} \vec{k}.$$

So,

$$\text{div} \vec{F}(x, y, z) = \left(\frac{1}{(x^2 + y^2 + z^2)^{3/2}} - \frac{3x^2}{(x^2 + y^2 + z^2)^{5/2}} \right)$$

$$+ \left(\frac{1}{(x^2 + y^2 + z^2)^{3/2}} - \frac{3y^2}{(x^2 + y^2 + z^2)^{5/2}} \right)$$

$$+ \left(\frac{1}{(x^2 + y^2 + z^2)^{3/2}} - \frac{3z^2}{(x^2 + y^2 + z^2)^{5/2}} \right)$$

$$= \left(\frac{x^2 + y^2 + z^2}{(x^2 + y^2 + z^2)^{5/2}} - \frac{3x^2}{(x^2 + y^2 + z^2)^{5/2}} \right)$$

$$+ \left(\frac{x^2 + y^2 + z^2}{(x^2 + y^2 + z^2)^{5/2}} - \frac{3y^2}{(x^2 + y^2 + z^2)^{5/2}} \right)$$

$$+ \left(\frac{x^2 + y^2 + z^2}{(x^2 + y^2 + z^2)^{5/2}} - \frac{3z^2}{(x^2 + y^2 + z^2)^{5/2}} \right)$$

$$= \frac{3(x^2 + y^2 + z^2) - 3(x^2 + y^2 + z^2)}{(x^2 + y^2 + z^2)^{5/2}}$$

$$= 0.$$

(c) We cannot apply the Divergence Theorem to the whole region within the box, because the vector field $\vec{F}$ is not defined at the origin. However, we can apply the Divergence Theorem to the region, W, between the sphere and the box. Since $\text{div} \vec{F} = 0$ there, the theorem tells us that

$$\int_{\substack{\text{Box} \\ \text{(outward)}}} \vec{F} \cdot d\vec{A} + \int_{\substack{\text{Sphere} \\ \text{(inward)}}} \vec{F} \cdot d\vec{A} = \int_W \text{div} \vec{F} \, dV = 0.$$

Therefore, the flux through the box and the sphere are equal if both are oriented outward:

$$\int_{\substack{\text{Box} \\ \text{(outward)}}} \vec{F} \cdot d\vec{A} = - \int_{\substack{\text{Sphere} \\ \text{(inward)}}} \vec{F} \cdot d\vec{A} = \int_{\substack{\text{Sphere} \\ \text{(outward)}}} \vec{F} \cdot d\vec{A} = 4\pi.$$

33. (a) At the point $(1, 2, 1)$, we have $\text{div} \vec{F} = 1 \cdot 2 \cdot 1^2 = 2$.

(b) Since the box is small, we use the approximation

$$\text{div} \vec{F} = \text{Flux density} \approx \frac{\text{Flux out of box}}{\text{Volume of box}}.$$

Thus

$$\text{Flux out of box} \approx (\text{div} \vec{F}) \cdot (\text{Volume of box}) = 2(0.2)^3 = 0.016.$$

(c) To calculate the flux exactly, we use the Divergence Theorem,

$$\text{Flux out of box} = \int_{\text{Box}} \text{div} \vec{F} \, dV = \int_{\text{Box}} xyz^2 \, dV.$$

Since the box has side 0.2, it is given by $0.9 < x < 1.1$, $1.9 < y < 2.1$, $0.9 < z < 1.1$, so

$$\text{Flux} = \int_{0.9}^{1.1} \int_{1.9}^{2.1} \int_{0.9}^{1.1} xyz^2 \, dz dy dx = \int_{0.9}^{1.1} \int_{1.9}^{2.1} xy \left. \frac{z^3}{3} \right|_{0.9}^{1.1} dy dx$$

$$= \frac{(1.1)^3 - (0.9)^3}{3} \int_{0.9}^{1.1} \left. \frac{xy^2}{2} \right|_{1.9}^{2.1} dx = \frac{(1.1)^3 - (0.9)^3}{3} \cdot \frac{(2.1)^2 - (1.9)^2}{2} \cdot \left. \frac{x^2}{2} \right|_{0.9}^{1.1}$$

$$= \frac{(1.1)^3 - (0.9)^3}{3} \cdot \frac{(2.1)^2 - (1.9)^2}{2} \cdot \frac{(1.1)^2 - (0.9)^2}{2} = 0.016053 \ldots .$$

Notice that you can calculate the flux without knowing the vector field, $\vec{F}$.

37. (a) The rate at which heat is generated at any point in the earth is div $\vec{F}$ at that point. So div $\vec{F} = 30$ watts/km^3.
   (b) Differentiating gives $\text{div}(\alpha(x\vec{i} + y\vec{j} + z\vec{k})) = \alpha(1 + 1 + 1) = 3\alpha$ so $\alpha = 30/3 = 10$ watts/km^3. Thus, $\vec{F} = \alpha\vec{r}$
      has constant divergence. Note that $\vec{F} = \alpha\vec{r}$ has flow lines going radially outward, and symmetric about the origin.
   (c) The vector grad T gives the direction of greatest increase in temperature. Thus, $-\text{grad } T$ gives the direction of
      greatest decrease in temperature. The equation $\vec{F} = -k \, \text{grad } T$ says that heat will flow in the direction of greatest
      decrease in temperature (i.e. from hot regions to cold), and at a rate proportional to the temperature gradient.
   (d) We assume that $\vec{F}$ is given by the answer to part (b). Then, using part (c), we have

$$\vec{F} = 10(x\vec{i} + y\vec{j} + z\vec{k}) = -30{,}000 \, \text{grad } T,$$

so

$$\text{grad } T = -\frac{10}{30{,}000}(x\vec{i} + y\vec{j} + z\vec{k}).$$

Integrating we get

$$T = \frac{-10}{2(30{,}000)}(x^2 + y^2 + z^2) + C.$$

At the surface of the earth, $x^2 + y^2 + z^2 = 6400^2$, and $T = 20^\circ C$, so

$$T = \frac{-1}{6000}(6400^2) + C = 20.$$

Thus,

$$C = 20 + \frac{6400^2}{6000} = 6847.$$

At the center of the earth, $x^2 + y^2 + z^2 = 0$, so

$$T = 6847^\circ C.$$

41. Check that div $\vec{E} = 0$ by taking partial derivatives. For instance,

$$\frac{\partial E_1}{\partial x} = \frac{\partial}{\partial x}[q(x - x_0)[(x - x_0)^2 + (y - y_0)^2 + (z - z_0)^2]^{-3/2}]$$

$$= q[(y - y_0)^2 + (z - z_0)^2 - 2(x - x_0)^2][(x - x_0)^2 + (y - y_0)^2 + (z - z_0)^2]^{-5/2}$$

and similarly,

$$\frac{\partial E_2}{\partial y} = q[(x - x_0)^2 + (z - z_0)^2 - 2(y - y_0)^2][(x - x_0)^2 + (y - y_0)^2 + (z - z_0)^2]^{-5/2}$$

$$\frac{\partial E_3}{\partial z} = q[(x - x_0)^2 + (y - y_0)^2 - 2(z - z_0)^2][(x - x_0)^2 + (y - y_0)^2 + (z - z_0)^2]^{-5/2}.$$

Therefore,

$$\frac{\partial E_1}{\partial x} + \frac{\partial E_2}{\partial y} + \frac{\partial E_3}{\partial z} = 0.$$

The vector field $\vec{E}$ is defined everywhere but at the point with position vector $\vec{r}_0$. If this point lies outside the surface S,
the Divergence Theorem can be applied to the region R enclosed by S, yielding:

$$\int_S \vec{E} \cdot d\vec{A} = \int_R \text{div } \vec{E} \, dV = 0.$$

If the charge q is located inside S, consider a small sphere S_a centered at q and contained in R. The Divergence Theorem for the region R' between the two spheres yields:

$$\int_S \vec{E} \cdot d\vec{A} + \int_{S_a} \vec{E} \cdot d\vec{A} = \int_{R'} \operatorname{div} \vec{E} \; dV = 0.$$

In this formula, the Divergence Theorem requires S to be given the outward orientation, and S_a the inward orientation. To compute $\int_{S_a} \vec{E} \cdot d\vec{A}$, we use the fact that on the surface of the sphere, $\vec{E}$ and $\Delta \vec{A}$ are parallel and in opposite directions, so

$$\vec{E} \cdot \Delta \vec{A} = -\|\vec{E}\| \|\Delta \vec{A}\|$$

since on the surface of a sphere of radius a,

$$\|\vec{E}\| = q \frac{\|\vec{r} - \vec{r}_0\|}{\|\vec{r} - \vec{r}_0\|^3} = \frac{q}{a^2}.$$

Then,

$$\int_{S_a} \vec{E} \cdot d\vec{A} = \int -\frac{q}{a^2} \|d\vec{A}\| = \frac{-q}{a^2} \cdot \text{Surface area of sphere} = -\frac{q}{a^2} \cdot 4\pi a^2 = -4\pi q.$$

$$\int_{S_a} \vec{E} \cdot d\vec{A} = -4\pi q.$$

$$\int_S \vec{E} \cdot d\vec{A} - \int_{S_a} \vec{E} \cdot d\vec{A} = 4\pi q.$$

Solutions for Section 20.3

Exercises

1. Vector. We have

$$\operatorname{curl}(z\vec{i} - x\vec{j} + y\vec{k}) = \begin{vmatrix} \vec{i} & \vec{j} & \vec{k} \\ \frac{\partial}{\partial x} & \frac{\partial}{\partial y} & \frac{\partial}{\partial z} \\ z & -x & y \end{vmatrix} = \vec{i} + \vec{j} - \vec{k}.$$

5. Using the definition of Cartesian coordinates,

$$\operatorname{curl} \vec{F} = \begin{vmatrix} \vec{i} & \vec{j} & \vec{k} \\ \frac{\partial}{\partial x} & \frac{\partial}{\partial y} & \frac{\partial}{\partial z} \\ (-x+y) & (y+z) & (-z+x) \end{vmatrix}$$

$$= \left(\frac{\partial}{\partial y}(-z+x) - \frac{\partial}{\partial z}(y+z) \right) \vec{i} + \left(-\frac{\partial}{\partial x}(-z+x) + \frac{\partial}{\partial z}(-x+y) \right) \vec{j}$$

$$+ \left(\frac{\partial}{\partial x}(y+z) - \frac{\partial}{\partial y}(-x+y) \right) \vec{k}$$

$$= -\vec{i} - \vec{j} - \vec{k}.$$

9. This vector field points radically outward and has unit length everywhere (except the origin). Thus, we would expect its curl to be $\vec{0}$. Computing the curl directly we get

$$\operatorname{curl}\left(\frac{\vec{r}}{\|\vec{r}\|}\right) = \begin{vmatrix} \vec{i} & \vec{j} & \vec{k} \\ \frac{\partial}{\partial x} & \frac{\partial}{\partial y} & \frac{\partial}{\partial z} \\ \frac{x}{(x^2+y^2+z^2)^{1/2}} & \frac{y}{(x^2+y^2+z^2)^{1/2}} & \frac{z}{(x^2+y^2+z^2)^{1/2}} \end{vmatrix}$$

The $\vec{i}$-component is given by $= \left(-\frac{1}{2} \cdot \frac{2yz}{(x^2 + y^2 + z^2)^{3/2}} - \left(-\frac{1}{2} \cdot \frac{2yz}{(x^2 + y^2 + z^2)^{1/2}} \right) \right) \vec{i}$

$$= \vec{0}$$

Similarly, the $\vec{j}$ and $\vec{k}$ components are also both $\vec{0}$.

13. This vector field shows no rotation, and the circulation around any closed curve appears to be zero, so the vector field has zero curl.

Problems

17. Yes. $\vec{F} = (1 + y^2)\vec{i}$ has constant direction and curl $\vec{F} = -2y\vec{k} \neq \vec{0}$. The flow lines of a vector field do not have to bend for it to have nonzero curl.

21. The curl is defined in such a way that if $\vec{n}$ is a unit vector and C is a small circle in the plane perpendicular to $\vec{n}$ and with orientation induced by $\vec{n}$, then

$$(\text{curl}\,\vec{G}) \cdot \vec{n} = \text{Circulation density}$$

$$\approx \frac{\int_C \vec{G} \cdot d\vec{r}}{\text{Area inside } C}$$

so

$$\text{Circulation} = \int_C \vec{G} \cdot d\vec{r} \approx \left((\text{curl}\,\vec{G}) \cdot \vec{n}\right) \cdot \text{Area inside } C.$$

(a) Let C be the circle in the xy-plane, and let $\vec{n} = \vec{k}$. Then

$$\text{Circulation} \approx (2\vec{i} - 3\vec{j} + 5\vec{k}) \cdot \vec{k} \cdot \pi(0.01)^2$$

$$= 0.0005\pi.$$

(b) By a similar argument to part (a), with $\vec{n} = \vec{i}$, we find the circulation around the circle in the yz-plane:

$$\text{Circulation} \approx (2\vec{i} - 3\vec{j} + 5\vec{k}) \cdot \vec{i} \cdot \pi(0.01)^2$$

$$= 0.0002\pi.$$

(c) Similarly for circulations around the circle in the xz-plane,

$$\text{Circulation} \approx -0.0003\pi.$$

25. Let $\vec{C} = a\vec{i} + b\vec{j} + c\vec{k}$. Then

$$\text{curl}(\vec{F} + \vec{C}) = \left(\frac{\partial}{\partial y}(F_3 + c) - \frac{\partial}{\partial z}(F_2 + b)\right)\vec{i} + \left(\frac{\partial}{\partial z}(F_1 + a) - \frac{\partial}{\partial x}(F_3 + c)\right)\vec{j}$$

$$+ \left(\frac{\partial}{\partial x}(F_2 + b) - \frac{\partial}{\partial y}(F_1 + a)\right)\vec{k}$$

$$= \left(\frac{\partial F_3}{\partial y} - \frac{\partial F_2}{\partial z}\right)\vec{i} + \left(\frac{\partial F_1}{\partial z} - \frac{\partial F_3}{\partial x}\right)\vec{j} + \left(\frac{\partial F_2}{\partial x} - \frac{\partial F_1}{\partial y}\right)\vec{k}$$

$$= \text{curl}\,\vec{F}.$$

29. By Problem 28, curl $\vec{F} = \text{grad}\,f \times \text{grad}\,g + f\,\text{curl}\,\text{grad}\,g = \text{grad}\,f \times \text{grad}\,g$, since curl grad $g = 0$. Since the cross product of two vectors is perpendicular to both vectors, curl $\vec{F}$ is perpendicular to grad g. But $\vec{F}$ is a scalar times grad g, so curl $\vec{F}$ is perpendicular to $\vec{F}$.

33. Investigate the velocity vector field of the atmosphere near the fire. If the curl of this vector field is non-zero, there is circulatory motion. Consequently, if the magnitude of the curl of this vector field is large near the fire, a fire storm has probably developed.

37. (a) Since $\vec{F}$ is in the xy-plane, curl $\vec{F}$ is parallel to $\vec{k}$ (because $F_3 = 0$ and F_1, F_2 have no z-dependence). Imagine computing the circulation of $\vec{F}$ counterclockwise around a small rectangle R at the point P with sides of length h parallel to $\vec{F}$ and sides of length t perpendicular to $\vec{F}$ as shown in Figure 20.4. Since $\vec{F}$ is perpendicular to C_2 and C_4, the line integral over these two sides is zero. Assuming that $\vec{F}$ is approximately constant on C_1 and C_3, its value on these sides is $F(Q)\vec{T}$ and $-F(P)\vec{T}$, respectively. Thus, since $\vec{F}$ is parallel to C_1 and C_3, the line integral over C_1 is approximately $F(Q)h$ and the line integral over C_3 is approximately $-F(P)h$. Finally

$$\text{curl}\,\vec{F}(P) \approx \frac{\text{Circulation around } R}{\text{Area of } R} \approx \frac{F(Q)h - F(P)h}{ht} = \frac{F(Q) - F(P)}{t}$$

$$\approx \text{Directional derivative of } F \text{ in the direction of } \overrightarrow{PQ}.$$

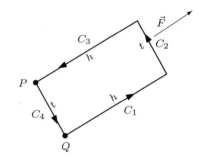

Figure 20.4: Path R used to find curl $\vec{F}$ at P

(b) Since $\vec{F} = F(x, y)\vec{T} = F(x, y)a\vec{i} + F(x, y)b\vec{j}$, with a, b constant, we have

$$\text{curl } \vec{F} = (bF_x - aF_y)\vec{k}.$$

Also $\vec{T} \times \vec{k} = (a\vec{i} + b\vec{j}) \times \vec{k} = b\vec{i} - a\vec{j}$, so

$$bF_x - aF_y = (\text{grad}F) \cdot (b\vec{i} - a\vec{j}) = \text{grad}F \cdot ((a\vec{i} + b\vec{j}) \times \vec{k}) = F_{\vec{T} \times \vec{k}},$$

where $F_{\vec{T} \times \vec{k}}$ is the directional derivative of F in the direction of the unit vector $\vec{T} \times \vec{k}$, which is perpendicular to $\vec{F}$. The right-hand rule applied to $\vec{T} \times \vec{k}$ shows that $\vec{T} \times \vec{k}$ is obtained by a *clockwise* rotation of $\vec{T}$ through $90°$.

Solutions for Section 20.4

Exercises

1. To calculate $\int_C \vec{F} \cdot d\vec{r}$ directly, we compute the integral along each of the sides C_1, C_2, C_3 in Figure 20.5. Now C_1 is parameterized by

$$x(t) = t, \quad y(t) = 0, \quad z(t) = 0 \quad \text{for } 0 \le t \le 5, \quad \text{so } r'(t) = \vec{i}.$$

Similarly, C_2 is parameterized by

$$x(t) = 5, \quad y(t) = t, \quad z(t) = 0 \quad \text{for } 0 \le t \le 5, \quad \text{so } \vec{r}'(t) = \vec{j}.$$

Also, C_3 is parameterized by

$$x(t) = 5 - t, \quad y(t) = 5 - t, \quad z(t) = 0 \quad \text{for } 0 \le t \le 5, \quad \text{so } \vec{r}'(t) = -\vec{i} - \vec{j}.$$

Thus

$$\int_C \vec{F} \cdot d\vec{r} = \int_{C_1} \vec{F} \cdot d\vec{r} + \int_{C_2} \vec{F} \cdot d\vec{r} + \int_{C_3} \vec{F} \cdot d\vec{r}$$

$$= \int_0^5 t(\vec{i} + \vec{j}) \cdot \vec{i} \, dt + \int_0^5 (5 - t)(\vec{i} + \vec{j}) \cdot \vec{j} \, dt + \int_0^5 ((5 - t) - (5 - t))(\vec{i} + \vec{j}) \cdot (-\vec{i} - \vec{j}) \, dt$$

$$= \int_0^5 t \, dt + \int_0^5 5 - t \, dt + \int_0^5 0 \, dt = \int_0^5 5 \, dt = 25.$$

To calculate $\int_C \vec{F} \cdot d\vec{r}$ using Stokes' Theorem, we find

$$\text{curl } \vec{F} = \begin{vmatrix} \vec{i} & \vec{j} & \vec{k} \\ \frac{\partial}{\partial x} & \frac{\partial}{\partial y} & \frac{\partial}{\partial z} \\ x - y + z & x - y + z & 0 \end{vmatrix} = -\vec{i} + \vec{j} + (1 - (-1))\vec{k} = -\vec{i} + \vec{j} + 2\vec{k}.$$

For Stokes' Theorem, the triangular region S in Figure 20.5 is oriented upward, so $d\vec{A} = \vec{k}\ dx\ dy$. Thus

$$\int_C \vec{F}\cdot d\vec{r} = \int_S \text{curl}\,\vec{F}\cdot d\vec{A} = \int_S (-\vec{i}+\vec{j}+2\vec{k})\cdot\vec{k}\ dx\ dy$$

$$= \int_S 2\ dx\ dy = 2\cdot\ \text{Area of triangle}\ = 2\cdot\frac{1}{2}\cdot 5\cdot 5 = 25.$$

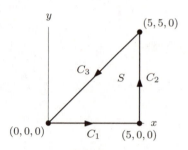

Figure 20.5

5. The boundary of S is C, the circle $x^2+y^2=1$, $z=0$, oriented counterclockwise and parameterized in polar coordinates by

$$\vec{r}(\theta) = \cos\theta\vec{i}+\sin\theta\vec{j}, \quad 0\le\theta\le 2\pi,$$

so,

$$\vec{r}\,'(\theta) = -\sin\theta\vec{i}+\cos\theta\vec{j}.$$

Hence

$$\int_C \vec{F}\cdot d\vec{r} = \int_0^{2\pi} (\sin\theta\vec{i}+0\vec{j}+\cos\theta\vec{k})\cdot(-\sin\theta\vec{i}+\cos\theta\vec{j}+0\vec{k})d\theta$$

$$= \int_0^{2\pi} -\sin^2\theta d\theta = -\pi.$$

Now consider the integral $\int_S \text{curl}\,\vec{F}\cdot d\vec{A}$. Here $\text{curl}\,\vec{F} = -\vec{i}-\vec{j}-\vec{k}$ and the area vector $d\vec{A}$, oriented upward, is given by

$$d\vec{A} = 2x\vec{i}+2y\vec{j}+\vec{k}\ dxdy.$$

If R is the disk $x^2+y^2\le 1$, then we have

$$\int_S \text{curl}\,\vec{F}\cdot d\vec{A} = \int_R (-\vec{i}-\vec{j}-\vec{k})\cdot(2x\vec{i}+2y\vec{j}+\vec{k})dxdy.$$

Converting to polar coordinates gives:

$$\int_S \text{curl}\,\vec{F}\cdot d\vec{A} = \int_0^{2\pi}\int_0^1 (-\vec{i}-\vec{j}-\vec{k})\cdot(2r\cos\theta\vec{i}+2r\sin\theta\vec{j}+\vec{k})rdrd\theta$$

$$= \int_0^{2\pi}\int_0^1 (-2r\cos\theta-2r\sin\theta-1)rdrd\theta$$

$$= \int_0^{2\pi} \left(\frac{2}{3}(-\cos\theta-\sin\theta)-\frac{1}{2}\right)d\theta$$

$$= -\pi.$$

Thus, we confirm that

$$\int_C \vec{F}\cdot d\vec{r} = \int_S \text{curl}\,\vec{F}\cdot d\vec{A}.$$

9. The circulation is the line integral $\int_C \vec{F} \cdot d\vec{r}$ which can be evaluated directly by parameterizing the circle, C. Or, since C is the boundary of a flat disk S, we can use Stokes' Theorem:

$$\int_C \vec{F} \cdot d\vec{r} = \int_S \text{curl}\,\vec{F} \cdot d\vec{A}$$

where S is the disk $x^2 + y^2 \leq 1$, $z = 2$ and is oriented upward (using the right hand rule). Then $\text{curl}\,\vec{F} = -y\vec{i} - x\vec{j} + \vec{k}$ and the unit normal to S is $\vec{k}$. So

$$\int_S \text{curl}\,\vec{F} \cdot d\vec{A} = \int_S (-y\vec{i} - x\vec{j} + \vec{k}) \cdot \vec{k}\, dxdy$$

$$= \int_S 1\, dxdy$$

$$= \text{Area of } S = \pi$$

Problems

13. Since

$$\text{curl}\,\vec{F} = \begin{vmatrix} \vec{i} & \vec{j} & \vec{k} \\ \frac{\partial}{\partial x} & \frac{\partial}{\partial y} & \frac{\partial}{\partial z} \\ y & -x & y-x \end{vmatrix} = \frac{\partial}{\partial y}(y-x)\vec{i} - \frac{\partial}{\partial x}(y-x)\vec{j} + \left(\frac{\partial}{\partial x}(-x) - \frac{\partial}{\partial y}(y)\right)\vec{k} = \vec{i} + \vec{j} - 2\vec{k},$$

writing S for the disk in the plane enclosed by the circle, Stokes' Theorem gives

$$\int_C \vec{F} \cdot d\vec{r} = \int_S \text{curl}\,\vec{F} \cdot d\vec{A} = \int_S (\vec{i} + \vec{j} - 2\vec{k}) \cdot d\vec{A}.$$

Now $d\vec{A} = \vec{n}\, dA$, where $\vec{n}$ is the unit vector perpendicular to the plane, so

$$\vec{n} = \frac{1}{\sqrt{3}}(\vec{i} + \vec{j} + \vec{k}).$$

Thus

$$\int_C \vec{F}\, d\vec{r} = \int_S (\vec{i} + \vec{j} - 2\vec{k}) \cdot \frac{\vec{i} + \vec{j} + \vec{k}}{\sqrt{3}}\, dA = \int_S \frac{0}{\sqrt{3}}\, dA = 0.$$

17. (a) The equation of the rim, C, is $x^2 + y^2 = 9$, $z = 2$. This is a circle of radius 3 centered on the z-axis, and lying in the plane $z = 2$.

(b) Use Stokes' Theorem, with C oriented clockwise when viewed from above:

$$\int_S \text{curl}(-y\vec{i} + x\vec{j} + z\vec{k}) \cdot d\vec{A} = \int_C (-y\vec{i} + x\vec{j} + z\vec{k}) \cdot d\vec{r}.$$

Since C is horizontal, the $\vec{k}$ component does not contribute to the integral. The remaining vector field, $-y\vec{i} + x\vec{j}$, is tangent to C, of constant magnitude $||-y\vec{i} + x\vec{j}|| = 3$ on C, and points in the opposite direction to the orientation. Thus

$$\int_S \text{curl}(-y\vec{i} + x\vec{j} + z\vec{k}) \cdot d\vec{A} = \int_C (-y\vec{i} + x\vec{j}) \cdot d\vec{r} = -3 \cdot \text{Length of curve} = -3 \cdot 2\pi3 = -18\pi.$$

21. (a) We have

$$\text{curl}\,\vec{F} = \begin{vmatrix} \vec{i} & \vec{j} & \vec{k} \\ \frac{\partial}{\partial x} & \frac{\partial}{\partial y} & \frac{\partial}{\partial z} \\ y & z & x \end{vmatrix} = \vec{i}(-1) - \vec{j}(1) + \vec{k}(-1) = -\vec{i} - \vec{j} - \vec{k}.$$

(b)   (i) Using Stokes' Theorem, with S representing the disk inside the circle, oriented upward, we have

$$\int_C \vec{F} \cdot d\vec{r} = \int_S \operatorname{curl} \vec{F} \cdot d\vec{A} = \int_S (-\vec{i} - \vec{j} - \vec{k}) \cdot \vec{k} \, dA = - \text{ Area of disk } = -4\pi.$$

(ii) This is a right triangle in the plane $x = 2$; it has height 5 and base length 3. Using Stokes' Theorem, with S representing the triangle, oriented toward the origin (in the direction $-\vec{i}$), we have

$$\int_C \vec{F} \cdot d\vec{r} = \int_S \operatorname{curl} \vec{F} \cdot d\vec{A} = \int_S (-\vec{i} - \vec{j} - \vec{k}) \cdot (-\vec{i} \, dA) = \int_S dA = \text{ Area of triangle } = \frac{1}{2} \cdot 3 \cdot 5 = \frac{15}{2}.$$

25. (a) It appears that $\operatorname{div} \vec{F} < 0$, and $\operatorname{div} \vec{G} < 0$; $\operatorname{div} \vec{G}$ is larger in magnitude (more negative) if the scales are the same.

(b) $\operatorname{curl} \vec{F}$ and $\operatorname{curl} \vec{G}$ both appear to be zero at the origin (and elsewhere).

(c) Yes, the cylinder with axis along the z-axis will have negative flux through it (ends parallel to xy-plane).

(d) Same as part(c).

(e) No, you cannot draw a closed curve around the origin such that $\vec{F}$ has a non-zero circulation around it because curl is zero. By Stokes' theorem, circulation equals the integral of the curl over the surface bounded by the curve.

(f) Same as part(e)

29. (a) $\vec{F}$ has only $\vec{i}$ and $\vec{j}$ components, and they do not depend on z. Thus $\vec{F}$ is everywhere parallel to the xy-plane, and takes the same values for every value of z.

(b) We have

$$\operatorname{curl} \vec{F} = \begin{vmatrix} \vec{i} & \vec{j} & \vec{k} \\ \dfrac{\partial}{\partial x} & \dfrac{\partial}{\partial y} & \dfrac{\partial}{\partial z} \\ F_1(x,y) & F_2(x,y) & 0 \end{vmatrix} = \left(\frac{\partial F_2}{\partial x} - \frac{\partial F_1}{\partial y} \right) \vec{k}.$$

(c) Since C is in the xy-plane, oriented counterclockwise when viewed from above, for an area element $d\vec{A}$ in S, we have $d\vec{A} = \vec{k} \, dx \, dy$. Thus Stokes' Theorem says

$$\int_C \vec{F} \cdot d\vec{r} = \int_S \operatorname{curl} \vec{F} \cdot d\vec{A} = \int_S \left(\frac{\partial F_2}{\partial x} - \frac{\partial F_1}{\partial y} \right) \vec{k} \cdot \vec{k} \, dx \, dy = \int_S \left(\frac{\partial F_2}{\partial x} - \frac{\partial F_1}{\partial y} \right) dx \, dy.$$

(d) Green's Theorem.

Solutions for Section 20.5

Exercises

1. Since $\operatorname{curl} \vec{F} = \vec{0}$ and $\vec{F}$ is defined everywhere, we know by the curl test that $\vec{F}$ is a gradient field. In fact, $\vec{F} = \operatorname{grad} f$, where $f(x, y, z) = xyz + yz^2$, so f is a potential function for $\vec{F}$.

Problems

5. Let $\vec{v} = a\vec{i} + b\vec{j} + c\vec{k}$ and try

$$\vec{F} = \vec{v} \times \vec{r} = (a\vec{i} + b\vec{j} + c\vec{k}) \times (x\vec{i} + y\vec{j} + z\vec{k}) = (bz - cy)\vec{i} + (cx - az)\vec{j} + (ay - bx)\vec{k}.$$

Then

$$\operatorname{curl} \vec{F} = \begin{vmatrix} \vec{i} & \vec{j} & \vec{k} \\ \dfrac{\partial}{\partial x} & \dfrac{\partial}{\partial y} & \dfrac{\partial}{\partial z} \\ bz - cy & cx - az & ay - bx \end{vmatrix} = 2a\vec{i} + 2b\vec{j} + 2c\vec{k}.$$

Taking $a = 1$, $b = -\frac{3}{2}$, $c = 2$ gives $\operatorname{curl} \vec{F} = 2\vec{i} - 3\vec{j} + 4\vec{k}$, so the desired vector field is $\vec{F} = (-\frac{3}{2}z - 2y)\vec{i} + (2x - z)\vec{j} + (y + \frac{3}{2}x)\vec{k}$.

9. Since $\operatorname{div} \vec{G} = 2x + 2y + 2z \neq 0$, there is not a vector potential for $\vec{G}$.

13. (a) Using the product rule from Problem 28 on page 430, we find

$$\operatorname{curl} \vec{E} = \operatorname{curl}\left(\frac{\vec{r}}{\|\vec{r}\|^p} \right) = \frac{1}{\|\vec{r}\|^p} \operatorname{curl} \vec{r} + \operatorname{grad}\left(\frac{1}{\|\vec{r}\|^p} \right) \times \vec{r}.$$

Now $\operatorname{curl} \vec{r} = \vec{0}$ and $\operatorname{grad}\left(\frac{1}{\|\vec{r}\|^p} \right)$ is parallel to $\vec{r}$, so both terms are zero. Thus $\operatorname{curl} \vec{E} = \vec{0}$.

(b) The domain of $\vec{E}$ is 3-space minus the origin if $p > 0$, and it is all of 3-space if $p \le 0$.

(c) Both domains have the property that any closed curve can be contracted to a point without hitting the origin, so $\vec{E}$ satisfies the curl test for all p. Since $\vec{E}$ has constant magnitude r^{1-p} on the sphere of radius r centered at the origin, and is parallel to the outward normal at every point of the sphere, the sphere must be a level surface of the potential function ϕ, that is, ϕ is a function of r alone. Further, since $\|\vec{E}\| = r^{1-p}$, a good guess is

$$\phi(r) = \int r^{1-p}\, dr,$$

that is,

$$\phi(r) = \begin{cases} \frac{r^{2-p}}{2-p} & \text{if } p \ne 2 \\ \ln r & \text{if } p = 2. \end{cases}$$

You can check that this is indeed a potential function for $\vec{E}$ by checking that $\operatorname{grad} \phi = \vec{E}$.

Solutions for Chapter 20 Review

Exercises

1. Scalar. $\operatorname{div}((2\sin(xy) + \tan z)\vec{i} + (\tan y)\vec{j} + (e^{x^2+y^2})\vec{k}) = 2y\cos(xy) + 1/\cos^2 y$.

5. We have

$$\operatorname{div} \vec{F} = \frac{\partial}{\partial x}(\cos x) + \frac{\partial}{\partial y}(e^y) + \frac{\partial}{\partial z}(x+y+z) = -\sin x + e^y + 1$$

$$\operatorname{curl} \vec{F} = \begin{vmatrix} \vec{i} & \vec{j} & \vec{k} \\ \frac{\partial}{\partial x} & \frac{\partial}{\partial y} & \frac{\partial}{\partial z} \\ \cos x & e^y & x+y+z \end{vmatrix} = \vec{i} - \vec{j}.$$

So $\vec{F}$ is not solenoidal and not irrotational.

9. C_2, C_3, C_4, C_6, since line integrals around C_1 and C_5 are clearly nonzero. You can see directly that $\int_{C_2} \vec{F} \cdot d\vec{r}$ and $\int_{C_6} \vec{F} \cdot d\vec{r}$ are zero, because C_2 and C_6 are perpendicular to their fields at every point.

13. (a) Direct method:

$$\operatorname{curl} \vec{F} = \begin{vmatrix} \vec{i} & \vec{j} & \vec{k} \\ \frac{\partial}{\partial x} & \frac{\partial}{\partial y} & \frac{\partial}{\partial z} \\ 0 & xz & -xy \end{vmatrix} = -2x\vec{i} - (-y)\vec{j} + z\vec{k} = -2x\vec{i} + y\vec{j} + z\vec{k}.$$

On the surface, $d\vec{A}$ has no $\vec{i}$-component, so the $\vec{i}$-component of $\operatorname{curl} \vec{F}$ does not contribute to the flux. Thus

$$\int_S \operatorname{curl} \vec{F} \cdot d\vec{A} = \int_S (y\vec{j} + z\vec{k}) \cdot d\vec{A}.$$

Since $y\vec{j} + z\vec{k}$ is perpendicular to S and $\|y\vec{j} + z\vec{k}\| = \sqrt{y^2 + z^2} = \sqrt{5}$ on S, we have

$$\int_S \operatorname{curl} \vec{F} \cdot d\vec{A} = \sqrt{5} \cdot \text{Area of } S = \sqrt{5} \cdot 2\pi\sqrt{5} \cdot 3 = 30\pi.$$

(b) Using Stokes' theorem, we replace the flux integral by two line integrals around the circular boundaries, C_1 and C_2, of S. See Figure 20.6.

$$\int_S \text{curl}\, \vec{F} \cdot ds = \int_{C_1} \vec{F} \cdot dr + \int_{C_2} \vec{F} \cdot d\vec{r}.$$

On C_1, the left boundary, $x = 0$, so $\vec{F} = \vec{0}$, and therefore $\int_{C_1} \vec{F} \cdot d\vec{r} = 0$. On C_2, the right boundary, $x = 3$, so $\vec{F} = 3z\vec{j} - 3y\vec{k}$. This vector field has $||\vec{F}|| = \sqrt{(3z)^2 + (-3y)^2} = \sqrt{9(z^2 + y^2)} = 3\sqrt{5}$. and $\vec{F}$ is tangent to the boundary C_2 and pointing in the same direction as C_2. Thus

$$\int_{C_2} \vec{F} \cdot d\vec{r} = ||\vec{F}|| \cdot \text{Length of } C_2 = 3\sqrt{5} \cdot 2\pi\sqrt{5} = 30\pi.$$

Figure 20.6

17. Since $\text{div}\, \vec{F} = 3x^2 + 3y^2$, using cylindrical coordinates to calculate the triple integral gives

$$\int_S \vec{F} \cdot d\vec{A} = \int_{\substack{\text{Interior} \\ \text{of cylinder}}} (3x^2 + 3y^2)\, dV = 3\int_0^{2\pi}\int_0^5\int_0^2 r^2 \cdot r\, dr\, dz\, d\theta = 3 \cdot 2\pi \cdot 5 \frac{r^4}{4}\Big|_0^2 = 120\pi.$$

21. If C is the rectangular path around the rectangle, traversed counterclockwise when viewed from above, Stokes' Theorem gives

$$\int_S \text{curl}\, \vec{F} \cdot d\vec{A} = \int_C \vec{F} \cdot d\vec{r}.$$

The $\vec{k}$ component of $\vec{F}$ does not contribute to the line integral, and the $\vec{j}$ component contributes to the line integral only along the segments of the curve parallel to the y-axis. Thus, if we break the line integral into four parts

$$\int_S \text{curl}\, \vec{F} \cdot d\vec{A} = \int_{(0,0)}^{(3,0)} \vec{F} \cdot d\vec{r} + \int_{(3,0)}^{(3,2)} \vec{F} \cdot d\vec{r} + \int_{(3,2)}^{(0,2)} \vec{F} \cdot d\vec{r} + \int_{(0,2)}^{(0,0)} \vec{F} \cdot d\vec{r},$$

we see that the first and third integrals are zero, and we can replace $\vec{F}$ by its $\vec{j}$ component in the other two

$$\int_S \text{curl}\, \vec{F} \cdot d\vec{A} = \int_{(3,0)}^{(3,2)} (x+7)\vec{j} \cdot d\vec{r} + \int_{(0,2)}^{(0,0)} (x+7)\vec{j} \cdot d\vec{r}.$$

Now $x = 3$ in the first integral and $x = 0$ in the second integral and the variable of integration is y in both, so

$$\int_S \text{curl}\, \vec{F} \cdot d\vec{A} = \int_0^2 10\, dy + \int_2^0 7\, dy = 20 - 14 = 6.$$

Problems

25. (a) The cube is in Figure 20.7. The vector field is parallel to the x-axis and zero on the yz-plane. Thus the only contribution to the flux is from S_2. On S_2, $x = c$, the normal is outward. Since $\vec{F}$ is constant on S_2, the flux through face S_2 is

$$\int_{S_2} \vec{F} \cdot d\vec{A} = \vec{F} \cdot \vec{A}_{S_2}$$
$$= c\vec{i} \cdot c^2\vec{i}$$
$$= c^3.$$

Thus, total flux through box $= c^3$.

(b) Using the geometric definition of divergence

$$\operatorname{div} \vec{F} = \lim_{c \to 0} \left(\frac{\text{Flux through box}}{\text{Volume of box}} \right)$$
$$= \lim_{c \to 0} \left(\frac{c^3}{c^3} \right)$$
$$= 1$$

(c) Using partial derivatives,

$$\operatorname{div} \vec{F} = \frac{\partial}{\partial x}(x) + \frac{\partial}{\partial y}(0) + \frac{\partial}{\partial z}(0) = 1 + 0 + 0 = 1.$$

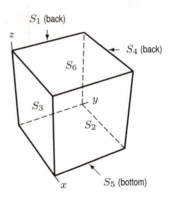

Figure 20.7

29. We close the cylinder, S, by adding the circular disk, S_1, at the top, $z = 3$. The surface $S + S_1$ is oriented outward, so S_1 is oriented upward. Applying the Divergence Theorem to the closed surface $S + S_1$ enclosing the region W, we have

$$\int_{S+S_1} \vec{F} \cdot d\vec{A} = \int_W \operatorname{div} \vec{F} \, dV$$
$$\int_S \vec{F} \cdot d\vec{A} + \int_{S_1} \vec{F} \cdot d\vec{A} = \int_W \operatorname{div} \vec{F} \, dV.$$

Since

$$\operatorname{div} \vec{F} = \operatorname{div}(z^2\vec{i} + x^2\vec{j} + 5\vec{k}) = 0,$$

we have

$$\int_S \vec{F} \cdot d\vec{A} = -\int_{S_1} \vec{F} \cdot d\vec{A}.$$

Only the $\vec{k}$-component of $\vec{F}$ contributes to the flux through S_1, and $d\vec{A} = \vec{k} \, dx \, dy$ on S_1, so

$$\int_S \vec{F} \cdot d\vec{A} = -\int_{S_1} (z^2\vec{i} + x^2\vec{j} + 5\vec{k}) \cdot \vec{k} \, dx \, dy = -\int_{S_1} 5 \, dx \, dy = -5 \cdot \text{Area of } S_1 = -5\pi(\sqrt{2})^2 = -10\pi.$$

33. We close the cylinder, S, by adding the circular disk, S_1, at the top, $z = 3$. The surface $S + S_1$ is oriented outward, so S_1 is oriented upward. Applying the Divergence Theorem to the closed surface $S + S_1$ enclosing the region W, we have

$$\int_{S+S_1} \vec{F} \cdot d\vec{A} = \int_W \operatorname{div} \vec{F} \, dV$$

$$\int_S \vec{F} \cdot d\vec{A} + \int_{S_1} \vec{F} \cdot d\vec{A} = \int \operatorname{div} \vec{F} \, dV.$$

Since

$$\operatorname{div} \vec{F} = \operatorname{div}(x^3\vec{i} + y^3\vec{j} + \vec{k}) = 3x^2 + 3y^2,$$

we have

$$\int_S \vec{F} \cdot d\vec{A} = \int_W (3x^2 + 3y^2)\, dV - \int_{S_1} \vec{F} \cdot d\vec{A}.$$

To find the integral over W, we use cylindrical coordinates. For the integral over S_1, we use the fact that $d\vec{A} = \vec{k}\, dx\, dy$, so only the k-component of $\vec{F}$ contributes to the flux.

$$\int_S \vec{F} \cdot d\vec{A} = \int_0^{2\pi} \int_{-3}^3 \int_0^{\sqrt{2}} 3r^2 \cdot r\, dr\, dz\, d\theta - \int_{S_1} (x^3\vec{i} + y^3\vec{j} + \vec{k}) \cdot \vec{k}\, dx\, dy$$

$$= \theta \Big|_0^{2\pi}\, z\Big|_{-3}^3\, \frac{3}{4}r^4 \Big|_0^{\sqrt{2}} - \int_{S_1} dx\, dy$$

$$= 2\pi \cdot 6 \cdot \frac{3}{4}(\sqrt{2})^4 - \text{Area of } S_1$$

$$= 36\pi - \pi(\sqrt{2})^2 = 34\pi.$$

37. Since

$$\operatorname{div}(3x\vec{i} + 4y\vec{j} + xy\vec{k}) = 3 + 4 + 0 = 7,$$

we calculate the flux using the Divergence Theorem:

$$\text{Flux} = \int_S (3x\vec{i} + 4y\vec{j} + xy\vec{k}) \cdot d\vec{A} = \int_W 7\, dV = 7 \cdot \text{Volume of box} = 7 \cdot 3 \cdot 5 \cdot 2 = 210.$$

41. We use Stokes' Theorem. Since

$$\operatorname{curl} \vec{F} = \begin{vmatrix} \vec{i} & \vec{j} & \vec{k} \\ \frac{\partial}{\partial x} & \frac{\partial}{\partial y} & \frac{\partial}{\partial z} \\ x+y & y+2z & z+3x \end{vmatrix} = -2\vec{i} - 3\vec{j} - \vec{k},$$

if S is the interior of the square, then

$$\int_C \vec{F} \cdot d\vec{r} = \int_S \operatorname{curl} \vec{F} \cdot d\vec{A} = \int_S (-2\vec{i} - 3\vec{j} - \vec{k}) \cdot d\vec{A}$$

Since the area vector of S is $49\vec{j}$, we have

$$\int_C \vec{F} \cdot d\vec{r} = \int_S (-2\vec{i} - 3\vec{j} - \vec{k}) \cdot d\vec{A} = -3\vec{j} \cdot 49\vec{j} = -147.$$

45. By the Divergence Theorem, since $\operatorname{div} \vec{F} = 0$, the flux through the cone equals the flux upward through the disk $r \le 4$ in the plane $z = 4$. The area vector of the disk is $\vec{A} = \pi 4^2 \vec{k}$. Since the flux is negative, $\vec{F} = -c(\vec{i} + \vec{k})$ with $c > 0$. Thus

$$\text{Flux} = \vec{F} \cdot \vec{A} = -c(\vec{i} + \vec{k}) \cdot \pi 4^2 \vec{k} = -16\pi c = -7.$$

so

$$c = \frac{7}{16\pi} \quad \text{and} \quad \vec{F} = -\frac{7}{16\pi}(\vec{i} + \vec{k}).$$

49. Since div $\vec{F} = 1 + 1 + 1 = 3$, the flux through the closed cylinder, S_1, with interior W, is

$$\int_{S_1} \vec{F} \cdot d\vec{A} = \int_W 3\,dV = 3 \cdot \text{Volume of cylinder} = 3\pi.$$

With the base of the cylinder oriented downward and the top of the cylinder oriented upward,

$$\int_S \vec{F} \cdot d\vec{A} = \int_{S_1} \vec{F} \cdot d\vec{A} - \int_{\text{Base}} \vec{F} \cdot d\vec{A} - \int_{\text{Top}} \vec{F} \cdot d\vec{A}.$$

Since $\vec{F}$ is parallel to the base, the flux through the base is 0. The flux through the top is contributed entirely by the $\vec{k}$ component. Since $z = 1$, we have

$$\text{Flux through top} = \int_{\text{Top}} \vec{F} \cdot d\vec{A} = \int_{\text{Top}} (x\vec{i} + y\vec{j} + \vec{k}) \cdot d\vec{A} = \int_{\text{Top}} \vec{k} \cdot d\vec{A} = \pi.$$

Thus

$$\int_S \vec{F} \cdot d\vec{A} = 3\pi - \pi = 2\pi.$$

53. (a) Can be computed. If W is the interior of the sphere, by the Divergence Theorem, we have

$$\int_S \vec{F} \cdot d\vec{A} = \int_W \text{div}\,\vec{F}\,dV = 4 \cdot \text{Volume of sphere} = 4 \cdot \frac{4}{3}\pi \cdot 2^3 = \frac{128\pi}{3}.$$

(b) Cannot be computed.

(c) Can be computed. Use the fact that $\text{div}(\text{curl}\,\vec{F}) = 0$. If W is the inside of the sphere, then by the Divergence Theorem,

$$\int_S \text{curl}\,\vec{F} \cdot d\vec{A} = \int_W \text{div}(\text{curl}\,\vec{F})dV = \int_W 0\,dV = 0$$

57. (a) We have

$$\text{curl}\,\vec{F} = \begin{vmatrix} \vec{i} & \vec{j} & \vec{k} \\ \dfrac{\partial}{\partial x} & \dfrac{\partial}{\partial y} & \dfrac{\partial}{\partial z} \\ \dfrac{-y}{x^2+y^2} & \dfrac{x}{x^2+y^2} & 0 \end{vmatrix} = 0\vec{i} + 0\vec{j} + \left(\frac{\partial}{\partial x}\left(\frac{x}{x^2+y^2} \right) + \frac{\partial}{\partial y}\left(\frac{y}{x^2+y^2} \right) \right)\vec{k}.$$

Since

$$\frac{\partial}{\partial x}\left(\frac{x}{x^2+y^2} \right) = \frac{1}{x^2+y^2} - \frac{x(2x)}{(x^2+y^2)^2} = \frac{x^2+y^2-2x^2}{(x^2+y^2)^2} = \frac{y^2-x^2}{(x^2+y^2)^2}$$

and similarly $\dfrac{\partial}{\partial y}\left(\dfrac{y}{x^2+y^2} \right) = \dfrac{x^2-y^2}{(x^2+y^2)^2}$, we have, provided $x^2 + y^2 \neq 0$,

$$\text{curl}\,\vec{F} = \vec{0}.$$

The domain of $\text{curl}\,\vec{F}$ is all points in 3-space except the z-axis.

(b) On C_1, the unit circle $x^2 + y^2 = 1$ in the xy-plane, the vector field $\vec{F}$ is tangent to the circle and $||\vec{F}|| = 1$. Thus

$$\text{Circulation} = \int_{C_1} \vec{F} \cdot d\vec{r} = ||\vec{F}|| \cdot \text{Perimeter of circle} = 2\pi.$$

Note that Stokes' Theorem cannot be used to calculate this circulation since the z-axis pierces any surface which has this circle as boundary.

(c) Consider the disk $(x-3)^2 + y^2 \leq 1$ in the plane $z = 4$. This disk has C_2 as boundary and $\text{curl}\,\vec{F} = \vec{0}$ everywhere on this disk. Thus, by Stokes' Theorem $\int_{C_2} \vec{F} \cdot d\vec{r} = 0$.

(d) The square S has an interior region which is pierced by the z-axis, so we cannot use Stokes' Theorem. We consider the region, D, between the circle C_1 and the square S. See Figure 20.8. Stokes' Theorem applies to the region D, provided C_1 is oriented clockwise. Then we have

$$\int_{C_1 \text{(clockwise)}} \vec{F} \cdot d\vec{r} + \int_S \vec{F} \cdot d\vec{r} = \int_D \text{curl}\, \vec{F} \cdot d\vec{A} = 0.$$

Thus,

$$\int_S \vec{F} \cdot d\vec{r} = -\int_{C_1 \text{(clockwise)}} \vec{F} \cdot d\vec{r} = \int_{C_1 \text{(counterclockwise)}} \vec{F} \cdot d\vec{r} = 2\pi.$$

(e) If a simple closed curve goes around the z-axis, then it contains a circle C of the form $x^2 + y^2 = a^2$. The circulation around C is 2π or -2π, depending on its orientation. A calculation similar to that in part (d) then shows that the circulation around the curve is 2π or -2π, again depending on its orientation. If the closed curve does not go around the z-axis, then curl $\vec{F} = \vec{0}$ everywhere on its interior and the circulation is zero.

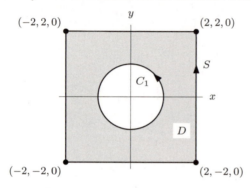

Figure 20.8

CHECK YOUR UNDERSTANDING

1. True. By Stokes' Theorem, the circulation of $\vec{F}$ around C is the flux of curl $\vec{F}$ through the flat disc S in the xy-plane enclosed by the circle. An area element for S is $d\vec{A} = \pm\vec{k}\, dA$, where the sign depends on the orientation of the circle. Since curl $\vec{F}$ is perpendicular to the z-axis, curl $\vec{F} \cdot d\vec{A} = \pm(\text{curl}\, \vec{F} \cdot \vec{k})dA = 0$, so the flux of curl $\vec{F}$ through S is zero, hence the circulation of $\vec{F}$ around C is zero.

5. True. div $\vec{F}$ is a scalar whose value depends on the point at which it is calculated.

9. False. The divergence is a scalar function that gives flux density at a point.

13. True. curl $\vec{F}$ is a vector whose value depends on the point at which it is calculated.

17. False. Since $\vec{F}$ can be written $\vec{F}(x, y, z) = x\vec{i} + y\vec{j} + z\vec{k}$, the divergence of $\vec{F}$ is 3.

21. True. By the Divergence theorem, $\int_S \vec{F} \cdot d\vec{A} = -\int_W \text{div}\, \vec{F}\, dV$, where W is the solid interior of S and the negative sign is due to the inward orientation of S. Since div $\vec{F} = 0$, we have $\int_S \vec{F} \cdot d\vec{A} = 0$.

25. True. The boundary of the cube W consists of six squares, but four of them are parallel to the xz or yz-planes and so contribute zero flux for this particular vector field. The only two surfaces of the boundary with nonzero flux are S_1 and S_2, which are parallel to the xy-plane.

29. True. The circulation density is obtained by dividing the circulation around a circle C (a scalar) by the area enclosed by C (also a scalar), in the limit as the area tends to zero.

33. False. The left-hand side of the equation does not make sense. The quantity $(\vec{F} \cdot \vec{G})$ is a scalar, so we cannot compute the curl of it.

37. False. For example, take $\vec{F} = z\vec{i}$ and $\vec{G} = x\vec{j}$. Then $\vec{F} \times \vec{G} = xz\vec{k}$, and curl$(\vec{F} \times \vec{G}) = -z\vec{j}$. However, $(\text{curl}\vec{F}) \times (\text{curl}\vec{G}) = \vec{j} \times \vec{k} = \vec{i}$.

41. True. By Stokes' theorem, both flux integrals are equal to the line integral $\int_C \vec{F} \cdot d\vec{r}$, where C is the circle $x^2 + y^2 = 1$, oriented counterclockwise when viewed from the positive z-axis.

45. True. By Stokes' theorem, $\int_S \text{curl}\, \vec{F} \cdot d\vec{A} = \int_C \vec{F} \cdot d\vec{r}$, where C is one or more closed curves that form the boundary of S. Since $\vec{F}$ is a gradient field, its line integral over any closed curve will be zero.

APPENDIX

Solutions for Section A

1. The graph is

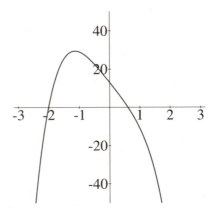

   (a) The range appears to be $y \leq 30$.
   (b) The function has two zeros.

5. The largest root is at about 2.5.

9. Using a graphing calculator, we see that when x is around 0.45, the graphs intersect.

13. (a) Only one real zero, at about $x = -1.15$.
    (b) Three real zeros: at $x = 1$, and at about $x = 1.41$ and $x = -1.41$.

17. (a) Since f is continuous, there must be one zero between $\theta = 1.4$ and $\theta = 1.6$, and another between $\theta = 1.6$ and $\theta = 1.8$. These are the only clear cases. We might also want to investigate the interval $0.6 \leq \theta \leq 0.8$ since $f(\theta)$ takes on values close to zero on at least part of this interval. Now, $\theta = 0.7$ is in this interval, and $f(0.7) = -0.01 < 0$, so f changes sign twice between $\theta = 0.6$ and $\theta = 0.8$ and hence has two zeros on this interval (assuming f is not *really* wiggly here, which it's not). There are a total of 4 zeros.
    (b) As an example, we find the zero of f between $\theta = 0.6$ and $\theta = 0.7$. $f(0.65)$ is positive; $f(0.66)$ is negative. So this zero is contained in $[0.65, 0.66]$. The other zeros are contained in the intervals $[0.72, 0.73]$, $[1.43, 1.44]$, and $[1.7, 1.71]$.
    (c) You've found all the zeros. A picture will confirm this; see Figure A.1.

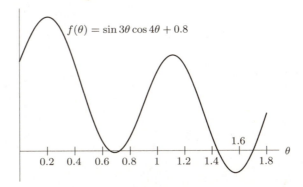

$f(\theta) = \sin 3\theta \cos 4\theta + 0.8$

Figure A.1

21.

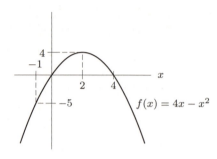

Bounded and $-5 \le f(x) \le 4$.

Solutions for Section B

1. $2e^{i\pi/2}$

5. $0e^{i\theta}$, for any θ.

9. $-3 - 4i$

13. $\frac{1}{4} - \frac{9i}{8}$

17. $5^3(\cos \frac{3\pi}{2} + i \sin \frac{3\pi}{2}) = -125i$

21. One value of $\sqrt[3]{i}$ is $\sqrt[3]{e^{i\frac{\pi}{2}}} = (e^{i\frac{\pi}{2}})^{\frac{1}{3}} = e^{i\frac{\pi}{6}} = \cos \frac{\pi}{6} + i \sin \frac{\pi}{6} = \frac{\sqrt{3}}{2} + \frac{i}{2}$

25. One value of $(-4 + 4i)^{2/3}$ is $[\sqrt{32}e^{(i3\pi/4)}]^{(2/3)} = (\sqrt{32})^{2/3}e^{(i\pi/2)} = 2^{5/3} \cos \frac{\pi}{2} + i2^{5/3} \sin \frac{\pi}{2} = 2i\sqrt[3]{4}$

29. We have

$$i^{-1} = \frac{1}{i} = \frac{1}{i} \cdot \frac{i}{i} = -i,$$

$$i^{-2} = \frac{1}{i^2} = -1,$$

$$i^{-3} = \frac{1}{i^3} = \frac{1}{-i} \cdot \frac{i}{i} = i,$$

$$i^{-4} = \frac{1}{i^4} = 1.$$

The pattern is

$$i^n = \begin{cases} -i & n = -1, -5, -9, \cdots \\ -1 & n = -2, -6, -10, \cdots \\ i & n = -3, -7, -11, \cdots \\ 1 & n = -4, -8, -12, \cdots. \end{cases}$$

Since 36 is a multiple of 4, we know $i^{-36} = 1$.
Since $41 = 4 \cdot 10 + 1$, we know $i^{-41} = -i$.

33. To confirm that $z = \frac{a + bi}{c + di}$, we calculate the product

$$z(c + di) = \left(\frac{ac + bd}{c^2 + d^2} = \frac{bc - ad}{c^2 + d^2}i \right)(c + di)$$

$$= \frac{ac^2 + bcd - bcd + ad^2 + (bc^2 - acd + acd + bd^2)i}{c^2 + d^2}$$

$$= \frac{a(c^2 + d^2) + b(c^2 + d^2)i}{c^2 + d^2} = a + bi.$$

37. True, since $\sqrt{a}$ is real for all $a \ge 0$.

41. True. We can write any nonzero complex number z as $re^{i\beta}$, where r and β are real numbers with $r > 0$. Since $r > 0$, we can write $r = e^c$ for some real number c. Therefore, $z = re^{i\beta} = e^c e^{i\beta} = e^{c+i\beta} = e^w$ where $w = c + i\beta$ is a complex number.

45. Using Euler's formula, we have:

$$e^{i(2\theta)} = \cos 2\theta + i \sin 2\theta$$

On the other hand,

$$e^{i(2\theta)} = \left(e^{i\theta}\right)^2 = (\cos \theta + i \sin \theta)^2 = (\cos^2\theta - \sin^2\theta) + i(2\cos \theta \sin \theta)$$

Equating real parts, we find

$$\cos 2\theta = \cos^2 \theta - \sin^2 \theta.$$

49. Replacing θ by $(x + y)$ in the formula for $\sin \theta$:

$$\sin(x + y) = \frac{1}{2i} \left(e^{i(x+y)} - e^{-i(x+y)}\right) = \frac{1}{2i} \left(e^{ix}e^{iy} - e^{-ix}e^{-iy}\right)$$

$$= \frac{1}{2i} \left((\cos x + i \sin x)(\cos y + i \sin y) - (\cos (-x) + i \sin (-x))(\cos (-y) + i \sin (-y))\right)$$

$$= \frac{1}{2i} \left((\cos x + i \sin x)(\cos y + i \sin y) - (\cos x - i \sin x)(\cos y - i \sin y)\right)$$

$$= \sin x \cos y + \cos x \sin y.$$

Solutions for Section C

1. (a) $f'(x) = 3x^2 + 6x + 3 = 3(x + 1)^2$. Thus $f'(x) > 0$ everywhere except at $x = -1$, so it is increasing everywhere except perhaps at $x = -1$. The function is in fact increasing at $x = -1$ since $f(x) > f(-1)$ for $x > -1$, and $f(x) < f(-1)$ for $x < -1$.

(b) The original equation can have at most one root, since it can only pass through the x-axis once if it never decreases. It must have one root, since $f(0) = -6$ and $f(1) = 1$.

(c) The root is in the interval $[0, 1]$, since $f(0) < 0 < f(1)$.

(d) Let $x_0 = 1$.

$$x_0 = 1$$

$$x_1 = 1 - \frac{f(1)}{f'(1)} = 1 - \frac{1}{12} = \frac{11}{12} \approx 0.917$$

$$x_2 = \frac{11}{12} - \frac{f\left(\frac{11}{12}\right)}{f'\left(\frac{11}{12}\right)} \approx 0.913$$

$$x_3 = 0.913 - \frac{f(0.913)}{f'(0.913)} \approx 0.913.$$

Since the digits repeat, they should be accurate. Thus $x \approx 0.913$.

5. Let $f(x) = \sin x - 1 + x$; we want to find all zeros of f, because $f(x) = 0$ implies $\sin x = 1 - x$. Graphing $\sin x$ and $1 - x$ in Figure C.2, we see that $f(x)$ has one solution at $x \approx \frac{1}{2}$.

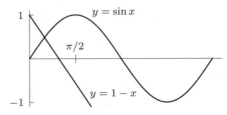

Figure C.2

Letting $x_0 = 0.5$, and using Newton's method, we have $f'(x) = \cos x + 1$, so that

$$x_1 = 0.5 - \frac{\sin(0.5) - 1 + 0.5}{\cos(0.5) + 1} \approx 0.511,$$

$$x_2 = 0.511 - \frac{\sin(0.511) - 1 + 0.511}{\cos(0.511) + 1} \approx 0.511.$$

Thus $\sin x = 1 - x$ has one solution at $x \approx 0.511$.

9. Let $f(x) = \ln x - \frac{1}{x}$, so $f'(x) = \frac{1}{x} + \frac{1}{x^2}$.
Now use Newton's method with an initial guess of $x_0 = 2$.

$$x_1 = 2 - \frac{\ln 2 - \frac{1}{2}}{\frac{1}{2} + \frac{1}{4}} \approx 1.7425,$$

$$x_2 \approx 1.763,$$

$$x_3 \approx 1.763.$$

Thus $x \approx 1.763$ is a solution. Since $f'(x) > 0$ for positive x, f is increasing: it must be the only solution.

Solutions for Section D

Exercises

1. The magnitude is $\|3\vec{i}\| = \sqrt{3^2 + 0^2} = 3$.
   The angle of $3\vec{i}$ is 0 because the vector lies along the positive x-axis.

5. $2\vec{v} + \vec{w} = (2 - 2)\vec{i} + (4 + 3)\vec{j} = 7\vec{j}$.

9. Two vectors have opposite direction if one is a negative scalar multiple of the other. Since

$$5\vec{j} = \frac{-5}{6}(-6\vec{j})$$

the vectors $5\vec{j}$ and $-6\vec{j}$ have opposite direction. Similarly, $-6\vec{j}$ and $\sqrt{2}\vec{j}$ have opposite direction.

13. Scalar multiplication by 2 doubles the magnitude of a vector without changing its direction. Thus, the vector is $2(4\vec{i} - 3\vec{j}) = 8\vec{i} - 6\vec{j}$.

17. In components, the vector from $(7, 7)$ to $(9, 11)$ is $(9 - 7)\vec{i} + (11 - 9)\vec{j} = 2\vec{i} + 2\vec{j}$.
   In components, the vector from $(8, 10)$ to $(10, 12)$ is $(10 - 8)\vec{i} + (12 - 10)\vec{j} = 2\vec{i} + 2\vec{j}$.
   The two vectors are equal.

21. The velocity is $\vec{v}(t) = e^t\vec{i} + (1/(1 + t))\vec{j}$. When $t = 0$, the velocity vector is $\vec{v} = \vec{i} + \vec{j}$.
   The speed is $\|\vec{v}\| = \sqrt{1^2 + 1^2} = \sqrt{2}$.
   The acceleration is $\vec{a}(t) = e^t\vec{i} - 1/((1 + t)^2)\vec{j}$. When $t = 0$, the acceleration vector is $\vec{a} = \vec{i} - \vec{j}$.

Notes

Notes

Notes

Notes

Notes

Notes

Notes

Notes